ADVANCES IN
BORON NEUTRON
CAPTURE THERAPY

The following States are Members of the International Atomic Energy Agency:

AFGHANISTAN
ALBANIA
ALGERIA
ANGOLA
ANTIGUA AND BARBUDA
ARGENTINA
ARMENIA
AUSTRALIA
AUSTRIA
AZERBAIJAN
BAHAMAS
BAHRAIN
BANGLADESH
BARBADOS
BELARUS
BELGIUM
BELIZE
BENIN
BOLIVIA, PLURINATIONAL
   STATE OF
BOSNIA AND HERZEGOVINA
BOTSWANA
BRAZIL
BRUNEI DARUSSALAM
BULGARIA
BURKINA FASO
BURUNDI
CAMBODIA
CAMEROON
CANADA
CENTRAL AFRICAN
   REPUBLIC
CHAD
CHILE
CHINA
COLOMBIA
COMOROS
CONGO
COSTA RICA
CÔTE D'IVOIRE
CROATIA
CUBA
CYPRUS
CZECH REPUBLIC
DEMOCRATIC REPUBLIC
   OF THE CONGO
DENMARK
DJIBOUTI
DOMINICA
DOMINICAN REPUBLIC
ECUADOR
EGYPT
EL SALVADOR
ERITREA
ESTONIA
ESWATINI
ETHIOPIA
FIJI
FINLAND
FRANCE
GABON
GEORGIA

GERMANY
GHANA
GREECE
GRENADA
GUATEMALA
GUYANA
HAITI
HOLY SEE
HONDURAS
HUNGARY
ICELAND
INDIA
INDONESIA
IRAN, ISLAMIC REPUBLIC OF
IRAQ
IRELAND
ISRAEL
ITALY
JAMAICA
JAPAN
JORDAN
KAZAKHSTAN
KENYA
KOREA, REPUBLIC OF
KUWAIT
KYRGYZSTAN
LAO PEOPLE'S DEMOCRATIC
   REPUBLIC
LATVIA
LEBANON
LESOTHO
LIBERIA
LIBYA
LIECHTENSTEIN
LITHUANIA
LUXEMBOURG
MADAGASCAR
MALAWI
MALAYSIA
MALI
MALTA
MARSHALL ISLANDS
MAURITANIA
MAURITIUS
MEXICO
MONACO
MONGOLIA
MONTENEGRO
MOROCCO
MOZAMBIQUE
MYANMAR
NAMIBIA
NEPAL
NETHERLANDS
NEW ZEALAND
NICARAGUA
NIGER
NIGERIA
NORTH MACEDONIA
NORWAY
OMAN
PAKISTAN

PALAU
PANAMA
PAPUA NEW GUINEA
PARAGUAY
PERU
PHILIPPINES
POLAND
PORTUGAL
QATAR
REPUBLIC OF MOLDOVA
ROMANIA
RUSSIAN FEDERATION
RWANDA
SAINT KITTS AND NEVIS
SAINT LUCIA
SAINT VINCENT AND
   THE GRENADINES
SAMOA
SAN MARINO
SAUDI ARABIA
SENEGAL
SERBIA
SEYCHELLES
SIERRA LEONE
SINGAPORE
SLOVAKIA
SLOVENIA
SOUTH AFRICA
SPAIN
SRI LANKA
SUDAN
SWEDEN
SWITZERLAND
SYRIAN ARAB REPUBLIC
TAJIKISTAN
THAILAND
TOGO
TONGA
TRINIDAD AND TOBAGO
TUNISIA
TÜRKİYE
TURKMENISTAN
UGANDA
UKRAINE
UNITED ARAB EMIRATES
UNITED KINGDOM OF
   GREAT BRITAIN AND
   NORTHERN IRELAND
UNITED REPUBLIC
   OF TANZANIA
UNITED STATES OF AMERICA
URUGUAY
UZBEKISTAN
VANUATU
VENEZUELA, BOLIVARIAN
   REPUBLIC OF
VIET NAM
YEMEN
ZAMBIA
ZIMBABWE

The Agency's Statute was approved on 23 October 1956 by the Conference on the Statute of the IAEA held at United Nations Headquarters, New York; it entered into force on 29 July 1957. The Headquarters of the Agency are situated in Vienna. Its principal objective is "to accelerate and enlarge the contribution of atomic energy to peace, health and prosperity throughout the world".

ADVANCES IN
BORON NEUTRON
CAPTURE THERAPY

INTERNATIONAL ATOMIC ENERGY AGENCY
VIENNA, 2023

COPYRIGHT NOTICE

For further information on this publication, please contact:

Physics Section
International Atomic Energy Agency
Vienna International Centre
PO Box 100
1400 Vienna, Austria
email: Official.Mail@iaea.org

© IAEA, 2023
Printed by the IAEA in Austria
June 2023

IAEA Library Cataloguing in Publication Data

Names: International Atomic Energy Agency.
Title: Advances in boron neutron capture therapy / International Atomic Energy Agency.
Description: Vienna : International Atomic Energy Agency, 2023. | Includes bibliographical references.
Identifiers: IAEAL 23-01601 | ISBN 978–92–0–132723–9 (paperback : alk. paper) | ISBN 978–92–0–132623–2 (pdf)
Subjects: LCSH: | Boron-neutron capture therapy. | Cancer—Treatment. | Radiotherapy.
Classification: UDC 661.65:615.849 | CRCP/BOR/002

FOREWORD

Boron neutron capture therapy is a combination cancer therapy utilizing an external beam of neutrons of appropriate energies and a 10B-containing pharmaceutical that preferentially concentrates in tumour tissues of the patient. The nuclear reaction between the neutron and the boron nucleus generates an alpha particle and a recoiling ^{7}Li nucleus that create a high degree of localized damage to the tumour cell. This concept, while simple and first proposed in 1936 by G. Locher, has proven challenging to implement and involves truly multidisciplinary teams. One difficulty in the past was that there were few neutron sources around the world of sufficient intensity and of the required energies for boron neutron capture therapy. The only neutron sources suitable were research reactors, which were distributed at universities and government laboratories around the world. Research reactors are not clinical environments, and, while many clinical trials were conducted and several centres reported encouraging results, the number of patients treated was small and the comparison of results from different centres was not simple. In 2001, the IAEA published Current Status of Neutron Capture Therapy (IAEA-TECDOC-1223), which summarized the state of the field based around reactor sources.

In the past two decades, the number of research reactors involved in boron neutron capture therapy has precipitously declined. However, during the same period, compact accelerator based neutron sources have been developed utilizing a variety of different accelerator technologies. Accelerators have the advantages that they can be installed directly in hospitals and clinics, providing a more suitable environment for treating patients, and that they can be registered as medical devices. In Japan, in 2020 the use of a cyclotron based neutron source in combination with a boron pharmaceutical and treatment planning device was approved for the treatment of unresectable, locally advanced, and recurrent carcinoma of the head and neck, funded through the national health insurance system.

These developments have led to renewed interest in boron neutron capture therapy around the world, to significant commercial investment, and to the realization that it was time for the IAEA to issue another report reflecting the many advances in the field. This report is the output of two virtual technical meetings on the subject, one in July 2020 attended by 106 participants from 20 Member States and the second in March 2022 attended by 111 participants from 18 Member States.

The IAEA acknowledges the valuable contributions and support of the numerous international experts who contributed to the drafting and review of this publication, in particular, W. Sauerwein (Germany) for his assistance in organizing the first consultancy and technical meetings and K. Igawa (Japan) for her assistance in coordinating and networking throughout this project. The IAEA officers responsible for this publication were I. Swainson, D. Ridikas, and A. Jalilian of the Division of Physical and Chemical Sciences and O. Belyakov of the Division of Human Health.

CONTENTS

1   INTRODUCTION

1.1   BACKGROUND

The principle and history of boron neutron capture therapy (BNCT) is well described in the introduction of the previous IAEA report covering this topic, published in 2001 [1]. The concept, first proposed in the mid-1930s, is that tumour cells may be destroyed if it is possible to concentrate a sufficient number of boron atoms within them and then expose the tissue to a neutron 'beam' of sufficient intensity that a treatment could be performed in a 'reasonable' time. The treatment relies on the fact that the ^{10}B nucleus has a very high affinity for neutron capture and promptly decomposes, predominantly via α emission; the recoiling nucleus and α particle have a short range of travel of approximately the size of a mammalian cell and they have a high linear energy transfer; i.e., they deposit significant ionizing energy over this range. The treatment also relies on finding a boron-containing pharmaceutical with a marked preference for uptake by cancer cells rather than normal cells. Normal tissues, with lower boron content, would receive much lower doses.

Although the concept is simple, practical implementation involves assembling a wide variety of different scientific and clinical expertise for the design (see Sections 2–5) and operation (see Section 6) of an accelerator based BNCT (AB-BNCT) facility. As a radiotherapy technique, BNCT is unique in that it requires the administration of a stable pharmaceutical (see Section 7) in conjunction with the application of a rather rare form of radiation, the neutron. It requires a relatively high flux of neutrons in order to provide treatment in a reasonable time frame, and in all such neutron sources, neutrons are not created in the energy range optimal for treatment but require partial moderation. Figure 1 shows a schematic diagram of an AB-BNCT system. A proton/deuteron beam emitted from an accelerator is steered, and its focus adjusted onto a neutron producing target through the beam transportation system. The fast neutrons emitted are moderated within the neutron irradiation system (in a component also commonly called the 'beam shaper assembly', Section 3.2) before being emitted into a treatment room in which a patient is precisely placed.

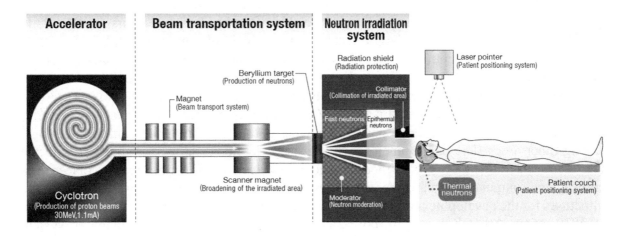

FIG. 1. Schematic diagram of an AB-BNCT system showing the principal components of the accelerator, beam transport, and the target and moderator inside the shielded neutron irradiation system (beam shaper assembly) emitting a therapeutic neutron 'beam' for a carefully positioned patient (courtesy of Osaka Prefecture, Japan).

The final neutron moderation takes place within the patient's hydrogen-rich tissues (Fig. 2). Analytical techniques are required to quantify the boron uptake in tissues (see Section 8). A detailed understanding of the relationship between different boron-containing pharmaceuticals and their effect on normal and tumour tissues when irradiated is required in order to optimize treatments (Section 9) and this knowledge is also indispensable to be able to calculate the neutron dose (see Sections 4 and 9–10) and to plan the treatments (Section 11) and report the doses received (Section 12).

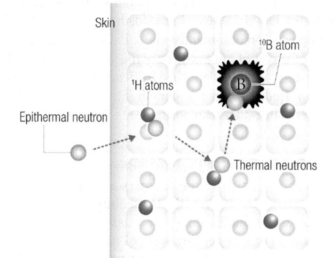

FIG. 2. Schematic diagram of the interaction of an individual epithermal neutron being moderated by the hydrogenous tissue to thermal energies and thereafter being captured by a ^{10}B atom, triggering the therapeutic α decay (courtesy of Osaka Prefecture, Japan).

Since the 1950s and until the publication of Ref. [1], research reactors had been almost the only neutron sources of sufficient flux available for use in BNCT and were used throughout the development of the field. Since Ref. [1] was published, while some research reactors remain in use for BNCT many have been shut down.[1] Fortunately, during the same period, low energy proton and deuteron accelerators have seen major developments such that they can now generate neutron fluxes sufficient to replace the role of such research reactors for BNCT. In the 2001 report [1], accelerator based neutron sources had only one paragraph dedicated to them as future possible neutron sources for BNCT. At that time, it was felt that a radiofrequency quadrupole proton accelerator with a Li target was the most promising technology. However, there are currently several different classes of accelerators being proposed and constructed for AB-BNCT with a variety of target materials (Section 2). In addition, in the last two decades much progress has been made in pharmaceuticals, boron measurements, and radiobiology, among other areas. The developments in modern accelerator technologies together with the fact that accelerators can be classified as medical devices, are far more readily placed in a clinical environment, and can, therefore, be approved for clinical use were some of the impetuses for the development of this new report. In particular, the approvals received in Japan in 2020, of an AB-BNCT system, dose calculation engine and treatment planning system together with a boron pharmaceutical for treatment of unresectable, locally advanced, and recurrent carcinoma of the head and neck under public health insurance has brought interest around the world in implementing similar facilities.

---

[1] See the statistics in the IAEA Research Reactor Database: https://nucleus.iaea.org/rrdb/

1.2 OBJECTIVE

The objective of this report is to describe many developments in the field of BNCT that have happened in the last two decades since publication of the last IAEA report on the topic [1], and to describe the recent developments associated with AB-BNCT that enable BNCT to be performed in a clinical environment.

1.3 SCOPE

This report focuses on developments in BNCT since Ref. [1] was published. Regarding neutron sources, it focuses on the developments in AB-BNCT and does not deal with the specifics of design of research reactor based BNCT facilities, for which it is considered that Ref. [1] remains relevant. However, reactor based facilities will find many useful discussions in the later sections concerning, e.g., developments in the fields of pharmaceuticals, boron measurement, radiobiology, as well as dose calculation, prescription, and reporting.

1.4 STRUCTURE

Section 2 provides a list of the low energy nuclear reactions involving accelerated hydrogen and deuterium ions that generate neutrons currently considered as compatible with a compact accelerator based neutron source. The relevant target materials and their physical properties are discussed. The possible combination of accelerator type and targets are considered and finally a list of the currently known facilities that are either operational, under commissioning, or in an advanced state of project development are given.

Section 3 describes the characteristics of a neutron 'beam' required for BNCT and goes on to describe the general features of the beam shaping assembly, which moderates, filters, shields and collimates the neutron 'beam'. It provides a summary of reference values for beam quality factors, which is supported by Appendix I.

Section 4 provides a description of the physical dosimetry requirements for characterizing the various radiation components emitted from the target and the beam shaping assembly into the treatment room. It provides a list of equipment and methods, as well as the uncertainties involved. It is supported by Appendices II and III.

Section 5 provides a description of the principles of facility design at an AB-BNCT facility, including a list of typical functional spaces, layouts showing their relationship, zoning of spaces, materials and ventilation requirements, and optional equipment that could either be at a BNCT clinic or in a nearby supporting hospital. This is supplemented in Annexes I–XI by several specific examples of recent AB-BNCT facilities with more specific details that designers of facilities may find useful to consider in new facility designs.

Section 6 gives an outline of what the operation and management of an AB-BNCT clinical facility may look like. It includes roles and numbers of staff that may be expected. It also provides an example clinical case approval flow chart. While details will vary from country to country, the expertise needed, processes, and regulatory requirements are likely to be similar in many countries.

Section 7 describes the pharmaceuticals and radiopharmaceuticals that have been developed for BNCT, concentrating on 4-borono-L-phenylalanine (BPA), the most commonly used pharmaceutical for BNCT. (This is supplemented by discussion on some of the earlier boron containing pharmaceuticals and results from clinical trials in Annex XII) The requirements for quantitative estimates of boron uptake with such an agent are discussed, through the example of 4-borono-2-[^{18}F]fluoro-L-phenylalanine (^{18}F-FBPA) used in PET imaging. Topics in this section include isomers, synthesis, metabolism, transport, kinetics, biodistribution, and clinical application. The two-step infusion method for BPA developed in Japan is discussed in Annex XIII, and development of future targeted pharmaceuticals for BNCT in Annex XIV. As BNCT involves the approvals of the pharmaceutical along with the accelerator and dose engine etc for clinical use, Annexes XV–XVIII describe some of the regulatory aspects with experience from Europe and Japan in obtaining authorizations for clinical use.

Section 8 covers the quantitative analytical methods (chemical, physical, nuclear) that are used to measure boron concentrations and boron distribution and imaging in tissues. It includes the inductively coupled plasma (ICP) analytical methods for concentration determination that are currently used in clinical practice as well as quantitative PET imaging of uptake of boron in tumours. A wide variety of other methods that are either experimental or in use in preclinical or fundamental studies are also described. It is supported by a table in Appendix IV that lists specific systems and parameters for ICP methods in use at various clinical facilities around the world.

Section 9 gives a broad overview of the field of radiobiology as applied to BNCT. It discusses some of the ideal considerations of boron-carrying pharmaceuticals and mechanisms of action when used in BNCT. It describes research into optimizing tumour targeting, radiotoxicity, and treatment methods. It discusses preclinical experiments required to support clinical trials. This is supplemented by Annex XVIII, which gives practical guidelines for such experiments at various types of AB-BNCT facilities. It ends with an outlook for the future is given, including the use of BNCT as a combination therapy. The section provides a bridge between many of the discussions in Sections 7–8 and the later sections.

Specialists, such as medical physicists and oncologists, are likely to find the last sections most relevant.

Section 10 describes general concepts involved in dose calculation for BNCT, provides some of the relevant sources of nuclear data that are required, and goes on to give an outline of the models used for dose calculation at the macroscopic and microscopic levels. It then describes how to relate BNCT doses to 'isoeffective' doses, using photon therapy as the reference.

Section 11 describes how some of the models and concepts outlined in Section 10 are put into practice in clinical use for a patient. This includes the computational approaches for dose calculation, quality assurance, and the definition of the monitor unit for the neutron fluence determination. It continues to describe the existing treatment planning systems in use at facilities around the world and how dose may be prescribed.

Section 12 describes dose reporting and volume specification for BNCT. It describes the need for harmonization and how that may be aided by providing a suggested approach within the context of the ICRU levels of dosimetry, highlighting some of the characteristics that would be unique to dose reporting from BNCT. It is supported by Annex XIX.

Section 13 describes an overview of clinical trials involving BNCT that have been undertaken for meningiomas, gliomas (specifically glioblastomas), head and neck cancers, and skin cancers. This is also supported in Annex XX by a case study of a patient treated with AB-BNCT for recurrent head and neck cancer.

2   ACCELERATOR BASED NEUTRON SOURCES FOR BNCT TREATMENT FACILITIES

In the last two decades there has been significant progress in the development of high intensity, accelerator based neutron sources for BNCT (AB-BNCT) and other applications. The aim is to inaugurate the era of hospital based BNCT facilities, moving progressively away from reactor based facilities which, according to the prevailing consensus, are more costly, more difficult to operate and license, and, more importantly, more difficult to install in hospitals. Although a reactor based BNCT facility was recently constructed next to a hospital in China [2], in Japan, it proved impossible to register a research reactor as a medical device and this was one of the impetuses that led to the concentration on AB-BNCT systems [3]. Today, there is an international consensus that AB-BNCT may change the prospects of BNCT in clinical practice due mainly to the relative ease of installing accelerators in a clinical environment. Hence, there is a worldwide quest to find technical solutions for such facilities that are reliable, safe, and affordable in order to promote the widest possible dissemination of AB-BNCT.

Currently, AB-BNCT facilities are being designed and constructed at medical centres around the world, and some are already being used in clinical trials and treatments. Notably, in Japan, one system was approved as a medical device in 2020 for the treatment of unresectable locally advanced or locally recurrent head and neck carcinoma [4].

In this section, the progress made in neutron sources in recent years will be assessed. Different options for AB-BNCT will be introduced according to the nuclear reactions employed, beam energies and currents, primary neutron spectra, target materials, types of accelerator technologies, and combinations of accelerators and reactions.

2.1   NEUTRON PRODUCTION

The neutron is a neutral particle that is generally stable when bound into a nucleus but is unstable when free, decaying with a half-life of a little over 10 minutes [5]. Neutrons are created with an energy that depends on the nucleus from which the neutron is liberated, the specific nuclear reaction liberating the neutron, and the energy of the interacting particle that triggers the neutron emission. For an AB-BNCT facility, the range of neutron energies involved is from a few tens of MeV down to a few meV, approximately 10 orders of magnitude. Neutron energy ranges are classified under names such as 'thermal', 'epithermal' and 'fast'. All neutron sources generate the majority of the neutrons in the fast energy range. There is no single definition of these terms, but for BNCT the definition is commonly taken as defined in Table 1. Different definitions may be used, but they need to be reported in any studies.

TABLE 1. DEFINITION OF NEUTRON ENERGY RANGES USED IN BNCT

Range name	Thermal	Epithermal	Fast
Neutron energy	< 0.5 eV	0.5eV–10 keV	> 10 keV

Note: Neutrons with energies slightly in excess of 10 keV (e.g., 20–30 keV) can be useful for epithermal BNCT as relative biological effectiveness does not undergo a step change at 10 keV.

The only alternative to reactors capable of producing sufficiently intense neutron beams for clinical BNCT is to use particle accelerators. Currently, proton or deuteron accelerators show the most promise. Non-spallation production of neutrons by compact accelerators was reviewed in a recent IAEA publication [6], along with cost considerations of various options.

For AB-BNCT, the required source intensity at the neutron production rate is greater than about 10^{13} s^{-1} [7], although the exact requirements depend on the primary neutron spectrum as well as the design of the beam shaping assembly (BSA) and beam collimator. Using appropriate nuclear reactions, such an intensity can be reached with proton or deuterons beams with either [7–9] low bombarding energies and high current or higher bombarding energies and lower current.

The incident particles travel within the target losing their energy and finally stop, forming a 'Bragg peak' towards the end of trajectory. The electronic stopping power is much larger than the nuclear stopping power and is the major contributor to energy loss of the incident particles (Fig. 3), where data taken from the 'Stopping Power for Light and Heavier Ions' database.[2]

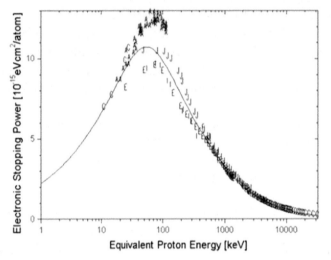

FIG. 3.Electronic stopping power of protons in Be (courtesy of Y. Kiyanagi, Nagoya University).

Figure 4 shows the heat deposition in a Be target at various proton energies as a function of distance from the target surface. The protons deposit a large energy density in a very narrow range where they finally stop and accumulate. As shown in Fig. 3, the electronic stopping power for protons in Be reaches a maximum around 100 keV. Around this energy the protons rapidly reduce their energy and form a 'Bragg peak' (Fig. 4). The protons gradually combine to form hydrogen gas bubbles within the target, which may cause blistering (see discussion in Section 2.2.2.).

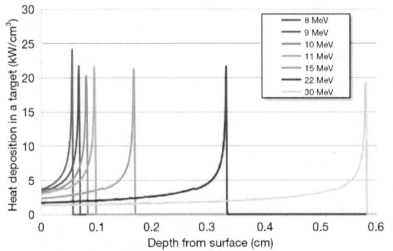

FIG. 4. Heat deposition in a Be target. This figure by Y. Kiyanagi is licenced under CC BY-NC-ND-4.0 [10].

---

[2] Stopping powers are available at https://www-nds.iaea.org/stopping/stopping_201410/index.html

2.1.1   Neutron production reactions

The neutron yields, defined as neutron production per unit current or charge (of a proton or deuteron beam) in neutrons per μC, from the candidate neutron producing reactions are shown in Fig. 5 as a function of the incident particle energy. Below, the following notation is used as an abbreviation to refer to the combination of a given accelerated particle and target: e.g., d-Be refers to a deuteron beam on a Be target and p-Li a proton beam on a Li target. Depending on the energies of the particle beams, more than one nuclear reaction may occur. The data concerning d-Be and d-Li are from Refs [9, 11–12], p-Li and p-Be from Ref. [13] and d-^{13}C from Ref. [14]. The ^{7}Li(p,n)^{7}Be reaction gives a much higher neutron yield at lower proton energies, E_p, than the ^{9}Be(p,n)^{9}B reaction. The intensity of the p-Be reaction becomes of practical use for AB-BNCT when $E_p \gtrsim 5$ MeV.

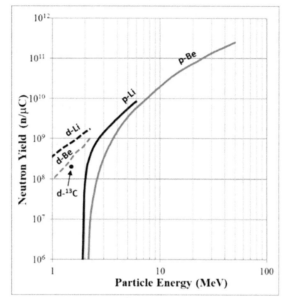

FIG. 5. *Neutron yields of various reactions as a function of incident particle energy for thick targets. Data for d-Be and d-Li are taken from Refs [9, 11–12], for p-Li from Ref. [13] and d-^{13}C from Ref. [14] (courtesy of Y. Kiyanagi, Nagoya University).*

Relevant parameters to be taken into account include:

a) Charged particle energy. As a general rule, the lower the particle energy required to produce neutrons, the smaller and cheaper is the accelerator. In addition, the Coulomb barriers of protons on common structural materials for accelerators, like Cu and Fe, are ~5 MeV. Hence, to minimize direct activation of accelerator components by the charged particle beam, a target energy of $\lesssim 5$ MeV would be preferred (but this is below the energy at which the p-Be reaction becomes significant). However, reaction can occur at energies much lower than this because of the tunnelling effect, but with significantly lower cross sections (reaction probabilities);

b) Highest possible neutron production yield at the lowest possible neutron energies. The latter has two important effects on design:

(i) The lower the neutron energy, the smaller and simpler the beam shaper assembly (BSA) may become to filter out the high energy neutrons and the lower the volume of activated material produced in the facility (see Section 3);

7

(ii) For activation other than through direct neutron capture, in general, more nuclear reaction channels open up with increasing neutron energy and the likelihood to produce activation increases.

To reduce the build-up of activation in a facility during its operational life and for decommissioning, it is important to choose materials that do not produce long lived radioisotopes [15].

The higher the energy of the particle incident on any given target, the higher both the yield and energy of the neutrons produced. The increase in neutron yield leads to a reduction in the required current for the accelerator. However, this gain is partially counterbalanced by the higher neutron energy, which leads to higher losses during the moderation of the neutrons down to the therapeutic epithermal range.

For neutron energies above the threshold of ~10.2 MeV, the reaction $^{16}O(n,p)^{16}N$ opens up as an activation product in cooling water, which is swept away from the target to a heat exchanger. With its very short half-life of ~7s, ^{16}N emits 6 MeV photons that can sometimes be a shielding problem.

c) High resistance of the target. Low energy incident particles have very short ranges in targets. This leads to high heat densities and may cause target blistering. This effect can be minimized by choosing very thin layers of the neutron producing materials deposited on materials that are resistant to blistering, in which the protons stop. As the energy of the charged particles increases, their range in the target becomes longer, and the target structure required to prevent blistering becomes simpler, so that the integrity of the target can be more easily maintained. This issue is discussed more fully in Section 2.2.2.

2.1.2   Low energy proton reactions

Both proton reactions, $^{9}Be(p,n)^{9}B$ and $^{7}Li(p,n)^{7}Be$, are endoenergetic; i.e., $Q < 0$, where $Q = (M_{init} - M_{final})c^2$ is the energy difference between the initial mass, M_{init}, and the final mass, M_{final}, multiplied by the speed of light squared, c^2. For both proton reactions, the maximum neutron energy, $E_{n,max}$, is given by $E_{n,max} \approx E_p - E_{th}$, where E_{th} is the threshold energy for the reaction, and E_p the energy of the accelerated proton beam (see Table 2) [8]. Conversely, the deuteron reactions are exoenergetic, $Q > 0$, and the maximum neutron energy is given by $E_{n,max} \approx E_d + Q$.

Protons reduce their energies continuously on their path through the target. Neutron sources using p-Be and p-Li reactions can use a thick target to increase the neutron yield. The minimum neutron energies, $E_{n,min}$, for p-Li and p-Be reactions indicated in Table 2 correspond to a target thickness which covers the energy range from the bombarding energy, E_p or E_d, down to the reaction threshold (except for p-Be at 30 MeV, see Table 2).

TABLE 2. LOW ENERGY NEUTRON PRODUCING REACTIONS

Reaction	E_{th} or Q (MeV)	E_p or E_d (MeV)	Total neutron yield (n/mA)	Percentage of neutrons with E_n < 1 MeV	$E_{n,max}$ (keV)	$E_{n,min}$ (keV)
ENDOENERGETIC REACTIONS						
^{7}Li(p,n)^{7}Be	1.880	1.89	6.3×10^9	100	67	0^c
		2.30	5.8×10^{11a}	100	573	0^c
		2.50	9.3×10^{11a}	100	787	0^c
		2.80	1.4×10^{12b}	92	1100	0^c
^{9}Be(p,n)^{9}B	2.057	2.50	3.9×10^{10}	100	574	0^d
		4.00	1.0×10^{12}	50	2117	0^d
		8.00	1.9×10^{13}	21^e	6136	0^d
^{9}Be(p,xn)^{9}B		30.0	1.34×10^{14}	9^f	28147	214^f
EXOENERGETIC REACTIONS						
^{9}Be(d,n)^{10}B	4.362	1.45^g	1.6×10^{11}	69	5763^h	225^i
		1.50^j	3.3×10^{11}	50	5815^h	15^k
^{13}C(d,n)^{14}N	5.237	1.50	1.9×10^{11}	70	6720^l	59^m
^{7}Li(d,n)^{8}Be	15.03	1.40	7.1×10^{11}	0	15765	12934^n

[a] Average of the values reported in Refs [14, 16].

[b] Ref. [17]

[c] Minimum neutron energies correspond to a target thick enough that the proton energy degrades down to the reaction threshold.

[d] Minimum neutron energies correspond to a target thick enough that the proton energy degrades down to the reaction threshold.

[e] Calculated with PHITS code/JENDL4.0.

[f] Corresponds to a target thickness of 5.5 mm so that the beam leaves the target with an energy of 5.76 MeV. All excited states in ^{9}B are taken into account. Calculated with MCNPX code/ENDF/B-7.

[g] 8 µm thin target: Ref. [18] and refs therein.

[h] Population of the ground state of ^{10}B.

[i] Strong population of the 5.11 MeV excited state in ^{10}B. In this case the highest neutron energy is 618 keV at 0^0 emission.

[j] Thick target. The incident deuteron has 1.5 MeV and it stops in the Be target.

[k] Lowest neutron energy for a thick target with a population of the 5.11 MeV excited state in ^{10}B and $E_d = 1.0$ MeV energy within the target.

[l] Population of the ground state of ^{14}N.

[m] Population of the 6.446 MeV excited state in ^{14}N.

[n] Lowest neutron energy for a very thin target. This reaction is not suitable due to its very high neutron energies.

2.1.2.1   Protons on lithium targets

The neutron energy spectra as a function of emission angle from the p-Li reaction are shown in Fig. 6 at different emission angles for $E_p = 2.8$ MeV [10], calculated from the Li-Yield code [16]. The peak energies appear around several hundred keV and then come down towards zero energy. There is a reaction resonance peak in the cross section at $E_p \sim 2.25$ MeV, which substantially increases neutron production: most accelerators using Li utilize an E_p slightly in excess of this value.

The residual nuclide from ^{7}Li(p,n)^{7}Be has a half-life of 53.2 days. It decays by electron capture to ^{7}Li either directly to the ground state or in 10.4% of the cases to an excited state, which emits a 478 keV γ ray upon decaying to the ground state. Beryllium-7 is produced at all incident proton energies in Li targets. This contrasts with the case in Be targets (see below).

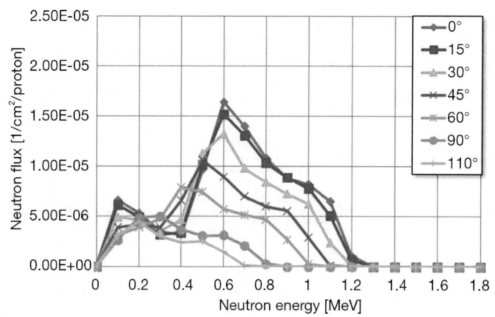

FIG. 6. Neutron energy spectra at various emission angles for the p-Li reaction at a proton energy of 2.8 MeV. This figure by Y. Kiyanagi is licenced under CC BY-NC-ND-4.0 [10].

2.1.2.2   *Protons on beryllium targets*

In the case of a Be target, a higher E_p is required to achieve the same yield as that from a ^{7}Li target, and the neutron spectrum shifts to higher energies. The angle dependent energy spectra of the ^{9}Be(p,n)^{9}B reaction are shown for a thick Be target at $E_p = 11$ MeV [19] in Fig. 7a. The spectrum for $E_p = 30$ MeV [15], calculated using the MCNPX code [20], is shown in Fig. 7b. The neutron intensity reaches a maximum at $E_n \sim 1$ MeV for the $E_p = 11$ MeV case (Fig. 7a), and at $E_n \sim 10$ MeV for the $E_p = 30$ MeV case (Fig. 7b). The maximum value of E_n for the $E_p = 11$ MeV case reaches up to ~ 9 MeV and for the $E_p = 30$ MeV case to ~ 28 MeV, as $E_{th} = 2.057$ MeV (Table 2).

Reaction channels other than ^{9}Be(p,n)^{9}B open up with increasing proton energy. Reactions such as ^{9}Be(p,xn), where x indicates the number of emitted neutrons, and ^{9}Be(p,chp), where chp can be a variety of light charged particles, can also open up. The high energy neutrons induce long lived activation in some cases. Also, tritium and ^{7}Be are produced through ^{9}Be(p,t)^{7}Be for $E_p \gtrsim 13$ MeV.

2.1.3   Low energy deuteron reactions

The deuteron is a good candidate particle for neutron production since there are several reactions with a high neutron yield that start at very low bombarding energies, $E_d \gtrsim 1.5$ MeV. For example, Fig. 5 shows that the reactions d-Li, d-Be and d-^{13}C give higher neutron yields per bombarding particle than the p-Li or p-Be reactions and start at much lower particle energies. This allows the use of a very low energy accelerator. The neutron yield is higher for ^{7}Li(d,n)^{8}Be than for ^{9}Be(d,n)^{10}B, but the neutrons are produced at very high energies (Table 2). The neutron yield of the d-^{13}C reaction for a thick target is 1.9×10^8 n/μC at $E_d = 1.5$ MeV [14]. This yield is somewhat smaller than the yield from the d-Be reaction in a thick target at the same bombarding energy, but somewhat larger than the neutron yield for d-Be with a thin target.

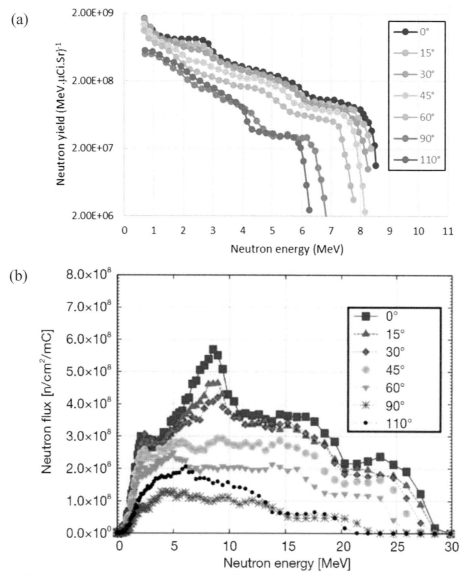

(a)

(b)

FIG. 7. (a) Neutron energy spectra as a function of emission angle for the p-Be reaction. Data are taken from Ref. [19]. (b) Neutron energy spectra at various emission angles for the p-Be reaction at a proton energy of 30 MeV. This figure by Y. Kiyanagi is licenced under CC BY-NC-ND-4.0 [10].

2.1.3.1   Deuterons on lithium targets

The neutron energy spectra of the d-Li reaction measured at $0°$ and $90°$ with $E_d = 2$ MeV are shown in Fig. 8 [21]. A broad peak appears at around a few MeV and extends out to more than 15 MeV due to the very large positive Q value of 15.03 MeV for this reaction (Table 2).

2.1.3.2   Deuterons on beryllium targets

The neutron energy spectra produced from the d-Be reaction measured at $0°$ with $E_d = 1.45$ MeV are shown in Fig. 9 [12, 22] for both thick and thin target cases. A peak appears at $E_n < 1$ MeV, then the intensity decreases rapidly and extends out to almost 6 MeV due to the large Q value of 4.362 MeV (Table 2). A thin Be target may be more appropriate for AB-BNCT as it suppresses the higher-energy components of the neutron emission [12, 18, 22]. However, the high energy component of the total neutron yield is in general much smaller than for the d-Li case (Fig. 8).

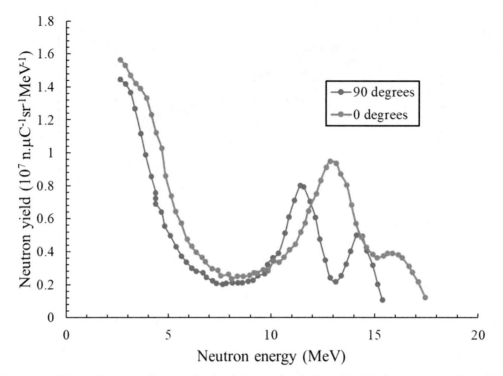

FIG. 8. Neutron yield as a function of energy for the d-Li reaction at E_d = 2 MeV at emission angles of 90° (blue) and 0° (orange). Data are from Ref. [21].

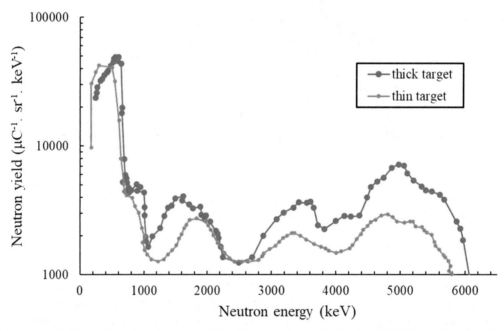

FIG. 9. Neutron yield as a function of energy of the d-Be reaction at 0° for 1.45 MeV for a thick (blue) and a thin (orange) target. Data are from Refs [18, 22].

2.1.3.3   Deuterons on carbon-13 targets

The neutron energy spectrum for the d-^{13}C reaction at E_d = 1.5 MeV is shown in Fig. 10 at 0° emission angle [14]. The shape resembles the d-Be case (Fig. 9). A peak appears at an energy less than 1 MeV and gradually decreases until about 3 MeV. Then it becomes flat and decreases again above 5 MeV. The intensity of neutrons with E_n < 1 MeV is about 70% of the total [12, 14].

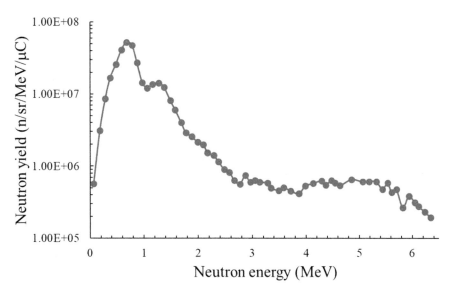

FIG. 10. Neutron yield as a function of energy of the d-^{13}C reaction at 0° for 1.5 MeV. Data are from Ref. [12].

2.1.3.4 Conclusions regarding low energy deuteron reactions

Based on these characteristics, both the d-Be (thin target) and the d-^{13}C reactions are good candidates for AB-BNCT, whereas the d-Li spectra contain a much higher energy neutron component without a significant gain in neutron intensity. While the main neutron producing reaction channels for d-Be and d-^{13}C lead to stable residual nuclei, they are accompanied by a smaller (d,t) channel. For d-Be and d-^{13}C reactions, the tritium yields are ~50% and ~10%, respectively, of their neutron yields, in both cases leading to stable residues.

The threshold for deuteron breakup is its binding energy of 2.224 MeV in the centre of mass frame; for a Be target this translates to 2.718 MeV in the laboratory frame. Hence, deuteron breakup under acceleration has no impact in the deuteron induced reactions at the energy regime proposed for BNCT.

2.1.4 Other neutron producing reactions

2.1.4.1 Photoneutrons

A photoneutron source using an electron accelerator with several tens of MeV has been proposed [23–24] but not yet planned for construction.

2.1.4.2 Fusion

The reactions d+d (Q = 3.27 MeV) and d+t (Q = 17.6 MeV) have fair yields even at very low accelerating energies (e.g., E_d ~ 120 keV), and produce quasi-monoenergetic neutrons at 2.45 MeV and 14.1 MeV, respectively. The smallest devices used to accelerate deuterons go by the name of neutron generators and are advantageous from the point of view of the low accelerating voltage required. The neutron yield of a d-t generator is about 50–100 times higher than for the d-d system. However, a d-t neutron generator requires sealed-tube operation with a secondary containment arrangement to prevent tritium leakage, as it decays with a 12.32-year half-life emitting β rays of 18.521 keV maximum energy. In addition, tritium gas is expensive and requires a special license for handling [25]. Therefore, a d-t system is not generally considered suitable as an in-hospital BNCT neutron source. However, considering the small amounts of tritium involved in these systems, a facility might not need more stringent safety requirements than other AB-BNCT facilities. For example, a tritium inventory below 400 GBq ($\sim 2 \times 10^{20}$ atoms) might be considered "UN2911 Class 7 Radioactive material, excepted package-instruments or articles" [6]. The neutron production rate obtained from a d-d/d-t neutron generator is at most 5×10^{10} s^{-1} [6, 25–26], short of the intensity needed for BNCT when using a single neutron tube (see the typical target source strength required in Section 2.1). Nevertheless, the utilization of a set of these inexpensive devices might be useful for research (e.g., small animal irradiation) and potentially clinical facilities (e.g., explanted-organ BNCT) [27]. There are also systems based on slightly larger, inexpensive, electrostatic accelerators (~ 300 keV, ~ 50 mA) and gas targets able to reach neutron production rates above 10^{13} s^{-1} (about 10^{12} mA$^{-1}\cdot$s^{-1}) [6] that could be useful for some BNCT applications.

2.2 TARGET MATERIALS

The physical properties of the target materials have to be carefully considered while designing appropriate neutron production targets for in-hospital use. A comparison of the characteristics of the low-energy neutron producing targets is given below:

- Beryllium targets for the p-Be reaction were the first used routinely in AB-BNCT facilities [28–31]. Beryllium has a relatively high melting point (1287 °C), good thermal conductivity for handling the heat load, and nearly no residual target radioactivity after irradiation at $E_p \lesssim 13$ MeV (see Section 2.1.2). However, careful handling is required during Be target manufacturing due to its toxicity in powder form;
- Lithium targets for the p-Li reaction have been adopted more recently [32–33]. In contrast to Be, Li has a low melting point (180.5 °C) and gradually becomes activated due to the ingrowth of ^{7}Be (see Section 2.1.2.1). As Li is chemically highly reactive and flammable when in contact with water or air, handling of Li is more complicated than Be. This disadvantage is compensated by the softer neutron spectrum than that of the p-Be reaction;
- A neutron source using a low energy accelerator coupled with d-Be and d-^{13}C reactions is under construction [8–9, 12], and development of the Be and C targets is in progress [34–35].

Table 3 collects the relevant thermal properties (melting points and thermal conductivities) of the targets for the options under consideration today. The most notable differences are that Li has a very low melting point and lower thermal conductivity than Be and C.

2.2.1 Thermal loading of the target

Heat removal is a significant issue for targets. To avoid too high a heat density, various methods have been adopted. One of the methods widely used is proton beam spreading to reduce the power density. Flat plate targets are used for p-Be neutron sources. For Li targets, other solutions have been developed to compensate for the low thermal conductivity and melting point. A cone shaped target was first demonstrated at the National Cancer Center in Japan [36–37] and a rotating Li target at Helsinki University Hospital [38]. However, even flat plate targets can be used [39].

Liquid Li targets are also considered since they can remove the heat flux more easily, increasing the potential proton current and neutron source strength [33]. However, efficient trapping of Li vapour has to be implemented in order to avoid short circuiting and contaminating the accelerator in case of an unwanted temperature excursion. Handling of a large amount of Li needs careful safety consideration and may require regulatory permission.

TABLE 3. THERMAL PROPERTIES OF NEUTRON PRODUCING TARGET MATERIALS

Material	Melting temperature (°C)	Thermal conductivity ($W \cdot m^{-1} \cdot K^{-1}$)
Li	180	84.7
Be	1287	201
C	3550	230

2.2.2 Blistering and mechanical issues

Blistering is another issue to be considered for the target. Depending on their energy and the target thickness, the protons (deuterons) may stop in the target just beyond the Bragg peak (Section 2.1) where they initially form 'injected interstitials', which reduces the thermal conductivity of the target. Eventually the protons/deuterons can recombine to form H_2/D_2 gas, and when the pressure becomes great enough the target can locally rupture to form blisters, further reducing the thermal conductivity. In order to avoid these problems, it is best to let the beam stop outside the Li or Be layer.

2.2.2.1 High energy protons

For protons of a few tens of MeV (e.g., 30 MeV [28]), the Be thickness required to produce the neutrons above the threshold energy is several mm (5.5 mm is reported in Ref. [28]). In such a case, the Be plate can be used as a structural material to separate the evacuated beam tube from atmospheric pressure. This makes the target structure simpler, namely, only two layers consisting of a Be plate and a backing containing a cooling water device. In such a case, the protons stop in the water layer behind the Be target plate, as the Bragg peak appears in the water layer and the high heat load is removed directly by the water flow. This is one of the advantages of using high energy protons with a Be target. However, the production of ^{7}Be and tritium (Section 2.1.2.2) in this energy range may be considered disadvantages.

2.2.2.2 Medium energy protons

The required Be thickness is thinner than for high energy protons and Be cannot be used as a structural material in this energy range (e.g., Ref. [31], 0.5 mm of Be for $E_p = 8$ MeV). Therefore, the solution of installing an antiblistering material immediately behind the neutron generating target has been adopted. Blistering by protons has been measured for several materials in target systems [40–41]. The characteristics of candidate anti-blistering materials and Cu are summarized in Table 4. Pd and Ta have almost the same blistering limit, withstanding proton fluences of greater than 200×10^{22} m^{-2}, V more than 120×10^{22} m^{-2}, and Cu is not very tolerant [40].

The thermal conductivity of V is the lowest among candidate anti-blistering materials, but the melting point is rather high, 1910 °C, and the half-life of ^{52}V is very short. Tantalum has a high blistering limit and a high melting point, but it produces a long half-life isotope [42]. In existing BNCT facilities, Pd has been adopted for Be [31] and Li [32] targets, and V has been adopted for a Be target used at a neutron source for neutron beam experiments [42].

A Li-target has been tested with a ~2 MeV proton beam [43], which suggested that a Li-target may be used without an anti-blistering material.

15

TABLE 4. CHARACTERISTICS OF ANTI-BLISTERING MATERIALS [40, 42]

Anti-blistering material	V	Ta	Pd	Cu
Proton blistering limit (10^{22} m^{-2})	> 120	> 230	200–300	0.4–0.1
Thermal conductivity (W·m^{-1}·K^{-1})	30.7	57.5	71.8	401
Melting point (°C)	1910	3017	1555	1084
Major activation product, half-life	^{52}V, 3.7 m	^{182}Ta, 114 d	^{103m}Rh, 56 m	^{64}Cu, 12.7 h

2.3 ACCELERATORS

Ref. [6] provides an overview of low energy neutron producing accelerators. There are many different types of accelerators ranging from low energy electrostatic machines, through radiofrequency quadrupoles (RFQs) and drift tube linacs (DTLs), to higher energy cyclotrons that are being adopted for BNCT facilities. This section introduces the accelerators presently in use at BNCT facilities.

2.3.1 Electrostatic accelerators

Electrostatic accelerators use static, high voltages to accelerate ions. The following designs are currently in use for BNCT applications:

- Dynamitron. A dynamitron is a single-ended machine, manufactured currently by IBA[3], enclosed in an SF$_6$-filled pressurized vessel, where the high voltage is generated through rectified RF power. A dynamitron has been used for BNCT with $E_p \sim 2.8$ MeV. Annex I describes the Nagoya facility based around a dynamitron;
- Vacuum Insulated Tandem Accelerator (VITA). The first facility to be developed was at the Budker Institute of Nuclear Physics in Novosibirsk, Russia [44–45]. The electrostatic voltage is sustained between equipotential metallic electrodes in vacuum (no insulating gas) and produced through rectified AC power. It delivers proton currents of up to 10 mA at 2.3 MeV. All tandem accelerators require change of charge from negative to positive particles using a stripper which has to be capable of reliable operation at high beam intensity. The VITA design is being commercialized by TAE Life Sciences[4]. Annexes II–IV describe facilities designed around such an accelerator;
- Single-ended conventional tube accelerator. Neutron Therapeutics Inc[5] markets a high current, single ended electrostatic accelerator with conventional accelerator tubes, enclosed in an SF$_6$ tank in which the high voltage is generated by a series of high voltage supplies powered by rotating and insulated shafts. A design based on a similar principle has been published in Ref. [46]. It delivers at least 30 mA, and up to 40 mA, of protons at 2.6 MeV. Annexes V and VI describe facilities designed around such an accelerator;
- Electrostatic quadrupole (ESQ) accelerator. An ESQ accelerator facility is under development in Argentina [8, 12]. It consists of a succession of discrete ESQs of alternating polarity with strong transverse focusing which produce a net focusing action to counteract the repulsive space charge effect [46]. The first example under construction is a single-ended ESQ with terminal voltage of up to 1.45 MV designed to work in air to avoid the need for a pressure vessel and insulating gas. The project aims at eventually delivering a 30 mA deuteron beam. For higher energies up to about 3 MeV, a folded tandem ESQ has also been proposed [46].

---

[3] IBA: www.iba-industrial.com/accelerators
[4] TAE Life Sciences: www.taelifesciences.com/
[5] Neutron Therapeutics: www.neutrontherapeutics.com/

2.3.2   Linear radiofrequency electrodynamic accelerators

For particle energies of up to about 3 MeV, the electrodynamic accelerator of choice is the RFQ. It is a resonating cavity working on the transverse electric TE_{210} mode. The RFQ principle was invented in 1970 [47]. It consists of a continuous succession of strongly transversally focusing quadrupoles which have, in addition, an accelerating longitudinal electric field [48]. It is machined with high precision from a single high quality piece of copper and has to be carefully temperature stabilized due to its high quality factor. Annex VII describes an RFQ based facility at the National Cancer Center, Japan. For energies above 3 MeV, the RFQ has to be supplemented with a drift tube linac (DTL) as a post accelerator. The DTL is a succession of cavities (drift tubes) whose lengths increase to keep pace with the increasing velocity of the accelerated particle. When the charged particle is inside the drift tubes, it is shielded from the electric field, but when it emerges it has to 'see' the correct polarity to be further accelerated; i.e., it has to be in phase with the RF electric field. Annexes VIII and IX describe facilities based around RFQ–DTL linacs. Both RFQs and DTLs produce a beam with a bunched time structure. Each of these machines needs sophisticated RF-power generating equipment, which is not readily available to all countries due to lack of domestic supply.

2.3.3   Cyclotrons

Cyclotrons accelerate ions in a spiral using a relatively simple RF structure and a fixed magnetic field. Charged particles from an ion source are injected into the centre of the cyclotron. The alternating electric field in the gap between the Dee-shaped electrodes is used to accelerate the ions, and they follow a spiral trajectory, gradually increasing their radius until they are extracted and directed towards a target. Annexes X and XI describe facilities based around a cyclotron.

Either positive or negative ions can be accelerated in a cyclotron:

- In the case of 'positive ion machines', protons are accelerated, and the beam is extracted at the beam extraction port by an electrostatic deflector. Due to beam loss, heat is generated locally, and wasteful radiation is generated that can activate cyclotron components;
- In recent years, 'negative ion machines' have come to dominate. A multi-cusp external ion source generates H^- currents of 10 mA or more. These ions are injected vertically into the centre of the cyclotron using focusing solenoids, a buncher and a carefully designed spiral inflector [49]. The proton beam is extracted by a conductive foil that strips the electrons from the H^- ion to leave a bare proton. The beam orbit due to charge conversion is diverted, and it is possible to extract almost 100% of the current from the acceleration orbit.

Commercial, 1 mA, 30 MeV negative ion cyclotrons developed by Sumitomo Heavy Industries[6] are used for clinical BNCT. With a Be target, it produces up to 10^{14} neutrons per second with neutron energies up to 28 MeV in the forward direction (Fig. 7, Table 2) to generate an epithermal neutron flux with an optimized BSA [50]. The first clinical trials in the world using AB-BNCT were performed on that system, and approval for manufacturing and selling it as a medical device was obtained in 2020 [51].

---

[6] SHI: www.shi.co.jp/industrial/en/product/medical/bnct/

2.4 CURRENT ACCELERATOR SYSTEMS WORLDWIDE

Figure 11 shows the world distribution of BNCT centres. Argentina, Belgium, China, Finland, Israel, Italy, Japan, Korea, Russia, Spain, and the UK are among the countries actively engaged in AB-BNCT projects.

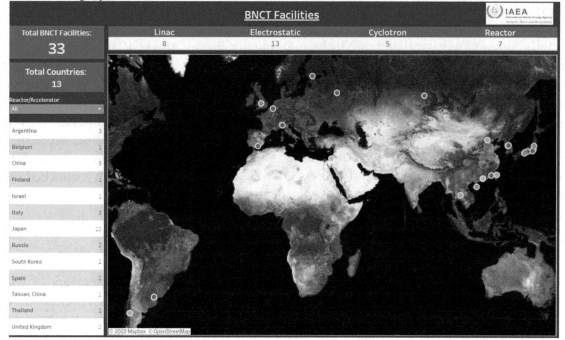

FIG. 11. Visualization of the IAEA BNCT database [52] showing the distribution of operational research reactors and the planned and operational accelerator based neutron facilities involved in BNCT. Colour code: red = electrostatic; orange = cyclotron; cyan = linac; green = reactor.

Japan is by far the country devoting the largest effort to this endeavour, and Japanese projects are shown separately in Fig. 12.

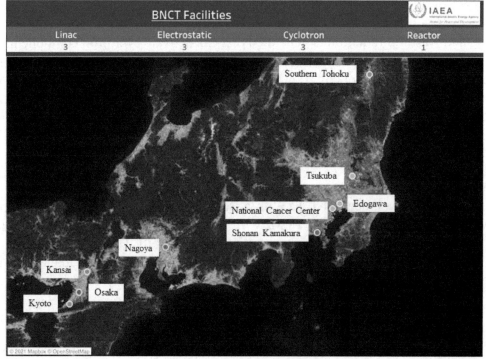

FIG. 12. Blow-up of the planned and operational Japanese BNCT facilities from Ref. [52]. Colour code: red = electrostatic; orange = cyclotron; cyan = linac. (Kyoto University also hosts the research reactor.)

18

Table 5 gives the present status and performance of the different accelerator projects underway for AB-BNCT facilities. Some have already been developed and some are under construction. Most of them still need a significant upgrade in terms of beam intensity. Below, a summary of the world situation with respect to accelerator type, energy and target materials is given:

- The lowest energy accelerator, the ESQ, uses deuterons with Be or ^{13}C targets;
- All the electrostatic accelerators with E_p in the range 2–2.8 MeV use Li targets;
- The RFQs with E_p < 3.5 MeV use Li targets;
- The higher energy RFQs and DTLs with E_p in the range 5–10 MeV use Be targets;
- The 30 MeV cyclotrons use Be targets.

- Limit out-of-field radiation dose. The role of the shielding around the BSA is to ensure that the radiation dose to the rest of the patient's body is kept within acceptable limits (Appendices II, III).

3.2 BEAM SHAPING ASSEMBLY

The structure of the BSA: its components, their sizes, and materials, depend on the energy distribution of the primary neutrons produced by the neutron source and the final required therapeutic spectrum (Section 2). It has to transform, as close as possible, the characteristics of the beam emitted from the target to those required by the facility designer.

Figure 13 shows a schematic image of a typical BSA that generates an epithermal neutron beam. It consists of the following components:

- A high-energy neutron filter installed after the target;
- Immediately downstream is the moderator used to further reduce the neutron energies to the therapeutic range: this is typically the largest component;
- Next, thermal neutron and gamma ray filters are installed just upstream of the beam collimator to remove thermal neutron and photon contamination, respectively;
- Neutrons, with appropriately adjusted energy spectrum, pass through a collimator, and are emitted from a beam aperture with appropriate directionality, spectrum and flux;
- Surrounding these components is shielding.

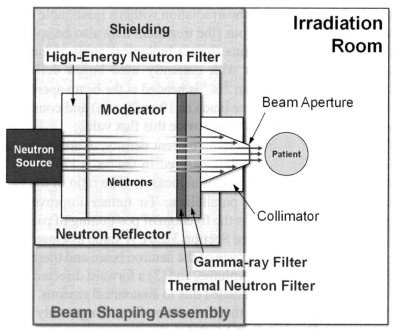

FIG. 13. Typical beam shaping assembly for an epithermal neutron beam. All the components enclosed by the red line are usually counted as part of the BSA (courtesy of H. Kumada, University of Tsukuba).

In the case of a BSA that generates a thermal neutron beam, the thermal neutron filter is not included in the BSA, but the size of the moderator required to further reduce the energy of neutrons may increase.

The experience gained concerning desirable neutron beam characteristics and BSA design from reactor based BNCT can also be applied to the design of the BSA for AB-BNCT. However, several additional restrictions for materials and handling may apply for an AB-BNCT device. These include safety and regulatory aspects for commercial or hospital based devices, which may differ from those applied to reactor based devices.

Next, the individual components of a BSA are described in more detail.

3.2.1 High-energy neutron filter

Accelerator-based neutron sources produce neutrons with a spectrum of energies, including 'fast neutrons'; i.e., those where $E_n > 10$ keV. Where present, these fast neutrons are not desirable for BNCT, and their flux has to be reduced in the treatment beam. A reference value of 2×10^{-13} Gy·cm^2 per unit epithermal neutron fluence was established by Ref. [1], where the fluence is defined as $\int \phi_{epi}(t) \cdot dt$, measured in units of cm^{-2}. There are clinical BNCT facilities with slightly higher fast neutron components which are delivering successful patient treatments, and therefore this has been revised (see Table 6 and Appendix I).

Generally, materials that have high cross-sections or resonance peaks in the high energy neutron range are used. In current AB-BNCT systems, Fe and Pb have been adopted as high-energy neutron filter materials [50, 55]. With these materials, it is possible to reduce the neutron energy to around 1 MeV by inelastic scattering, and also through (n,2n), (n,3n) [50] and (n,xn) reactions, although some reactions lead to long lived activation products. For the different combinations of particle and targets described in Section 2.4, the following observations can be made:

- Currently designed p-Be systems all use protons with E_p in the range 5–30 MeV (Table 5), and the maximum neutron energy emitted from the target in the forward direction is $E_p - E_{th}$, where $E_{th} = 2.057$ MeV (Table 2, Section 2.1.2.2). Thus, a BSA designed for a Be-based neutron source has to include a high energy neutron filter, and its thickness depends on E_p;
- In a p-Li system, neutrons are created by irradiation of the Li target with protons of $E_p \sim 2.5$ MeV. The energy of the primary neutrons is in the order of a few hundred keV (Section 2.1.2.1). Since there are no neutrons with energy in the MeV range, a high-energy neutron filter may not be required in the BSA;
- For d-Be and d-^{13}C systems, there is generally no need for a high-energy neutron filter [12, 18].

3.2.2 Moderator

Generally, the energy of neutrons that have passed through a high-energy filter is still too high to use for patient irradiation. The primary role of the moderator in a BSA for an AB-BNCT system is to reduce and adjust the energy of the neutrons to the therapeutic range required to be emitted from the beam aperture. For the reasons mentioned above, the moderator for a p-Li system can be relatively small.

The design of a moderator has to consider a balance between the neutron flux and the spectrum at the beam aperture. Further reduction of the fast neutron component, increasing the low energy component of the epithermal range, or the production of a predominantly thermal beam can all be achieved by increasing the size of any given moderator. However, increasing the size of the moderator results in the reduction of therapeutic neutron flux leaving the beam aperture. This is because it increases the opportunity for neutron absorption within the moderator or loss through the reflectors. Therefore, the choice of moderator materials may need to be changed to more efficient materials. A moderator may consist of a single material or a combination of different materials. To be feasible as components of a moderator, these materials have neither to decompose in a high radiation field nor produce moisture. A high physical density is also advantageous. Furthermore, it would be preferable if the half-life of the neutron activation products from these materials were short.

Moderation occurs via predominantly elastic collisions between neutrons and the nuclei of the moderator [72]. It is most efficient when the nuclei are approximately the same mass as the neutron. However, for BNCT requiring an epithermal spectrum, the moderator only has to be effective at relatively high energies to tune the spectrum down from the fast regime and into the epithermal range without producing too many thermal neutrons. For this reason, the moderator materials generally differ from those seen in reactor cores. Epithermal moderators use moderate weight elements (e.g., Mg, Ca, Al, F) in the form of ceramics that have resonances in their scattering cross sections in the upper epithermal range [6], thereby reducing the required moderator size. As examples, Fig. 14 shows scattering, absorption, and γ production cross sections for the main isotopes of Al, O, and F. The following are examples of materials used:

- MgF_2, AlF_3 and CaF_2 have been successfully applied in AB-BNCT systems [73];
- Fluental, developed for reactor based BNCT by VTT[7] may also be used in an AB-BNCT device;
- For d-Be and d-^{13}C systems, two combinations of materials have been proposed, respectively: Al–AlF_3 and Al–PTFE (Teflon).

D_2O has been used in the BSA of reactor based BNCT facilities. D_2O, a liquid, is useful because the neutron spectrum at the beam aperture can be shifted easily by changing the volume of the material in a moderator's vessel. However, from the point of view of regulation and handling, it may be difficult to use as a moderator in a hospital based AB-BNCT device, as it gradually becomes tritiated over time, and as it is a fluid, there is the risk of a spill or leak.

Monte Carlo calculations have shown that systems using low-energy deuteron beams ($E_d \sim 1.45$ MeV) together with simple optimized AlF_3-based BSAs could deliver treatments of brain tumours comparable to the best present-day reactors [8–9, 12, 18].

3.2.3 Neutron reflector

Neutrons scatter on their journey through the moderators and filters. Thus, to obtain as many neutrons as possible at the beam aperture, the BSA is surrounded by neutron reflectors to scatter neutrons back into the BSA. Generally, materials that have high cross sections or reaction resonance peaks in the high energy neutron range are used. Lead is usually used as the reflector of the BSA for typical epithermal BNCT systems. The choice of a heavy reflector material ensures no additional beam moderation occurs, only redirection of neutrons. Materials such as Be and graphite can be applied as reflectors for thermal neutron sources.

3.2.4 Thermal neutron filter

During generation of an epithermal neutron beam, the thermal neutron component needs to be minimized, and a thermal neutron filter is installed in the BSA, usually downstream of the moderator or else in the collimator upstream of the beam aperture. Materials, such as Cd, which have a high absorption cross-section for low-energy neutrons may be used, although it is a γ emitter. Materials containing ^6Li are often used since it has a high capture cross section and the reaction does not produce γ rays. A reference number for the ratio of thermal flux to epithermal flux for an AB-BNCT beam is 0.05 or less (as was proposed in Ref. [1] for research reactors). However, the proportion of thermal neutrons that is acceptable in a beam depends on the location and dimension of the tumour. For a deep-seated tumour a low thermal component is profitable, while for large tumours extending to the skin surface (i.e., some head and neck malignancies) a thermal component might be useful.

---

[7] VTT, Finland. https://www.vttresearch.com/en

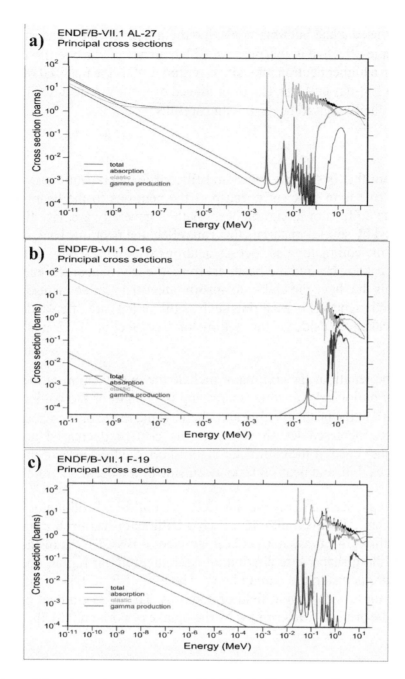

FIG. 14. Comparison of the principal cross sections of (a) ^{27}Al, (b) ^{16}O, and (c) ^{19}F. Although the elastic cross sections of ^{16}O and ^{19}F are similar over a wide energy range, in the range E >10 keV, the nuclear resonances give ^{19}F a significantly greater value. Similarly, ^{27}Al has several strong resonances that are beneficial in this range. As the neutron energy drops outside of the resonance range, the cross sections (probabilities) for absorption (capture) increase proportional to 1/v of the neutron, as is typical for most isotopes. Data are from the IAEA Evaluated Nuclear Data File. Reproduced from Ref. [6].

3.2.5 Gamma ray filter

Neutrons passing through the BSA may activate its various components and produce a substantial quantity of photons that contribute to an undesirable radiation dose deposited in tissues. While some of the photons are from prompt (n,γ) processes, the γ background also gradually builds up from longer term activation of the upstream components. Thus, a γ ray filter needs to be installed at the end of the neutron beam path in a BSA with the aim of reducing the γ component of dose to 2×10^{-13} Gy·cm^2 per epithermal neutron fluence, in line with that suggested for reactor sources in Ref. [1].

Lead and Bi are usually used since both materials, having a high Z value, can efficiently attenuate γ rays. Although Pb is more effective in this process, Bi has an advantage over Pb in terms of neutron permeability. Thus, when a higher neutron intensity is required, Bi is the material of choice. However, ^{210}Po, an α emitter with a 138-day half-life, can be produced by neutron irradiation of Bi. The presence of such an α emitter at a facility has to be dealt with carefully.

3.2.6 Collimator

Neutrons, with a spectrum that is suitably tuned and filtered for therapeutic use, are delivered to a collimator, whose main role is to efficiently transport the neutrons to the beam aperture. For this purpose, the centre of the collimator along the beam path is generally a cavity filled with air. This cavity space is surrounded by several materials that can reflect the neutrons back into the beam path. The downstream part of the collimator also focuses neutrons within the cavity to the beam aperture, and, to prevent neutron leakage beyond the beam aperture, the cavity near the end of the collimator is surrounded by materials that have the ability to absorb rather than reflect neutrons. The collimator is typically made of polyethylene or Pb. Materials such as LiF or B_4C are often added to polyethylene to reduce the neutron leakage outside of the collimator (see Section 4.2.3 and Appendix III for discussion).

The design criteria for the length of the collimator include the optimization of the neutron flux, the current-to-flux ratio, and biological shielding. Increasing the length of the collimator improves the 'current-to-flux ratio (J/ϕ)' at the beam aperture and, thereby, the range of therapeutic neutrons in the patient's body. However, increased length comes at the cost of decreased neutron flux at the aperture's exit. On the other hand, if the collimator is too short, the thickness of the shielding wall at the beam aperture becomes thin, and neutron leakage outside of the beam aperture may increase.

In conventional radiotherapy, such as X ray therapy (XRT), a multi-leaf collimator is commonly used to shape the radiation beam. However, although such a collimator has been proposed for use in a reactor based BNCT facility [74], it has not yet been applied in BNCT. One of the reasons is that it is impossible to make a sharp shape dose distribution around a tumour region due to the inevitable diffusion of thermal neutrons inside the human body. The aim of BNCT is to infiltrate target cells with boron and irradiate these with a wide field of neutrons. The dose gradient between cancerous and healthy tissue has to be primarily determined by the uptake of boron not by the neutron field (see Section 12 and Annex XIX for discussion).

3.2.7 Beam aperture

An optimized neutron beam, whose spectrum has been tuned by filters and the moderator, and subsequently focused by the collimator is released from the beam aperture to irradiate the patient. Typically, the aperture is circular. Often, attachments to the beam aperture with different diameters are prepared in advance. The diameter depends on the size and shape of a target region. For the irradiation for head-and-neck cancers, malignant brain tumours, and malignant melanomas, apertures of 10–15 cm in diameter are generally used. For the irradiation of the breast region, as used for mesothelioma and lung cancers, beam apertures with a larger diameter are applied. The beam aperture is usually made with materials that are good for shielding neutrons (typically, LiF-spiked polyethylene) and photons. Developments allowing extensions to the beam collimator/aperture to allow more flexible patient positioning are underway at reactor and accelerator based sites [75–77].

A shutter or portable lead shielding may be placed at the aperture when the system is not operating to reduce γ dose to staff working in the irradiation room arising from the BSA's activated components.

3.2.8 Shielding

Radiation has to be confined as much as possible to the beam path and within the BSA. This requires additional shielding materials at extra cost. The shielding design depends on optimal selection of materials and their arrangement to fit the characteristics of the radiation at the specific location. Neutron energy can be reduced by using hydrogen rich materials such as polyethylene that further moderate the spectrum. The thermalized neutrons can finally be captured by materials containing Li and ^{10}B. These materials are used in particular at the end of the BSA in order to reduce the neutron leakage outside of the beam defining aperture. Gamma rays (photons) are shielded by using Pb or Bi.

Generally, concrete construction is used around the BSA and treatment room. Concrete contains light elements in the cement (hydrous phases) and heavier elements in the light and coarse aggregate. It therefore acts as both a structural engineering component for the facility and a formable composite neutron and γ shield (see Section 5.2.1.2).

3.3 SUMMARY OF REFERENCE BEAM QUALITY FACTORS

The values in Table 6 are presented in the context of the main themes of this publication. They are therefore relevant to the use of epithermal neutron beams derived from accelerators and used for the treatment of relatively deep-seated targets where the boron carrier is BPA. For different clinical cases and different boron compounds, a different set of beam reference values may be appropriate. The reference values in Table 6 are specifically not suggested as 'requirements' or 'recommendations' that have to be achieved by a system provider or facility.

TABLE 6. REFERENCE NEUTRON BEAM QUALITY FACTORS

Beam quality component	Symbol or definition	Reference value	Section	Ref.
Therapeutic epithermal flux[a]	ϕ_{epi}	$\geq 5 \times 10^8$ cm^{-2}·s^{-1} [e]	3.1	[1, 78[e]]
Thermal to epithermal flux ratio	ϕ_{th} / ϕ_{epi}	≤ 0.05	3.1	[1]
Beam directionality[b]	J / ϕ_{epi}	≥ 0.7	3.1	[1]
Fast neutron dose per unit epithermal fluence[c]	$D_H / \int\phi_{epi}(t)\cdot dt$	$\leq 7 \times 10^{-13}$ Gy·cm^2	3.2.1	Table 27
Gamma dose per unit epithermal fluence[c,d]	$D_\gamma / \int\phi_{epi}(t)\cdot dt$	$\leq 2 \times 10^{-13}$ Gy·cm^2	3.2.5	[1]

[a] ϕ_{epi} refers to the flux of epithermal neutrons in the energy range typically defined for BNCT (Table 1).
[b] Much lower values (e.g., 0.3) can be used for treatment of melanomas with more thermalized beams.
[c] These are doses per unit fluence of epithermal neutrons, where epithermal fluence is defined as $\int\phi_{epi}(t)\cdot dt$, in units of cm^{-2}.
[d] The range reported in reactor based BNCT facilities is $1–13 \times 10^{-13}$ Gy·cm^2.
[e] Ref. [78] reports $\int\phi_{epi}(t)\cdot dt = 5.3 \times 10^{11}$ cm^{-2} used clinically in $t = 17$ min, corresponding to $\phi_{epi} = 5.2 \times 10^8$ cm^{-2}·s^{-1}.

The value suggested in this publication as a reference for the fast neutron dose per unit epithermal fluence is different from that in the original IAEA TECDOC-1223 [1]. The reasons for this change are explained in Table 27 of Appendix I and originate from clinical experience that has been gathered in the intervening years.

4 PHYSICAL DOSIMETRY AND DETERMINATION OF NEUTRON FIELD PARAMETERS

This section describes the radiation fields involved in BNCT and the sources of dose to patients. It describes the challenges involved with physical dosimetry at BNCT centres, the methods used and some of the estimated measurement uncertainties. Dosimetry for BNCT is much more complex than for standard radiotherapy where collimated photons mainly produce electrons releasing all their kinetic energies by ionization. BNCT relies on the capture of a neutron by ^{10}B which generates an α particle and a recoiling ^{7}Li nucleus via two channels: in 93.7% of the cases the ^{7}Li nucleus is in an excited state (^{7}Li*), which decays promptly to the ground state accompanied by a 478 keV γ photon, and in 6.3% of the cases it enters the ground state directly (Fig. 15) [79]:

$$^{10}\text{B} + {}^1\text{n} \rightarrow {}^{11}\text{B} \quad
\begin{array}{ll}
\xrightarrow{6.3\%} & \alpha\ (1776\ \text{keV}) + {}^7\text{Li}\ (1015\ \text{keV}) \\
\xrightarrow{93.7\%} & \alpha\ (1472\ \text{keV}) + ({}^7\text{Li})^*
\end{array}$$

$$\downarrow$$

$$\alpha\ (1472\ \text{keV}) + {}^7\text{Li}\ (840\ \text{keV}) + \gamma\ (478\ \text{keV})$$

FIG. 15. The boron neutron capture reaction.

As BNCT is based on a neutron field produced by a nuclear reaction and subsequent moderation, one needs to measure or to estimate by simulations not only the number of neutron capture reactions on ^{10}B, but also the neutron capture reactions on ^{1}H and ^{14}N, the associated γ production of these neutron capture reactions and their interaction with tissues. In addition, an important contribution comes from elastic collisions of fast and epithermal neutrons on ^{1}H, ^{12}C, and ^{14}N present in human tissue, as the neutron field undergoes further moderation within the body. The neutron capture reactions on ^{10}B and on ^{14}N produce nuclear recoil products (^{4}He, ^{7}Li, and ^{14}C), releasing a large amount of energy (> 200 keV/μm), which is the main contribution to the high linear energy transfer (LET) of BNCT (see Section 10). Neutron capture cross sections strongly depend on neutron energy. The principal neutron cross-sections for ^{10}B are shown in Fig. 16, where it can be seen that absorption is the dominant process for epithermal and thermal neutrons.

The choices of particles, energies, and targets for neutron production (Section 2) in compact accelerators create an important degree of freedom for the optimization of the neutron field produced and, therefore, the BNCT dose to tumours while minimizing the secondary dose to healthy tissues. Epithermal neutron energies can then be optimized during the design to treat tumours at a given depth. The BSA around the targets defines the neutron field[8] or 'beam' used in BNCT (Section 3). BNCT dosimetry is therefore strongly related to the upstream choice of accelerator, target, and BSA.

Before performing treatment with an AB-BNCT system, it is necessary to confirm the characteristics of the treatment 'beam', or field, by appropriate radiation measurements. In-air, in-phantom, and whole-body position dosimetry are needed during a system acceptance test or commissioning (Fig. 17). The results measured during commissioning are used as quality assurance (QA) reference values during treatment. Dosimetry is used for standardization of the radiation sources during commissioning of treatment planning system. During treatment, calibrated 'beam' monitor values are used to deliver the prescribed dose. The calibration of the 'beam' monitor is performed at the time of commissioning. Treatment is performed after the confirmation that the measured value does not

---

[8] The term 'neutron beam' is commonly used when describing the accelerator and BSA. However, from the point of view of metrology, what is measured is the distribution of flux, often simultaneously with the energy spectrum, of neutrons through space; i.e., in a neutron field.

deviate from the reference value set by the QA irradiation performed before the treatment. Thus, physical dosimetry is important in ensuring delivery of the prescribed dose during BNCT.

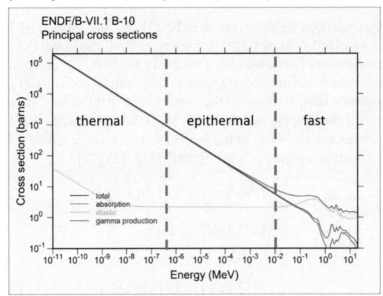

FIG. 16. Neutron cross sections for ^{10}B as a function of energy displaying a very strong rise in the probability of capture as the neutron energy drops. The ranges thermal, epithermal, and fast are shown as defined in the common terminology of BNCT (Section 2). Data originate from the IAEA Evaluated Neutron Data File.

Since neutrons have different radiation qualities in the thermal, epithermal, and fast energy regions, it is desirable to discriminate and measure them in each energy region. In addition, γ rays are mixed in the irradiation field, and it is necessary to discriminate the γ ray contribution from that of the neutrons. The thermal neutron flux, after further moderation in the body, required at the tumour position is of the order ~10^9 cm^{-2}·s^{-1}, and it is necessary to select an appropriate measurement method.

In the following sections, the physical dosimetry required for BNCT treatment is introduced. This consists of measurements in-air, in-phantom, whole body exposure (as shown in Fig. 17), real time beam monitor, and QA.

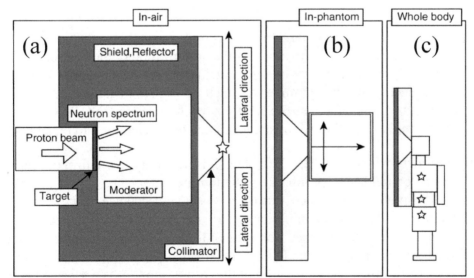

FIG. 17. Schematic layout of physical dosimetry in an AB-BNCT system showing (a) in-air, (b) in-phantom and (c) whole body measurements (courtesy of H. Tanaka, Kyoto University).

4.1 RADIATION COMPONENTS

The following four radiation components in the BNCT irradiation field are considered in dosimetry. Each component of the total dose results from ionization energies arising from different nuclear reactions (Table 7).

TABLE 7. THE FOUR PRINCIPAL DOSE COMPONENTS IN BNCT

Symbol	Common name(s)	Originating reactions
D_B	Boron dose	The $^{10}B(n,\alpha)^7Li$ reaction (Fig. 15);
D_N	Nitrogen dose Thermal neutron dose	Mainly the $^{14}N(n,p)^{14}C$ reaction yielding a recoiling ^{14}C nucleus and 583 keV protons
D_H (D_n, D_f)	Hydrogen dose Fast neutron dose Proton recoil dose	The (n,n') moderation reaction occurs mainly with ^1H in the body. Energy is released by the recoiling protons produced by interaction with fast and epithermal neutrons.
D_γ	Photon dose	Prompt γ from radiative capture within the patient mainly from the $^1H(n,\gamma)^2H$ reaction and γ contamination of the incident radiation field.

Note: In the literature, D_H has several notations, including D_n and D_f, where n = neutron and f = fast.

The principles of neutron detection, various types of detectors, instruments, and dosimeters are described in the IAEA Safety Series Report 115 [80]. The neutron energy range of the BNCT irradiation field is continuous from a few meV up to a few MeV with a flux as high as 10^9 cm^{-2}·s^{-1}. It is difficult to measure the neutron spectrum directly at all energies. The neutron energy spectrum is, therefore, evaluated by a combination of Monte Carlo simulations and measured values.

4.2 IN-AIR MEASUREMENTS

The following sections discuss the requirements for neutron energy spectra measurements at various points along the system. In-air measurements have two main purposes. First, to verify that the BSA is performing as specified by the manufacturer. Second, to validate the neutron 'beam' model for treatment planning and dosimetric calculations (see Section 11.1.2).

4.2.1 Neutron energy spectrum at the target

In an AB-BNCT system, a nuclear reaction between a charged particle and a target nuclide generates neutrons (Section 2.1.1). Neutron transport calculations are performed using Monte Carlo simulations when designing the system. It is desirable to use a neutron energy spectrum from evaluated nuclear data and an appropriate nuclear reaction model. Calculations can provide neutron energy and angular distributions from the target.[9] Calculations are also backed by measurements using various methods. There is a new possibility to characterize the neutron spectrum based on the 3D nuclear recoils produced by neutrons [12]. The Bonner sphere spectrometer (BSS) can be used for this purpose: it utilizes a response function that combines polyethylene spheres of several sizes with a thermal neutron detector [81]. A nested moderator spectrometer has also been developed and tested at other neutron producing accelerator facilities [82].

4.2.2 Neutron energy spectrum at the beam port

Fast neutrons generated by the target are moderated to epithermal neutrons suitable for BNCT using the moderator in the BSA (Section 3.2.2). As the patient is irradiated with a therapeutic 'beam' or

---

[9] When a nuclear reaction is not included in the evaluated nuclear data used for neutron transport calculation, it is desirable to confirm the neutron energy spectrum from the target by direct measurement and to verify the validity of the nuclear reaction model, usually using the time-of-flight method. However, this may not be practicable in all accelerator types used for AB-BNCT.

neutron field, defined by the BSA to match the size of the treatment area (see Section 12 and Annex XIX), it is important to evaluate the neutron spectrum at the collimator outlet. Fast and thermal neutrons are mixed with the therapeutic epithermal neutron field, and it is desirable to discriminate and measure each component of the spectrum. Among the choices are:

- A BSS can be used [83] at the beam port in conjunction with unfolding techniques [84–86] to obtain the neutron spectrum. Depending on the size of the polyethylene sphere, the BSS perturbs the neutron field to a greater or lesser extent [87]. When using a BSS in a narrow irradiation field, such as at the exit of a collimator, corrections may be required;
- Multi-foil activation methods [50] can also be used. A list of activation foils useful across a wide range of energies is given in Table 1 of Ref. [88] and Table 2 of Ref. [89];
- Based on previous work on single moderator, directional spectrometers [90–91], a spectrometer with emphasised response in the epithermal domain has been designed and prototyped especially for BNCT [92];
- Recombination chambers, based on high pressure ionization chambers, have been applied for many years in high energy radiation fields. New cylindrical pen-like recombination chambers of the T5 and G5 types have been constructed specially for BNCT, capable of measuring all four BNCT dose components [93–94];
- Microdosimetry with a tissue-equivalent gas proportional counter can measure the quality of the therapeutic beam [95]. Since the neutron energy spectrum is not measured directly, evaluation in combination with Monte Carlo simulation is desirable.
- The adaptation of new detectors able to perform fast neutron spectroscopy and measure angular distribution is expected in the future [96–97].

4.2.3 Out-of-field leakage

Measuring the dose outside the irradiation field is important in assessing the exposure of the patient in the whole-body position [29]. Due to the high dose rate of neutrons and γ rays near the collimator, neutron survey meters typically used for radiation protection cannot be used. It is desirable to evaluate the neutron dose by combining the measured value from the activation foil method with Monte Carlo simulation. Gamma ray dose can be measured by thermoluminescence dosimeters (TLDs), glass dosimeters, and ionization chambers. Since some glass dosimeters and TLDs are also sensitive to neutrons, it is necessary to discriminate the dose contribution from neutrons.

The out-of-field leakage is another important quantity to be taken into consideration in the BSA design, and it needs to be evaluated and reported. The out-of-field leakage is the unintended radiation dose leaking through the BSA to the patient outside of the treatment volume and outside the primary neutron beam. The leakage radiation in BNCT consists of neutrons and photons. The out-of-field leakage dose, also called non-target radiation dose, is of concern because it causes an unnecessary risk of harm to the patient. Even relatively low (≤ 1 Gy) doses have been associated with increased risk of secondary cancer, cardiac toxicity, skin cancer, eye lens cataract, implanted medical device malfunction, and adverse effects on the foetus [98].

So far, specific recommendations for out-of-field leakage dose reporting or limits have not been established for BNCT. International guidelines and standards for light ion beam therapy techniques are helpful as a basis in specifying the specific parameters and limits for BNCT. Requirements for other radiotherapy modalities cannot be directly applied in BNCT due to the fundamental differences in the techniques and characteristics of each radiation modality. BNCT patient dose differs from that in other types of radiotherapies. Definition of the prescribed dose or dose delivered at the equipment reference point, the denominator to calculate the leakage dose ratio, depends on several parameters. The dose distribution in a tumour region is not uniform due to the heterogeneity of the neutron and photon dose distributions in the patient. Furthermore, boron concentration during the treatment is also different for each patient, boron compound, and the way the compound is administered (BPA infusion is either stopped prior to or continued throughout the irradiation, see Annex XIII). Thus, when reporting the prescribed dose in the out-of-field leakage evaluation, common values ought to be applied for parameters of prescribed dose such as the boron concentration, tumour-to-blood boron concentration ratio, the RBE and CBE factors, and the geometry in which the prescribed dose is defined (the phantom type, its location related to the neutron beam and the dose evaluation depth in the phantom). The requirements defined by the International Electrotechnical Commission (IEC) standard for the basic safety and essential performance of light ion beam medical electrical equipment and guidelines by the American Association of Physicists in Medicine (AAPM) for out-of-field leakage are described in Appendix II, and an example of the parameters for out-of-field leakage reporting in BNCT are given in Appendix III. The aim is to include BNCT specific dosimetry requirements, including the out-of-field leakage parameters, in international standards in the near future.

4.2.4 Neutron spatial distribution

Measuring the spatial distribution of neutrons in the treatment 'beam' contributes to the validation of the calculated neutron distribution that is output by the treatment planning system (see Section 11). The distribution can be verified by irradiating activation foils side by side, although the results are not available instantaneously as they have subsequently to be individually counted, e.g., in a high purity germanium spectrometer. Another method for measuring the 2D spatial distribution of neutrons is to irradiate a metal foil and then expose an imaging plate to it [99]. Absolute value correction is possible by measuring the saturated radioactivity of a metal foil such as Cu. Attempts have also been made to measure the neutron distribution using Gafchromic film [100]. Paired ionization chambers can also be used to assess the incident neutron distribution.

4.3 IN-PHANTOM MEASUREMENTS

The majority of BNCT tumour doses are due to neutron capture by ^{10}B (D_B defined in Section 4.1). Therefore, it is important to measure the depth distribution of thermal neutron fluence in a phantom that simulates a human body. A water phantom with a rectangular or cylindrical acrylic wall is used. Design of the phantom depends on the purpose of the measurement. For dose calibration, it is usual to use a sufficiently large water phantom that provides full backscattering conditions. Other phantom designs are used to represent patient treatment conditions and for QA. Solid phantoms are also used.

An activation method using either diluted or metallic Au or Mn foils may be used for the measurement of thermal neutron fluence [101]. When using Au foils, thermal neutron fluence can be discriminated from higher energy neutron fluence, by measuring uncovered foils together with foils covered by Cd. Scanning a region with a small scintillator coupled with a fibre can also be used to map the thermal neutron distribution [63]. TLDs and glass dosimeters can be used to determine γ ray doses in phantoms. However, as they are also neutron sensitive, suitable correction has to be made.

Since the contribution to dose from fast neutrons (D_H defined in Section 4.1) cannot be ignored near the surface of the phantom, it ought to be evaluated. The paired ionization chamber technique provides one option. Another approach is the MIMAC-FastN spectrometer, which has an active phantom mode able to count the number of neutron capture reactions at the tumour level with a known amount of ^{10}B nuclei [102]. This spectrometer permits the coupling of external high density polyethylene covers to adapt to the tumour depth. Recently, In foils have been applied for fast neutron dose verification in phantom [29].

It is necessary to measure the distribution of γ rays, which are non-therapeutic radiation components of BNCT. Measurements of the lateral distribution at the appropriate depth within the phantom are also useful in confirming beam symmetry. The neutron and γ ray dose distributions can be discriminated and measured using paired ionization chambers using a combination of a tissue-equivalent wall chamber and a graphite- or Mg-wall ionization chamber.

4.4   WHOLE BODY EXPOSURE

Whole body dose assessment is important to confirm the radiation safety of the device. In addition to the evaluation of neutron and γ ray doses in the lateral direction, described in Section 4.2.3, a phantom simulating a human body may be installed to evaluate the dose in the body. An activation foil, TLD, or glass dosimeter is installed at a position corresponding to each organ, and the neutron and γ ray doses are evaluated by comparing with the Monte Carlo simulation after irradiation [103]. Self-powered neutron detectors have been used for on-line determination of thermal neutron flux in patients, tissues, and reference positions during irradiations [104–106].

4.5   REAL TIME BEAM MONITOR

During treatment using AB-BNCT, a beam monitor calibrated in a similar manner to that for X ray treatment is required to determine the prescribed dose. Two approaches are possible.

4.5.1   Electric current of the accelerated charged particle beam

Before treatment of patients can be authorized, a linear relationship between the number of accelerated protons or deuterons (Section 2.1) incident on the target (as measured by the current) and the neutrons output has to be confirmed. Then, the prescribed dose may be delivered by using the indicated integrated current of charged particles [50]. It is advisable to monitor the health of the target, as changes in its condition may affect the linearity. The current meter measuring the current of charged particles has to be calibrated and its stability confirmed. This approach has been used at the National Cancer Center, Japan (see Annex VII).

4.5.2   Real-time neutron monitor

The location of a real time neutron monitor needs to be chosen so that it does not change the treatment 'beam' and is not affected by reflections from the patient. Therefore, it is usually installed inside or around the moderator. Fission chambers or BF_3 counters, which are resistant to large neutron fluxes and can discriminate them from γ ray events, are, in general, selected for this role. A small scintillator with fibre may also be a candidate in cases where radiation damage can be resolved. Before treatment, it is necessary to confirm the linearity between the number of counts recorded by the monitor and the neutron output. Typically, two independent monitor channels would be installed. Section 3.12 of Ref. [1] provides a useful summary of some practical features of a neutron monitor for BNCT.

4.6   MONITOR UNIT

When determining the prescribed dose, the concept of a monitor unit (MU), similar to that used in XRT, is required. In the case of XRT, an ionization chamber is installed upstream of the treatment beam, and the MU is calibrated using the absorbed dose in a water phantom (Section 4.3).

For the case of BNCT:

- When using the amount of charge (integrated current) of charged particles (Section 4.5.1) as the MU, the thermal neutron fluence distribution in the water phantom per unit charge is derived by measurement. Since the dose per unit charge is also derived in the treatment planning system, the prescribed dose can be delivered by setting the value of the integrated current for the treatment;

- When the count number of a real-time neutron monitor (Section 4.5.2) is used as the MU, this measurement has to relate to a useful dose quantity measured in the phantom. Often the thermal neutron fluence distribution in the phantom per unit count of the neutron monitor is measured. Since the treatment planning system evaluates the dose per count, the prescribed dose can be delivered by setting the total number of counted neutrons.

More details concerning the role of the MU in the treatment planning system, including IEC requirements for redundancy, are given in Section 11.1.5.

4.7 UNCERTAINTIES, TRACEABILITY

Reference [107] reviewed dosimetry methods for BNCT and gave recommendations for beam characterization, beam monitoring, and calibration, as well as frequency of measurements. It is important to evaluate the measurement uncertainties in the determination of doses contributed by neutrons and γ rays in BNCT dosimetry. Sources of uncertainty are extensively discussed in Ref. [107]. They depend on the measurement methods, and can include contributions from counting statistics, systematic effects, and the calibration of equipment and sources. For example, it has been suggested that the measurement uncertainty of thermal neutron fluence using the Au activation method lies in the range from ±5% to ±7% [101, 107], where this measurement uncertainty includes the uncertainty in the efficiency of the high purity germanium detector used for measuring saturated radioactivity. It is reported from two different centres in Japan that the measurement uncertainty of the γ ray absorbed dose is ±12–20% for a TLD enclosed in quartz glass [101, 108]. TLDs are calibrated in a standard ^{60}Co γ ray field that is traceable to a national primary standard. Even when using twin ionization chambers, it is possible to calibrate the electrometer output and absorbed dose in a standard ^{60}Co γ ray field. Measurement uncertainties from ±2.5% to ±9% have been reported for γ ray absorbed dose measured by this method [107]. For neutron dose measurements using this method, uncertainty estimates range from ±17% to ±24% [107]. Uncertainties in the measurement of the fast neutron absorbed dose have been determined as ±5 to ±30% at different centres [109].

A summary of the typical uncertainties in measurements and their effects on determined doses is given in Table 8. The uncertainties associated with these physical dosimetric methods are used in the quality assurance of the treatment planning system (TPS), as described in Section 11.1.4.

National standards for BNCT are required to establish the reliability of the measurements. Currently, the national standards laboratories typically have neutron sources with maximum fluxes of 10^6 to $10^7 \, cm^{-2} \cdot s^{-1}$, and there is not yet a national standard field for a $10^9 \, cm^{-2} \cdot s^{-1}$ epithermal neutron flux that is typical of a BNCT irradiation field. The Bureau International des Poids et Mesures is aware of the gap, and it is hoped that such a standard field will be developed in the future. In the meantime, some form of traceability to the existing standards is helpful.

The National Institute of Advanced Industrial Science and Technology of Japan (better known by its initials AIST) is considering steps to maintain the traceability for BNCT by collaborating with some BNCT facilities. Some BNCT facilities are considered to be calibration facilities and have been experimentally investigated. AIST is also developing calibration techniques for BNCT, using their accelerator based neutron source whose intensity has been enhanced using a thick target and moderators.

An overview of neutron dosimeters is given in Ref. [80], and methods of dosimetry used in BNCT applications are summarized in Table 9.

TABLE 8. UNCERTAINTIES IN MEASURED RADIATION QUANTITIES

Component	Targeted uncertainty (1σ)	Example reports and typical uncertainties
Thermal neutron fluence, $\int\phi_{th}(t)\cdot dt$	5%	Au wire: 7% [101] Activation foils: 4–10 [109] Activation foils: 2–3.4% [110]
Photon dose, D_γ	5%	Paired ionization chambers: 2.5–9% [107] Paired ionization chambers: 4.4–20% [109] Paired ionization chambers: 30.84% [110] TLD: 12% [101] TLD: < 15% [50] TLD: 20% [108]
Fast neutron dose, D_H	10%	Paired ionization chambers, foils: 5–30% [109] Paired ionization chambers: 17–24% [107] Paired ionization chambers: 9.74% [110]

Note: The targeted uncertainties were defined in Section 4.4.4.3 of Ref. [107].

TABLE 9. MEASUREMENT METHODS FOR PHYSICAL DOSIMETRY OF GAMMA RAYS AND NEUTRONS

Application	Measurement location	Reactor based methods	Accelerator based methods	Section
In-air measurements	Neutron energy spectrum at the target	Not applicable	Time of flight [111] 3d nuclear recoil detector [12] BSS [81]	4.2.1
	Neutron energy spectrum at the beam port	Multi-foil activation methods [112–113] BSS [114] Tissue-equivalent proportional counter [115] Spherical-type activation detector [116] MIMAC-FastN [96]	Multi-foil activation methods [50] BSS [83] Single moderator spectrometer [90] Tissue-equivalent proportional counter [95] MIMAC-FastN [96–97]	4.2.2
	Neutron/γ ray dose in the lateral direction	Activation foil methods, TLDs and paired ionization chambers [29, 107]	Activation foil methods, TLDs and paired ionization chambers [29, 107]	4.3
		Activated metal foil with imaging plate [117] Gafchromic film [100]	Activated metal foil with imaging plate [117] Gafchromic film [100]	4.2.4
In-phantom measurements		Activation foil methods [118–119] TLDs [120] Paired ionization chamber [109, 121–123] A small scintillator with optical fibre [63] MIMAC-active phantom [102] Self-powered flux detectors [104–106]	Activation foil methods [101] TLDs [120] Paired ionization chamber [109, 121–123] A small scintillator with optical fibre [63] MIMAC-active phantom [102] Self-powered flux detectors [104–106]	4.3
Whole body exposure		Human simulated phantom with activation foil methods and TLDs [103] Self-powered flux detectors [104–106]	Human simulated phantom with activation foil methods and TLDs [103] Self-powered flux detectors [104–106]	4.4
Real time beam monitor	Electric current of the charged particle beam	Not applicable	Charged particle electric monitor [50]	4.5.1
	Real time neutron monitor	Fission chamber [124–125] Ionization chamber [124] Small scintillator with optical fibre [126] Self-powered flux detectors [104–106]	Fission chamber [124–125] Ionization chamber [124] Small scintillator with optical fibre [126] Self-powered flux detectors [104–106]	4.5.2

Note: Self-powered flux detectors were developed for highly thermalized beams but could be adapted for certain epithermal fluxes.

5   FACILITY DESIGN

The purpose of this section is to provide an overview of the functions and spaces required, and some examples of 'treatment facility designs' for AB-BNCT facilities inside or near a hospital campus. Detailed description of peripheral devices for BNCT is excluded in this section because those depend on the treatment protocol and procedure in each facility. Annexes I–XI give more design details about many facilities operating or under construction.

Clinical studies using BNCT with neutron 'beams' from research reactors have been conducted for recurrent malignant gliomas and head and neck cancers [127]. To investigate its efficacy for other malignancies or to treat more patients, BNCT treatment facilities need to move from their 'traditional' setting in research reactors to medical institutes [128], and AB-BNCT provides a possible future treatment modality to enable this (Sections 1–2). Several new AB-BNCT facilities have been planned and constructed in recent years worldwide [7], and the first clinical trials have been performed [128]. Although each type of accelerator used in AB-BNCT has some unique requirements, such as space, many of the requirements for the other areas of the facility and functions within it, like radiation safety, supporting laboratories, and patient care, do not differ significantly. Proximity to a hospital allows full use of hospital infrastructure, including personnel, availability of laboratory spaces, and various imaging modalities (CT, MRI, PET). The IAEA has a publication describing considerations around radiation technology facilities [129] and has publication on setting up a conventional radiotherapy program which includes information on facility design that may also be considered when designing a BNCT facility [130].

The following functional spaces and major equipment need to be installed:

- Neutron delivery and radiation protection spaces:
    o   Accelerator room – accelerator;
    o   Control room – accelerator control system;
    o   Treatment room – neutron beam aperture, patient imaging/positioning system, robotic (optional) couch or chair.
- Laboratory spaces:
    o   Dosimetry room – standards, foils, phantoms;
    o   Boron laboratory – Inductively Coupled Plasma (ICP) or other methods.

The following functional spaces need to be located somewhere in the BNCT facility or in a participating hospital:

- Radioactive waste storage;
- Patient care spaces:
    o   Treatment planning room;
    o   Treatment simulation room – treatment simulation system with X ray, X-CT;
    o   Patient pre-fixation room;
    o   Pre-treatment room (physical examination and administration of boron drug);
    o   Rest recovery room (post-treatment);
    o   Outpatient clinic.

The following facilities are optional depending on anticipated need:

- Radiobiology laboratory;
- Conference room;
- PET-CT system.

Some of the functional spaces and equipment may be provided at neighbouring major medical centres.

5.1 PLANNING THE FACILITY

The facility in which the accelerator will be installed needs to be designed and constructed under the supervision of the appropriate regulatory agencies, and the necessary licenses obtained at each stage of the project. The IAEA publication 'Regulatory Control of the Safety of Ion Radiotherapy Centres' may be a useful checklist when considering design of a facility, as there is a lot of common ground with a BNCT facility [131].

5.1.1 Radiation safety

Radiation safety of the facility needs to fulfil the 'As Low As Reasonably Achievable' (ALARA) principle. This takes into account the dose rates and shielding requirements of the BNCT facility and requires minimizing the radiation risk during operation from non-therapeutic radiation doses to the:

- Patients;
- Workers engaged in BNCT;
- General public.

In addition, residual activation of the workspaces needs to be considered to ensure that doses to workers during routine daily maintenance and annual shutdown maintenance are acceptable and in accordance with ALARA. Occupancy factors in various spaces can be used to design and control processes to ensure that worker exposure limits remain within regulatory limits. Everything inside the accelerator hall and, in particular, the treatment room has the potential to become activated due to neutron exposure. Therefore, optimisation/minimisation of the residual dose rates needs to be undertaken using the following approach:

- Selecting component materials that are less susceptible to neutron activation;
- Removing or relocating components to positions where they are less likely to become activated;
- Providing neutron shielding around components to minimise activation due to neutron exposure;
- Providing γ shielding around activated components to reduce ambient dose to both staff and patients.

A centralized building radiation monitoring system may be required to monitor the radiation fields in those areas in which the hazards are higher (e.g., the accelerator room, treatment room, and neighbouring spaces).

5.1.2 Classification of areas

Radiation facilities are typically divided into different areas (a process commonly known as zoning) according to the possible risk from radiation, which allows oversight proportional to the risks. National regulatory bodies assign the exposure limits for these areas. They are typically termed:

- 'Controlled', where there is a risk of receiving a relatively high dose within a short period of time, and work is conducted by qualified and trained personnel. Visiting workers typically work in such zones under special permissions and control after receiving specific training;
- 'Supervised', where there is a risk of receiving a slightly lower dose within a short period of time;
- 'Uncontrolled', where the general public can often be present, but where there may still be other hazards (e.g., chemical).

Typically, the accelerator room and treatment room are classified as controlled areas, where the radiation level is actively monitored. Other areas near the treatment room, like the control room, are typically classified as supervised areas. Access controls are put in place to prevent unauthorized or unqualified people from accessing zones with greater potential hazards and to secure radioactive materials. Boundaries between these zones have to be clearly demarcated with appropriate signage. Depending on local regulations and the risk of contamination, crossing between certain zones may also require contamination monitoring (e.g., hand/foot monitors) and changes of clothing. Space for storage of clean and used clothing may be required at these points, as well as provision for washing hands or other areas of exposed skin.

5.1.3 Emergency response

In the planning for the operation of an AB-BNCT facility, it is necessary to consider and implement several emergency response strategies in case of disaster. This planning will impact the facility design as it will need to comply with national and possibly local fire regulations for buildings (e.g., consideration of fire compartments, distance between fire exits, sensor and alarm placement etc.). Another concern for such a facility would be target failure, e.g., after a loss of mechanical integrity, or loss of cooling, etc. The plan has to describe what each worker has to do in the event of a given emergency. In addition to scenarios such as a facility fire, which will be common concerns for all facilities, the types of accident scenario involved in emergency response planning often include site specific incidents, which may include responses to earthquakes, flooding, hurricanes, tornados, etc.

Management of a clinical emergency is discussed separately in Section 6.5.

5.1.4 Other considerations

The design of the facility may need to consider access by disabled workers (e.g., wheelchair access), which can impact corridor widths, require provision of ramps, etc. In planning for the security of a facility an alarm system and video monitoring may need to be considered. Some countries and individual institutions have specific regulations governing the storage, handling, and distribution of hydrogen (or deuterium) gas that need to be considered with respect to the ion source.

5.1.5 Decommissioning of the facility

It is important for the BNCT facility to consider residual radiation inside the facility and identify possible risks. It is possible to simulate in advance the residual radiation after several years of typical operation of the BNCT facility, from which one may discover particular risk areas, using the same transport codes that are used in designing the BSA (e.g., FLUKA, MCNP, PHITS etc.). Regular measurement of the activation of components and structures within the facility may be needed, and follow-up to limit or reduce the build-up may be required to reduce longer term decommissioning costs [132–133]. If lithiated plastic is used for shielding in the treatment room walls, tritium produced during a facility's operational life may need to be considered at its end of life.

It may be a regulatory requirement to have a preliminary plan for decommissioning, an estimate of the budgetary requirement to execute that plan, and possibly funds segregated specifically for this purpose. The IAEA has issued a publication on decommissioning particle accelerators [134].

5.2 FACILITY RESOURCES

Table 5 shows a list of AB-BNCT facilities including those that are either in clinical use or are in the process of preparation for clinical use. One plan for each of the common accelerator types outlined in Section 2 is shown below. The figures mostly display the spaces classified as 'neutron delivery and radiation protection spaces' and 'laboratory spaces.' Many of the 'patient care spaces' are in the neighbouring spaces or clinics. Some of the common features and considerations for design are then discussed. A brief description of some facility layouts is given below, and more extensive individual facility descriptions are given in Annexes I–XI.

The first few Annexes cover detailed facility descriptions based around various kinds of electrostatic accelerators. Annex I provides an example of a facility designed around a Dyamitron electrostatic accelerator. Annex II gives details of the facility at the Budker Institute, Russia where the first VITA was developed. Annexes III and IV, respectively, describe the TAE Life Sciences VITA based facilities at Xiamen, China and CNAO, Italy. Figure 18 shows the layout of the AB-BNCT system at the hospital in Helsinki, Finland. The single-ended electrostatic accelerator system is manufactured by Neutron Therapeutics Inc. Annexes V and VI describe two facilities based around this source.

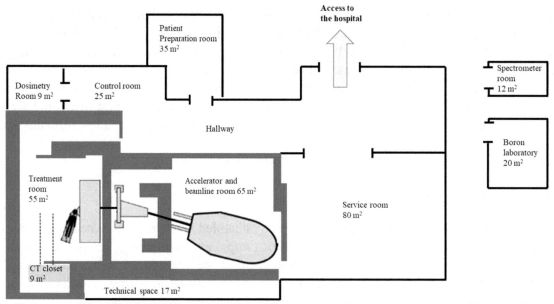

FIG. 18. Layout of the Helsinki BNCT station with the electrostatic nuBeam neutron source of Neutron Therapeutics Inc.

Figure 19 shows the layout of the RFQ system installed at the National Cancer Centre, Tokyo, Japan. The system is located next to medical linacs for radiation therapy. The accelerator lies on the floor above the treatment room, delivering the beam downwards. The 2.5 MeV RFQ system is manufactured by Cancer Intelligence Care systems [135]. Annex VII provides more details on this facility.

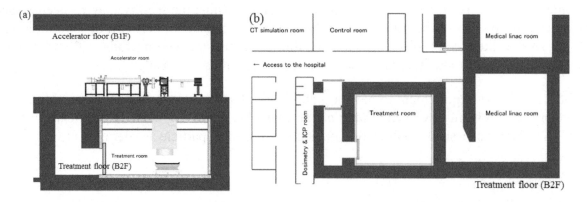

FIG. 19. Layout of BNCT system of Cancer Intelligence Care Systems, Inc. in the National Cancer Center Hospital (a) Elevation showing the accelerator floor above the treatment floor (b) Plan view at the level of the treatment floor (courtesy of M. Nakamura, Cancer Intelligence Care Systems, Inc.).

Figure 20 shows the layout of the AB-BNCT system using an 8 MeV RFQ–DTL that is installed at the iBNCT facility in Tsukuba, Japan. See Ref. [136] for further information. Annexes VIII and IX describe facilities based around RFQ–DTL linacs in detail.

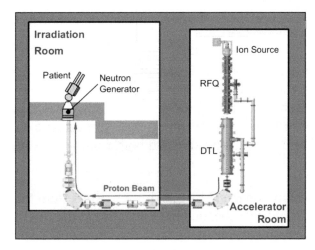

FIG. 20. Layout of the demonstration iBNCT system at Tsukuba University (courtesy of H. Kumada, Tsukuba University).

Figure 21 shows the layout of the AB-BNCT system installed at Kansai BNCT research centre, located on the campus of Osaka Medical College, Japan. The 30 MeV cyclotron system is manufactured by Sumitomo Heavy Industries (see Ref. [137] for further information). Annexes X and XI describe facilities based around such a source.

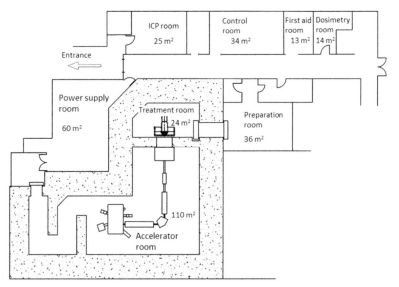

FIG. 21. Layout of the NeuCure system by Sumitomo Heavy Industries, Ltd. at Kansai BNCT Medical Center (courtesy of T. Mitsumoto, Sumitomo Heavy Industries).

5.2.1   Neutron delivery and radiation protection

This section describes in more detail some of the main functional spaces that are typically found in an AB-BNCT facility. As AB-BNCT facilities are neutron producing accelerators, a review of some of the IAEA guidance on designing a cyclotron facility may also be useful to designers [138], as they have many design considerations in common. Ref. [80] discusses neutron dosimetry and monitoring.

5.2.1.1   Accelerator room and beamline rooms

The size and service requirements for the accelerator depend on the manufacturer's specifications. Ancillary devices such as cabinets for the power supplies, RF (if required), and machine protection systems may be placed in the accelerator room alongside the accelerator or in dedicated technical spaces outside of that room.

The cooling water for the accelerator and especially the neutron producing target may become activated, depending on the choice of target, and the choices of accelerated particle and its energy. This activation needs to be simulated during the design stage to ensure proper provisions are made. The heat exchanger and water filtration can become a significant source of radiation for higher energy accelerators and may need to be placed and shielded both from people and sensitive electronics (see Section 2.1.1).

The accelerator room is typically separated by one or more walls from the treatment room. The BSA and its immediate shielding is either embedded in or attached to the final wall (Section 3). Typically, the accelerator does not have a direct 'line of sight' to the target and BSA. This reduces activation of accelerator components from neutrons scattered backwards from the target. Putting a bend in the beam path can also give flexibility in facility layout and save space. This flexibility includes the potential to place the accelerator on a different floor above the treatment room, rather than the same floor (Fig. 20). Bending and extending the beam path may require additional investment in bending and quadrupole focussing magnets.

In order to prevent personnel being trapped in the accelerator room during operation, the system needs to be equipped with a set of one or more 'last person out' buttons that feed into a suitable programmable logic controller. This interlocks the room and issues a signal that allows the use of the accelerator. The accelerator room is equipped with at least one emergency stop button.

Despite careful selection of materials (selected on the basis of cross sections for activation and half-life of activated products), the neutron 'beam' may activate materials in the facility. It is important that the shielding design considers the number of patients expected to be treated per day, as this determines the build-up of residual activation in the treatment room over time due to the accumulated neutron fluence. To minimize residual activity and unnecessary radiation exposure, the interior surfaces may need to be covered with lithiated or boronated plastic covers that efficiently slow down and absorb neutrons. Regulations (including contamination control and fire) have to be considered when choosing the interior surface material to cover the bare concrete (see Section 5.2.2.3 below).

Where space is not a critical issue, standard concrete compositions are typically used to absorb secondary γ rays. During construction, it may be advisable to perform quality control on each batch of concrete to ensure it meets the expected strength and density, as shielding efficiency is dependent on composition. Where space is an issue, heavy concrete is often used to absorb secondary γ rays; e.g., to encapsulate the BNCT treatment room. Heavy concrete is concrete mixed with aggregates such as ilmenite (iron ore) to increase the γ shielding. Its cost is generally greater than that of standard construction grades of concrete.

5.2.1.2   Activation of concrete shielding and planning for decommissioning

Relatively short-lived activation occurs, chiefly from neutron capture of alkali metals (e.g., production of ^{22}Na, ^{134}Cs) within concrete. Longer lived activation comes from other metals (e.g., Mn) that are frequently present in the aggregates.

Therefore, the proportions and compositions of concrete components need to be considered and, if possible, controlled [139–140]:

- Care is advisable in selecting the source of the fine and coarse aggregate (sand and gravel);
- The iron ore in heavy concrete can contain impurities that become activated (e.g., Mn, Co);
- Fly ash, often added as a cost effective substitute for Portland cement and to improve the properties of concrete, has uncontrolled composition and may become activated;
- Rebar can become activated (e.g., due to the presence of Co). It may be possible to push it deeper into the concrete wall to reduce activation. Non-metallic rebar could be considered, at additional expense.

The main issue for long-term activation is the Eu content of concrete that originates in the limestone. It is present at the low ppm level but because of its large neutron capture cross section and long half-life of its activation products it is frequently the determinant for whether concrete is classed as radioactive waste at the end of life of the facility [139]. The concrete itself can be shielded by boronated or lithiated polyethylene or rubber to reduce the build-up of long term activation, if required.

5.2.1.3   Service and maintenance

Sufficient space has to be left to enable easy access for service and maintenance of the accelerator. The amount of space required is very specific to the type of accelerator. Provision has to be made to change the neutron producing targets under routine conditions and in case of target failure. The hazards associated with this process depend on the target material, the approach taken for changing it, the choice of accelerated particle and its energy (see Section 2). Specialized equipment may be needed. Since this work involves the most highly irradiated components in the facility, not only the direct dose but also contamination control has to be carefully considered for this operation.

5.2.1.4   Control room

The control room needs to fulfil the requirements for a radiotherapy control room [141]. It needs not only to control the neutron delivery to the patient but also to allow patient monitoring and communication with the patient during treatment. It has to include an emergency stop button in order to interrupt the treatment prematurely if needed. It is convenient to place the control room close to the treatment room and to give easy access to the patient in case of an emergency. Usually, a set of warning lights showing the operational state of the accelerator and the dose rate inside the treatment room is visible from the control room.

5.2.1.5   Decay rooms and radioactive waste

Shielded space for temporary radioactive waste storage may be required in an AB-BNCT facility unless it is part of a larger hospital which already has such dedicated space. Materials to be disposed of include activated components, and contaminated personal protective equipment such as gloves. In case of handling radioactive materials, a risk assessment needs to be done before the work procedure to avoid unexpected exposure. A certain amount of space is required to store used targets. The amount of space required and its location in the facility depend on the nature of the target as well as its expected service lifetime. Final disposal to a licenced radioactive waste facility may be required for some components. Rooms storing activated materials and waste are usually placed under a negative pressure differential with respect to neighbouring spaces as radioactive gases may be evolved.

5.2.2 Patient care

The following sections deal with spaces in equipment directly involved with patients.

5.2.2.1 Patient preparation room

A patient preparation room where physical examination, boron infusion, blood sampling, etc. can be performed is required. Where an AB-BNCT facility is located inside a hospital, existing hospital spaces and resources can be used for this purpose. After the treatment, the preparation room may be also utilized for patient recovery. In a high throughput facility, several such rooms may be required.

5.2.2.2 Treatment planning

Treatment planning is an important part of BNCT treatment. If the facility is built in a hospital, this can be done in the existing hospital spaces, where patient information is stored and available. Otherwise, dedicated space equipped with a computer and adequate data connections is required. Treatment planning is discussed in more detail in Section 11.

5.2.2.3 Treatment room

The treatment room needs not only to be shielded but also has to meet applicable standards for medical facilities and radiation safety requirements [141]. Back reflection of thermalized neutrons from the walls needs to be considered and modelled during design. Back reflection perturbs the therapeutic neutron field and may give rise to additional non-therapeutic neutron dose to the patient. For this reason, a rule of thumb of assuming a minimum distance of ca. 3 m from the BSA in all directions may be a useful starting point when beginning the design process for a new facility. The room also requires sufficient space to handle the patient positioning device and any ancillary equipment that may be required. These requirements may suggest a minimum footprint of 30 m^2. The treatment room has to be large enough to accommodate any treatment posture provided in the treatment planning, easy access for patients and quick access in case of an emergency for medical staff.

The patient couch, beam alignment, and monitoring equipment need to be designed and constructed of materials bearing in mind the likelihood of activation and tolerance for neutron radiation damage. Covering equipment with flexible neutron absorbing material, such as borated or lithiated plastic, can inhibit activation due to neutron exposure. Its efficacy can be tested before patient use. This approach is suitable for modest size items, such as pillows, cushions, and restraints. Similarly, the treatment room's interior surfaces are often covered with suitable neutron absorbing materials. This is to reduce the production of long term activation and prompt γ radiation, thereby reducing nontherapeutic γ dose to the patient. There is a desire in the AB-BNCT field to develop a live monitoring system for boron concentrations in patients, such as PG-SPECT scanning, which would operate during BNCT treatment (Section 8.2.1) using the 478 keV prompt γ ray from the boron neutron capture reaction (Fig. 15). If desired, provision may be made for accommodation of such a system during facility design (perhaps similar to that for the conventional CT system in Fig. 18). For these reasons, the decision of where to use lithiated and boronated shielding in the treatment room requires careful thought. The floor of the treatment room is generally covered with a suitable material that follows national regulations to prevent loose concrete dust coming into contact with clothing, shoes, or skin. Epoxy or equivalent materials are candidates for this application. (Some of the considerations regarding concrete compositions are discussed in Section 5.2.1.2.) A 'maze' may be introduced to the design to reduce the neutron and γ fluxes on the exit door (see Figs 18, 19(b)). A maze design requires more space than a facility without one, but may enable the construction of a thinner, more easily operated door.

Another potential source of dose is ^{41}Ar production due to neutron capture by ^{40}Ar that is naturally present in air at just under 1% concentration. Therefore, the treatment room is actively ventilated with fresh, temperature-controlled air and is usually under a negative pressure differential with respect to neighbouring spaces. The exhausted air from the facility may (depending on the level of activation and national regulations) need to have the ventilation volume and radioactivity actively monitored.

Similar to the requirements for the accelerator and beamline rooms (Section 5.2.1.1), in order to prevent personnel becoming accidentally trapped inside the treatment room during a treatment, the system needs to be equipped with 'last person out' buttons feeding signals to a programmable logic controller that interlocks the room and allows the use of the accelerator. These interlocks also have to ensure that the irradiation is terminated if the door is opened during irradiation. Irradiation cannot be resumed automatically after the door is closed again, but the process of authorization of start-up has to be followed again. A fail-safe interlock system is required that can function in case of power failures or of failure one or more components of the system [141]. The room will need to be equipped with adequate communication devices such as cameras and an intercom.

A continuous ambient dose equivalent monitor for γ radiation is usually installed inside the treatment room. A signal outside of the treatment room can be used to warn when the neutron radiation is on or if the residual dose rate inside the treatment room remains too high for personnel to enter the room safely. A set of signals that displays the operational state of the accelerator and the dose rate inside the treatment room is usually located immediately outside the treatment room.

5.2.2.4 *Patient positioning system*

Patient positioning is important for BNCT facility design, as the dose is delivered to the patient in accordance with the treatment planning system. A patient positioning system included in a BNCT facility assists in reproducible patient positioning to achieve a safe and reliable treatment. Depending on the specific treatment procedures and the facility design, patient positioning can be done either inside or outside the treatment room. The most important requirement for the patient positioning system is to ensure the accuracy required clinically. For BNCT, there is, at the time of publication, no consensus on the optimum technology for patient position verification.

One option in facility design is to have a patient/beam simulation room near the treatment room as part of the patient setting system; e.g., in the preparation room in Fig. 21. This enables patient positioning without exposing the patient or staff to activated components in the treatment room. However, in this case, the reproducibility of patient positioning between the simulation and the treatment rooms has to be validated regularly. During treatment, the patient is often close to the irradiation port, making certain treatment postures difficult. In this case, the patient/beam simulation room can be also used for a feasibility check of the patient posture and setting before the actual treatment.

5.2.3 Laboratory spaces

The following sections deal with analytical and technical spaces and equipment needed to support the operation.

5.2.3.1 *Dosimetry*

The neutron 'beam' requires daily quality assurance and regular measurement. A room where dosimetric tools, such as water phantoms, ionization chambers and dose rate meters can be stored is needed. If the neutron flux is measured by neutron activation foil methods, a detector, such as a high purity germanium spectrometer, is needed. Depending on the dosimetry equipment and methods being used, up to 5×5 m^2 could be reserved for this function. Typical dosimetry methods and the necessary tools are described in more detail in Section 4.

5.2.3.2   *Boron laboratory*

In order to make accurate dose calculations, it is important to measure the boron concentration in patients' blood. Therefore, a boron laboratory is located usually inside or close to the BNCT facility. Details concerning techniques and instruments for measurement of the concentration of boron in blood (and other tissues) are described in Section 8. As various techniques are still under development for boron determination, flexibility of design for future changes needs to be kept in mind when designing this space. In Figs 19 and 21 the boron laboratory is labelled the 'ICP room' referring to a set of some of the major techniques for boron determination described in Section 8.1.1.

5.2.4   Option: Dedicated neutron analytical beamline

For some proposed neutron analytical techniques, including prompt gamma activation analysis and prompt alpha neutron spectroscopy (Section 8.2.2), a separate thermal neutron beam may be required in the facility design. Whereas a typical research reactor typically has multiple beam tubes, the current AB-BNCT systems typically have a dedicated BSA with a single moderator and beam port. To provide 'live' boron concentration analysis with these techniques would require an additional smaller port that could be operated while BNCT was being performed and which produced a thermalized neutron spectrum in a space with a lower background than the treatment room.

5.2.5   Option: Radiobiology laboratory

It would be suitable if the radiobiology experiment rooms in the BNCT facility are designed for Good Laboratory Practice level, although this requires considerable effort to achieve and maintain [142–143].

5.2.6   Option: Diagnostic imaging system

PET-CT may provide valuable information to determine whether BNCT is performed on a patient (see Section 7.6.1). Thus, a BNCT medical centre that is independent of the main hospital may require access to a PET-CT system or consider including one within its facility. Guidance on how to design and set-up a clinical PET-CT facility is provided in Ref. [144].

6   OPERATION AND MANAGEMENT OF A FACILITY

BNCT is becoming a part of general cancer treatment using accelerators located either at or in close proximity to a larger medical facility such as a hospital. The maximum number of BNCT treatments that can be performed per week depends in part on the characteristics of the accelerator based neutron source and irradiation system. Because BNCT is typically delivered in a one or two fractions, the theoretical patient capacity of a single treatment room is much higher than one used for conventional or hadron radiotherapy. With that in mind, this section provides an overview of the likely requirements of staffing for the operation and management of the BNCT facility. The IAEA publication on setting up a conventional radiotherapy programme may also be consulted [130], and Ref. [145] gives an overview of how to plan staffing at such a centre, which may be useful in planning a BNCT centre. It includes an Excel sheet to assist in this process.[10]

6.1   THE ROLE OF EACH PROFESSIONAL

Table 10 lists the personnel listed below considered to be the minimal requirement for a BNCT centre. Specific titles, roles, and responsibilities may vary by institutional practice and national regulation. Annexes III, V, IX and X give actual staffing plans at various BNCT facilities.

TABLE 10. MINIMAL PROFESSIONS INVOLVED, THEIR ROLES, AND NUMBERS REQUIRED PER PATIENT FOR OPERATION OF AN AB-BNCT CENTRE

Position	Roles	No. per patient
Radiation oncologist	Overall supervision Judgement of the indication for suitability of BNCT Dose prescription Preparation of treatment plan and dose evaluation	1
Medical physicist[a]	QA/QC for BNCT Dose verification Treatment planning of BNCT and its implementation	1–2
Radiation therapist	Preparation of patient positioning including imaging Simulation of neutron irradiation Operation of BNCT irradiation	2
Accelerator operator/engineer	Controlling the accelerator parameters General maintenance	1
Clinical laboratory scientist	Measurement of boron concentration in blood	1
Radiation protection officer[b]	Staff dose recording and reporting Radiation safety Reporting to regulatory agency	1
Nurse	Care of BNCT patients Infusion of patient with BPA Taking blood samples from the patient	1

[a] In some centres, the medical physicist may be assisted by a medical dosimetrist.
[b] Necessary if the centre is set up independently of the hospital. Otherwise, the role may be shared.

---

[10] http://www-pub.iaea.org/MTCD/Publications/PDF/P-1705_CD-Rom.zip

In the future, radiation oncologists at a BNCT centre and closely collaborating physicians involved in BNCT treatment may be offered training and certification by the national Society of Neutron Capture Therapy. In addition, each specialist physician may be certified by the pertinent medical society (e.g., radiation oncology society, hepatology society, etc.) for BNCT. Furthermore, it is preferred that the medical physicists be certified as specialized in radiation oncology or to have experience in the practice of general radiation oncology.

The IAEA publication SSG-46 describes the roles and responsibilities of the different groups with respect to radiation protection and safety of medical use of radiation [146]. The roles of radiation oncologists, medical physicists, radiological technologists, and nurses in general radiation oncology practice have also been discussed by the American Society for Radiation Oncology [147]: they need to be modified in part according to the very specific requirements of BNCT. BNCT can deliver an extremely large dose to the tumour, but the surrounding normal tissue also usually receives a fairly large dose during the single fraction. Therefore, risks of acute and/or late adverse effects exist. Close collaboration with specialist physicians and long term follow-up of the patient are both indispensable. Table 11 shows who may be involved in the operation of the centre, possibly based at the hospital.

TABLE 11. CLOSELY COLLABORATING PROFESSIONALS IN BNCT TREATMENT

Position	Roles
Specialists in the treated organs (head and neck surgeon, neurosurgeon, dermatologist, urologist, digestive surgeon, orthopaedic surgeon etc…)	Judgement of the indication of BNCT close collaboration with the radiation oncologist; Treatment plan creation collaboration with the radiation oncologist and medical physicist; BNCT implementation collaboration with related staff in the centre.
Nuclear medicine physician	^{18}F-FBPA PET implementation; Estimation of concentration of ^{10}B in tissue.
Radiopharmacist	QA and oversight of ^{18}F-FBPA manufacturing under GMP.
Pharmacist	QA and oversight of BPA compounding (e.g., with fructose), if compounded locally

Figure 22 shows how the groups may be coordinated during the treatment of a case.

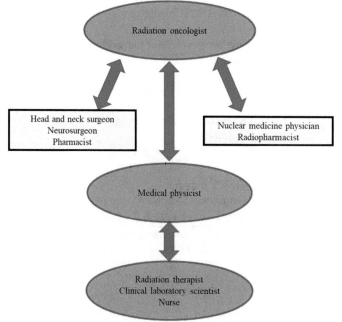

FIG. 22. Treatment planning cooperation diagram for each professional. Red indicates the minimum core professionals involved in operating a BNCT centre. The square boxes indicate other professionals that need to be involved, perhaps based at the main hospital or a specialized clinic.

6.2 PATIENT FLOW FOR A HEAD AND NECK CANCER PATIENT FOR AN INDEPENDENT CENTER WITHOUT AN IN-PATIENT UNIT

Figure 23 shows an example flow chart for the treatment of a head and neck cancer at a BNCT centre.

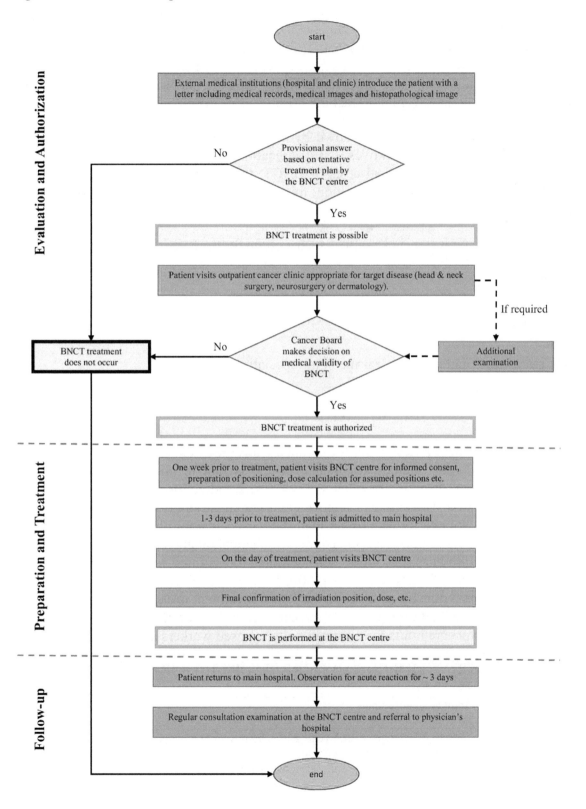

FIG. 23. Typical flowchart for head and neck cancer treatment by BNCT.

6.3 CASE REVIEW CONFERENCE, INDIVIDUAL CLINICAL DEPARTMENTS AND THE CANCER BOARD

Case conferences attended by core members of the BNCT centre are important as a venue to perform careful examination and discussion of an individual case from the perspective of radiation oncology and radiology. During case conferences, new clinical and basic research issues may come to light. These case conferences are also places of education for inexperienced staff. Case conferences in individual clinical departments are also indispensable because they consider the indication of BNCT from the viewpoint of each participating physician's field. They are also a place for education by participating radiation oncologists; e.g., in a teaching hospital.

In most countries, a group, which may be known by various names such as a Cancer Board or Multidisciplinary Team, assesses and approves BNCT indications for individual patients (Fig. 23). If public health insurance covers the medical costs of BNCT, this group's review and decision is a common condition of financial support. In addition, this group can be positioned as an educational place to expand understanding of BNCT.

Below is an example of the typical membership of such a group for the case of head and neck cancer:

- Radiation oncologist;
- Head and neck surgeon;
- Medical oncologist;
- Pathologist;
- Diagnostic radiologist (including nuclear medicine specialist);
- Palliative medicine physician.

6.4 RADIATION EXPOSURE MANAGEMENT

ICRP Report 103 [148] includes a description of planned exposure situations; i.e., those involving deliberate operation of radiation sources. Medical exposure is that incurred by patients for the purposes of medical or dental diagnosis or treatment; by carers and comforters; and by volunteers subject to exposure as part of a programme of biomedical research (where carers and comforters are persons who willingly and voluntarily help (other than in their occupation) in the care, support and comfort of patients undergoing radiological procedures for medical diagnosis or medical treatment) [146, 149]. The activation products, half-lives, considerations concerning release of patients after BNCT, and doses to others are discussed in Refs [150–152] (see also Section 9.7.3). The exposure received by medical personnel directly involved in the BNCT treatment is classified as occupational exposure as is that of nurses and others involved in the care and assistance of patients after treatment in the ward. Some of the exposure limits for occupationally exposed workers are listed in Table 12.

TABLE 12. DOSE LIMITS RECOMMENDED BY ICRP REPORT 103 FOR OCCUPATIONALLY EXPOSED WORKERS [148]

Occupational exposure	Individual dose limits
Whole body	20 mSv/y averaged over 5 years and ≤ 50 mSv/y in any one year
Lens	20 mSv/year averaged over 5 years and ≤ 50 mSv/y in any one year (as updated in Ref. [153])
Skin	500 mSv/y
Hand and foot	500 mSv/y

6.5 MANAGEMENT OF A CLINICAL EMERGENCY

The principal boron drug used in BNCT, BPA, is an amino acid of low molecular weight (see Section 7.4). It has relatively safe toxicity profiles and is less likely to cause serious reactions such as anaphylaxis than many other conventional drugs. But an emergency can occur at any time. A large volume of fluid administration may result in cardiac stress. For example, a patient of 60 kg weight would require a 1 L infusion if BPA is administered at 500 mg/kg, because Steboronine (SPM-011) is a 300 mg/mL solution of BPA (see Annex XVII). Preparations for a clinical emergency are always necessary, and an emergency cart needs to be just beside the patient during BNCT procedures. From the start of BPA administration to the end of neutron irradiation, it is desirable to check the vital signs shown in Table 13. However, since neutron irradiation during vital sign monitoring may cause malfunction or failure of related equipment, the necessity of monitoring needs to be judged considering the individual patient's conditions. If they are monitored during neutron irradiation, the equipment needs to be protected from damage by elevated neutron and γ fields using appropriate shielding, such as acrylic blocks and lead blocks, respectively. The order in which the shielding components are placed is also important: the outer shielding for neutrons, with γ shielding closer to the equipment.

TABLE 13. VITAL SIGNS REQUIRING ACTIVE MONITORING DURING TREATMENT

Items that need to be checked continuously
Electrocardiogram
Heart rate
Respiratory rate
Percutaneous arterial oxygen saturation (SpO$_2$)
Items to be checked every 15 minutes
Blood pressure
Body temperature
The presence of crystals and the colour of urine

6.6 CONSIDERATIONS FOR INTERNATIONAL PATIENTS

As alluded to later, in Section 13, BNCT has been used in clinical situations where no further standard treatments are available for a given condition. For these patients, the other alternative will be symptomatic treatment under the auspices of palliative care. Such patients exist in every country in the world. However, BNCT facilities are not widely available, and it is unlikely they will be for the next decade or so. Until then, patients requiring BNCT will have to travel overseas from their home country to BNCT facilities. Overseas treatments provide additional challenges which, if inappropriately handled, may give a negative experience that may hinder future development of BNCT.

The purpose of the following three subsections is to elucidate possible issues that may be faced and how they can be addressed. The following terms and their definition will be used:

- Departure country: Home country of the patient;
- Destination country: The country where BNCT treatment is performed.

6.6.1 Prior to leaving for the destination country

The following steps need to be considered in preparation for overseas treatment:

- Assess medical suitability for BNCT:
  - Patients and caregivers need to be reviewed to ascertain that the condition the patient has is suitable for BNCT;
  - This needs to be done as an interdisciplinary discussion, either through live video-consultation or other means of communication, with clinicians in the destination country;
- Assess physical suitability to travel:
  - Patients would have to be assessed to be able to tolerate the overseas trip;
  - Assessment would include:
    - Prognostication, as patients with short prognoses of three months or less may have a better quality of life by NOT undergoing the arduous trip;
    - Fitness to travel:
      - Ability to tolerate the entire flight, including conditions during take-off and landing;
      - Ability to tolerate land transfer;
- Palliative medicine input:
  - This allows exploration of goals of care, and ensures that BNCT is in alignment with the goals of care, while providing an avenue for discussion of further management options in the event BNCT is deemed unsuitable;
  - This includes a discussion of action plans in the event of deterioration and/or death in the foreign land;
- Financial discussion:
  - This may be done at this stage as finances may be a concern or reason for not choosing BNCT, and it is necessary to clarify whether insurance will pay for the treatment.
- If the patient has an indication suitable for BNCT and is deemed suitable to travel, then the following logistical issues need to be addressed:
  - Details of stay in the destination country, including:
    - Dates and venues of appointments;
    - Overall duration of stay;
    - Itinerary during stay;
    - Estimated costs (medical and ancillary);
  - Ancillary information:
    - Suggestions of places of accommodation nearby;
    - Land transportation contacts;
    - Emergency contacts;
  - Indemnification:
    - Until such time that there are entities that can provide a seamless service between the departure country and the destination country, it is necessary to stress to patients that this process is of their own accord and the hospital cannot be held responsible for anything that happens during it;
  - Discussion with airline companies:
    - Especially if there are special needs – e.g.; the patient is unable to lie flat;
  - Letters for immigration authorities:

- To permit and explain possession of items that may be deemed contraband, such as opioids for symptom relief and needles for parenteral medications;
- (OPTIONAL) Discussion about medical tourism
    - As there are periods of time between initial consultation, procedures (e.g., CT simulation), and actual BNCT treatment, some patients and their loved ones may want to spend meaningful time being tourists. This is important as these groups of patients may otherwise have a limited remaining lifespan and creating beautiful memories, in what may be their last trip with loved ones, will create a positive experience for all.

6.6.2 In the destination country

The preparations in the preceding section ought to have addressed all the possible issues:

- Contact numbers need to be provided, including:
    - For the hospitals, in case a medical emergency arises;
    - For the respective embassy, in case there is an unexpected turn of events requiring urgent evacuation, including death;
- (OPTIONAL) Medical tourism as planned during the periods between initial consultation and the day of BNCT, if conditions permit;
- Upon completion of BNCT the following need to be addressed in preparation for return:
    - Information concerning the BNCT treatment rendered, which may include treatment plans and dose–volume histograms;
    - Letters to airlines and immigration, in particular to highlight any new changes that may have occurred in the interim period.

6.6.3 After return to the departure country

The following need to be addressed after return to the departure country:

- Continued follow-up with a clinician familiar with BNCT with regular:
    - Physical assessments for the management of side effects, if any;
    - Quality of life assessments;
    - Radiological examinations;
- Follow-up with palliative medicine:
    - To address symptoms and issues that have arisen due to treatment rendered;
    - To address symptoms and issues that have arisen due to ultimately inevitable disease progression.

A schematic of the workflow for referral of a patient overseas is given in Fig. 24.

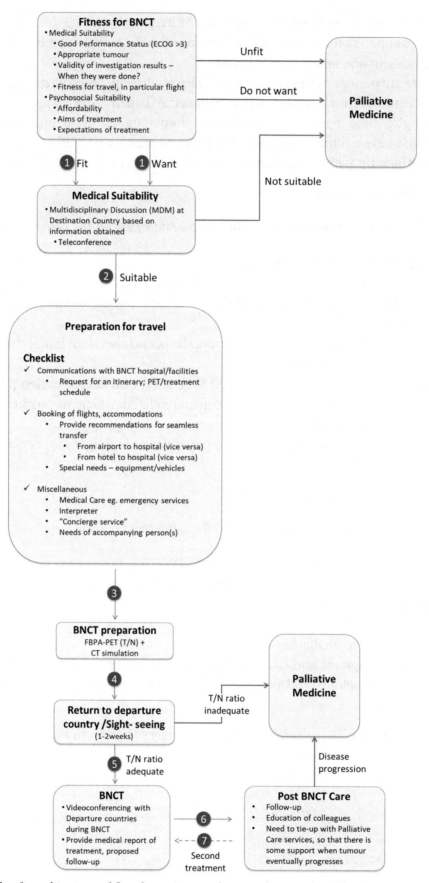

FIG. 24. An example of a multi-step workflow for patients under consideration for referral overseas for BNCT treatment (courtesy of D. Quah, National Cancer Center Singapore).

7 PHARMACEUTICALS AND RADIOPHARMACEUTICALS

This section deals with the boron containing pharmaceuticals that are currently used in BNCT as well as some that were investigated historically. It describes some ideal properties of such agents, as well as some of the information known about their kinetics, biodistribution, and metabolism. It also discusses the ^{18}F-labelled derivatives of these pharmaceuticals used for diagnostic purposes.

7.1 REQUIREMENTS OF A BORON AGENT

BNCT provides a tool to selectively destroy malignant cells when a sufficient amount of [10]B (~20–35 µg/g cell) is selectively delivered to the tumour with sufficient thermal neutron fluence (~10^{12} cm^{-2}) from an external radiation source [154–155].

Boron agents used in BNCT would ideally:

- Be able to maintain the [10]B-concentration in tumour tissues to a level at which an antitumour effect can be anticipated during neutron irradiation;
- Have systemic toxicity low enough to ensure safety. Furthermore, while achieving higher uptake into the tumour tissue than the normal tissue, the ratios 'concentration in tumour tissue / concentration in normal tissue' (T/N) and 'concentration in tumour tissue / concentration in blood' (T/B) have to be high;
- Be rapidly cleared from normal tissues and blood after neutron irradiation;
- Comply with the guidelines for neoplastic agents of the International Council for Harmonisation of Technical Requirements for Pharmaceuticals for Human Use[11] (see Annex XVII for a detailed example).

In other words, the tumour selectivity of ideal boron drugs minimizes their effect on normal tissues. The radiobiological aspects of an ideal pharmaceutical are described further in Section 9.3 and of those used in clinical trials in Section 9.4.

The only boron isotope that contributes to BNCT is [10]B. Since the natural abundance of [10]B is only 19.8%, in order to reduce the pharmaceutical dose of boron agents to patients, a high level of enrichment of [10]B is required. Isotopically enriched precursor chemicals are commercially available, and pharmaceuticals with a [10]B enrichment of 99% or higher are routinely manufactured (Annex XVII). Furthermore, the efficacy of BNCT depends not only on the concentration but also on the localization of boron agents at the cell level [156] (see Sections 9.6.1 and 10.2.2).

7.2 EARLY CLINICAL TRIALS

Early clinical trials of BNCT for brain tumours were carried out with boron delivery agents, such as borax, p-carboxyphenylboronic acid (PCPB), and sodium decahydrodecaborate (GB-10) (Fig. 25) in the 1950s and early 1960s in the United States of America. However, the results were disappointing (see Annex XII), as patients succumbed from recurrent disease. A retrospective analysis indicated that the unsatisfactory results stemmed from inadequacies in both the boron compounds and in the neutron 'beams' that were then available [157].

---

[11] International Council for Harmonisation of Technical Requirements for Pharmaceuticals for Human Use https://www.ich.org/

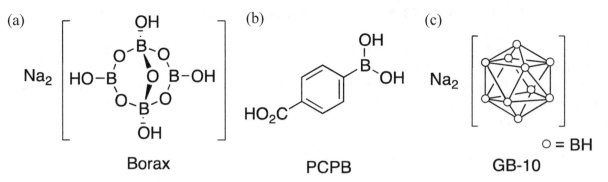

FIG. 25. Boron compounds used in early clinical trials in the USA (a) borax (b) p-carboxyphenylboronic acid (PCPB), (c) sodium decahydrodecaborate (GB-10).

7.3   MERCAPTO-UNDECAHYDRO-*CLOSO*-DODECABORATE

Mercapto-undecahydro-*closo*-dodecaborate (BSH), an icosahedral, dianionic, water soluble boron anion cluster composed of 12 boron atoms (Fig. 26a), was developed in the late 1960s [158]. In 1975, BSH was the first boron agent to demonstrate a significant antitumour effect in BNCT of malignant brain tumours [159] from treatments that began in 1968. Since the first successful BNCT for a malignant brain tumour with BSH, over 400 brain tumour cases were treated using BSH in the early stages of BNCT [160–161].

Although BSH has a low selectivity for cancer cells, it has high water solubility and low toxicity. In a healthy brain, water soluble substances in blood are sparingly taken up by normal brain tissue due to the presence of the blood–brain barrier. However, in brain tumours, disruption to the blood–brain barrier permits the entry and accumulation of water soluble compounds in brain tissue. Thus, BSH is considered to be accumulated in brain tumours.

7.4   4-BORONO-L-PHENYLALANINE

Although BSH was the boron agent initially used in BNCT for malignant brain tumours, due to the restrictions outlined in Section 7.1, 4-borono-L-phenylalanine (BPA) is currently the most used boron agent in BNCT as it has a high affinity for tumour cells. BPA used in BNCT is structurally similar to the essential amino acid phenylalanine and has a boric acid residue on the benzene ring of L-phenylalanine (Fig. 26b). All information on both accelerators and irradiation plans described in this publication is based on the properties of BPA, as this is the standard agent currently in use.

The development of the application of BPA in BNCT is outlined below:

- 1989: the potential for treating melanomas with BPA was first demonstrated [162];
- 1999: BPA was used for treatment of malignant brain tumours [163];
- 2001: the first extracorporeal liver BNCT treatment of diffuse metastases using BPA was performed [164];
- 2004: the first BNCT treatment of head and neck cancers with both BPA and BSH in combination using epithermal neutrons was performed [165]. Within a couple of years, the effectiveness of BNCT with BPA alone was confirmed in the treatment of locally recurrent head and neck cancer [166–167];
- 2020: BPA in use with AB-BNCT was approved for treating locally recurrent head and neck cancer in Japan (Annex XVII).

7.4.1 Structural and optical isomers

BPA has three structural isomers, which differ by the position of the boric acid residue relative to the alanine functional group on the aromatic ring (see Fig. 26b): the *o*-isomer (2-position), the *m*-isomer (3-position), and the *p*-isomer (4-position), the latter isomer being depicted in Fig. 26b). A distribution test of these three structural isomers in the Greene's melanoma cell transplantation model of Syrian hamsters found that the *p*-isomer yields a higher boron concentration in tumours than the other isomers [168–169]. Therefore, the *p*-isomer (4-borono-phenylalanine) is the structural isomer used in current clinical research and clinical trials.

BPA is a chiral molecule with both L- and D-optical isomers. Distribution tests using the BALB/c mouse Harding–Passey melanoma cell transplantation model and the KHJJ breast cancer cell transplantation model have shown that the L-isomer delivers boron to the tumour tissues better than the D-isomer [170–171]. The L-type large neutral amino acid transporter 1 (LAT-1) has been reported to be involved in the mechanism of tumour uptake of BPA in a system using brain tissue excised from patients with brain tumours [172] and in a system using African clawed frog oocytes [173] (see Section 7.6.4). Based on these observations, the 4-borono-L-phenylalanine isomer is widely used in BNCT clinical research and as a drug substance in ongoing clinical trials. Throughout this publication, unless specified otherwise, BPA can be taken to refer to the specific isomer 4-borono-L-phenylalanine.

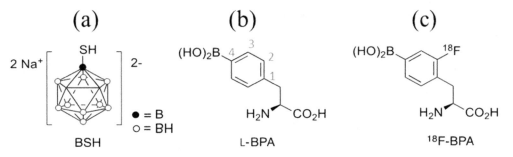

FIG. 26. Boron compounds widely used in clinical studies (a) BSH, (b) p-BPA, where the numbers in blue refer to individual C atoms on the aromatic ring, (c) the ^{18}F-labelled derivative of BPA, ^{18}F-FBPA.

7.4.2 Dosage form

It is desirable to select the optimal dosage form of the boron drug to be administered in BNCT in accordance with the route of administration. Since BPA is currently intravenously administered, an aqueous solution is desirable. This aqueous solution would preferably have the same pH as that under physiological conditions and a clinically acceptable osmotic pressure.

BPA, which is a zwitterion, has little charge at pH 7.4 (representative of physiological conditions) and is therefore poorly soluble in water. To improve the solubility of BPA under physiological conditions, anion complexes of BPA with various carbohydrates were prepared, and it was found that high water solubility of BPA was achieved with the use of D-fructose as a solubilizer [174]. The pH of BPA–fructose (hereinafter, BPA–fr) solution is 7.94, which is close to the pH of physiological conditions [175]. The problem with a BPA–fr solution is its instability: it turns brown due to the Maillard reaction initiated by the condensation of the amino group with the carbonyl group within 2–3 days after preparation, and the quantitative value of BPA also decreases. Hence, it is better to prepare BPA–fr solution on site just before BNCT [176]. Formulations with other dissolution additives that would not be affected by Maillard reactions are therefore desirable. In this regard, formulations with D-sorbitol can be stored for long periods in a refrigerated state.

7.4.3 Administration method

The optimal method of administration of a boron agent for BNCT depends upon the characteristics of the boron drug. In the current practice of BNCT, BPA is usually administered intravenously, and studies are being conducted on the administration method (see Annex XIII). A concentration in blood of 20 ppm or more of ^{10}B is regarded as required for an effective BNCT treatment. Measuring the ^{10}B-concentration in blood just before radiation (see Section 8) is useful for accurate calculation of the tumour dose that will result from BNCT. Blood BPA concentration drops relatively quickly when the administration is discontinued. Although different tumours are expected to have different intratumoural concentrations, it is reasonable to assume that the intrastromal concentration of tumour cells will decrease with changes in blood concentration. Because target tumour cells are unlikely to take up BPA homogeneously, the effect of BNCT on such tumour cells depends on the boron concentration in the extracellular stroma. In the future, depending on the disease for which BNCT is to be applied, the current administration methods may need to be reconsidered.

7.5 FUTURE PROSPECTS FOR BORON PHARMACEUTICALS

The development of boron agents for BNCT is challenging as the required high boron concentration and selectivity for tumours also requires them to have low toxicity. It is also necessary to consider the imaging properties via PET or MRI of the boron agent (or close analogues thereof) during the molecular design in order to estimate the pharmacokinetics and treatment plan of each patient (see Sections 8.1.2 and 8.2.4.1). When future ^{10}B delivery agents are developed, ^{18}F-labelled analogues will also be required to be developed for PET imaging. BPA is currently the only boron agent that meets all requirements. A review of future potential theranostic agents for BNCT is available in Ref. [177].

Figure 27 shows the molecular weight-dependent classification of boron agents that have been reported so far. In general, lower molecular weight drugs, such as BPA, are excreted rapidly from the kidney, whereas higher molecular weight drugs, so-called 'nanocarriers', possess long circulation times in the blood. A new class of BSH-encapsulating 10% distearoyl boron lipid liposomes has been developed, where the liposome shell itself possesses cytocidal potential in addition to that of its encapsulated agents [178]. These liposomes display excellent efficacy of boron delivery to tumours. For further improvement of BNCT, the use of drug delivery systems is one of the methods to be considered. Various nanocarriers can deliver their contents to tumours in a manner that is essentially independent of their contents. Many nanocarriers are highly dependent on tumoural blood vessel architecture for effective accumulation. Conjugation of tumour-specific ligands to the surface of nanocarriers is a possible approach for targeting specific tumours. This approach is probably essential for targeting different subpopulations of tumour cells and subcellular sites that differ biochemically, physiologically, and pharmacologically from one another. Annex XIV discusses developments of future ^{10}B-pharmaceuticals for BNCT.

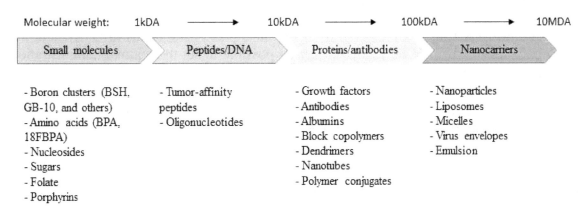

FIG. 27. Classification of boron agents by molecular weight.

An amplified epidermal growth factor receptor (EGFR) gene is expressed by high grade gliomas and their cell surfaces therefore contain increased numbers of EGFRs. In the past, monoclonal antibodies (mAbs) have been used as carriers to improve selectivity and achieve a high concentration of ^{10}B in cancer cells: boronated monoclonal antibodies have been obtained by chemical derivatization to epidermal growth factor [179–180], but the specificity of the bioconjugate decreased significantly, probably due both to conformational changes and to steric hindrance, which impaired binding to the receptor. An in silico pipeline for selection of the best candidates from boron containing ligands obtained from the literature and the DrugBank[12] has been developed. It also evaluates the most suitable residues to be boronated, i.e., those that maintain the availability to recognize their specific target proteins on tumour cells [181]. The pipeline was applied to Cetuximab, a chimeric mAb capable of inhibiting the EGFR. It was found that four boronated residues for each chain resulted in the best performance as measured by retaining the native protein folding and guaranteeing the high binding specificity of Cetuximab to EGFR. This potentially allows increasing the preferential ^{10}B uptake on tumour cells to 10^6 times that on normal cells [182–183].

7.6   4-BORONO-2-[^{18}F]FLUORO-L-PHENYLALANINE

In BNCT, it is important to quantify the accumulation of the ^{10}B delivery agent in the target tumours in order to select appropriate candidate patients, to optimize neutron irradiation protocols, and to maximize the cancer-cell killing effect. It is also important to verify that the accumulation of the boron agent in surrounding normal tissue is low to avoid adverse effects and to result in better treatment outcome. [^{10}B]4-borono-L-phenylalanine (^{10}B-BPA) is now utilized as a ^{10}B delivery agent in BNCT.

For the development of radiotracers to image and follow biological and/or pathological processes, the tracer would ideally contain a radioisotope of an element already contained in the molecule to be traced; e.g., ^{11}C-choline is an exact analogue for the choline molecule, responsible for detection and follow up of various cancers via tumour cell mitochondrial transport. However, due to synthesis restrictions and half-life issues, various ^{18}F-choline agents have been developed and used [184]. In BPA-based BNCT, 4-borono-2-[^{18}F]fluoro-L-phenylalanine (^{18}F-FBPA) can be utilized to estimate ^{10}B-concentration quantitatively at the ppm level in the target tumours and surrounding normal tissue [185–188]. This is described in Section 8.1.2. Research is currently on-going to assess its usefulness in identifying patients who would benefit from BNCT.

---

[12] https://go.drugbank.com/

7.6.1 Positron emission tomography imaging

The use of ^{18}F-FBPA (Fig. 26c) is currently being anticipated as a means to understand the pharmacokinetics of BPA. In the future, it may be possible to set the optimal administration method and irradiation timing for each patient by combining with an ^{18}F-FBPA PET diagnosis. The sections immediately below describe its synthesis, and examples of the tests undertaken to determine its metabolism, biodistribution, and kinetics. Similar tests would be required for any future ^{18}F-derivatives of future ^{10}B-pharmaceuticals developed for BNCT.

7.6.2 Synthesis

Unlike the nucleophilic route of fluorination that is used in the modern production of 2-deoxy-2-[^{18}F]fluoro-D-glucose (^{18}F-FDG), which uses ^{18}F-fluoride produced in a ^{18}O-enriched water target, ^{18}F-FBPA is currently produced by electrophilic fluorination of BPA using carrier-added ^{18}F-F$_2$ gas [189–190] produced in gas targets, either neon or ^{18}O-O$_2$:

- Neon target: ^{18}F-FBPA was originally synthesized by using ^{18}F-F$_2$ produced via the ^{20}Ne(d,α)^{18}F nuclear reaction in a cyclotron target. The labelling agent ^{18}F-CH$_3$COOF was synthesized for the fluorination reaction. The activity yield of ^{18}F-F$_2$ was limited (Table 14), which permitted ^{18}F-FBPA studies for only two to three patients after a single synthesis;
- Oxygen target: In order to improve the activity yield and the molar activity of ^{18}F-FBPA, the production of ^{18}F-F$_2$ using the ^{18}O(p,n)^{18}F nuclear reaction with ^{18}O-O$_2$ as the target was introduced [190]. ^{18}F-F$_2$ was used as a labelling agent. This removes the requirement for a cyclotron capable of accelerating deuterons. The activity yields and molar activities were improved (Table 14) with this route [191].

More information on targets for ^{18}F production is available in Refs [192–193].

Until a nucleophilic route of fluorination is developed, cyclotrons currently producing ^{18}F-FDG may need to invest in a gas target to produce ^{18}F-FBPA for a BNCT centre. The Xiamen BNCT centre is involved with a clinical study of ^{18}F-FBPA produced via a nucleophilic fluorination route [194], and a nucleophilic synthesis route has been published [195].

TABLE 14. ACTIVITY YIELDS AND MOLAR ACTIVITIES OF ^{18}F-FBPA FROM TWO ROUTES OF ELECTROPHILIC PRODUCTION

Target reaction	Activity yield (MBq)	Molar activity (MBq/μmol)	Ref.
^{20}Ne(d,α)^{18}F	1000	20–130	[189]
^{18}O-O$_2$ (p,n)^{18}F	5300	257	[190]

7.6.3 Tracer kinetics: uptake by tumours

Extensive studies have been undertaken to determine whether ^{18}F-FBPA–fr PET accurately traces the pharmacokinetics of ^{10}B-BPA–fr both qualitatively and quantitatively due to the differences in:

- Chemical structure between the boron agent and its ^{18}F-labelled derivative;
- Therapeutic dose of ^{10}B-BPA–fr (mg/kg) versus the tracer dose of ^{18}F-FBPA–fr (μg/kg);
- Administration protocol between ^{10}B-BPA–fr (slow bolus iv. followed by drip infusion) and ^{18}F-FBPA–fr (single bolus iv.).

7.6.3.1 Preclinical studies

Examples of the results and knowledge obtained from preclinical work undertaken during the development of ^{18}F-FBPA include the following:

- The distribution of ^{10}B-BPA–fr and ^{19}F-FBPA–fr in rats after therapeutic dose administration (~500 mg/kg) showed no significant difference in boron concentration by inductively coupled plasma atomic emission spectroscopy (ICP-AES) in blood, tumour tissue, and other normal tissue using the same administration protocol [196] (Fig. 28);
- The ^{10}B-concentration after intravenous injection of a mixture of ^{18}F-FBPA (1.0–2.6 mg/kg, tracer dose) and ^{10}B-BPA–fr (14–80 mg/kg, therapeutic dose) was estimated by means of a γ-counter and ICP-AES, respectively. The ^{10}B-concentrations calculated from ^{18}F-radioactivity were comparable with those measured by ICP-AES in Greene's melanoma [197]. The time course of ^{10}B-BPA–fr (therapeutic dose) and ^{18}F-FBPA–fr (tracer dose) was measured ex vivo by ICP-mass spectroscopy and a γ-counter, respectively, in F98 glioma-bearing rats. The concentrations of ^{10}B-BPA–fr and ^{18}F-FBPA–fr both reached their maxima one hour after administration. This indicated that both ^{10}B-BPA–fr and ^{18}F-FBPA–fr follow a similar time course of cellular uptake [198];
- The correlation between ^{10}B-BPA–fr concentration and ^{18}F-FBPA accumulation in RGC-6 glioma-bearing rats was studied [199]. On day 20 after tumour cell implantation, PET imaging was performed one hour post intravenous injection of an ^{18}F-FBPA solution (30.5 ± 0.7 MBq, 1.69 ± 1.21 mg/kg). One hour after this scan, an intravenous injection of ^{10}B-BPA–fr (167.32 ± 18.65 mg/kg) was given. The rats were dissected one hour post injection. The absolute concentration of ^{10}B in autopsied tissues (tumour, brain, lung, liver, kidney, spleen, small intestine, large intestine, pancreas, and blood) was measured by ICP-optical emission spectrometry (ICP-OES). A significant correlation was found between the accumulation levels of ^{10}B-BPA–fr and ^{18}F-FBPA;
- A similar intracellular microdistribution of both ^{10}B-BPA and ^{18}F-FBPA was found in human glioblastoma T98G cells [200]. By using double-tracer micro-autoradiography, the highest ^{18}F-FBPA accumulation was found in S phase melanocytes and the lowest in non-S phase and non-melanocyte [201]. ^{10}B-BPA accumulation was found in areas of increased ^{3}H-thymidine metabolism [170]. These studies indicate that ^{18}F-FBPA accumulation is influenced by DNA synthesis and melanin metabolism.

In summary, it has been demonstrated that the pharmacokinetics of ^{18}F-FBPA–fr is similar to that of ^{10}B-BPA–fr at the cellular and animal levels. Similar studies would be required in the development of future ^{18}F-agents.

7.6.3.2 Clinical studies

In humans, ^{18}F-FBPA PET imaging was performed on patients with high grade glioma [187–188]. Then, ^{10}B-BPA–fr (therapeutic dose) was administered before surgery. The absolute concentration of ^{10}B in autopsied tumour was measured by prompt gamma spectroscopy. The estimated ^{10}B-concentrations based on the ^{18}F-FBPA PET parameters were well correlated with absolute ^{10}B-concentration by prompt gamma spectroscopy. Although data are still limited, it appears that the pharmacokinetics of ^{18}F-FBPA–fr and ^{10}B-BPA–fr are very similar in humans. This holds for both cancers and normal tissues.

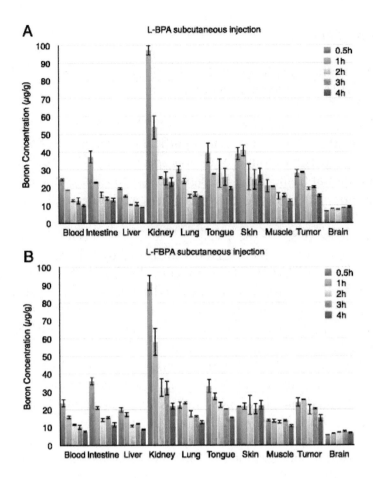

FIG. 28. Transition of the boron concentrations in blood, normal tissue, and tumour tissue. (A) Boron concentrations after L-BPA subcutaneous injection. (B) Boron concentrations after [^{19}F]-L-FBPA subcutaneous injection. This figure by the authors of Ref. [196] is licensed under CC BY-4.0.

7.6.4 Cellular transport mechanisms

The transport mechanisms of ^{10}B-BPA and ^{18}F-FBPA and their fructose complexes (–fr) have been studied and the following gives a precis of what is known. Both compounds are transported via the L-type neutral amino acid transporter (LAT), which operates across the blood–brain barrier, and two of the transporter components, LAT-1 and LAT-2, are known to be involved [171]. Several studies have revealed that ^{10}B-BPA is predominantly transported through LAT-1 [172, 202–203]. Since ^{18}F-FBPA is a ^{18}F-labelled phenylalanine derivative, it was expected that it would also be transported via LAT-1. ^{18}F-FBPA uptake by System-L to A172, T98G, U-87MG cancer cells has been shown to be linearly correlated with both LAT-1 specific and total LAT gene expression [204]. ^{18}F-FBPA uptake has been shown to be strongly correlated with ^{14}C-BPA uptake in seven human cancer cell lines (Fig. 29), and the use of a LAT inhibitor inhibited both ^{18}F-FBPA and ^{14}C-BPA in human cancer cells [205].

In the subtype specificity analysis, transport of ^{18}F-FBPA via LAT-1 expressed on human cancer cell membranes dominated that via LAT-2 expressed on normal cell membranes. The value of the Michaelis constant (K_m) of ^{18}F-FBPA is lower for LAT-1 than for LAT-2 (Table 15) suggesting higher specificity to LAT-1 than LAT-2 [206]. This preference is also present for ^{10}B-BPA (Table 15) [173]. In summary, these studies indicate that both ^{18}F-FBPA and ^{10}B-BPA are transported preferentially via LAT-1 [207, 208], but with greater specificity in the case of ^{18}F-FBPA.

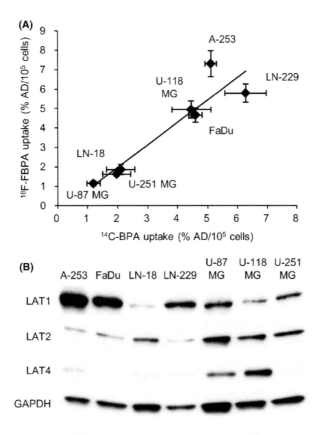

FIG. 29. (A) Correlation between ^{14}C-4-borono-L-phenylalanine (^{14}C-BPA) and 4-borono-2-^{18}F-fluoro-phenylalanine (^{18}F-FBPA) uptake levels in 7 tumour cell lines ($r = .93$, $P < .01$). Data are represented as the mean ± SD. (B) Expression levels of amino acid transporters (LAT1, LAT2 and LAT4) in 7 tumour cell lines. %AD/10^5 cells, percentage of the administered dose per 10^5 cells. This figure by the authors of Ref. [205] is licensed under CC BY-NC.

TABLE 15. MICHAELIS CONSTANTS OF ^{10}B-BPA AND ^{18}F-FBPA FOR LAT-1 AND LAT-2

	K_m (μM)		Ratio
	LAT-1	LAT-2	LAT-2:LAT-1
^{18}F-FBPA	196.8 ± 11.4	2813.8 ± 574.5	14
^{10}B-BPA	20	88	4.3

7.6.5 Cellular metabolism

Since PET visualizes ^{18}F distribution in the body, it is important to determine the fraction of unmetabolized ^{18}F-FBPA in the tissue. Some examples of what is currently known at the cellular level are given below:

- In FM3A mammary carcinoma, the majority of the ^{18}F activity in the tumour over 6 hours post injection was found to be in the form of ^{18}F-FBPA, and the protein-bound fraction of tissue was negligible;
- In B16 melanoma, a significant proportion of ^{18}F-activity was found to be in the form of ^{18}F-melanin in the protein-binding fraction [209];
- In plasma, the unmetabolized fraction of ^{18}F-FBPA was more than 96% at 60 min post injection in tumour-bearing mice [210]. In five healthy human subjects, the fraction of un-metabolized ^{18}F-FBPA in plasma was 97.68 ± 1.57% and 96.03 ± 1.64% at 20 and 30 min, respectively. These studies indicated little metabolic change of ^{18}F-FBPA during the first hour of administration [207].

7.7 IMAGING NORMAL HUMANS WITH 4-BORONO-2-[¹⁸F]FLUORO-L-PHENYLALANINE

In a study on healthy humans, seven whole-body ¹⁸F-FBPA PET scans were performed during one hour, and average ¹⁸F-FBPA distributions of 13 organs were measured [203]. The maximum mean ¹⁸F-FBPA concentrations were reached 2–6 minutes post injection in all organs except the brain and bladder. In normal brain, the average ¹⁸F-FBPA concentration plateaued 24 min after injection. Immediately after injection, high and moderate ¹⁸F-FBPA uptakes were observed in the kidney and pancreas, respectively. The ¹⁸F-FBPA accumulation in the other organs was very low throughout the scans, showing rapid excretion of the tracer via the urinary system. Maximum ¹⁸F-FBPA concentration was observed at 6.0, 2.4, 6.4, 5.5, 4.6, 4.2, 3.5, 3.6, 3.2, 3.1, and 2.4 min after injection for submandibular glands, bone marrow, parotid glands, thyroid gland, lung, heart, liver, spleen, pancreas, kidney, and intestine, respectively (Fig. 30). The ¹⁰B-concentration was calculated based on:

- ¹⁸F activity of each organ (Bq/ml) 60 min after ¹⁸F-FBPA injection;
- Molar activity of ¹⁸F-FBPA (Bq/mole);
- Calibration factors among the PET, well counter, and dose calibrator.

In humans, most organs showed ¹⁰B-concentration to be less than 20 ppm when 30g (500 mg/kg) of ¹⁰B-BPA was administered.

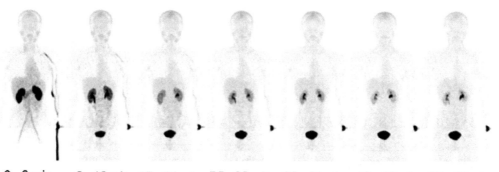

0 - 8 min   8 - 16 min   17 - 24 min   25 - 33 min   33 - 41 min   42 - 49 min   50 - 58 min

FIG. 30. Series of whole-body PET images of ¹⁸F-FBPA uptake in a representative subject. High and moderate uptakes of ¹⁸F-FBPA immediately after injection were observed in the kidney and pancreas, respectively. The accumulation of ¹⁸F-FBPA in other organs was very low (reproduced from Ref. [203] copyright (2016) The Japanese Society of Nuclear Medicine, with permission courtesy of Springer Nature).

7.7.1 Dosimetry

Radiation dosimetry of ¹⁸F-FBPA was first investigated in mice [188]. Subsequently, it was investigated in adult humans by dynamic whole body PET scanning, which showed that the effective dose of ¹⁸F-FBPA was similar to other more commonly used ¹⁸F diagnostic tracers: 2-deoxy-2-[¹⁸F]fluoro-D-glucose (¹⁸F-FDG), *O*-(2-¹⁸F-fluoroethyl)-L-tyrosine (¹⁸F-FLT), and 6-¹⁸F-fluorol-L-dopa (¹⁸F-F-DOPA) (Table 16) [211]. The effective dose per unit activity of ¹⁸F-FBPA in paediatric patients was larger than that in adult patients (Table 16) [208].

It is essential to demonstrate that there is a meaningful correlation between ¹⁸F-FBPA and BPA uptake if PET scans taken with ¹⁸F-FBPA are to be used to contribute to the dose calculation for BNCT treatment. However, the cellularity of the tumour (proportion of cancerous to non-cancerous cells) is not known from this technique due to its low spatial resolution (see Section 8.1.2). What is measured is the gross tumour:normal tissue (T/N) ratio and not the concentration ratio between cancerous and normal cells.

TABLE 16. EFFECTIVE DOSES PER UNIT ACTIVITY FOR SELECTED ^{18}F TRACERS

	Doses (μSv/MBq)				
	^{18}F-FBPA	^{18}F-FDG	^{18}F-FLT	^{18}F-F-DOPA	Ref.
Patients	23.9, $n = 6$	19–29	16.5	19.9	[207]
Pediatrics	31, $n = 3$				[208]
Adults	15, $n = 6$				[208]

7.8   ANALYTICAL METHODS OF MEASURING ACCUMULATION

There are two major analytical methods to evaluate ^{18}F-FBPA accumulation in the tumour and normal tissue. One is a kinetic analysis based on dynamic PET data using the time–activity curve of blood. The other is a single-time-point image analysis based on static PET data at a certain time point.

7.8.1   Kinetic analysis

In one study, a four-compartment mathematical model (Fig. 31) was used to describe the pharmacokinetics of ^{18}F-FBPA based on dynamic PET scanning for 120 min and the image-derived activity of whole blood [187]. By using the kinetic transport constants of the model for ^{18}F-FBPA, the magnitude of accumulation and the time-dependent change in [10]B-BPA–fr accumulation after constant infusion for thirty minutes was predicted. In two patients with glioblastoma multiforme, the predicted accumulation of [10]B-BPA was consistent with the concentration measured in biopsied tissue. From this, the optimal timing for effective BNCT was determined to be 60–90 min after initiation of continuous [10]B-BPA drip infusion [187].

The concentration of [10]B-BPA in biopsied samples has been shown to be consistent with that mathematically estimated by dynamic ^{18}F-FBPA PET imaging and from the arterial input function of ^{18}F-FBPA as measured by arterial blood sampling [187–188]. It has also been concluded that ^{18}F-FBPA can be used to predict the absolute concentration of [10]B-BPA before BNCT treatment. By applying kinetic analysis to ^{18}F-FBPA PET data, it has been demonstrated that [10]B-BPA tissue concentration can be predicted at any point in time [212].

7.8.2   Single-time-point image analysis

The Standard Uptake Value (SUV) for ^{18}F-FBPA is defined as the ratio of the ^{18}F-FBPA concentration within one region to its mean concentration in the whole body.

$$\text{SUV} = \frac{[^{18}\text{F-FBPA}]_{\text{region}}}{[^{18}\text{F-FBPA}]_{\text{whole body}_{\text{IsoE}}}} \tag{1}$$

where $[^{18}\text{F-FBPA}]_{\text{whole body}}$ is defined as the total activity divided by the body weight of a patient and where it is assumed that ^{18}F-FBPA is distributed homogenously throughout the whole body. The merit of using the SUV approach is that there is no need of a reference tissue. In one study, SUVs of ^{18}F-FBPA based on a PET image were taken forty minutes after ^{18}F-FBPA injection. The SUV_{max} (maximal PET count in region of interest, ROI) or SUV_{mean} (mean PET count in the ROI) were used as practical parameters [213].

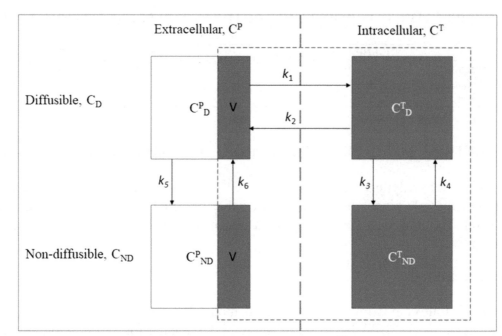

FIG. 31. Four-component pharmacokinetic model from Ref. [187] for ^{18}F-FBPA–fr in glioblastoma multiforme. The model consists of a vascular, extracellular space, C^P, and an intracellular, tissue space, C^D, each of which contain diffusible and non-diffusible spaces. The blue shaded areas represent activity detected in the ROI of a PET scan. It builds on earlier three-component models by the addition of the C^P_{ND} component that describes binding of ^{18}F-FBPA–fr and appearance of its metabolites in whole blood (k_5) from outside of the ROI. Parameter k_6 is assumed to be zero. V is the estimated vascular fraction.

As described below, most of the recent clinical studies of ^{18}F-FBPA PET imaging have employed tumour/normal (T/N) or tumour/blood (T/B) count ratios by defining relevant regions of interest on the PET image. It is usually possible to define the normal reference tissue, but it can sometimes be difficult because of the deformity of normal organs after surgical intervention. The blood count of ^{18}F-FBPA has usually been measured by setting the ROI of the PET camera on the blood pool. For purposes of comparison, ^{18}F-FBPA activity has been directly measured in venous blood using the PET count obtained by setting the ROI as the left cardiac chamber (see Ref. [207]): the PET image-derived ^{18}F-FBPA activity was 20% lower than that measured by blood sampling. Therefore, one could normalize the PET image-derived T/B ratio by dividing by 0.8 to yield the T/B ratio measured directly by blood sampling [207].

7.9 CLINICAL APPLICATIONS OF IMAGING

^{18}F-FBPA PET studies have been clinically applied in patients with brain tumours [187–188, 214–219], malignant melanoma [186, 220], and head and neck cancers [165, 167, 221]. A few examples are given below.

7.9.1 Brain tumours

BNCT for malignant brain tumours has been conducted by combining the use of ^{10}B-BPA with ^{18}F-FBPA PET [216], by using an epithermal neutron beam with combined use of both ^{10}B-BSH and ^{10}B-BPA as ^{10}B-carriers [217], and by combining with radiotherapy [218–219].

Research suggests that the kinetics are similar for the tracer dose of ^{18}F-FBPA–fr and the therapeutic dose of ^{10}B-BPA–fr. For example, two patients with glioblastoma multiforme were studied by means of ^{18}F-FBPA–fr PET [187]. ^{18}F-FBPA–fr accumulated in tumours that had been identified earlier by MR imaging. The T/N uptake ratio of 3.1:1 for ^{18}F-FBPA–fr was similar to that determined from ^{10}B analysis of brain in a patient that had been infused with BPA–fr before debulking surgery.[13]

In Ref. [219], based on histopathological examination, lesions were classified as tumour recurrence, radiation necrosis with a small portion of viable tumour cells, or complete radiation necrosis. A significantly greater ^{18}F-FBPA accumulation was found in radiation necrosis with viable tumour cells than that seen in complete necrosis cases. This indicated heterogeneous distribution of ^{18}F-FBPA in the necrotic tissue. Additional studies, including stereotactic tissue sampling based on three-dimensional ^{18}F-FBPA PET/MR or PET/CT, are needed to clarify radiation necrosis with viable tumour cells.

7.9.2 Head and neck tumours

BNCT has been applied to non-resectable head and neck cancer using ^{10}B-BPA and thermal neutrons from a research reactor, where complete remission of parotid gland cancer was demonstrated [165], followed by other research with favourable clinical outcomes [221–223].

7.9.3 Other tumours

BNCT has been applied to malignant melanoma using ^{10}B-BPA with thermal neutrons from a research reactor [186], and by means of ^{10}B-BPA–fr as the ^{10}B-carrier combined with ^{18}F-FBPA or ^{18}F-FBPA–fr PET for primary lesions and metastatic lesions [220].

7.10 SUMMARY

Because of the high selectivity of ^{18}F-FBPA for LAT-1, expressed predominantly on cancer cells, ^{18}F-FBPA is a useful PET tracer for visualizing cancer [224]. ^{18}F-FBPA PET/CT has distinguished between viable tumours and inflammation in an animal model [213], and between recurrence of glioblastoma (Fig. 32) and radiation necrosis (Fig. 33) in patients [215]. ^{18}F-FBPA PET/MR has been proven to improve tumour localization in head and neck tumours [225].

Because of the pharmacokinetic similarity between ^{10}B-BPA–fr and ^{18}F-FBPA–fr, the gross tissue concentration of ^{10}B can be estimated by means of ^{18}F-FBPA–fr PET when ^{10}B-BPA–fr is to be used as a boron carrier for BNCT. Dynamic ^{18}F-FBPA PET imaging and direct measurement of blood ^{18}F activity provides detailed information for patient-specific optimization of BNCT procedures. In practice, an estimate of ^{18}F-FBPA accumulation in the gross tumour and normal tissue can be made for each patient by taking the SUV of tumour relative to that of normal reference tissue. However, as stated above, the cellularity of the tumour and the homogeneity of boron uptake are not known from this technique due to its low spatial resolution, so that it is the gross T/N concentration ratio that is determined.

---

[13] However, the gross T/N concentration ratio of 3.1:1 ought not to be confused with the 3.5:1 ratio of intratumoural to whole blood concentration ratio used in previous clinical studies.

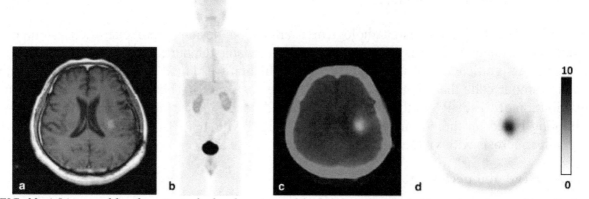

FIG. 32. A 54-year-old male patient who has been treated for left frontoparietal oligo-astrocytoma with radiation therapy. (a) T1 MRI image showed a contrast-enhanced lesion in the left frontoparietal area. ^{18}F-FBPA PET scan images were obtained 5 months after radiation therapy: Maximum intensity projection image of the whole body (b), PET/CT fusion image (c), and PET image (d) demonstrated increased accumulation of the tracer in the lesion and high PET metabolic indices (SUVmax = 6.8, SUVmean = 4.5, metabolically active tumour volume (MTV) = 28.5 mL, and total lesion uptake of FBPA = 128.8 g) (reproduced from Ref. [215] copyright (2018) The Japanese Society of Nuclear Medicine, with permission courtesy of Springer Nature.)

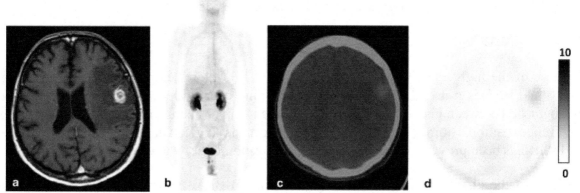

FIG. 33. A 72-year-old male patient with a history of left frontal lobe metastasis of lung adenocarcinoma. (a) T1 MRI image showed a contrast-enhanced lesion in the left frontal area. An ^{18}F-FBPA PET/CT scan was performed 24 months after stereotactic radiotherapy: Maximum intensity projection image of whole body (b), PET/CT fusion image (c), and PET image (d) revealed mild uptake in the lesion with low PET metabolic indices (SUVmax = 1.46, SUVmean = 1.04, MTV = 6.03 mL, and total lesion uptake of FBPA = 6.28 g). The lesion appears less active than the one in Fig. 32, suggesting a necrotic lesion, poorly characterized by MRI (reproduced from Ref. [215] copyright (2018) The Japanese Society of Nuclear Medicine, with permission courtesy of Springer Nature.)

8   BORON CONCENTRATION DETERMINATION AND IMAGING

This section describes the various techniques in clinical use and many under experimental development for boron determination and imaging. It describes some of the equipment necessary, known limitations, as well as some of the known advantages and disadvantages. The ability to measure very small amounts of boron in samples of biological origin is important for BNCT. Biodistribution studies are fundamental in clinical applications to understand the behaviour of boron compounds in tumours, always considering the biological variability that exists between individuals of the same species [226] and intertumoral heterogeneity [227–228]. The information obtained in these biodistribution studies is used to determine the boron concentration ratios [229]: T/B and T/N tissue ratios are then used to calculate the neutron irradiation time, considering a pharmacokinetic model that predicts the concentration in blood during irradiation [230].

It is important to distinguish between 'boron determination' and 'boron concentration determination'. The former determines the mass of boron present in the sample. The latter determines how much boron per unit mass of the sample is present. For BNCT the concentration of boron and its spatial distribution need to be determined. The knowledge of the concentration of ^{10}B in various tissues, both healthy and cancerous, is fundamental to developing a treatment plan. These data, combined with knowledge of the neutron fluence, permit the calculation of the absorbed dose produced by the boron neutron capture reaction (Fig. 15) during neutron irradiation at any point within the patient (see Section 10).

To calculate the absorbed dose, it is not sufficient to know only the total mass of boron, dm_B, present in the volume element dV considered, but it is also necessary to know the mass, dm_T, of tissue contained in dV. C_B, the concentration of boron in dV, is calculated as:

$$C_B = \frac{dm_B}{dm_T} \tag{2}$$

If dV has linear dimensions much larger than the ranges of the α particle and ^{7}Li nucleus emitted in the boron neutron capture reaction (i.e., mm compared to μm), charged particle equilibrium (Section 10.1.1.4) conditions are realized in dV, and the boron dose, D_B (Table 7), can be calculated through the collision KERMA (Section 10.1.1), K_c, and neutron fluence (Table 6), $\int\phi(t)\cdot dt$:

$$D_B = K_c = \int\phi(t)\cdot dt \ \times \ F_n \tag{3}$$

with

$$F_n = \frac{1}{dm_T} \times 1.602 \times 10^{-13} N_B \, \sigma \, Q_{ch} = 1.602 \times 10^{-13} \frac{C_B}{10} \, N_A \, \sigma \, Q_{ch} \tag{4}$$

where:

- F_n is the KERMA factor (in Gy·cm^2) associated with the ^{10}B(n,α)^{7}Li reaction;
- Q_{ch} (in MeV) is the reaction's positive Q-value transferred to charged particles and deposited within a few microns of where they were produced;
- $N_B = \frac{dm_B}{10} N_A = \frac{C_B}{10} N_A \, dm_T$ is the total number of ^{10}B nuclei contained in dV;
- σ is the nuclear capture cross section;
- N_A is Avogadro's number.

In the end, the boron dose (in Gy) in dV as a function of its concentration (in μg/g or ppm) and the neutron energy E is equal to:

$$D_B = 1.602 \times 10^{-13} \int\phi(t)\cdot dt \; \frac{C_B}{10} \; N_A \, \sigma \, Q_{ch} \tag{5}$$

What makes it particularly difficult to determine the concentration of boron is the need to know at the same time both the mass dm_B of boron, measurable through an appropriate signal that depends on the technique used, and the mass of tissue from which that signal is emitted. Often, the signal from boron has good intensity, but the mass of the sample is so small that it is difficult to measure precisely. In such cases, the technique may permit verifying the presence and spatial localization of boron (e.g., a specific carrier accumulates boron in tissues, within the cytoplasm or nucleus of cells or intracellular spaces) but not its concentration. Such a technique cannot be used to evaluate the boron dose.

8.1 TECHNIQUES USED IN CLINICAL PRACTICE WITH ACCELERATORS

The measurement of boron concentration in vivo still represents a major challenge for BNCT. In addition, it has to be taken into account that, during treatment, the dose also depends on the time t, as well as on the position $\mathbf{r}$; i.e., the dose is a function of $\mathbf{r}$ and t through $N_B(\mathbf{r}, t)$ and $\phi(\mathbf{r}, t)$; the former quantity relates to the metabolism in the patient and the latter relates to the stability of the flux distribution from the neutron source. Fortunately, the dose depends on the product of these two quantities and the cross section according to the equation

$$dD_B(\mathbf{r}, t) \propto N_B(\mathbf{r}, t) \, \sigma \, \phi(\mathbf{r}, t) \, F_n \, dt \tag{6}$$

In principle, the spatial distribution of the product $N_B(\mathbf{r}, t) \, \sigma \, \phi(\mathbf{r}, t)$ could be measured online during the treatment using, for example, a PG-SPECT technique (see Section 8.2.1 and Ref. [231]) or a Compton camera technique [232], thus obtaining an image of the spatial distribution of the dose taking into account the real values of boron concentration, cross section, and neutron flux. Unfortunately, the current methods used in the clinic, based on inductively coupled plasma (ICP) techniques, measure boron concentration in the patient's blood. Based on this value, boron concentrations in the tumour and healthy tissues are assumed. There is an urgent need for a method that is able to directly measure boron concentrations in the tissues subjected to neutron irradiation. PET imaging is a step in this direction but is currently available only for BPA, based on the use of ^{18}F-FBPA/^{18}F-FBPA–fr, and only gives concentrations at the gross tumour level.

8.1.1 Inductively coupled plasma optical emission spectroscopy and mass spectroscopy

An inductively coupled plasma (ICP) is created by a set of electrical coils, external to the sample, through which current is applied whose phase alternates at radiofrequencies. This induces an alternating electromagnetic field (inductive coupling) to any fluid passing within the coils. Sufficient power is applied to heat and ionize the sample to form a plasma (i.e., a fully ionized, but neutral gas). Inductive argon plasma spectrometry has very good characteristics regarding detection limit, precision, and accuracy.

Two ICP methods are widely used in routine BNCT applications:

- Optical emission spectroscopy (ICP-OES) is sensitive enough to determine boron levels in samples with low boron concentration. There is a lot of literature supporting the use of the ICP-OES to measure boron in BNCT [233–235];
- Mass spectroscopy (ICP-MS) has even lower detection limits (about three orders of magnitude better than ICP-OES) and the isotopy of boron can be analysed [236].

These methods provide information on mean boron concentration in the sample, with no information on boron localization. Both ICP measurement systems consist of two main parts:

(i) ICP source;
(ii) Detection system.

Figure 34 shows a simplified diagram of ICP instruments.

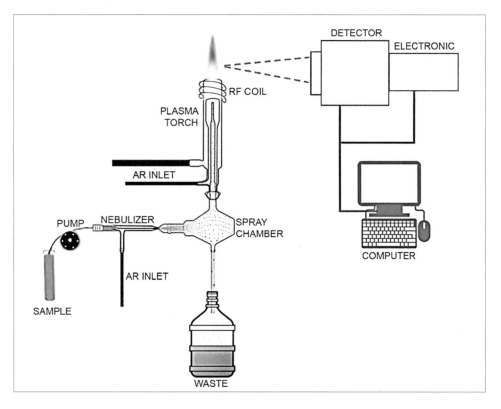

FIG. 34. Major components and layout of a typical ICP instrument (courtesy of S. Nievas, CNEA).

8.1.1.1   Sample pre-treatment

Sample introduction can be a significant source of random and systematic error in the measurement of samples by ICP methods: all samples need to be in an aqueous state. It is important that both the dilution and digestion processes are as efficient as possible to provide representative aqueous solutions of the original sample.

There are different sample measurement protocols in BNCT treatments. Whole blood samples, once extracted from the patient, are conditioned with an anticoagulant. Some groups perform a dilution of whole blood with surfactant [229] or with surfactant and nitric acid [237]. Other groups have developed digestion methods of whole blood samples that use trichloroacetic acid or wet ashing by microwave digestion (nitric acid + hydrogen peroxide) [235]. Still others perform a separation of the blood plasma from the cells and platelets of each sample, and the blood plasma becomes the sample measured. All the aforementioned procedures aim to obtain an optimal liquid sample to be introduced into an ICP to feed the measurement process.

8.1.1.2   Introduction of samples

The next step is the production of an aerosol using a nebulizer, which is used to convert liquid samples into finely divided drops through the pneumatic action of an Ar gas flow. As the drop sizes used in ICP-MS are so small, the concentration of matrix components needs to be kept below 0.2% to reduce sample-specific matrix effects and potential blockage of the nebulizer, whereas the nebulizers of ICP-OES are designed to handle up to 20% dissolved solids. The most common pneumatic nebulizers used in commercial ICPs are the concentric and cross-flow designs. The concentric design is more suitable for clean samples, while the cross-flow is generally more tolerant of samples that contain higher concentrations of solids or particles.

A spray chamber is used for the selection of only the smallest aerosol droplets (less than 10 μm in diameter) to obtain a small droplet size distribution. In addition, the spray chamber smooths any pulse in the sample flow produced by the nebulization process, coming from a peristaltic pump. Larger droplets collide with the walls and drain away. The sample introduction efficiency for ICP is extremely low, less than 2%. The resulting aerosol is transferred to the base of the plasma by the sample injector [238].

The torch unit of an ICP is used to create and sustain a plasma. Due to the extreme temperatures (up to 10,000 °C) of the Ar plasma used in modern ICP systems, the excitation of atoms is practically complete for most elements. The torch system is composed of three concentric quartz tubes. The sample and the Ar gas used to aspirate it pass through the internal tube into the region surrounded by the ICP coils where they are turned into a plasma. The Ar plasma then passes through the middle tube, and a separate flow of Ar passing through the outer tube is used to cool the quartz torch.

8.1.1.3   Detection systems

Stability of the plasma during measurement is critical to reduce variations in data acquisition, as the plasma serves as an ion source for spectrometers where the signals are subsequently recorded by a detection system [239]. ICP-MS can determine the isotopic abundance (^{10}B:^{11}B) via the mass/charge ratio (m/z) of B^+ ions; characteristic optical emission lines of B^+ are measured by ICP-OES and are used to determinate the presence of boron independent of its isotopic composition [236].

8.1.1.4   Advantages and disadvantages

Both ICP techniques have advantages in common, as they:

- Provide high sample throughput;
- Yield multi-elemental analysis;
- Have wide dynamic analytical range;
- Require low sample volumes.

Specific advantages include:

- ICP-OES requires simpler sample preparation, has greater tolerance to complex matrices, and the instruments are cheaper to purchase and operate than ICP-MS instruments;
- ICP-MS is a more sensitive technique and is also sensitive to the isotopic composition of boron.

Both ICP techniques have the disadvantages that:

- They do not measure boron concentration directly in the tissues of interest, although they can be combined with other measurements (e.g., PET imaging);
- Their spectra can suffer from interferences, but these can be eliminated or minimized:
  o ICP-OES: Specific spectral interferences can be eliminated using another uncontaminated spectral line. Complex matrix interferences can be smoothed out using internal standards [236];
  o ICP-MS: Isobaric overlaps [240] are relatively easy to reduce using alternative isotopes or through mathematical equations related to their relative natural abundances [241].

8.1.1.5 *Treatment protocol*

On the day of BNCT treatment, prior to irradiation, patients receive the full infusion of the boron drug. Blood samples are drawn at different intervals. Boron concentration in blood is assumed to be proportional to that in normal tissue when calculating boron dose in both normal tissue and tumour [227]. The estimate of boron concentration in the tumour is based on the boron concentration in the blood sample, which is compared with a calibration curve at the time of its measurement by ICP. These values are used in BNCT treatment planning (Section 11).

Table 29 in the Appendix IV shows some of the ICP measurement methods used in different countries. Differences between the procedures used could result in variations in the determination of boron concentrations.

8.1.2 Quantitative imaging

Positron Emission Tomography (PET) is an imaging modality to measure the concentration of positron-emitting radioisotope in the human body in vivo. The images derived from PET are usually fused with those from another imaging modality such as X ray computed tomography (PET/CT) or magnetic resonance imaging (PET/MR) to improve the understanding of the localization of the accumulation of the positron-emitting radiopharmaceutical. ^{18}F-FBPA is the labelled analogue of BPA and is used to provide a measure of boron concentration in tumours. Its synthesis, cell transport, uptake, and metabolism are described in Section 7.6.1.

PET imaging has several advantages for quantitative measurement of radioactivity in the body:

- When compared to γ rays typically used in SPECT imaging (see Section 8.2.1):
  o The 511 keV annihilation γ rays are of higher energy so that signals from structures deeper in the body can be collected;
  o Because of coincidence counting, PET imaging can reduce noise due to scattering events and random events, which may result in better quality images.
- Differential tissue attenuation of the γ rays (due to variable thicknesses or densities of tissues encountered between the tumour and detector) can be corrected by a transmission

scan using an external ring source of ^{68}Ga–^{68}Ge or by using the co-registered CT or MR images;

- Activity measured in specific regions of the body at specific times after tracer injection can be converted to absolute concentration of the administered tracer by cross calibrating among the PET scanner, dose meter, and well scintillation counter, and by knowing the molar activity of the injected tracer.

Positron emission tomography imaging has the disadvantage that it cannot be conducted during BNCT treatment. The other major limitation of PET imaging is the low spatial resolution compared with CT or MRI. The radioactivity measured by PET is a function of the volume of the target tissue. When the volume is less than 1 ml, in general, the radioactivity contained in the nodule is underestimated. This is called the 'partial volume effect' [242]. Figure 35 shows the relationship between nodule size and radioactivity.

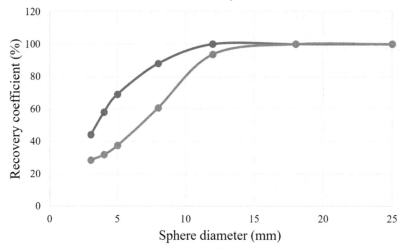

FIG. 35. Partial volume effect in small nodules as a function of size from corrected positron emission mammography (blue curve and data points) data, and PET data (orange curves and data points). Data are taken from Ref. [246].

Another factor of underestimation is tissue density or cancer cell density [243]. For example:

- When a lung cancer contains an air fraction, such as in adenocarcinoma with ground-glass appearance, regional activity in the ROI is low even though the cancer tissue contains a high concentration of ^{18}F-FDG;
- In brain tumours, when residual cancer cells contain high concentrations of ^{18}F-FDG but cell density is low, it is sometimes difficult to identify remaining cancer cells at the microscopic level.

These limitations are being reduced in the current PET scanner technologies by improving the spatial resolution, reducing noise, and optimizing image reconstruction and analytical methods. Because of the short travel range of α and ^{7}Li particles (see Table 20 in Section 10) produced by the boron neutron capture reaction, the specific location of ^{10}B-BPA within the cell's internal structures affect the cell-killing effect of BNCT [244]. However, the travel range of the positron prior to annihilation (ca. 0.54 mm for ^{18}F) contributes to the limited spatial resolution of current PET scanners [245], so that it is difficult to estimate the degree of intracellular localization of ^{18}F-FBPA in the clinical setting.

Both ICP-OES and ICP-MS may provide the atomic concentration of boron in blood after administration of a ^{10}B-carrier compound. By means of ^{18}F-FBPA PET, the ratio of concentrations of ^{10}B in tissue to that in blood can be estimated in each patient. If ^{10}B-BPA–fr is utilized as the ^{10}B-carrier in BNCT, the ^{10}B tissue concentration can be predicted by combining these two measures as follows:

$$C_{\text{tissue}} = C_{\text{blood}} R_{\text{tissue/blood}} \tag{7}$$

where:

- C_{tissue} is the ^{10}B-concentration in tissue;
- C_{blood} is the ^{10}B-concentration in blood measured by ICP-OES or ICP-MS;
- $R_{\text{tissue/blood}}$ is the tissue to blood activity ratio measured by ^{18}F-FBPA PET.

Most institutions employ ICP-OES to measure the atomic concentration of boron in blood (see Table 29 in Appendix IV). Since isotopic composition is not measurable by means of ICP-OES, the isotopic composition of ^{10}B is taken into account as follows:

$$cC_{\text{blood}} = C_{\text{blood}} \, IC_{^{10}\text{B}} \tag{8}$$

where:

- cC_{blood} is ^{10}B-concentration in blood;
- C_{blood} is atomic concentration of boron in blood measured by ICP-OES;
- $IC_{^{10}\text{B}}$ is the known isotopic composition of ^{10}B in the carrier molecule ^{10}B-BPA–fr.

In an ^{18}F-FBPA PET measurement, $R_{\text{tissue/blood}}$ is obtained as follows:

$$R_{\text{tissue/blood}} = \frac{A_{\text{tissue}}}{\frac{A_{\text{blood}}}{S} F} \tag{9}$$

where

- $R_{\text{tissue/blood}}$ is the ^{10}B-concentration ratio of tissue to blood;
- A_{tissue} is the PET count from the tissue of interest;
- A_{blood} is the PET count from the blood pool, e.g., as measured in the left ventricle;
- F is the fraction of unmetabolized ^{18}F-FBPA at the time of the PET scan;
- S is the correction factor for underestimation of the image-derived blood count.

The parameter S is a PET scanner dependent factor related to its spatial resolution. If, for example, the PET count of the left ventricle is underestimated by 80%, then $S = 0.8$ [207] and the measured A_{blood} is corrected for by dividing measured value by $S = 0.8$ (see also Section 7.8.2). Since the origins of measured activity are unmetabolized ^{18}F-FBPA and ^{18}F-labelled metabolites, the measured A_{blood} is corrected by using the fraction of unmetabolized ^{18}F-FBPA in blood (F). The mean fraction of unmetabolized ^{18}F-FBPA in blood at 20 min and 50 min after administration has been previously determined to be 97.68 ± 1.57 % and 96.03 ± 1.64 %, respectively [207]. The activity of tissue, A_{tissue}, is underestimated for nodules less than 15-mm diameter due to the partial volume effect, the degree of underestimation being dependent on their size (Fig. 35). In an 8mm-diameter nodule, the measured activity was reduced to 60% of the true value; in a 3 mm-diameter nodule, activity was reduced to 25% [246]. The tissue activity can be corrected for the partial volume effect as follows:

$$cA_{\text{tissue}} = A_{\text{tissue}} \, \text{PVE} \tag{10}$$

where

- cA_{tissue} is the corrected tissue radioactivity for the partial volume effect in ^{18}F-FBPA PET/CT scans;
- A_{tissue} is the tissue activity measured by ^{18}F-FBPA PET/CT;
- PVE is the factor describing the partial volume effect.

For the case of pulmonary nodules, another source of underestimation is movement when breathing during the PET scan. Correction for this underestimation is installed in current PET scanners.

In summary, ^{10}B-concentration in tissue (cC_{tissue}) can be estimated by using ICP-OES and ^{18}F-FBPA PET as follows:

$$cC_{tissue} = \frac{A_{tissue}\,PVE}{\frac{A_{blood}}{S}F} \qquad (11)$$

Early PET studies in patients with high grade gliomas demonstrated that the absolute ^{10}B-concentration per unit weight of resected tumour tissue correlated well with ^{18}F-FBPA PET estimates per unit volume [186–187] (see Section 7.6.3.2). As a worked example of the above equations, Fig. 36 shows an ^{18}F-FBPA PET/CT image from a patient with malignant melanoma in the lung. Local accumulation of ^{18}F-FBPA was found in a large tumour in the left upper lobule (Fig. 36a) and in a small tumour in the left lower lobule (Fig. 36b). The activity of ^{18}F was determined to be 6.0 and 4.0 SUV_{mean} units for the tumours in panels a and b, respectively. Activity in the left ventricle was determined to be 1.1 SUV_{mean} units. The value of S was 0.80 for this PET/CT scanner. The unmetabolized fraction, F, was 95% during PET scanning. The isotopic composition of ^{10}B-BPA–fr was 99% ^{10}B. The diameters of the tumours in panel a and b were > 30 mm and 10 mm, respectively. PVE was 1.0 (no partial volume effect) and 1.25 (1/0.80) for tumours in panels a and b of Fig. 36, respectively. The atomic concentration of boron in blood was determined to be 10 ppm by ICP-OES. The ^{10}B-concentration in the gross tumour tissues in panels a and b could then be calculated to be 46.4 ppm and 38.7 ppm, respectively.

Tumour tissue consists of various cell components such as cancer cells, stroma cells, immune cells, inflammatory cells, and blood vessels. In BNCT, the ^{10}B-concentration in each single cancer cell directly affects the therapeutic effect. This can be estimated by measuring gross total ^{10}B content in the tumour, the fractional volume of cancer cells in the bulk of the tumour, and the cancer cell density in the fraction. Current PET scanners are not able to measure with precision the amount of ^{18}F-FBPA within cancer cells because of their limited spatial resolution. The homogeneity of boron uptake is also unknown and, therefore, so is the ^{10}B-concentration within cancer cells. The concentrations derived from ^{18}F-FBPA PET are therefore a lower bound on the concentrations in cancerous cells when the boron uptake is assumed to be homogeneous. By combining cancer cell specific radiopharmaceuticals such as ^{18}F-FBPA and diffusion MRI sensitive to tissue cellularity, ^{10}B content per cancer cell may be more accurately estimated by pre-BNCT imaging studies.

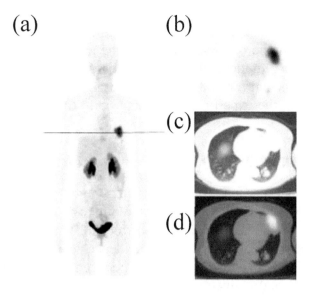

FIG. 36. ¹⁸*F-FBPA PET in a patient with malignant melanoma in lung. (a) Whole-body view of* ^{18}*F-FBPA distribution 40 min after* ^{18}*F-FBPA injection. A tomographic image of (b)* ^{18}*F-FBPA PET, (c) X ray CT, and (d) fused image of* ^{18}*F-FBPA PET and X ray CT (courtesy of J. Hatazawa, Osaka University).*

8.2 TECHNIQUES NOT IN CLINICAL USE AT ACCELERATOR FACIITIES

The most commonly used techniques in the BNCT field are summarized in Table 17 and described below; some are only intended for research, while others, such as Single Photon Emission Computed Tomography (SPECT) and Magnetic Resonance Imaging (MRI), aim to become useful in clinical practice, although they are still at the research and development stage.

TABLE 17. DETECTION LIMITS OF VARIOUS TECHNIQUES USED FOR ^{10}B-CONCENTRATION MEASUREMENTS IN BIOLOGICAL SAMPLES

Technique		Detection limit	Imaging
Inductively Coupled Plasma Spectroscopy	ICP-MS	1–3 ppb in biological materials, 0.15 ppb in saline waters, 0.5 ppb in human serum [247]	No
	ICP-OES	15–30 ppb [236]	No
Prompt Gamma Neutron Activation Analysis	PGNA	0.04–10 ppm for sample masses of 20–500 mg with thermal neutron fluence of ~10^9–10^{12} cm^{-2} [248, 249]	No
Prompt Alpha Neutron Spectrometry	PANS	0.5 ppm for sample masses of 500 mg with thermal neutron fluence of ~5×10^{11} cm^{-2} [250]	No
Neutron autoradiography*	NCR	~0.01–0.1 ppm with thermal neutron fluence of ~10^{12}–10^{14} cm^{-2} [251]	Yes
Prompt Gamma Single Photon Emission Computed Tomography	PG-SPECT	10 ppm with thermal neutron fluence of ~10^{12} cm^{-2} [252]	Yes
Secondary Ion (Neutral) Mass Spectrometry	SIMS	0.1–10 ppb [253]	Yes
	laser-SNMS	~ ppb, with lateral resolution of ~200 nm [253]	Yes
Laser-Induced Breakdown Spectroscopy	LIBS	~ ppm (< 20 ppm) [254]	Yes
Magnetic Resonance Spectroscopy	MRS	~20 ppm of ^{10}B in vivo at 1.5 T [255]	Yes

* Also known as Neutron Capture Radiography

These techniques can demonstrate the selective uptake of boron carriers and show the boron spatial distribution at macroscopic (organs and tissues) or microscopic (cellular and subcellular) levels. While knowledge of the boron concentration at the macroscopic level is important for the calculation of absorbed dose in tumours and healthy tissues in BNCT, measurements at the microscopic level are fundamental for understanding the biological effectiveness of different boron carriers; they represent the experimental starting point of microdosimetry calculations (see Section 10.2.2).

8.2.1 Prompt gamma analysis methods

Conventional Single Photon Emission Computed Tomography (SPECT) is similar to PET imaging in that a ring of detectors is used and a three-dimensional image of the intensity of localised radioactivity can be mathematically reconstructed. Unlike PET, only one γ ray is directly emitted per decaying nucleus. Therefore, coincidence detection cannot be used, and the spatial resolution is lower. There is a major difference between SPECT for BNCT (hereinafter referred to as PG-SPECT[14]) and conventional SPECT imaging, which requires an injected radiopharmaceutical (e.g., a ^{99m}Tc-labelled compound). In PG-SPECT, the source of γ rays is intrinsic and arises from the prompt gamma at 478 keV emitted from the recoiling ^{7}Li* nucleus created by the boron neutron capture reaction. This characteristic γ ray is produced in 93.7% of neutron capture events by ^{10}B (Fig. 15). In theory, this can be measured in real time during BNCT to estimate the local boron dose. However, the realization of PG-SPECT is not straightforward and such a system has yet to be applied in clinical BNCT because the intense primary neutron radiation produces a strong secondary γ ray background field. Moreover, although the neutron capture reaction has a high cross section, ^{10}B is present at only a few tens of ppm, even in a tumour.[15]

As shown above, the number density of ^{10}B can be predicted by ^{18}F-FBPA-PET [216], before performing BNCT. However, the true local dose due to the boron neutron capture reaction is not straightforward to determine, because it has to be determined during BNCT. Currently, neither boron number density nor local dose can be known in real time during BNCT treatment.

The local boron dose, $D_B(\mathbf{r}, t)$, is theoretically proportional to $N_B(\mathbf{r}, t) \, \sigma \, \phi(\mathbf{r}, t)$ as described in Section 8.1. However, its value changes depending not only on $N_B(\mathbf{r}, t)$, the number density of ^{10}B at time t and position $\mathbf{r}$, but also on the neutron flux, $\phi(\mathbf{r}, t)$. The cross section of the ^{10}B(n,α)^{7}Li reaction, σ, is a strong function of energy over the thermal to epithermal neutron energy region (Fig. 16). It is desired to know both $D_B(\mathbf{r}, t)$ and $N_B(\mathbf{r}, t)$, a differential value that changes continuously during BNCT and is directly related to the T/N ratio. The boron dose, D_B, is an integral value, and its value from the start of irradiation until a given time t can be estimated from $\iint N_B(\mathbf{r}, t) \, \sigma \, \phi(\mathbf{r}, t) \mathrm{d}t \mathrm{d}\mathbf{r}$ (see Eq. (6)). However, as ^{10}B metabolism occurs during BNCT and causes real-time changes in $N_B(\mathbf{r}, t)$, it is hard to estimate D_B experimentally.

In PG-SPECT, a three-dimensional image has to be reconstructed from measured projection data. During a normal diagnostic procedure, dual modality imaging (combining SPECT with CT or MRI) is carried out in an ideal environment, i.e., free of background, where 360° detection is possible. In such cases, the image reconstruction procedure is well established. However, during BNCT a patient is usually positioned very close to the wall from which the neutron 'beam' is emitted. Additional equipment and/or a jig to fix the patient in position are

---

[14] The terms PG-SPECT (prompt gamma SPECT) and BNCT-SPECT are both used in the field to describe the same technique.
[15] The absorption cross section of ^{10}B(n,α)^{7}Li is known to be very large compared to those of other nuclides in a human body (by around a factor of one thousand).

attached around the tumour or the patient, who is usually placed in a chair or on a bed. Due to these constraints, the count rate and signal-to-noise ratio from PG-SPECT is not high. In general, a SPECT instrument needs to be positioned no more than a few cm from the surface of the human body. Therefore, 360° projection data acquisition is difficult.

Compared to the need for knowledge of the local boron dose, the need for real-time measurement of boron concentration is not considered so urgent. Nevertheless, in Osaka University, Japan, development of a SPECT system (named 'T/N-SPECT') to determine real-time boron concentration is underway [256]. To determine the three-dimensional boron concentration distribution, supplementary image data by MRI are required. A PG-SPECT system needs to be able to discriminate the characteristic 478 keV prompt gamma arising from the de-excitation of the excited $^7Li^*$ nucleus (Fig. 15) from the background 511 keV positron annihilation peak in a high γ background and to deliver 'live' updates on boron concentration during the one-hour patient irradiation. Some of the design requirements for such a system are discussed in Refs [231, 256] and below.

8.2.1.1 Determining boron concentration using prompt gammas

In the past, a simple technique was proposed [257]: 478 keV γ rays were measured from an adjacent room through a hole in the wall (i) to see all of the irradiated region including the tumour and (ii) to avoid the high background. However, this is a compromise. Figure 37 shows a conceptual arrangement of PG-SPECT. The patient is placed near the exit of the neutron 'beam'. The SPECT system is positioned very close to the patient, which may lead to a substantial increase in background. This would allow the system to view the tumour directly and for this purpose it needs to be close to the tumour with a heavily shielded collimator.

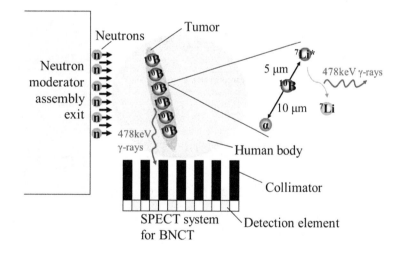

FIG. 37. Conceptual design of PG-SPECT. ^{10}B atoms undergo capture of epithermal and thermal neutrons. The SPECT detector element requires collimation ahead of it to determine the direction of γ emission (reproduced from Ref. [252] copyright (2022) with permission of Elsevier).

First, a segmented γ ray detection device needs to be specified: there are many candidate materials, including CdTe/CZT, LaBr₃(Ce) and GAGG(Ce), with good performance values for detection efficiency and energy resolution. PG-SPECT requires a collimator to shield the background radiation for good statistical accuracy (good signal-to-noise ratio) and to achieve the required spatial resolution.

Figure 38 shows an example of a pulse height spectrum of a CdTe detector calculated by MCNP [258]. The challenge of discerning the 478 keV characteristic γ ray is clear from this figure. Furthermore, although a strong peak due to hydrogen capture γ rays is observed at 2.22 MeV, the annihilation γ ray peak at 511 keV is not seen in this simulation, as the relevant process were not included. (In real applications, a strong 511 keV peak appears.) By measuring the intensity of the hydrogen capture γ rays, the boron concentration can be deduced [256], as the neutron capture cross sections of ^{1}H and ^{10}B show similar energy dependences that are inversely proportional to the velocity of the neutron in the thermal and epithermal energy regions. This '1/v' variation in the cross section is a characteristic of many absorption cross sections in this neutron energy range (see Fig. 14). The ratio of the measured intensities of the 478 keV and 2.22 MeV γ rays is the same as the ratio of the two nuclear reaction rates, which is, in turn, the same as the ratio of the number densities of ^{10}B and ^{1}H multiplied by the cross section ratio of both neutron capture reactions in the '1/v' region. If the hydrogen number density is known, which is easily established through MRI, the ^{10}B-concentration can be estimated.

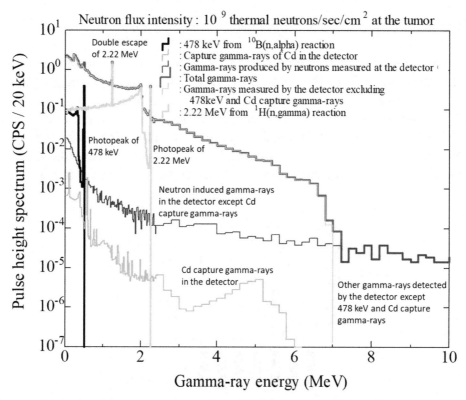

FIG. 38. Simulated pulse height spectrum measured of a CdTe detector in PG-SPECT (reproduced from Ref. [252] copyright (2022) with permission of Elsevier).

8.2.1.2 Designs of prompt gamma systems

Table 18 shows the examples of designs of PG-SPECT developed so far [252, 257, 259–262]. Basic research is supporting these advances in PG-SPECT [263–264]. Since each PG-SPECT system has unique design conditions, comparisons are not simple. Both semiconductor (such as CdTe and CZT) and scintillator detectors (such as LaBr₃(Ce) and GAGG(Ce)) have been used. The system described in Ref. [260] used BGO but concluded that CZT would be preferable.

TABLE 18. EXAMPLES OF PG-SPECT DETECTOR AND COLLIMATOR SYSTEMS

Detector element		Collimator			
Material	Size	Material	Diameter (mm)	Length (cm)	Ref.
CZT	$1 \times 1 \times 1$ cm	Pb	4×4	20	[259]
BGO	5 (Φ) mm $\times$ 5 cm	Heavy metal	5.4 (Φ)	32.1	[260]
CdTe	Not mentioned	W	4×4	20	[257]
LaBr$_3$	1"(Φ) $\times$ 1"	Pb	5 (Φ)	30	[261]
GAGG	$3.5 \times 3.5 \times 30$ mm	W	3.5 (Φ)	26	[252]

Note: Φ indicates a diameter. CZT = cadmium zinc telluride; BGO = bismuth germanate; GAGG = gadolinium aluminium gallium garnet

The implementation of PG-SPECT will be important but difficult, as the system has to be positioned in strong neutron and secondary γ ray fields in which relatively weak signals from the 478 keV characteristic γ rays have to be detected with acceptable accuracy, appropriate spatial resolution, and within a limited time. The development of PG-SPECT systems is likely to be a field of active research for the foreseeable future.

8.2.2 Other prompt nuclear spectroscopic methods

This section outlines two methods for the measurement of boron concentration that are based on on-line spectrometry of two boron neutron capture reaction products (Fig. 15): photons, in the case of PGNA (Prompt Gamma Neutron Activation Analysis), and α particles, in the case of PANS (Prompt Alpha Neutron Spectrometry).

8.2.2.1 *Prompt gamma neutron activation analysis*

Prompt gamma neutron activation analysis [248] is a non-destructive technique used to study the composition of materials; it is based on analysis of the spectrum of the prompt γ ray emitted in a radiative capture reaction

$$n + {}^{A}_{Z}X \rightarrow {}^{A+1}_{Z}Y + \gamma$$

In addition to the (n,α) boron neutron capture reaction, the basis of BNCT (Fig. 15), there is also an (n,γ) capture reaction. There are two possible candidate γ rays:

(a) Theoretically, the (n,γ) reaction could be used for PGNA. However, while the cross section of the (n,α) reaction is 3840 barns for a thermal neutron, the (n,γ) cross section is only ~0.4 barns for ^{10}B and about two orders of magnitude less than this for ^{11}B. This, combined with the high energy of the emitted γ ray (over 4 MeV), makes it difficult to use this reaction to detect boron in biological matrices at typical BNCT concentrations (tens of ppm);

(b) The PGNA technique can use the same characteristic γ ray as used in PG-SPECT: i.e., the 478 keV γ ray emitted by de-excitation of ^{7}Li*. Figure 39 shows an operating PGNA facility at a research reactor designed for BNCT applications.

The first applications of PGNA to the detection of boron in matrices of various kinds date back to the 1960s; the application of the technique to BNCT was introduced in the early 1980s and then applied in various BNCT research centres [249, 265–266]. The basic instrumentation used in PGNA includes a beam of thermal or cold neutrons, a sample holder, and a detector for gamma spectrometry (usually a high purity Ge detector).

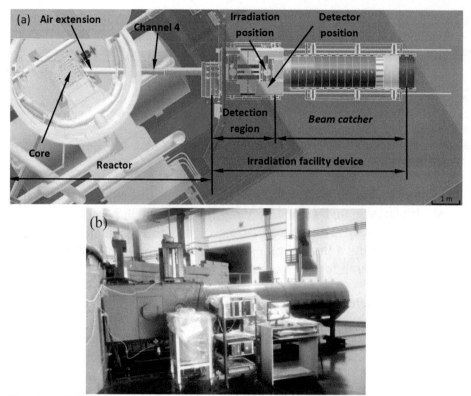

FIG. 39. (a) Top view of the horizontal plane of the engineering drawing of the prompt gamma facility for BNCT at the RA-3 nuclear reactor (CNEA, Argentina). This design concept was based on numerical modelling and design [267]. (b) Irradiation facility device built and positioned at the end of channel 4 of the RA-3 nuclear reactor. This operating facility has experimentally reached a thermal neutron flux of $(3.0 \pm 0.2) \times 10^7$ cm$^{-2} \cdot$s^{-1} at the sample position, with a thermal / suprathermal neutron components ratio of around $\sim 1 \times 10^4$. The current detection limit is of ~ 3.2 ppm of ^{10}B in samples of 250 mg of mass in 600 s [268]. The detection limit of this facility is being improved (courtesy of J. Quintana, CNEA).

Currently designed AB-BNCT centres do not include an additional neutron beam, although in principle future centres may. Requirements for a PGNA neutron beam include:

- Good collimation to ensure spatial uniformity across the sample;
- A well-defined and characterized energy spectrum;
- Low contamination by γ rays, epithermal, and fast neutrons;
- A stable source intensity if there is no neutron fluence monitor.

The high hydrogen content (about 10% by weight) in biological samples can give rise to two potential complications:

- The cross section for scattering from ^{1}H is very much larger than other nuclei. In the event of a heterogeneous distribution of hydrogen in a sample, the flux can be depleted in regions of the sample that are locally rich in hydrogen (a self-shielding effect);
- The scattering cross section of ^{1}H varies between ca. 20 and 80 barns in the thermal neutron energy regime, dependent on the specific bonding and dynamics of a given biological molecule [269].

As the 478 keV photons are emitted during recoil of the excited ^{7}Li* nucleus during its slowing down in matter (Fig. 15), the γ spectrum does not exhibit a Gaussian peak, but a Doppler-broadened peak of width $\Delta E = 2\beta E = 26.88$ keV centred at 478 keV; the shape of the curve in this region depends on the material in which the boron is contained; an example is shown in Fig. 40. The area under the broad curve is proportional to the boron content.

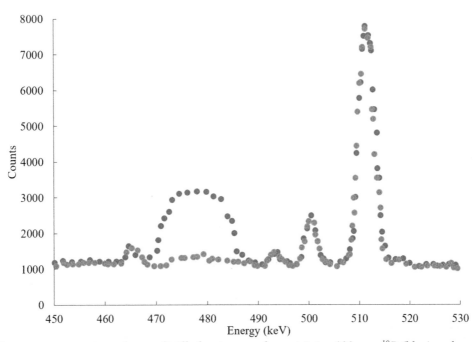

FIG. 40. Photon energy spectrum from a distilled water sample containing 100 ppm ^{10}B (blue) and a pure distilled water sample (orange) obtained using prompt gamma spectroscopy. The area under the blue peak centred at 478 keV is proportional to the boron content. Note the flat shape to the peak due to the recoil of the emitting particle. Data are taken from Ref. [270].

In general, the preparation of a PGNA sample is quite simple: if it is liquid (such as blood, urine…), it is enclosed in a vial, weighed, and irradiated. The sample needs to have a well-defined geometric shape even if it is solid [270]. The minimum detectable concentration values depend on the available neutron flux, measurement time, and sample mass. For a neutron flux of $1.7 \times 10^7 \, cm^{-2} \cdot s^{-1}$, the minimum detectable concentrations vary between 0.04 and 10 μg/g for sample masses between approximately 20 and 500 mg, and the collection times from between a few minutes to a few thousand minutes [249]. The sample mass required for measurement can be reduced by lowering the γ background. To this end, it is also important that the material used in the shielding is free of boron.

The hydrogen present in biological samples can be used as a fluence monitor to compare measurements made on different samples and times using the ^{1}H(n,γ)^{2}H capture reaction that emits a γ ray of 2.22 MeV. This is useful to compensate for fluctuations in source intensity and sample placement. Moreover, if it can be assumed that all analysed samples have the same known hydrogen content, the 2.22 MeV peak area can be used to evaluate the boron concentration in the samples (see discussion in the section above).

Since the first introduction of PGNA in BNCT, its potential for measuring boron in vivo as well as in small samples has been assessed. The first applications in vivo were dedicated to skin melanoma by studying the dependence of the ^{10}B/H ratio on the concentration of boron in the tumour (see Fig. 41) and the shape and size of the tumour itself [257, 266, 271–274]. By use of standards, the counting ratio can be converted into ppm ^{10}B. Over the years, PGNA has been applied for the measurement of boron concentration in various biological tissues and blood, as well as for pharmacokinetic studies both in animal models, and in patients treated with BNCT mediated by BPA or BSH.

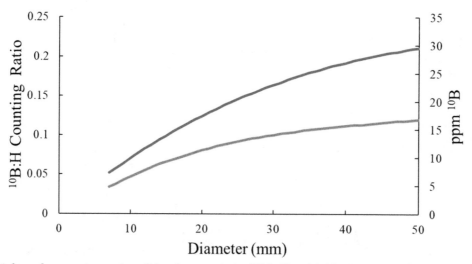

FIG. 41. A fit to the counting ratios of the characteristic 478 keV and 2.22 MeV γ rays due to neutron capture by ^{10}B and ^1H for a concentration of 20 ppm of ^{10}B in a tumour. The upper blue curve represents a 10 mm thick tumour and the lower orange curve a 5 mm thick tumour. The secondary vertical axis gives the values of ^{10}B-concentration determined from the calibration. Data are taken from Ref. [273].

As in the case for ICP methods (Section 8.1.1), PGNA measurement of the concentration of ^{10}B in blood is used to estimate the concentration of ^{10}B in tumour and healthy tissues. Naturally, to be used in the clinic the measurement system needs to be able to provide reliable and low error responses in a short time (a few minutes) compared to the treatment times; the limit of detection needs to be 1 ppm or better for sample masses of the order of 0.4–1.0 g.

There are two important points of note:

(a) In the case of samples with a non-homogeneous distribution of boron (as in the case of tumour tissue mixed with healthy tissues), the response represents an average value of the whole sample; therefore, it does not allow determination of the concentration of each individual tissue present in a sample;
(b) A BNCT centre wishing to use this technique requires a thermal moderator for PGNA instrumentation in addition to the standard epithermal BSA design used for patient treatment (see Section 5.2.4).

8.2.2.2 *Prompt alpha neutron spectrometry*

Prompt Alpha Neutron Spectrometry (PANS), like PGNA and PG-SPECT, exploits the boron neutron capture reaction to measure the concentration of ^{10}B present in samples of various kinds. However, for PANS it is the energy spectrum of the α particle (Fig. 15) that is measured directly from solid frozen biological samples. The technique [247, 250, 275–277] has been specially developed for the analysis of heterogeneous samples in which a tumour has infiltrated into normal tissues [164]; for this reason, PANS is commonly used in conjunction with biological staining and neutron autoradiography.

The instrumentation required for PANS measurement includes a thermal neutron 'beam', not necessarily collimated, a solid state charged particle detector, and a cryostat to section the sample to be analysed. Sample and detector need to be under vacuum during the measurement.

Biopsies taken from treated animals or patients are divided into smaller samples of about 1 cm³ and frozen in liquid nitrogen. Then, three consecutive slices are cut for the measurement [250]:

- The first section cut (70 μm thick) is deposited on a 100 μm thickness mylar disk for charged particle spectroscopy (Fig. 42);
- The second section cut (10 μm thick) is deposited on a glass plate for morphological analysis by standard hematoxylin–eosin staining;
- The third slice (70 μm thick) is deposited on a solid-state nuclear track detector for imaging of the macroscopic boron distribution by neutron autoradiography (Section 8.2.3).

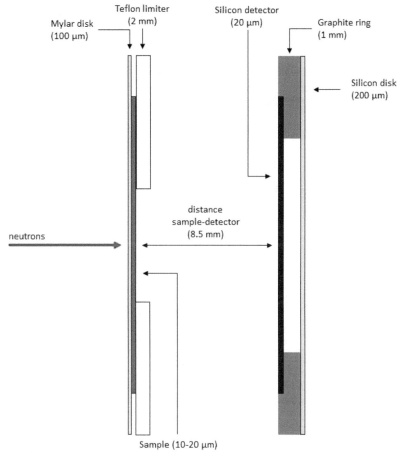

FIG. 42. Sketch of the setup for charged particle spectrometry (not drawn to scale).

It is important to underline that with this technique the surface of the sample to be analyzed is fixed with a collimator whose diameter is chosen so that it is smaller than the detector. The maximum sample thickness is equal to the range of the charged particles in the material under examination. In the detector, in practice, the charged particles release 'residual energy', E_{res}, equal to the difference between the energy with which they are produced in the reaction, E_0, and the energy, E_{diss}, they lose in the tissue before leaving it.

$$E_{res}(x) = E_0 - E_{diss}(x \rightarrow 0) = E_0 - \int_x^0 \frac{dE}{dx}\,dx \tag{12}$$

The use of a sample with a thickness much smaller than the range, R_0, would allow detection of particles with an energy practically equal to the emission energy in the reaction, and the typical Gaussian peak would be formed on the detector. However, for measurement of the concentration of ^{10}B present in the sample, which is needed for BNCT, it is necessary to know the mass of the portion of the sample from which the particles come. The use of samples that are very thin does not allow measurement of this mass with adequate precision. Therefore, samples with a thickness larger than R_0 are prepared for concentration measurement (Fig. 43); the corresponding spectra have a typical absorption shape like the one shown in Fig. 44.

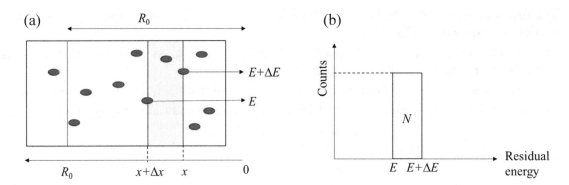

FIG. 43. Sample thickness and α particle energy. (a) The maximum range of α particles in tissue is R_0. Those α particles produced between x and x + Δx travelling perpendicular to the detector reach the detector with energies between E and E + ΔE. (b) The integral of the curve in the detector of events between E and E + ΔE is proportional to the number of such events, N.

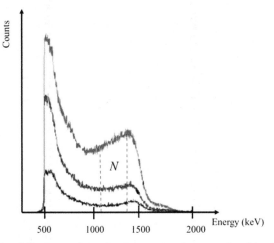

FIG. 44. Typical spectra obtained from two hepatic tissue samples, one healthy (blue) and one with a tumour (red): for comparison, the background of measurement is also shown (black); the two vertical dashed lines identify the area of the spectrum used for the evaluation of boron concentration.

The dose from ^{10}B needs to be calculated in the fresh tissue in its normal conditions of hydration; the sample used for the measurement, once cut and deposited on the mylar, loses its water content, and the measurement is carried out in the dehydrated state. Assuming that only water, and not boron, is lost during the dehydration process of the sample, it can be demonstrated that the concentration, C_f, in the fresh sample can be calculated from the N counts (Figs 43(b), 44, 45) obtained in the energy range ΔE using the equation

$$C_f = \frac{N}{\eta \sigma \phi S_d} \cdot \frac{A_w}{N_A} \left[\frac{\Delta E}{\Delta(\varrho x)} \right]_d \cdot \frac{m_{td}}{m_{tf}}$$ (13)

— the counted events derive from a sample mass $m_d = S_d[\Delta(\rho x)]_d$;
— $\frac{m_{td}}{m_{tf}}$ represents the ratio between the masses of the fresh and dehydrated sample;
— ϕ is the thermal neutron fluence at the measurement point,
— σ the thermal neutron cross section of the capture reaction ^{10}B(n,α)^{7}Li;
— η the detector counting efficiency.
— The ratios $\frac{m_{td}}{m_{tf}}$ and $\left[\frac{\Delta E}{\Delta(\varrho x)} \right]_d$ are measured experimentally and depend on the type of tissue to be analyzed;
— A_w is the atomic mass number of ^{10}B;
— N_A is Avogadro's number.

The above equation can be used for healthy and tumour homogeneous samples. If dealing with heterogeneous samples where tumour is mixed with healthy tissue, it provides a mean value over the entire sample (as does PGNA).

In the case of PANS, two consecutive slices can be used to separate the contribution due to healthy tissue from that produced by the tumour: a histological image is used to measure the percentages of the two types of tissue contained in the sample, and neutron autoradiography is used to confirm the different absorption due to boron in the two tissue types.

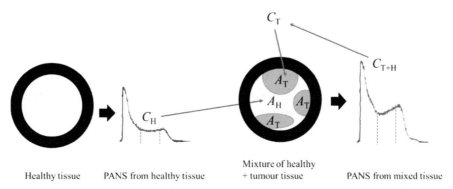

Healthy tissue   PANS from healthy tissue   Mixture of healthy + tumour tissue   PANS from mixed tissue

FIG. 45. Schematic diagram of the procedure used to assess concentration of boron in tumour and healthy tissue. C_H, C_{T+H}, represent concentrations of boron in healthy and mixed tissue as determined by PANS measurement. The area of healthy (A_H) and tumour (A_T) tissue can be determined. From this C_T can be calculated.

By defining A_{T+H} as the area of the sample analyzed, A_T, as the area of the tumour contained in the sample, and C_H as the concentration of boron in healthy tissue (obtained with a measurement on a tumour-free sample), the concentration C_T in the tumour is given by

$$C_T = \frac{C_H}{\frac{A_T}{A_{T+H}}}\left(\frac{C_{T+H}}{C_H} - \left(1 - \frac{A_T}{A_{T+H}}\right)\right) \tag{14}$$

This technique has the undoubted advantage of clearly separating the concentration of boron in tumour from that in healthy tissue, with both histological and neutron autoradiography findings based on the same reaction used in therapy. Moreover, the measured values refer to thicknesses of the biological sample on the order of a micron (a thickness equal to about one tenth of the useful range of the α particles used in BNCT, see Table 20 in Section 10). The spectral analysis guarantees that the analyzed signal is the correct one.

With a thermal neutron flux of $\phi = 10^9$ cm$^{-2}\cdot$s^{-1} it is possible to measure concentrations of ^{10}B lower than 0.5 ppm within times of the order of 10 minutes. The overall error associated with the technique is of the order of 20% due to the 17% error associated with the measurement of the ratio $\frac{\Delta E}{\Delta(\varrho x)}$. Everything else contributes with an error of about 5%. Accuracy can therefore be improved by reducing the error associated with the measurement of $\frac{\Delta E}{\Delta(\varrho x)}$ [250].

Prompt alpha neutron spectrometry was conceived and developed in the preclinical research phase of BNCT applied to diffuse liver metastases and was subsequently applied in the treatment of patients [164]; it was later used to measure the concentration of boron in different types of tissue such as lung [278] and bone [278–279] treated with different boron carriers such as BPA and GB-10 [280]. PANS has been found to be in good agreement with the results from other techniques such as ICP-MS and neutron autoradiography [281].

8.2.3 Neutron autoradiography with nuclear track detectors

The success of BNCT strongly depends on the intratumour delivery of ^{10}B. Knowledge of the localization of ^{10}B at the tissue and even cellular levels is essential to evaluate the probability of a successful therapy and to understand the observed post-irradiation effects [247]. Only a small number of techniques allow precise studies of boron microdistribution: neutron autoradiography with nuclear track detectors (NTDs) is an attractive option due to its high resolution and relatively low cost. Neutron autoradiography is also known by a variety of other names including 'alpha autoradiography', 'neutron capture radiography', 'neutron capture autoradiography', 'neutron-induced autoradiography', 'alpha tracking method'. The term 'autoradiography' is also used to refer to the emulsion-based technique [282]. Neutron autoradiography with NTDs was applied for the first time for BNCT by Fairchild [283] using liquid photographic nuclear track emulsion, and later developed using solid state NTDs [251, 284–286]. Since then, different approaches have been used to determine boron microdistributions in several in-vitro and in-vivo biological models. This methodology was also applied to samples from patients subjected to BNCT and biodistribution studies [287–288].

Solid state NTDs are insulating materials (organic or inorganic) that register the damage produced by ions along their trajectories, making a permanent alteration to the material. In the case of BNCT, the latent damaged zones on the NTD are produced by exposing a B-loaded biological matrix that is in contact with the NTD to a thermal neutron flux: the α particles and ^{7}Li nuclei emitted by the neutron capture reaction that can reach the detector create latent tracks. Latent tracks can then be revealed by a chemical attack (etching) with an appropriate solution (alkaline solutions are used in the case of many polymeric NTDs). The etching velocity close to the ion trajectory (V_T) is higher than the etching velocity in the bulk unirradiated material (V_b). This preferential attack velocity leads to etch pits (or tracks) that can be observed by electron or optical microscopy, depending on the etching conditions [289]. Thus, a track will be 'observable' whenever $V = V_T/V_b > 1$. The etching parameters such as the concentration of the reagents, the temperature, and the etching time determine the morphology of the nuclear tracks.

Nuclear track detectors are insensitive to visible light, γ radiation, and thermal neutrons, so they are especially suitable for use in mixed radiation fields. The polymers cellulose nitrate (Kodak), pollyallildiglycol carbonate (CR-39) and polycarbonate (e.g., Lexan) have been extensively used as NTDs for boron imaging of samples coming from BNCT protocols. KOH and NaOH solutions are used as etching reagents. Pollyallildiglycol carbonate is the most sensitive detector material in common use [290], but also registers the recoil protons coming from the ^{14}N(n,p)^{14}C reaction. The proton pits produced from this reaction represent an unwanted background which needs to be filtered out. Desensitization of the CR-39 detector can be effected by choosing 'PEW solutions' ($KOH + C_2H_5OH + H_2O$), as they present a higher V_b than etching reagents based on NaOH solutions, for a given temperature [291].

When analysing biological samples (e.g., cell cultures, tissue sections), boron microdistribution can be assessed by mapping the nuclear tracks and correlating their positions with the histological image. The following analytical expression relates track density per unit area, ρ, and boron concentration in the sample [292]:

$$\rho = C(B)\frac{N_v \sigma_B \phi}{4}\left(R^{\alpha}\cos^2\theta_{\alpha} + R^{Li}\cos^2\theta_{Li}\right) \tag{15}$$

where:

— $C(B)$ is the mass concentration of B;
— N_v is the number of atoms per unit volume;
— σ_B is the neutron capture cross section for ^{10}B;
— ϕ is the thermal neutron fluence;
— R^α, R^{Li} are the ranges of the α particles and Li fragments in the sample;
— θ_α and θ_{Li} are critical angles of ions entering the detector (the angle between the ion's trajectory and the axis normal to the detector surface).

Critical angles, θ_α and θ_{Li}, refer to the possibility of particles being recorded in the NTD and can be calculated in terms of V; ions entering the NTD with angles larger than the critical angle are not expected to be preferentially etched [292].

Several approaches to the neutron autoradiography technique have been described by different groups, allowing both qualitative and quantitative analyses. Depending on the neutron fluence and the etching conditions, overlapping tracks may be deliberately generated and exploited or prevented. Thus, the collective optical effect on the NTD surface, in terms of shades of grey, can be analyzed to describe the boron distribution [293–296]. Quantification can be performed by combining these data with a gross boron measurement determined from alpha spectrometry [276, 278]. Other approaches seek to determine boron concentration ratios by analysing profile intensities [297]. Moreover, optical density analysis can be performed on these high fluence images, and an absolute boron concentration value can be obtained [298].

Conversely, if nuclear tracks are separated enough to allow individual counting, track density can be translated to ^{10}B-concentration using a calibration system, thus allowing a quantitative analysis. The material used as a reference standard needs to comply with certain requirements:

• Homogeneous distribution of boron atoms in the material;
• Minimal number of background tracks produced by the material;
• Easy handling;
• Proportionality of track density to ^{10}B-concentration and thermal neutron fluence [299].

The material also needs to behave in a similar manner to biological samples, especially in terms of energy deposition (charged particle stopping power and range) [300]. Aqueous solutions [299, 301], drying filter paper sheets [291], blood [302], agarose [303], tissue homogenates [304], pulverized bone [305], etc. have been used as reference standards. When performing a calibration, it is essential to validate the system with other techniques, such as ICP-OES and ICP-MS [306], PGNA [307], and PANS [50]. Intercomparison between neutron autoradiography techniques from different laboratories is of great relevance for benchmarking [308].

Due to the short path of the α particles and ^7Li nuclei (Table 20 in Section 10), biological samples need to be put in close contact with the NTD surface:

• For in-vitro studies, cells can be grown on a substrate that is then put in contact with the NTD. This strategy is useful for an overall measurement of all cells within a sample [309], but when analysing boron localization at the cellular level, it is necessary to ensure that the sample consists of isolated cells [310] or a monolayer [311];
• For in-vivo studies, the tissue specimens need to be frozen in isopentane or liquid nitrogen in order to avoid migration of boron atoms from their original positions. Therefore, the tissue samples are sectioned in a cryostatic microtome and then put in contact with the NTD. The assembly 'section+NTD' can be freeze dried [312] or maintained at a low temperature

during irradiation [313], but irradiations are usually performed at ambient temperature. In such a case, the water trapped in the section is evaporated; this mass loss results in a reduction of the section thickness. The evaporation process changes the number of particles reaching the detector, producing an amplification effect [314]. This issue has been already addressed in Section 8.2.2.2 when describing PANS [250] and it has been extensively studied in the frame of autoradiography techniques [315].

Different methods have been explored in order to improve the spatial resolution of neutron autoradiography. In particular, it can be enhanced by using the same sample for the histological analysis and neutron autoradiography [316]. For this purpose, a biological sample has to be carefully processed in order to obtain accurate results and tissue needs to be thin enough to allow a broad histological analysis. For thin samples, track density increases up to a saturation value that is related to the range of the α particles and Li nuclei [317]. The optimal thickness depends on the tissue characteristics [318]. The use of reference marks to correlate the regions of interest in both the histological and autoradiographic images becomes necessary, when using one sample for both methods as the sample is removed from the detector before chemical etching. The design and characterization of reference marks has been well studied [156, 310]. Different filtering and segmentation procedures on the digital images have been proposed to characterize nuclear tracks, such as circularity, incidence angle, and area/radius relationship [301, 319].

A better spatial correlation and a more precise knowledge of boron atom positions could be achieved if tissue structures and nuclear tracks were simultaneously observed. High resolution quantitative radiography seeks to preserve the biological sample during the etching process, by means of the use of thin detector films [320]. Another technique used for some polymeric detectors consists of interposing a UV-C exposure step between the irradiation with thermal neutrons and the staining and histological analysis. The V_b in the NTD can increase considerably because of photodegradation. If a biological sample is put in contact with the detector, it partially protects the NTD from UV-C action. Thus, the etching solution will attack the NTD surface at different velocities depending on the sample topography, and an imprint of the biological material will be formed. Both imprint and tracks are revealed in the same process (see Fig. 46 and Refs [321–322]). The feasibility of use of this technique with polycarbonate detectors for BNCT has been demonstrated [323]. The methodology has been recently optimized [324], as a fading effect on nuclear tracks was observed in polycarbonate due to the UV-C action [325].

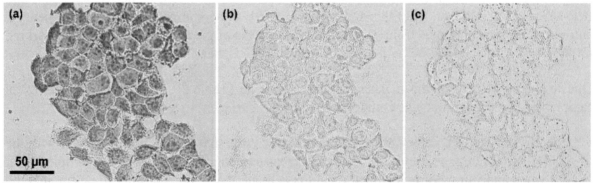

FIG. 46. Example of autoradiography + UV-C. (a) MCF-7 cells cultured on Lexan and stained with hematoxylin, and (b) its corresponding images of imprints, and (c) nuclear tracks (courtesy of A. Portu, CNEA).

Neutron autoradiography is a powerful and versatile methodology. The different approaches vary in complexity and resolution and need to be selected depending on the objective of the study. The information obtained from autoradiographic analyses has provided essential information to numerous studies: evaluation of boron compounds [326–328], comparison of response among different cell lines [329], evaluation of in vivo models [330], correlation with BNCT efficacy in radiobiological studies (e.g., [331–332]), dosimetric evaluation in non-uniform boron distribution scenarios [333], and microdosimetry analysis [334].

8.2.4 Quantification of boron distribution by nuclear magnetic resonance

Magnetic resonance imaging (MRI) requires liquid helium cooled high field magnets to create a very homogenous magnetic field that aligns the nuclear spins of the patient inside the scanner. An oscillating magnetic field is then applied at a frequency tuned to the resonance of a target nucleus, causing the nuclear spins to be perturbed from their natural precession. Excited atoms emit signals in the radiofrequency range as they gradually relax back to their aligned positions and these signals are recorded in a receiving coil. By varying the local magnetic field with gradient coils across the patient, scans can be performed that allow spatial information to be extracted. The contrast in an MRI arises mainly from differences in the spin–spin and spin–lattice relaxation times (T_1 and T_2) of water protons in tissue, due to their interaction with macromolecules, paramagnetic metal ions, and biological membranes. T1 describes the rate at which nuclear spins return to their equilibrium positions and T2 that at which they go in and out of phase with one another [335]. Nuclear magnetic resonance (NMR) detection of BNCT agents can be pursued by both imaging (MRI) and spectroscopy (MRS) modalities [336], using ^1H, ^{19}F, and ^{10}B or ^{11}B (Table 19).

TABLE 19. PROPERTIES OF NUCLEI THAT CAN BE USED IN MRI FOR BNCT

Nucleus	Natural abundance (%)	Spin	Relative gyromagnetic ratio	Quadrupole moment (fm^2)
^1H	99.9885	1/2	1	0
^{10}B	19.9	3	0.107456	8.45
^{11}B	80.1	3/2	0.320897	4.059
^{19}F	100.0	1/2	0.941286	0

8.2.4.1 Proton magnetic resonance imaging and spectroscopy

In the last few decades, MRI has become one of the key modalities in clinical settings thanks to its superb spatial resolution and its outstanding ability to differentiate soft tissues [337]. Although less sensitive than nuclear and optical modalities, the high spatial resolution (< 100 μm) of ^1H-MRI provides morphological details, and the lack of ionizing radiation makes it safer and easier to use than techniques involving radioisotopes. Moreover, due to the non-ionizing radiofrequency pulses used, the time window of observation is significantly larger than techniques employing ionizing radiation, and scanning can be repeated without harmful effect.

Proton magnetic resonance spectroscopy (^1H-MRS) allows the non-invasive, in vivo detection and quantification of brain metabolites [338]. Proton MRS has been proposed as a non-invasive and quantitative method to evaluate BPA concentration in vivo [339–341]. Detection of BPA is based on its aromatic protons, which lie in the range 6.5–9.0 ppm of an MR spectrum. Due to the almost complete absence of signals arising from human brain metabolites, the aromatic proton signals of BPA (four signals centred at ~7.4 ppm) are easily detectable without interference. The detection limits of BPA have been determined in vitro to be 1.4 mM (in a 1.5 T field) and 0.7 mM (in a 3.0 T field) when using a typical clinical voxel size of 8 ml and a clinically compatible MRS measurement time of about 10 min [255]. Based on these results, it was estimated that the in vivo detection limit at 1.5 T was approximately 2 mM BPA (~20 mg/kg of ^{10}B in the brain). Figure 47 shows a metabolite map derived from imaging the aromatic region during/after FBPA infusion, demonstrating that several voxels show its presence in tumour tissue, whereas no signal was detected in normal brain tissue [342].

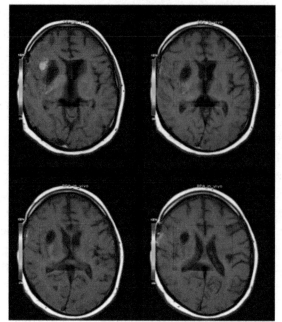

FIG. 47. Metabolite map formed by MRI using the aromatic region of the molecular structure of FBPA from a patient during/after FBPA infusion. The colour bars represent concentration, with relatively high uptake in the region of the tumour (red) and low uptake in normal tissue (courtesy of M. Timonen, University of Helsinki).

The main advantages of this technique are that it allows BPA determination during intravenous infusion and that the technique is available in the majority of high field clinical MRI scanners. The disadvantage is the low sensitivity that fixes the detection limit threshold to 15–20 ppm.

BNCT is frequently applied after the surgical removal of malignant glioma. The remaining residual tumour forms a narrow lining in the wall of the resection cavity which contains haemostatic agents and coagulated blood with fluid–tissue and fluid–air interfaces [247]. This could generate interferences in the detection of BPA via ^{1}H-MRS. For this reason, unoperated tumours are more suitable for MRS based BPA detection.

Contrast agents significantly improve the endogenous contrast by decreasing T_1 and T_2 of water protons in the tissues where they distribute. Their use is fundamental in oncological studies to obtain a precise determination of the extension of the tumour mass, even those of relatively small size (< 0.5 cm). The contrast agents used in clinical settings are polyaminocarboxylate complexes of the Gd^{3+} metal ion. The ligands are multidentate, forming very stable complexes, limiting release free Gd^{3+} ions, which are highly toxic [343–344]. An indirect boron quantification using these agents can be used once conjugated to the boron containing compound by measuring the MRI signal intensity enhancement against a previous calibration. This enhancement is directly proportional to the local concentration of the Gd-containing probe. Dual probes, possessing both a Gd complex for MRI detection and ^{10}B atoms for BNCT, have been reported [345].

Unfortunately, the conjugation of BPA to a Gd complex (Gd-DTPA, molecular weight 574 Daltons) has been reported to cause a dramatic change in the biodistribution: the intratumour boron concentration was significantly reduced with respect to BPA alone, as a consequence of decreased affinity for BPA receptors [346–347]. However, nanosized delivery agents (i.e., liposomes, micelles, polymers, proteins, etc.) can simultaneously deliver both BNCT and MRI agents whose biodistribution and receptor affinity is not affected by loading with small sized agents (see Section 7.5 and Refs [348–349]). Moreover, Gd-DOTA has also been encapsulated in the aqueous core of a liposome and the biodistribution measured by MRI before neutron irradiation. Significant antitumour effects were observed in mice that had been injected with BSH-encapsulating 10% distearoyl boron lipid liposomes: even at a pharmaceutical dose of 15 mg B/kg; the tumour completely disappeared three weeks after thermal neutron irradiation. Another pre-clinical study has been reported based on the use of bioconjugated Fe_3O_4–$GdBO_3$ composite nanoplatforms as next generation theranostic agents [350]. They have advantages not only of simplicity, non-toxicity, and low cost of the starting materials, but also the intrinsic possibility to monitor the delivery status.

An alternative approach under development is based on the preparation of a dual probe MRI–BNCT, consisting of a carborane linked to a Gd-DOTA monoamide complex for MRI and an aliphatic chain for binding to low-density lipoproteins [351–353]. The therapeutic dose of boron was measured by MRI on cells and animal models. The intracellular concentration of boron atoms was calculated from the relaxation rates, and a good correlation with the concentration of boron atoms measured by ICP-MS was obtained. The covalent bond between carborane and the MRI contrast agent also ensures correct detection after in vivo injection and cellular uptake. The boron delivery system has been tested on melanoma, breast, and mesothelioma tumour models, and the results appear very promising, providing patients an improved therapeutic option, exploiting low-density lipoproteins transporters.

8.2.4.2 Fluorine-19 magnetic resonance imaging

Fluorine-19 MRI can be used to investigate the biodistribution of fluorinated drugs. The stable ^{19}F nucleus has near 100% natural abundance (Table 19) with an NMR sensitivity ca. 20% less than that of ^{1}H. The endogenous ^{19}F-MRI signal from a human body is negligible, as the concentration of fluorine is below the detection limit. This gives ^{19}F-MRI a high signal-to-noise ratio and specificity when a fluorinated compound is administered as an exogenous contrast agent or drug [354].

Fluorine-19 MRI in combination with ^{1}H-MRI can selectively map the biodistribution of ^{19}F-FBPA, as demonstrated by in vivo experiments on the C6 rat glioma model. Correlation between ^{19}F-MRI measurements on rat brain and ^{19}F magnetic resonance spectroscopy on blood samples showed maximum uptake of ^{19}F-FBPA in the C6 glioma model 2.5 h post infusion. The improved uptake of the ^{19}F-FBPA–fr complex in C6 tumour-bearing rats after L-DOPA pre-treatment was also observed using ^{19}F-MRI [355].

The routine use of this interesting approach is limited by the requirement for fluorine probes on human clinical scanners together with long acquisition times due to the long T1 of fluorine.

8.2.4.3 Boron-10 and ^{11}B nuclear magnetic resonance

Both natural boron isotopes, ^{11}B (80% natural abundance) and ^{10}B (20% natural abundance), are detectable by NMR (Table 19). For clinical applications, the boronated molecules are $\geq 95\%$ ^{10}B enriched. Thus, ^{10}B-NMR detection is preferred in spite of its lower sensitivity and spectral resolution. The unique relaxation properties of the spin 3 nucleus (Table 19) might cause the T_2 of ^{10}B to be longer (1.75 ms) than that of ^{11}B for molecules of the same size. This helps in a higher recovery of the signal after a spin echo sequence. For this reason, both boron isotopes have been proposed in MRI–BNCT applications.

Boron-10 and ^{11}B NMR based spectroscopic methods have the advantage of not requiring any chemical modification of the boron carrier used in BNCT treatment, but with significantly lower sensitivity than PET, SPECT, or MRI, which does not allow boron quantification at the tumour site during BNCT treatments. One interesting example of in vivo MRI measurement of ^{10}B-enriched BSH biodistribution has been described [356]. However, since these initial studies were reported, there have been no more publications on this topic.

8.2.5 Elemental imaging of boron in biological samples

Secondary Ion Mass Spectrometry (SIMS), laser post-ionization Secondary Neutral Mass Spectrometry (laser-SNMS), Electron Energy Loss Spectroscopy (EELS), and Laser-Induced Breakdown Spectroscopy (LIBS) are complementary tools for efficient boron identification, in preclinical investigations or clinical translation. For multiple elemental imaging, LIBS is a flexible tool, particularly to detect boron. High resolution techniques such as laser-SNMS and EELS can provide information at the cell scale, while LIBS might be useful for ex-vivo imaging of an entire organ. Laser-SNMS includes a laser post-ionisation stage that can substantially increase the measurement sensitivity over that of SIMS, by accessing the dominant neutral fraction of the ablated sample material. This section describes the main principle of these different methods.

8.2.5.1 Secondary ion mass spectrometry

Secondary ion mass spectrometry creates images of the concentration of secondary ions released from the surface of a sample under high vacuum and subjected to bombardment from a primary focussed, raster-scanned beam of low energy ions (Fig. 48). The ions that are sputtered from the sample are then analyzed by mass in a mass spectrometer.

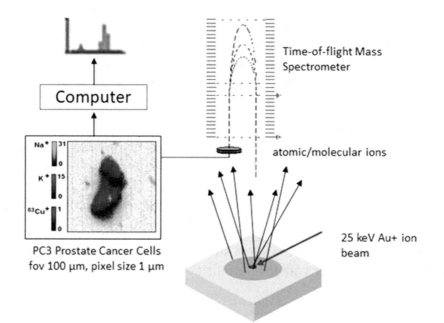

FIG. 48. SIMS set-up used at the University of Manchester circa 2010.

Secondary ion mass spectrometry has played an important role in the development and understanding of the uptake and biodistribution of boron compounds. The work of the group from Cornell [357] was critical in showing the potential benefit of longer infusions of BPA for targeting disease which has spread from the main tumour mass.

As an example of the capabilities of SIMS measurements, the images in Fig. 49 are derived from human tumour biopsy tissue which was then incubated (ex-vivo) in BPA before imprinting to provide tissue layers for analysis. The sodium and potassium images show the well-maintained cellular structures, which are more clearly visible in the less densely packed brain-around-tumour images (the densely packed cells of the main tumour bulk are harder to visualise). The cell membranes are revealed in the sodium images while the potassium is (as expected) mainly intra-cellular (Fig. 49b,c). Figure 49a,d) show the overlay of boron (green) and gives an indication of the degree to which BPA has accumulated into the cells. SIMS has developed considerably as a technique since these images were obtained, and now has much higher resolution capability in the form of the so called nanoSIMS [358]. Later work reported by Ref. [253] used this to provide a higher resolution, more quantitative analysis of the samples shown in Fig. 49.

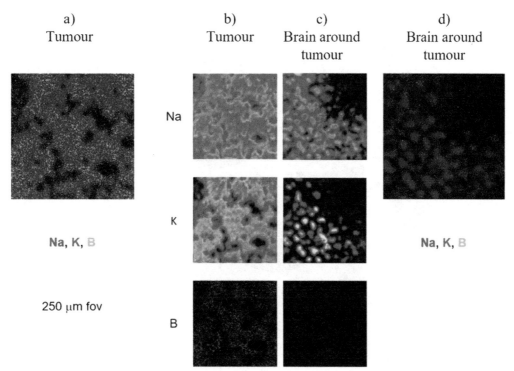

FIG. 49. Example SIMS images for tumour (a,b) and brain around tumour (c,d) for human biopsy samples which have then been incubated in BPA. Panels (b,c) show individual element distributions for Na, K, and B, and (a,d) show a composite elemental distribution with the colour code shown in the legend (courtesy of N. Lockyer, University of Manchester and G. Cruikshanck, University of Birmingham).

8.2.5.2 *Laser post-ionization secondary neutral mass spectrometry and electron energy loss spectroscopy*

In laser-SNMS analysis, a pulsed ion gun bombards and sputters the sample surface, yielding initially neutral particles. Subsequently, a focused laser beam is used to photoionize the neutral particles before their extraction and mass analysis. Such instrumentation combines a high useful yield and high background discrimination. Laser-SNMS works in vacuum conditions on cryo-fixed samples [359–360] and can detect boron at low concentration (~ppb) with high lateral spatial resolution (~200 nm) permitting the identification of physiologic structures [247].

To image boron distributions at the cellular level, EELS might also be used. In EELS analysis, the sample is exposed to an electron beam and the characteristic energy loss of the incoming electrons due to inelastic scattering with other electrons within the sample is recorded. However, boron imaging using EELS is complex in biological samples and the signals need to be carefully extracted to discriminate the boron signal from the typically high concentrations of phosphorus [361]. This technique possesses a very high spatial resolution and a sensitivity of around a few tens of ppm [361]. However, it requires cryo-fixations, generation of ultra-thin cryo-sections (< 100 nm thick), and cryo-analysis of the sample (< 110 K) [362].

8.2.5.3   Laser-induced breakdown spectroscopy imaging

Laser-induced breakdown spectroscopy (LIBS) imaging is an all-optical technique allowing multi-elemental imaging and quantification with a maximal spatial resolution of 10 μm [254]. This technique is currently used in preclinical experiments for boron detection, and in a clinical trial for metals detection (clinical trial NCT03901196)[16]. A single laser pulse, focused ~50 μm under the surface of the sample, is used to simultaneously ablate a very small amount of matter and excite it to produce a plasma. The emissions from the plasma are collected through optical fibres connected to one or more spectrometers (Fig. 50) [363–366].

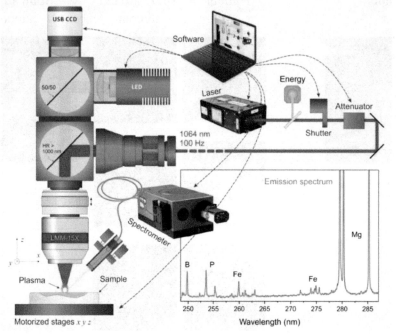

FIG. 50. LIBS set-up and representative emission of a sample containing boron (courtesy of V. Motto-Ros, ILM, Lyon).

Spectrometers can collect the main emissions from boron at 208.8 nm or 249.8 nm with a sensitivity close to a ppm for a single laser pulse. In a single-shot ablation, several elements might be observed and identified, depending on the collection range of spectrometers. As an example, some phosphorous emission peaks (214.9 or 253–255 nm) are next to the main boron emission peaks, but resolvable (see Figs 50 and 51). As phosphorous is present in high concentrations in biological samples, it is often used as a marker of the cell density, and phosphorous may be used to delineate the tissue and tumour area [367–368]. To illustrate boron imaging, organs of mice have been sampled and analyzed after intravenous administration of boron-rich compound to examine simultaneously both boron and phosphorous (Fig. 51).

Laser-induced breakdown spectroscopy was first designed for geological investigations (soil analysis as an example) and has been developed for biological [369–373] and medical uses [374–375] with a specific focus on the detection of metallic elements. Analysis of biological samples requires only minor sample preparation: any flat sample surface may be analyzed whatever its nature (e.g., Formalin-Fixed Paraffin-Embedded (FFPE) samples or frozen samples). The method works under ambient laboratory conditions, although for specific experiments, an Ar or He gas flow might be used to increase the signal of the element of interest. Most elements can be detected, with their limits of detection related to both the element itself

---

[16] https://clinicaltrials.gov/ct2/show/NCT03901196

and the nature and hardness of the sample (Fig. 52). Theoretically, isotopes can be identified; however, in biological samples, ^{10}B and ^{11}B cannot currently be discriminated.

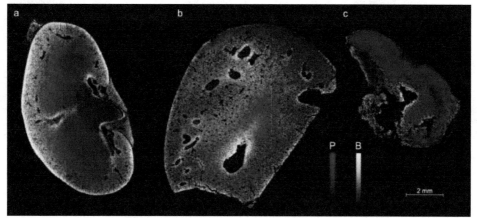

FIG. 51. LIBS imaging of a mouse kidney (a), liver (b) and tumour (c) section after intravenous administration of a boron rich compound. Boron and phosphorus are represented in yellow and blue false colours, respectively (courtesy of L. Sancey, IAB Grenoble).

The main advantages of LIBS imaging are the ease of use (once the instrument settings and sample preparation have been optimized), the size of the samples that can be analyzed (more than 10 × 10 cm), and the speed of analysis. Currently, for biological sample analysis, the systems can work at 100 Hz, even for FFPE samples made of very soft matter, leading to large data sets (> 1 M pixels). It can be used as a relative or quantitative method [373]. The spectra from different samples might be directly compared for relative or semi-quantitative applications. However, for quantitative measurements, a calibration curve needs to be established using standards of the element of interest prepared at known concentrations. In practice, the main difficulty may be to produce representative standards, constituted of biological matter.

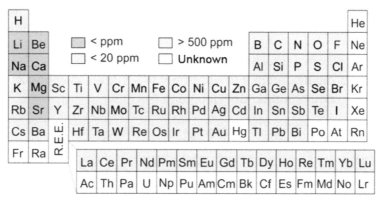

FIG. 52. Periodic table of the elements and theoretical limits of detection for LIBS shown in the colour coded legend of the figure (courtesy of B. Busser, Institut universitaire de France).

8.2.5.4 Photoemission electron microscopy

The BNCT group in Okayama University, Japan, is developing a visualization technique for boron distribution in a cancer cell utilizing PhotoEmission Electron Microscopy (PEEM) combined with X ray absorption spectroscopy at the boron K-edge at Hiroshima Synchrotron Radiation Center. The spatial resolution of the system is currently ~1 μm but could be improved to ~100 nm. PEEM may be useful to determine the microdistribution of boron compounds taken up in different tissues within a tumour and also within individual cells. Figure 53 shows the results of preliminary feasibility tests undertaken before biological measurements began.

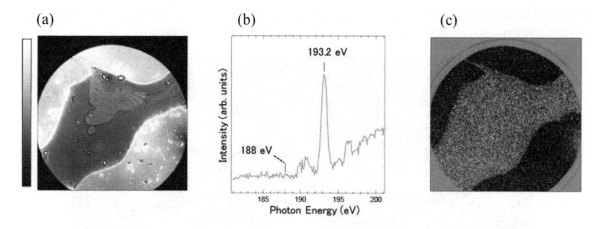

FIG. 53. (a) A PEEM image of a BPA microdroplet coagulating on a Si surface measured with a Hg lamp. The field of view is 150 µm. (b) A local X ray absorption spectrum from PEEM images measured with photon energies around the boron K-edge. (c) A boron distribution image obtained as a difference image between two PEEM images measured with photon energies of 193.2 eV and 188 eV (courtesy of W. Takanori, Okayama University).

8.3 SUMMARY OF BORON CONCENTRATION MEASUREMENT METHODS IN CLINICAL USE

In a BNCT clinical environment, PET imaging is currently only available for treatment with BPA, and it is not possible to perform during treatment. ICP techniques and PGNA are used to measure boron concentration in blood samples from which one can deduce the concentrations in tumours and healthy tissues, and calculate doses. However, these techniques present some critical issues:

- It is not possible to take into account the variability of partitioning from one patient to another;
- The neutron flux used to calculate the dose is that evaluated by the treatment planning system (Section 11) and not the one really present in the patient during treatment.

It would be important to have a technique able to measure online the dose-rate distribution present in the patient during treatment. For now, only PG-SPECT (Section 8.2.1) seems to have this potential, and this requires further development.

9 RADIOBIOLOGY

As described in previous sections, the boron neutron capture reaction (Fig. 15) between a thermal neutron and a ^{10}B nucleus gives rise to the formation of an α particle and a recoiling ^{7}Li nucleus, both characterised by having high linear energy transfers (LET)[17]. High LET particles are densely ionizing and have a high relative biological effectiveness (RBE)[18] over short distances in tissue (5–9 μm, see Table 20 in Section 10), confining most of the damage to tumour tissue where ^{10}B atoms are preferentially localized [376–378]. Theoretically, high LET particles with a high RBE would be able to overcome radioresistance to photon radiotherapies (such as X rays) [3, 379].

BNCT involves mixed field irradiation associated with several different radiation dose components, each with different physical properties and RBEs. Although the neutron capture cross sections for the elements in normal tissue are several orders of magnitude lower than that of ^{10}B, hydrogen, and nitrogen are in high concentrations and their neutron capture reactions Table 7 contribute significantly to total absorbed dose. The background dose affects tumour and normal tissue similarly.

As a radiation therapy, BNCT protocols ideally maximize the boron radiation component that is specific to the tumour and minimize the non-selective background dose [378]. Maximizing/optimizing the delivery of boron to tumour is the most effective way to do this. Conversely, increasing exposure to neutrons increases the non-specific background dose with no net gain in the therapeutic ratio.

As a nuclear medicine therapy, BNCT is ideally suited to treat undetectable micrometastases as it involves biochemical rather than geometric targeting, because of preferential incorporation of boron carriers in tumour cells [380] with minimum damage to healthy tissues in the treatment volume [381]. In principle, the Planning Target Volume (PTV, see Section 12.4) can be enlarged to include potentially undetectable target cells in the organ to be treated, outside the gross tumour [222]. In addition, since dose deposition mainly depends on boron localization, with BNCT it is less necessary to adjust for organ motions [3, 278] than in the case of other particle therapies.

BNCT is a non-mutilating, organ preserving treatment [382–383]. Compared to standard radiotherapy, it contributes to quality of life as it is usually administered in a single session (or at most two), while possessing the same level of therapeutic efficacy and toxicity. It needs to be borne in mind that BNCT is a local treatment and is not suited for the treatment of systemic disease.

9.1 BIOLOGICALLY EFFECTIVE DOSE

The biologically effective dose of BNCT depends on the RBE of the different dose components. The RBE in turn depends on the LET of the radiation. For a given radiation component, the RBE will depend on the tissue type, the tumour or normal tissue response endpoints under evaluation, and also on the level of effect at which they are assessed. In the case of the boron dose, D_B, the short ranges of α particles and recoiling ^{7}Li ions (approximately the diameter of a cell) confer a central role to the microdistribution of ^{10}B in BNCT's biological efficacy [384].

---

[17] The linear energy transfer (LET) is the amount of energy that an ionising radiation deposits in the matter traversed per unit distance.

[18] The relative biological effectiveness (RBE) is the the ratio of absorbed doses of a reference radiation such as X or γ rays and the radiation under study required to produce the same biological effect.

If ^{10}B is close to a sensitive target such as DNA, the effect will be maximized. The compound biological effectiveness (CBE) is defined for each boron compound and depends on the:

- Intrinsic biological efficiency of α particles and ^{7}Li ions;
- Microdistribution of boron in tissue;
- Geometry of the target (shape and volume);
- End point, tissue, dose rate, etc., as for RBE in general.

The administration protocol may influence the microdistribution of ^{10}B of a particular carrier within the tumour or normal tissues [377–378], and so affect the CBE value.

9.2 PHOTON EQUIVALENT DOSE: CONCEPTS AND OUTLINE OF APPROACHES TO CALCULATION

The dosimetry of BNCT involves contributions from different types of radiation with different biological efficacy. Within this context, the absorbed dose alone fails to explain the therapeutic and toxic effects of BNCT. A widespread aim is to prescribe safe and effective treatment protocols. To do this, it is necessary to compare different protocols within a BNCT centre or between centres and consider the data on outcomes with BNCT and conventional photon radiation [385], for which the values of absorbed dose are insufficient for this analysis. Comparison of BNCT with photon therapies needs a model for dose calculation that yields an adequate 'photon-equivalent' value. This issue will be addressed in detail in Section 10 and will be only briefly mentioned in this section.

The usual procedure for calculating photon-equivalent doses is to sum the main contributions of the different radiations (i.e., the absorbed doses) to form the total absorbed dose, each weighed by a constant (dose and rate independent) factor [378]. The single numbers used in the clinic are the RBE and CBE factors (see also Section 10.3.2) from cell survival experiments for a given endpoint and cellular system, using γ rays produced from accelerators or a ^{60}Co source as the reference radiation, converting each contribution to a term of photon-equivalent dose. The various absorbed dose components contributing to the total radiation dose are usually assumed to act independently, although there is some evidence that there may be synergistic effects between high LET and low LET components of the total dose [386]. This effect is not simple and depends on the relative mix of high- and low-LET radiation and the neutron energy spectrum [387–388].

While this approach to calculate photon-equivalent doses based on fixed RBE and CBE factors (often derived in one system and used for another) has been contributory, there is ample room and need for improvement to avoid overdosing dose-limiting tissues or underdosing tumours. The scientific community has proposed novel approaches that are currently under consideration. Sato and collaborators [389–390] proposed a model that would predict the biological effectiveness of newly developed ^{10}B-compounds based on their intra- and intercellular distributions and stressed the impact of the dose dependence of RBE on biological effect. The structure of the CBE factor was analyzed [391], and it has been proposed that the effect of dose from the boron neutron capture reaction can be predicted from the nucleo-cytoplasmic ratio or the cell size [244]. Ref. [392] claimed that the usual procedure for photon-equivalent dose calculation in BNCT may lead to unrealistically high calculated doses to tumour that cannot explain observed responses. A new approach for calculating photon isoeffective doses in BNCT has been introduced (see Section 10.3.4). The formalism of the photon isoeffective dose, proposed in Ref. [392], is expected to describe more precisely the biological effect of BNCT, provided that the radiobiological data required are known accurately. This new formalism permits determination of photon isoeffective dose for tumours and also for unacceptable complications in dose-limiting normal tissue [385], and its use is under validation by different groups [390]. Recently, a simplification of the photon isoeffective dose formalism has been proposed [393]. Ongoing studies are expected to provide input for the parameters required by the photon isoeffective dose formalism [394]. It is possible to evaluate and compare the performance of different beams by analyzing the radiobiological figures of merit of uncomplicated tumour control probability [395]. It has been shown that the physical characteristics of the beam are not sufficient to select the clinical BNCT 'beam' with the best performance for a particular clinical scenario [396]. The therapeutic potential is evaluated by analyzing the classical in-air free-beam characteristics and calculating a radiobiological figure of merit related to therapy outcome. The suitability considers the safety of the 'beam', by evaluation of the peripheral dose absorbed by out-of-beam organs.

Although it is recognized that the use of fixed RBE and CBE factors is not a suitable strategy to calculate mixed-field dose, and that in vitro studies have limitations in terms of extrapolation to an in vivo scenario, cell survival curves are still a valid tool to assess the potential of BNCT to kill tumour cells [309]. However, for studies performed at different laboratories to be intercomparable and more representative of in vivo conditions, Ref. [309] recommends the use of adherent rather than suspended cells. Adherent cells are more radiosensitive than suspended cells, and these differences preclude a comparative analysis of studies conducted with differing configurations.

Whatever model is used to express BNCT dose in photon-equivalent units, the capacity to design adequate treatment planning strategies and perform accurate dosimetry relies on sound, robust, and inter-comparable radiobiological data. Guidelines to produce these radiobiological data to feed computational models are needed. Furthermore, in many cases, radiobiological data produced in keeping with specific guidelines are part of the requirements of regulatory agencies for approval of clinical trials.

Despite the efforts to report the photon-equivalent dose as accurately as possible, a certain degree of uncertainty cannot be ruled out and different approaches are still under discussion, pending validation. In this context, it is always appropriate to also quote the total absorbed doses, including the breakdown of the total absorbed dose into its different components for the neutron dose employed, the spectrum of the neutron source, and boron concentration values in blood and clinically relevant tissues when available. Otherwise, important information for potential future use will be lost. Dose reporting in the clinical context is extensively discussed in Section 12.

9.3   RADIOBIOLOGICAL CONSIDERATIONS FOR AN IDEAL BORON CARRIER

As ^{10}B has a natural abundance of 19.9%, BNCT agents need to be enriched in ^{10}B [397] to increase the probability of thermal neutron capture. Potential systemic toxicity is an issue of concern that might limit the maximum dose of boron administered, and a boron carrier needs to be non-toxic at therapeutic dose levels. An ideal boron carrier will accumulate preferentially in tumour cells, will target all tumour cells homogeneously, and will deliver ^{10}B to the tumour efficiently close to a sensitive target. Preferential accumulation of boron in the tumour contributes to the therapeutic advantage of BNCT for tumour versus normal tissue. While Section 7.1 gives a brief outline of an ideal boron carrier from a pharmaceutical point of view, in this section the desirable aspects from the point of view of radiobiology are reviewed.

The BNCT literature typically stresses the importance of a high ($\geq 3/1$) T/N tissue and T/B boron concentration ratios. However, translational studies established somewhat different guidelines for potential therapeutic efficacy (e.g., Ref. [398]), i.e., no manifest toxicity, absolute boron concentration in tumour >20 ppm, and T/N and T/B ratios ≥ 1. The achievement of high mean T/N and T/B-concentration ratios is clearly an asset. However, the ultimate optimization of BNCT needs the majority of all the clonogenic tumour cells to be targeted, irrespective of their position in the tumour, their metabolism and degree of proliferation or differentiation. As tumours are often heterogeneous, targeting all the appropriate tumour cell populations is a well-known challenge. Tumour heterogeneity is now thought to be one of the major difficulties in the treatment of solid tumours [399–400]. Many tumours contain phenotypically and functionally heterogeneous cancer cells. Tumour cells poorly loaded with boron are less responsive or refractory to BNCT and can lead to therapeutic failure, e.g., Ref. [163]. The heterogeneous boron microdistribution within tumours, influenced by the microenvironment and the characteristics of the boron carrier, limits BNCT's therapeutic effect on local tumours [401]. It is well known that tumour blood vessels are characteristically dilated, saccular, and tortuous, have large inter-endothelial cell junctions and multiple fenestrations, and lack a normal basement membrane. Aberrant blood vessels affect blood flow, impairing convective fluid transport and distribution of blood-borne boron carriers [402]. Therefore, merely increasing pharmaceutical dose might not be effective in delivering boron to inaccessible areas.

Absolute boron content in tumour tissue is of utmost importance in the therapeutic efficacy of BNCT. It needs to be high enough, approximately 10^9 atoms ^{10}B/cell (i.e., 20–50 ppm ^{10}B) to allow sufficient boron neutron capture reactions to occur in a single cell for the effect to be lethal. D_B increases approximately linearly with the concentration of boron in tissues, provided there are no self-shielding effects due to high boron concentrations. Changes in D_B will result in changes in the relative contribution of the four dose components (Table 7) at any point in the treatment volume. Also, at a given T/N ratio, high absolute ^{10}B tumour concentrations are an asset because they allow for shorter irradiation times to achieve the same total dose and reduce the background dose that affects both healthy tissue and tumour non-selectively [378].

An ideal boron carrier rapidly clears from normal tissue and blood yet persists in tumour for sufficient time to permit BNCT treatment. The ideal carrier's microdistribution leaves ^{10}B atoms near therapeutically sensitive targets like DNA. The short ranges of α and recoiling ^{7}Li nuclei mean that the boron microdistribution relative to the subcellular target is of critical radiobiological significance [331, 398, 403]. Developing an ideal boron compound to satisfy all these requirements is a major challenge. Within this context, the international community devotes much effort and resources to search for the 'ideal' boron compound that would potentially replace the 'imperfect' compounds that have been used in humans; i.e., BPA, BSH, and GB-10. Only BPA and BSH have been used significantly in clinical trials. Although these individual agents delivered as a single intravenous injection/infusion before BNCT have shown therapeutic potential for several pathologies (e.g., Refs [161, 222, 404–409]), improvements are possible and needed. Other boron compounds have not yet been subjected to a clinical biodistribution study. Any promising new boron carrier identified in cell culture studies has to go through several stages, i.e., studies of biodistribution in experimental tumour models, in vivo toxicity, and subsequently in vivo radiobiological studies. Clinical biodistribution studies may follow if the preliminary results are promising enough. Such studies are very costly, do not benefit participants directly, and are subject to stringent regulatory requirements [410–411]. If all these studies are promising, the boron compound may enter a clinical trial. In the USA, only five in 10,000 medicinal compounds under development enter clinical trials, and only one receives approval for treatment. The process 'from bench to bedside' typically lasts a decade and costs may exceed a billion US Dollars [412].

The development and validation of improved boron carriers for BNCT are pivotal to the advancement of BNCT. However, because this is a lengthy and costly process, optimization of the delivery of ^{10}B-compounds that have been or are currently used in humans arises as an excellent short and medium term strategy. It can help to bridge the gap between research and clinical application. Concomitant international efforts to develop and assess novel boron carriers will also benefit from the knowledge derived from these studies. Improvement in the delivery of boron compounds that have been or are currently used would be the most effective way to make progress in the short and medium term [403, 413–415].

9.4 RADIOBIOLOGICAL CONSIDERATIONS UNDERLYING THE BORON CARRIERS EMPLOYED IN CLINICAL BNCT STUDIES

The main radiobiological features of the boron carriers employed in BNCT clinical studies are outlined in this section.

BPA is incorporated into the cell via the L-amino-acid cell transporter system, particularly LAT-1 (Section 7.6.4, Ref. [202]). LAT-1 is upregulated in various cancers, where it is thought to contribute to tumour growth by increasing amino acid supply [173]. The amino acid transporters $ATB^{0,+}$, as well as LAT-1, could contribute significantly to the tumour accumulation of BPA at clinically useful doses. Conversely, LAT-2 is expressed in normal tissues and would be detrimental to selective tumour targeting with BPA [173]. Several translational and clinical studies showed that BPA targets different types of tumours selectively, including brain tumours (e.g., Refs [378, 408, 416–417]), possibly because of increased metabolic activity and upregulation of L-type amino acid transport expression in tumour cells at the cell membrane. Thus, BPA would target tumours selectively based on a 'cell by cell' mechanism. BPA would accumulate in metabolically active basal cells within the epidermis and in tumour parenchyma [418]. This would explain the high biological efficacy of BPA-BNCT in terms of tumour response, albeit coupled to significant toxicity in terms of mucositis in normal and, above all, in the more sensitive field cancerized mucosa. The negative aspect associated with preferential incorporation of BPA in cells with enhanced protein expression is that BPA heterogeneously targets tumour cell populations with different LAT-1 expressions. For example, quiescent cancer cell populations may uptake BPA less effectively [419–420]. Tumour cells that are poorly loaded with boron will be refractory to BNCT [421–422]. Because it traverses the intact brain barrier, BPA was initially considered unsuitable for BNCT of brain tumours. However, this may be essential for effective BNCT of glioblastoma multiforme and other highly infiltrative brain tumours [378].

As described in Section 7.3, BSH does not cross the intact blood–brain barrier in the normal brain but accumulates in brain tumours mainly by passive diffusion because the blood vessels in intracranial tumours lack a functional blood–brain barrier [423–426]. Some studies suggest that BSH would accumulate more in mostly necrotic areas [427] and would be more effective as a boron carrier to treat quiescent sub-populations [420]. The handling and storage procedures for BSH to avoid the formation of oxidation products that might be toxic are somewhat complex. BSH in combination with BPA has been used to treat recurrent head and neck tumours with good results [165, 428].

The sodium salt of *closo*-$B_{10}H_{10}^{2-}$ (GB-10) is a largely diffusive, low molecular weight agent that, like BSH, does not traverse the intact blood–brain barrier [429]. Because of this, GB-10 was originally proposed as a boron carrier for BNCT of brain tumours as well as for BNCT enhanced fast neutron therapy of extra-cranial tumours such as non-small cell lung cancer [430]. While GB-10 was approved for use in humans, it has only been used in biodistribution studies in a reduced number of brain tumour patients during surgery and was employed clinically in two patients within the context of a BNCT clinical trial for brain tumours in 1961 [431]. It has also been used in a pharmacokinetic study of boron concentration in blood samples of subjects with glioblastoma multiforme or non-small cell lung cancer [432]. It is non-toxic at high concentrations and readily soluble. Administered alone or jointly with BPA, it has been explored in experimental tumour models that do not involve the blood–brain barrier [403]. As it is largely diffusive, GB-10 does not preferentially target the basal layer and GB-10-BNCT therapy induces only reversible, mild mucositis in dose-limiting precancerous tissue [405] or dermatitis in exposed skin [433]. Although GB-10 is not taken up preferentially by tumours in tissues not protected by the blood–brain barrier, GB-10-BNCT would exert a selective, therapeutic effect on tumour via selective damage to the aberrant tumour blood vessels, while preserving 'normal' blood vessels in healthy tissue or 'less aberrant' blood vessels in field cancerized tissue [405]. Homogeneity of boron targeting, achieved by administering more than one boron compound (as described below), or by using a largely diffusive compound like GB-10, has resulted in improved complete remission rates for larger tumours [405].

Microdistribution data from scanning transmission ion microscopy analysis revealed that a significant factor affecting the difference between CBE factors for BSH and BPA, despite similar gross boron concentrations, is the comparatively low uptake of BSH in the mucosal epithelium vs. lamina propria compared to that of BPA [434]. The preferential gross boron uptake by tumour tissue does not explain differences in CBE values between tumour tissue and normal tissue. Differences in CBE values would be due to variations in boron microdistribution. In turn, the microdistribution of boron is determined by features such as cell/tissue metabolism and cell/tissue geometry/architecture coupled to differences in the properties of the boron carriers that condition their uptake. Section 8 describes several techniques capable of determining microdistribution.

Although clinical trials with BPA and BSH have shown encouraging results [414], there is an unquestionable need for improvement in terms of specificity and efficiency of targeting tissues and cells. Values of CBE and RBE for BPA and BSH in normal tissues of a variety of organs have been recently reviewed [435].

9.5 MECHANISMS OF ACTION

Several groups are seeking to identify the mechanisms of action of BNCT. Translational research is essential to progress in this field. Understanding the mechanisms of action of BNCT will contribute to optimize the therapeutic efficacy of BNCT and reduce associated toxicity. Although many of the mechanisms involved in the effects of BNCT have not yet been elucidated, examples of contributory findings are given below.

Cell death induced by BNCT is triggered by the release of charged particles following the capture reaction of a thermal neutron by a ^{10}B atom. These high-LET particles create a high density of ionization tracks along their trajectories and induce complex, irreparable DNA double strand breaks (DSB) by direct damage [436]. Conversely, the low LET radiation dose component induces mostly reparable DNA single strand breaks (SSB), where the effect is dominated by indirect damage, but with a minor but non-negligible contribution from direct damage. However, it has been reported that BNCT-induced DNA damage is partially repaired by DNA ligase IV [437].

An increased ^{10}B isotope concentration was found to increase the DSB/SSB ratio [438]. The excellent anti-tumour effect of BNCT may result from the unrepaired DSBs: the DSB marker, γH2AX, was reported to be elevated 20 hours post-treatment with BPA-BNCT [439]. Poly(ADP-ribose) (a product generated by a nuclear enzyme that signals DNA damage by catalyzing the binding of ADP-ribose units to proteins), DSB and SSB markers of DNA were also elevated. The persistent staining of γH2AX and poly(ADP-ribose) in the BPA-BNCT treated group suggests accumulated DSB damage after BNCT [440]. BNCT does not induce significant changes in free radical production, supporting a predominantly direct effect on DNA [441]. Intranuclear ^{10}B localization enhances the efficiency of cell killing via direct damage to the DNA [442]. The molecular mechanisms of DNA damage response and repair in the mixed n–γ field of BNCT (Fig. 54) have recently been discussed [443].

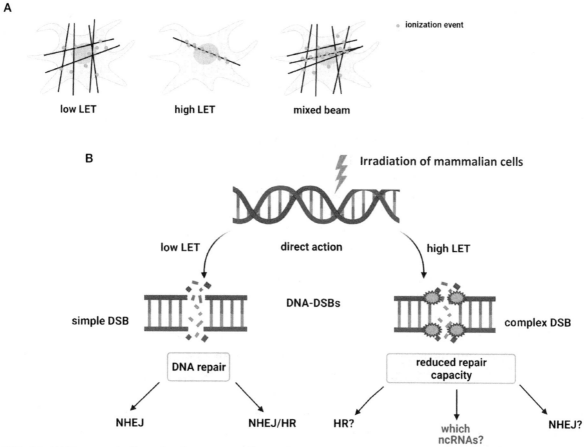

FIG. 54. (A) Types and effect of radiation according to linear energy transfer (LET): Low-LET radiation produces sparse ionization along its track, homogeneously within a cell. High-LET radiation causes dense ionization along its track. Mixed beam— both effects observed within a single cell. (B) Radiation-induced DNA damage: DSBs induced by low-LET radiation are repaired by non-homologous end-joining pathway (NHEJ) alone or NHEJ and homologous recombination (HR). Mechanisms of repair of complex DSBs induced by high-LET radiation are not fully determined. Created with BioRender.com. (This figure by the authors of Ref. [443] is licensed under CC BY).

Attempts have been made to understand the increased metastatic capability of tumours in the case of high grade gliomas following BNCT [444–445]. The high biological dose rate to tumours delivered by BNCT might be associated with a higher rate of specific mutations that could lead to a more aggressive pattern of dissemination. In addition, these authors suggested that transcription factors that regulate cellular response to ionizing radiation could be involved in bystander responses. The most important of these is NFκB. An altered balance in signalling pathways in bystander cells may increase survival rates and perhaps encourage the development of a more aggressive phenotype.

Multiple pathways related to cell death and cell cycle arrest may be involved in the treatment of melanoma by BNCT [446]. In B16F10 melanoma cells, Ki67 expression was significantly reduced after BNCT treatment, without affecting normal melanocytes. These findings agree with earlier studies that demonstrated that decreasing cyclin D1 triggers cell cycle arrest only in tumour cells (human and murine melanomas) after BNCT treatment. BNCT inhibited melanoma proliferation, altered the extracellular matrix by decreasing collagen synthesis, induced apoptosis by regulating Bcl-2/Bax expression, increased the levels of TNF receptor and cleaved caspases 3, 7, 8 and 9 in melanoma cells. Apoptosis is triggered by intrinsic and extrinsic pathways in BNCT-treated melanoma cells. BNCT has been shown to induce cell apoptosis via the mitochondrial pathway and to cause cell cycle arrest in the G_2/M phase, with changes in the expression of associated proteins [446]. It reduces the cells in G_2/M and G_0/G_1 phases, indicated by the reduction of cyclin D1, and increases the quantities of fragmented DNA in melanoma cells [447]. Reference [448] showed that, 24 and 48 h after BNCT, the cell cycle was arrested in G_2/M, and cell necrosis was the main cause of cell death. Up-regulation of p53 protein occurred 24 h after BNCT, which appears uncorrelated with the p53-dependent G_1/S cell cycle arrest. These studies agree with findings of Ref. [449] that BNCT induces G_2/M phase arrest and that cells in the G_2/M phase are the most sensitive to ionizing radiation.

Oral squamous cell carcinoma (SCC) cells with mutant-type p53 are more resistant than those with wild-type p53 to BNCT's cell-killing effect [450]. It has been concluded that a functional p53 is required for the induction of apoptosis related to G_1 arrest and that the lack of G_1 arrest and related apoptosis may contribute to the resistance of cells with mutant-type p53. Further work showed that at a physical dose affecting the cell cycle, BNCT inhibits oral SCC cells in p53-dependent and -independent manners [450]. It has also been reported that BNCT triggers cell death in glioblastoma cells independent of p53 mutation status [451]. In T98G cells, both cell killing and apoptosis happened followed BNCT.

Apoptosis or programmed cell death, marked by the fragmentation of nuclear DNA, has been found to contribute to the effect of BNCT. It is a normal physiological process that eliminates DNA-damaged or superfluous cells, and when halted (as by gene mutation for example) may result in uncontrolled cell growth and tumour formation. Overall, induction of apoptosis constitutes the main contribution to the effect of BNCT in glioma cells in vitro. Apoptosis triggered by BNCT activates Bax and downregulates Bcl-2 [452]. The pro-apoptotic effect of BNCT plays an important role in the reduction of cell proliferation. BNCT induces apoptosis in vivo, and, while the apoptotic rate is low, the proportion of apoptotic cells increased slightly 6 h after BNCT [420, 453]. Apoptosis is a form of cell death induced by BNCT [450], and both chromosome aberrations and apoptosis would be involved in the effect of BNCT [454]. However, Ref. [455] reported that apoptosis one day post-treatment did not have a significant role in BNCT-induced hamster cheek pouch tumour control and that one of the mechanisms involved in BNCT-induced tumour response in this model would be an inhibitory effect on DNA synthesis. Ref. [456] reported that differences in apoptotic cells pre- and post-BNCT in human oral SCC xenografts were minimal.

The DNA damage pattern and the repair pathways that are activated by BNCT in thyroid follicular carcinoma cells has been examined using, respectively, γH2AX foci and the expression of Ku70, Rad51, and Rad54, main effector enzymes of non-homologous end joining and homologous recombination repair pathways [457]. The findings were consistent with an activation of the homologous recombination repair mechanism in thyroid cells. A melanoma cell line showed a different DNA damage pattern and the activation of both repair pathways.

The importance of considering microdosimetry when attempting to explain results that apparently have no suitable explanation by considering only average quantities, such as absorbed dose or LET, has been stressed [384, 458]. Within this context, the importance of studying nitrogen concentration of cells, cell medium, and flasks is also pointed out. It has been suggested [458] that correction factors that express dose-effect parameters as a function of boron (and nitrogen) microdistribution are a suitable approach to establish a common framework with macroscopic dosimetry and treatment planning.

Recent studies have examined the impact of BNCT on the proteomic profile. Proteome analyses were performed of human squamous tumour SAS cells after BNCT mediated by BPA [459]. These analyses showed that proteins with a role in endoplasmic reticulum, DNA repair, and RNA processing exhibited dynamic changes shortly after BNCT and may be involved in the regulation of cellular response. Proof of principle that proteasome changes post-treatment can be followed based on urine samples has been provided [460]. It has been stressed that the potential value of evaluating baseline molecular profiles (metabolome and proteome) and genetic profiles to screen for those patients that will benefit the most from therapy and will suffer the least adverse effects [461]. With this aim in mind, the challenge is to encourage synergy between biochemists and clinicians. In the future, this strategy will enable monitoring of therapeutic efficacy and identification of disease and therapy related pathways and will contribute to the development of precision/personalized medicine in BNCT, based on evaluating the eligibility of patients for specific therapies [462]. It would be a valuable contribution to explore, within the context of BNCT radiobiological studies in animal models, the characterization of molecular profiles, tumour features, and biomarkers useful to monitor efficacy, side effects and recurrence. Proteomics would be an important tool to further optimize BNCT [463] and may play a role in predicting responders and non-responders.

Recent studies showed the abscopal effect of BNCT. The abscopal effect,[19] originally described in 1953 [464], refers to the inhibitory action on development and growth of non-targeted tumours, distant from the irradiation area after radiotherapy [465–466]. The abscopal effect would be mediated by radiation-induced immune responses [467], which could cause immunogenic cell death and cross-prime tumour-specific T cells that may act in a manner similar to an in situ tumour vaccine [468]. Radiotherapy would induce and enhance the endogenous anti-tumour innate and adaptive tumour response [469]. The mechanisms involved in the abscopal effect are still unclear [465], and several contributory effects are likely [469–470]. Proof of principle in an ectopic model of colon carcinoma in BDIX rats that the positive local response of a tumour to BNCT can induce an abscopal effect has recently been provided [471]. Ongoing studies suggest that BNCT combined with immunotherapy (Bacillus Calmette-Guerin) induces robust local and abscopal therapeutic effects [472]. Very recently it has been shown that the boron neutron capture reaction in peripheral blood mononuclear cells resulted in the modification of these cells to anti-tumour phenotype in a model of murine mammary (EMT-6) tumours [473]. These authors described the immunomodulatory effects of BNCT when boron-rich compounds are delivered systemically.

While the mechanisms involved in BNCT-induced effects have only been partially elucidated and require further study, the mechanisms specifically involved in the effect of AB-BNCT require special attention henceforth.

9.6   APPLIED AND TRANSLATIONAL RESEARCH

Translational research in adequate experimental models is necessary to optimize BNCT for different pathologies. The international community employs experimental models to study those pathologies that respond poorly to standard therapies and would potentially benefit from more effective and selective therapies. Translational BNCT studies in in vivo animal models

---

[19] 'ab-scopus', away from the target

are crucial to design clinical protocols for existing or new targets of BNCT. Fortunately, and very importantly, there has been an impressive rise in radiobiological research in BNCT since the publication of the last IAEA publication on this topic [1]. Different international groups have contributed to the advancement of BNCT with applied and translational research. Examples include translational studies to evaluate BNCT as a new approach for clear cell sarcoma treatment employing a lung metastasis model of clear cell sarcoma [474] and a review of how translational studies in animals have led to clinical advances in BNCT [409]. Translational studies in a newly developed model of oral precancer in the hamster cheek pouch suggested a potential new application of BNCT, i.e., its inhibitory effect on tumour development from field cancerized tissue [475]. This is an issue of great clinical relevance because locoregional recurrences are frequently the cause of therapeutic failure in the treatment of head and neck cancer.

9.6.1 Optimization of tumour targeting with boron

The development of boron compounds for BNCT (e.g., Ref. [397]) is described in Annex XII and Section 7, and the development of new ones is also discussed in Annex XIV. This section focuses on different administration strategies that seek to optimize delivery and microdistribution of boron carriers that have been/are approved for their use in humans [403, 415]. The optimization of tumour targeting with boron employing compounds approved for use in humans allows for a more direct and less costly extrapolation to a clinical environment [403]. Within this context, translational research is pivotal. Novel boron carriers will also eventually benefit from administration strategies that improve targeting of boron. In addition, some examples of more general strategies (not directly related to biodistribution and microdistribution of boron) to improve the therapeutic efficacy of BNCT employing boron compounds that have been authorized for their use in humans are outlined.

Recurrence following BNCT is caused mainly by the non-homogeneous uptake of boron compounds with poor microdistribution in some regions of the tumour. Aside from the potential development of novel boron agents, the best way to improve response and cure rates would be to optimize BPA dosing, either alone or in combination with BSH [415]. A variety of methods to improve the delivery and distribution of BSH and BPA in experimental models and human subjects has been tried with varying degrees of success, including combined administration of BPA and BSH [476]. Combined administration of boron compounds that differ in properties and uptake mechanisms may make the targeting of tumours more homogeneous and overcome the potential toxicity that would arise from larger doses required when the compounds are given alone. Improved homogeneity of tumour cell targeting by boron would improve the response of larger, more heterogeneous tumours, as shown by an experimental model of oral SCC treated with BNCT mediated by GB-10 + BPA [405, 477]. In addition, the use of the combined administration of two boron compounds with different mechanisms of action in BNCT would enhance therapeutic efficacy. Such is the case of BNCT mediated by GB-10 + BPA, combining the vascular and cellular targeting mechanisms of BPA and GB-10, respectively [405]. Combinations of agents may perform better than any single agent [476–478].

Additional examples of techniques used to improve accumulation of boron in brain tumours are the use of a hyperosmotic agent such as mannitol to disrupt the blood–brain barrier, intracarotid administration of the boron agent and convection enhanced delivery [397, 479–480]. The identification of a new pathway controlling glucose uptake in high grade gliomas presents an opportunity to reposition existing drugs and design novel drugs [481].

Slow infusions of BPA would improve targeting of infiltrating tumour cells by boron in the brain [411, 482–483]. A two-stage injection protocol (180 mg·kg^{-1}·h^{-1} for 2 h, followed by 90 mg·kg^{-1}·h^{-1} for 30 min) to maintain a stable blood BPA concentration during BNCT has been used [237]. Neutron irradiation under continuous BPA infusion contributed to solving the problem of heterogeneous distribution of BPA [484]. Annex XIII describes this two-step BPA infusion process developed in Japan.

Pre-loading with an amino acid analogue, such as L-3,4-dihydroxyphenylalanine (L-DOPA), structurally similar to BPA, which enters the cell through the LAT system (see Section 7), would increase accumulation of subsequently administered BPA via an antiport (exchange) mechanism [485]. In mouse and human cell lines, preloading with L-BPA, L-DOPA, and L-tyrosine enhances the uptake of ^{18}F-FBPA [486]. L-phenylalanine preloading reduces the boron component of the irradiation dose to the normal brain in BNCT for brain tumours by reducing the accumulation of L-BPA in the normal brain relative to tumour tissue [487]. Amino acid transport control, which involves co-loading BPA with L-type amino acid esters, would be useful to enhance intracellular accumulation of boron [488]. The increase is due to activation of a System-L amino acid exchanger between L-tyrosine and BPA. Intracellular hydrolytic enzymes metabolise the amino acid esters to amino acids, increasing the concentrations of intracellular amino acids and stimulating exchange transportation [488]. In the case of BPA, the intracellular free pool of boron is affected by the presence of phenylalanine. These notions may contribute to the development of improved regimens for the administration of BPA, especially the placement of patients on low phenylalanine diets prior to BNCT [442].

Mild temperature hyperthermia would improve boron carrier delivery by increasing blood flow [489]. Thermally sensitive liposomes, which maintain a stable drug payload at physiological temperatures but are designed to be highly permeable under mild hyperthermia, improved the tumour-specific delivery of BPA [490].

Sequential-BNCT involves the sequential application of BPA-BNCT and GB-10-BNCT with an interval of 24–48 h between applications and was developed based on radiobiological findings [491]. The first application reduces interstitial fluid pressure and void space is created because of cell death. This favours intratumoural delivery of GB-10 in a second application, which improves tumour targeting homogeneity by boron. Combining boron compounds is, therefore, a benefit, and a brief interval between the applications avoids tumour repopulation. A second application would favour targeting of tumour cell populations that did not respond to the first. Sequential-BNCT significantly enhanced tumour response (partial + complete remission) compared to a single dose-matched application of BNCT mediated by GB-10 + BPA. Sequential-BNCT did not exacerbate mucositis in precancerous tissue, which limits the dose that can be applied to the tumour.

The need to fix the defective vascular system in tumours to allow for adequate tumour targeting by boron has been identified [421]. The structure of aberrant tumour blood vessels causes deficient distribution of blood-borne agents. Transient normalization of defective tumour blood vessels was performed in the oral cancer model in the hamster cheek pouch. Thalidomide has been used as an anti-angiogenic drug to try to improve the distribution of boron in the tumour. BPA was administered in the normalization window. It was expected that the boron concentration in tumours would increase due to blood vessel normalization prior to the administration of BPA. However, gross concentration measurements by ICP-MS failed to show the expected increase in tumour concentration. However, microdistribution studies and radiobiological BNCT experiments revealed that a normalized vascular system improved targeting homogeneity of boron in tumours [421], enhancing tumour response. Rather than increase total uptake of boron, normalization would lead to improved distributions of boron to a larger proportion of tumour cells.

Further seeking to favour delivery and distribution of boron in tumours, electroporation associated with the administration of GB-10 was shown to enhance the therapeutic efficacy of BNCT by improving gross uptake of boron and microdistribution of GB-10 in the model of oral cancer in hamster without an increase in toxicity [296, 492]. BNCT with electroporation can be clinically useful [424]. Sonoporation was shown to enhance the efficiency of BPA-mediated BNCT for oral SCC in nude mice by modulating the microlocalization of BPA and BSH in tumours and increasing their intracellular levels [488].

Another possible approach to enhance BPA uptake in tumours with no concomitant increase in normal tissue uptake is to use high intensity focused ultrasound as shown in nude mice bearing intra-oral xenografts of a human SCC cell line [493] and in patients with high-grade gliomas [494]. The use of pulsed ultrasound has been shown to transiently disrupt the blood–brain barrier, improving the uptake of BPA and BSH as well as optimizing their microdistribution within the tumour [495–497].

^{10}BSH-entrapped water-in-oil-in-water emulsion has been developed and evaluated as a selective boron carrier for BNCT in hepatocellular carcinoma treatment [498–499]. Utilizing the hepatic artery, the application of the ^{10}BSH-entrapped water-in-oil-in-water emulsion as a novel intra-arterial boron carrier for BNCT proved to be feasible. The compound would be targeted to the tumour by enhanced permeability and retention effects. A complementary intracellular mechanism is sought to target cancer cells via binding of the compound to certain ligands such as monoclonal antibodies [500].

It has been shown in an orthotopic human oral SCC bearing animal model that low dose γ irradiation increases BPA accumulation in tumour as well as the T/N and T/B concentration ratios of boron, enhancing BNCT efficacy and extending the overall survival rate [501].

9.6.2 Strategies to improve efficacy and reduce radiotoxicity

In view of the high cost and complex regulatory approval process for new boron compounds, researchers have been investigating a wide variety of approaches to improve delivery and enhanced effectiveness of currently available boron compounds. These approaches are reviewed in this section.

Undifferentiated thyroid carcinoma is amenable to treatment by BNCT [404], and a histone deacetylase inhibitor such as sodium butyrate can act as a radiosensitizer of BPA-BNCT for poorly differentiated thyroid carcinoma [502]. Sodium butyrate increased the percentage of necrotic and apoptotic cells and the accumulation of cells in G2/M phase 24 h post-BNCT. Biodistribution studies of BPA showed that treatment with sodium butyrate increased tumour concentration of boron, contributing to enhanced tumour response to BPA-BNCT.

Solid tumours, in particular human tumours, are believed to have a high proportion of quiescent (Q) cells. The presence of these cells would be partly due to a microregional deficiency in the concentrations of oxygen, glucose, and other nutritional factors in the tumours caused by a poor and heterogeneous tumour vascular supply. Q tumour cells have lower radiosensitivity than proliferative (P) tumour cells in solid tumours in vivo, irrespective of the p53 status of tumour cells. Thus, more Q tumour cells survive after radiotherapy than P cells. The sensitivity difference between total and Q cells is widened in the presence of a ^{10}B-carrier, especially in the case of BPA. The control of Q tumour cells is an unmet challenge and has a great impact on the outcome of radiation therapy [503]. Translational studies showed that an acute hypoxia-releasing treatment such as administration of nicotinamide or bevacizumab administration, may be promising to enhance the reduction in lung metastases induced by BPA-BNCT [489]. The control of the chronic hypoxic Q cell population of the primary solid tumour has been shown as pivotal to the control of local tumours. Control of the acute hypoxic total tumour cell population of the primary solid tumour will condition the control of metastases. BSH-BNCT combined with a hypoxic cytotoxin, tirapazamine, with or without mild temperature hyperthermia, improved local tumour control, while BPA-BNCT in combination with both tirapazamine and mild temperature hyperthermia as well as nicotinamide is believed to decrease the number of lung metastases [503]. Control of the acute hypoxia-rich total tumour cell population of the primary solid tumour may impact the control of metastases, as illustrated in Fig. 55. The vascular targeting agent ZD6126 (*N*-acetylcochinol-*O*-phosphate) in the rodent SCC VII carcinoma model, in combination with BNCT, was better at enhancing the sensitivity of the Q cells than the total tumour cells (Fig. 56), reducing the marked difference in sensitivity between the Q and total tumour cells caused by the use of a ^{10}B-carrier for BNCT [504].

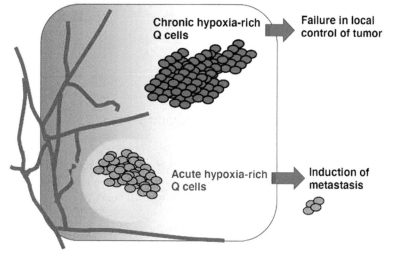

FIG. 55. It is postulated that control of chronic and acute hypoxia-rich Q-cells has the potential to impact control of tumours as a whole and metastases, respectively, after BNCT (courtesy of S. Masunaga, Osaka Prefecture University and M. Masutani, Nagasaki University).

Oxygen pressure during incubation with a ^{10}B-carrier had a critical impact on ^{10}B uptake of cultured tumour cells [505]: uptake of BPA decreased linearly as a function of oxygen level reduction from 20% through to 1%. Hypoxia with < 10% O_2 significantly decreased mRNA expression of LAT-1 in glioblastoma cell lines: the reduction of uptake of BPA in glioblastoma in such conditions may be caused by reduced expression of this transporter protein. The linear reduction in uptake of BPA caused by hypoxia in glioblastoma cells suggests that approaches to overcome local tumour hypoxia may improve the success of BNCT. The reduction of uptake of BPA by cancer cells due to local hypoxia is likely to negatively influence the efficacy of BNCT [506].

Cancer stem cells are linked to tumour resistance to treatment in general and to BNCT in particular. They are considered to be highly resistant to DNA damage and part of the Q tumour population [507]. Although cancer stem cells were initially described as residing in a perivascular niche around tumour vasculature [508], they have also been proposed to exist in a secondary niche within cancers that is further away from vasculature and more hypoxic [509]. One approach to overcome resistance is to increase the boron uptake in stem cells. Research in this area is lacking. However, one such attempt has been made [510]. Since CD133 is frequently expressed in the membrane of glioma stem cells, these authors developed a bioconjugate nanoparticle that targets human CD133-positive glioma stem cells and would be a potential boron agent in BNCT. Very importantly, it has recently been shown that glioma stem-cell like cells may take up BPA and be amenable to targeting by BNCT [511].

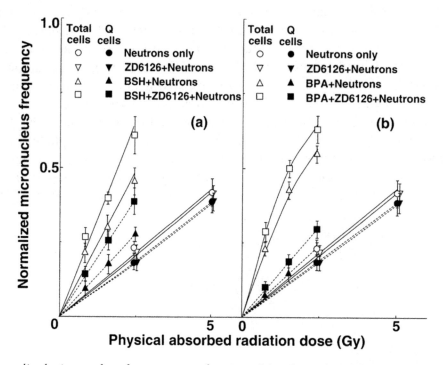

FIG. 56. Normalized micronucleus frequency as a function of the physical absorbed radiation dose in total and quiescent (Q) tumour cells after irradiation with reactor thermal neutron beams following (a) BSH or (b) BPA administration in combination with or without ZD6126, following ZD6126 administration only, or without any drug administration (reproduced from Ref. [512] with permission of Elsevier).

A novel strategy of selective boron delivery into tumours via oligopeptide transporter PEPT1, upregulated in various cancers has been proposed. This strategy uses the dipeptides of BPA and tyrosine (BPA-Tyr and Tyr-BPA) [513]. Kinetic analyses have indicated that BPA-Tyr and Tyr-BPA are transported by oligopeptide transporters, PEPT1 and PEPT2. PEPT1 is the dominant oligopeptide transporter in tumour cell lines. Furthermore, using Tyr-BPA and BPA-Tyr, boron was taken up by PEPT1-expressing pancreatic cancer AsPC-1 cells via a PEPT1-mediated mechanism. Oligopeptide transporters, especially PEPT1, are promising candidates for molecular targets of boron delivery. BPA-containing dipeptides may be used for developing novel boron carriers targeting PEPT1. BNCT using the dipeptide-based boron compounds may become an alternative therapeutic option that can be applied to patients that are expected to respond poorly to BPA-BNCT because of low expression of LAT-1 responsible for BPA accumulation in tumour cells.

In an attempt to improve the uptake of BSH, a BSH/micelle was developed that was stable in a normal physiological environment but released BSH in response to a high level of redox potential in cancer cells [514]. Furthermore, intracellular delivery of BSH was promoted by the BSH/micelle through the endosome escape function of micelles, enhancing the therapeutic efficacy of BNCT.

The use of theranostic compounds for BNCT is actively being explored [177]. For example, it has been shown that water-soluble aza-BODIPYs (such as the aza-BODIPY/^{10}B-BSH conjugate) can be used as theranostic vectors for boron complexes, opening a new perspective for compound development for BNCT applications [515]. An approach that combines the boron carrier for therapeutic purposes with an imaging tool might contribute to non-invasive determination of the ^{10}B-concentration in a specific tissue [463]. Radiobiological BNCT studies employing theranostic compounds will gain momentum in the coming years [177].

Regardless of the treatment protocols employed, the use of radiosensitizers and radioprotectors is a strategy that has received attention and continues to be explored with encouraging results in certain conditions. One such example is histamine (approved for use in humans), employed as a radioprotector, that was shown to reduce BNCT-induced mucositis in experimental oral precancer without jeopardizing therapeutic efficacy [381].

9.6.3 Research to identify the best candidates/treatment strategies

Translational studies have shown a significant correlation between tumour temperature, certain immunohistochemical markers, and the T/B-concentration ratio of boron [516]. These methods might be useful to design screening protocols to identify patients that might benefit the most from BNCT, thus optimizing the efficacy of BNCT in selected patients.

Genetic background and environmental factors may identify patients with different risks for oral cancer and mucositis after treatment. BNCT-induced mucositis in field cancerized tissue limits the dose that can be applied to tumours and affects quality of life. Mucositis caused by BNCT increases as a function of the aggressiveness of the carcinogenesis protocol employed. This finding ought to be considered during BNCT treatment planning for head and neck cancer patients that differ in tumour stage and aggressiveness of the field-cancerized tissue. The study of different oral cancer patient scenarios would help to develop personalized therapies, identifying factors that could negatively affect outcome, reducing toxicity and associated costs and eventually improving therapeutic outcome [517].

Within the context of potentially successful BNCT and prolonged survival, an issue that emerges and warrants consideration is the risk of secondary cancer in healthy organs of patients treated with BNCT. This issue has been examined using a radiation computational phantom and Monte Carlo software [518].

Experimental models are employed to test the therapeutic potential of boron compounds. Biodistribution studies are essential to design preclinical, clinical-veterinary, and clinical research protocols. They identify boron compounds and administration protocols that may be useful and can be used to optimize the time of neutron irradiation after the administration of the boron carrier. Currently, there is no practical, online, non-invasive way to determine boron concentration in tissue during BNCT (see Section 8). Dose calculations are, therefore, based on values in blood, tumour, and normal tissue from biodistribution studies [398]. At most, blood samples can be taken just before and, in some cases, during irradiation to measure concentrations of boron in blood and infer them in tissue, assuming the T/B-concentration ratios established in previously performed biodistribution studies hold true [519]. In experimental models, dose calculations are based on the mean concentration values of boron obtained from biodistribution studies in separate sets of animals [416]. Large intra- and inter-tumour, as well as inter-subject variations in gross boron contents have been observed [429, 477]. These variations need to be taken account of in the calculation and prescription of dose to avoid either underdosing the tumour or exceeding the radiotolerance of healthy tissues. More recent improvements (see Sections 7.6–7.6.1 and 8.1.2) include the development ^{18}F-FBPA, an ^{18}F-labelled derivative of BPA that has been developed to predict ^{10}B accumulation in tumours and normal tissues by PET imaging [224].

Although biodistribution studies are essential and provide valuable guidelines, radiobiological studies in experimental models to determine the efficacy of administration protocols and of boron carriers are vitally important. The combined biological effects of a heterogeneous boron distribution, a particular microdistribution of ^{10}B, and a mixed field irradiation with high- and low-LET components can only be investigated in studies involving irradiation [414, 520]. A boron carrier ruled from biodistribution studies may be useful in actual BNCT studies [403]. Even when a boron compound is taken up selectively by tumour it can target them homogeneously, and when used as a boron carrier for BNCT, it can elicit a selective effect on tumours. This selective action in tumours would be based on preferential effects on the aberrant tumour vasculature, while sparing both the normal blood vessels in normal tissue and the only slightly aberrant blood vessels in field-cancerized epithelial tissue. Such is the case of GB-10 in an experimental model of oral cancer in the hamster cheek pouch. Therefore, selective tumour lethality may not arise from selective uptake of a boron carrier but from a selective effect on aberrant tumour blood vessels [405]. This new paradigm arose out of in vivo BNCT studies and demonstrates the importance of radiobiological studies to determine therapeutic efficacy of boron carriers and administration protocols.

Translational research also serves to confer new value on boron compounds for a particular application that were previously classified as 'not useful for BNCT' based on their poor performance in biodistribution studies and in BNCT; e.g., boric acid was dismissed as a 'non-selective' boron compound for BNCT, but its therapeutic potential as a boron carrier for BNCT was 're-discovered' in hepatic VX2 tumour-bearing rabbits [521] and for hepatocellular carcinoma in a rat model [522]. Autoradiography showed that boric acid was selectively targeted to tumours and their blood vessels; histopathological examination showed that radiation damage to tumour-bearing liver was mainly seen in the tumour regions after BNCT.

As vascular damage rather than killing of tumour cells may be the main method of tumour control in certain cases, and as targeting of heterogeneous tumour cell populations plays a crucial role in the BNCT's biological effect, the need to evaluate the therapeutic efficacy of BNCT in vivo is clear. Furthermore, in vitro experiments are generally insufficient to determine biodistribution of boron in vivo and the radiobiology of BNCT for a particular set of conditions (pathology, site, boron compound, etc.). For example, boronated liposomes were shown to be non-toxic in vitro but caused massive haemorrhaging in vivo, conceivably due to the induction of apoptosis in the cells of the endothelial lining of blood vessels in the tumour [523]. In addition, the very important role of blood vessels in boron carrier uptake and effect of BNCT cannot be assessed in vitro. To understand organ response, ideally, the response of a whole organ would be evaluated (e.g., brain, spinal cord). Small animals would receive a lethal dose for such experiments conducted with epithermal neutrons, so ideally large animals would be used. Large animal experiments are costly, difficult to perform, and need to comply with stringent regulations. Surrogate models consist of small animals protected by an adequate shield to enable exposure of the target area while minimizing exposure to the animals' bodies, both in the case of irradiation with epithermal [405] and thermal [421] neutron beams (Fig. 57).

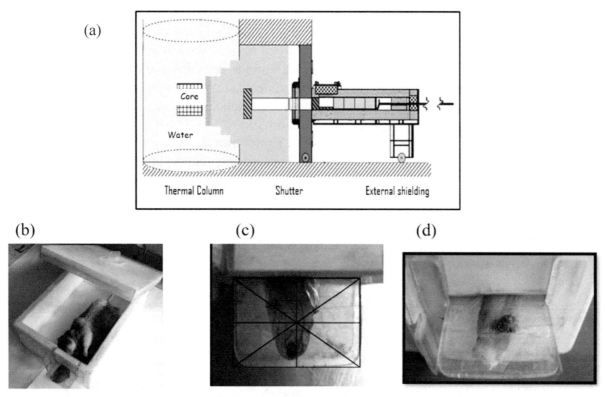

FIG. 57. A shielded set-up for small animal BNCT irradiation. (a) A tunnel in the graphite thermal column permits placement of samples into a near-isotropic neutron field. (b) A shield made of lithium carbonate, enriched in ^6Li was constructed to protect the body of the animal from the thermal neutron flux while exposing the tumour bearing pouch (c) and (d) (courtesy of A. Schwint, CNEA).

The important role of the tumour microenvironment (TME) in therapeutic outcome can only really be explored in in vivo models. In regions of hypoxia, low blood flow, and deficient nutrition, tumour cells will enter a quiescent (Q) state, exhibiting low radiosensitivity and reduced drug distribution. These conditions contribute to tumour resistance to treatment [402]. A pivotal feature of the microenvironment that promotes tumour immune escape and ensuing treatment failure is the lack of tumour antigen recognition and of antitumour T-cells. Although radiation was reported to augment microenvironmental immunosuppressive effects it is also accepted that radiotherapy also induces immune activation (Fig. 58) [402]. Inflammation and cancer can be interrelated [524]. The TME orchestrated by inflammatory cells is an indispensable participant in cancer initiation, promoting cellular proliferation, survival, and migration [525]. These aspects related to the microenvironment have a significant impact on response to therapy and eligibility for treatment and require in vivo studies.

Another example of the contribution of translational and clinical-veterinary studies to the knowledge of BNCT radiobiology is the low dose BNCT studies performed at the RA-1 reactor in Argentina for the treatment of SCC in feline patients [526]. The results were surprisingly good, considering the low absorbed doses administered. Given the sensitivity of SCC to fast neutrons alone [527], the contribution of the significant fast neutron component in the neutron spectrum at RA-1 might have played a role in tumour response. Likewise, comparative BNCT studies in the hamster cheek pouch oral cancer model at the thermalized epithermal B1 neutron beam at the RA-6 reactor [405] and the thermal neutron facility at the RA-3 reactor (with a negligible fast neutron component) in Argentina [528] exhibited the same trend. Within this context, the ideal neutron spectrum will depend on the pathology to be treated, even considering similar tumour volumes and localization.

Advances in cancer biology led to defining sixteen hallmarks of the TME that empower tumour cells for sustained growth and survival [529]. Radiation therapy can perturb the functions of these sixteen TME hallmarks to overcome sustained tumour growth and cause various modes of tumour cell death (Fig. 58). Such killing of tumour cells depends on the ability of the radiation dose and the fraction to affect these sixteen characteristic hallmarks, resulting in apoptosis, clonogenic death, and immunogenic death. Hence, each radiation dose and fraction can selectively perturb certain characteristic hallmarks of TME, so that radiation dose and fraction act like targeted drugs affecting several signal transduction pathways that dictate the mode of tumour cell death. For example, an increase in proapoptotic signaling (perturbing apoptosis evasion) with concomitant shutting down of pro-survival factors (perturbing oncogene addiction) in response to a radiation dose/fraction will result in the killing of tumour cells, and such killing can further be enhanced with the right signal transduction targeted drug combined with specific radiation dose [530].

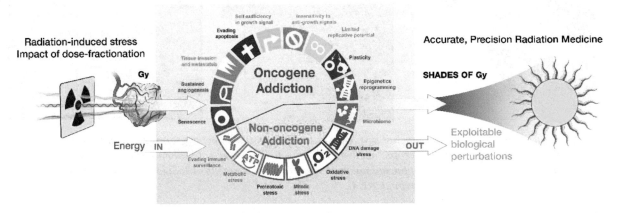

FIG. 58. Impact of radiation dose and fractions on sixteen TME characteristics. Each dose and fraction lead to unique biological perturbations, similar to drugs with a unique action mechanism, triggering several signal transduction pathways (courtesy of M. Ahmed, National Institutes of Health, U.S. Government Works).

Multiple translational studies at research reactors have been performed in different tumour models using a variety of boron compounds, different routes of administration, and different administration strategies. Radiobiological studies have contributed to clinical-veterinary studies in dogs and cats with spontaneous tumours. Dogs with spontaneous brain tumours have been used to study the biodistribution of BSH and implications for its use in BNCT [531]. Studies of spontaneous tumours may be more relevant than experimental tumours [531], as there may be differences in vascular permeability and blood flow, kinetics, and tumour morphology. The efficacy of B- and Gd-mediated neutron capture therapy has been compared for canine oral melanoma and osteosarcoma [532] and the possible application of BSH-BNCT to spontaneous canine osteosarcoma has been studied [533]. Clinical-veterinary BNCT studies in cats with spontaneous SCCs were performed at the RA-1 Nuclear Reactor [526, 534] and similar studies in dogs with advanced head and neck cancer (Fig. 59) were performed [535] in preparation for a clinical BNCT trial for head and neck cancer at the RA-6 reactor in Argentina. All these studies have contributed to the knowledge and advancement of BNCT and have provided evidence of the potential role of BNCT in veterinary medicine.

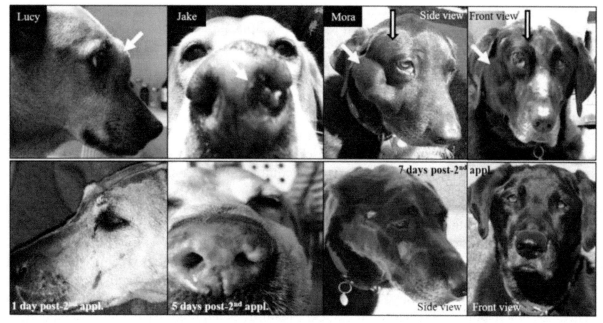

FIG. 59. Examples of dog patients pre-BNCT (top panel) and post-BNCT (lower panel). Effective and fast tumour response as early as 1–7 days after the second application of BNCT led to a prompt and impressive improvement in quality of life that persisted for several months. (This figure from the authors of Ref. [535] is licenced under CC BY.)

9.7 CURRENT STATUS OF RADIOBIOLOGY FOR ACCELERATOR BASED SYSTEMS

As described in Section 13, the efficacy of reactor based BNCT has been confirmed for certain malignancies. However, the BNCT community worldwide is working on the development of AB-BNCT [8, 536–538].

9.7.1 Biological studies reported for accelerator facilities

The fact that AB-BNCT is under study worldwide (e.g., [8, 539–541]) poses the need for radiobiological in vitro and in vivo translational studies employing accelerators as the neutron source rather than nuclear reactors, and studies are scarce. One of the first efforts was performed employing an accelerator based neutron source constructed for BNCT at the Budker Institute of Nuclear Physics, Novosibirsk, Russian Federation [542]. The activity in the bodies of mice induced by exposure to an AB-BNCT epithermal neutron source has been reported [32], and dose calculations to assess the efficacy of AB-BNCT have been performed [543]. Differences in RBE arising from different neutron source technologies have also started to be examined. Several cell lines were examined in an AB-BNCT system with a Be target, which concluded that the RBE of an epithermal neutron beam was 2.2–2.6, smaller than the value of 3.0 obtained at the Kyoto University Research Reactor [3]. The RBE of an epithermal neutron beam for AB-BNCT system with a Li target was reported to be 1.7–1.9 [544]. More basic and clinical studies are needed to move to an era when more patients with refractory cancers will be treated [128].

9.7.2 Planning preclinical tests prior to clinical trials

There is a need for careful consideration of an appropriate radiobiological program for a BNCT facility. A limited set of radiobiological measurements for example using cell cultures may be necessary to demonstrate safety of the beam and equivalence with already established facilities.

A much more extensive program is required if the centre is developing new administration regimes or new boron compounds or exploring new clinical indications. In such cases, animal experiments will very likely be necessary.

In some countries, it may not be possible to irradiate animals in the same room as humans are treated, and complexities handling animals outside a dedicated animal facility would need to be managed. Small rodents such as mice of certain specific pathogen free levels might be required for preclinical tests in hospitals. In contrast, middle and larger size animals are difficult to use, because of regulations on activation levels (see Section 9.7.3) as well as medical laws. These factors mean that development of BNCT needs a combination of clinical and research facilities.

Full experimental planning before clinical use of the system or a new clinical trial ought to be discussed in detail with the regulatory authorities of each country before preclinical tests are set up. For a hospital centre treating patients within established indications for BNCT, regulators may require very limited new radiobiological data or none, with the facility being characterized with physical dosimetric measurements (Section 4). For a research facility, testing new compounds and/or new clinical indications and/or developing new drug administration schedules much more sophisticated preclinical capability is required. An example of a possible preclinical test plan could be the following:

(a)   Plan for clinical trials for BNCT;
(b)   Based on this, build preclinical evaluation plans to discuss with the regulation authorities. For cell based and xenograft experiments, it is preferred to include specific cancer types for planned clinical trials;
(c)   Determine the neutron beam's RBE (hydrogen dose) specific to each type of AB-BNCT system, if not already established;
(d)   Safety evaluation study:
    (i)   Cytotoxicity can be measured with cancer and normal cells in the absence and presence of boron compounds at multiple concentrations. The effects can be compared with those of γ or X ray irradiation to calculate RBE values;
    (ii)   Genotoxicity can be measured employing the frequencies of micronuclei formation or sister chromatid exchanges as the endpoints, using cancer and normal cell lines;
    (iii)   Assessment of in vivo effects using mice, such as survival, skin lesions, mucosal lesions, haematological status, and intestinal crypt regeneration [545] after whole body irradiation or local irradiation in the absence and presence of boron compounds. The effects can be compared with those of γ or X ray irradiation to calculate RBE values. The observation periods can be set to investigate both short and long term effects;
    (iv)   If medium or larger size animals can be used, especially for newly developed boron compounds, they may provide further useful information in terms of safety evaluation;
(e)   Efficacy evaluation study:
    (i)   Cancer cell killing can be measured with clonogenic cell survival assay with several cancer cell lines in the absence and presence of boron compounds;
    (ii)   Effects on tumour growth can be evaluated using xenograft models of mice with target cancer types. Tumour growth delay can be measured in comparison with local X ray irradiation for xenograft models. It ought to be noted that dose limitation for subcutaneous xenograft models may depend on the skin doses calculated with the predetermined N/B (Normal tissue/Blood) boron concentration ratio and pharmacokinetic values;

(f)   The final evaluation with preclinical tests ought to be carried out using drugs manufactured under Good Manufacturing Practice [20] and conducted in laboratories operating under near-Good Laboratory Practice [142–143]. In hospitals, after irradiation with the BNCT system, the long-term observation of mice would be often difficult. Therefore, it becomes necessary to check their activation levels and transfer the irradiated animals to outside facilities. Consultation with regulation authorities for full planning is unavoidable.

9.7.3   The activation of cells and animals

For animal models, activation levels could be a concern and they ought to be lower than that prescribed in the IAEA's GSR Part 3 [149], as regulated by each facility, when the animals are to be transferred to outside facilities for biological tests. The induced radionuclides in mice after irradiation with an AB-BNCT system with a Li target were ^{24}Na, ^{38}Cl ,^{42}K, ^{56}Mn, ^{80m}Br and ^{82}Br [32]. The activation levels of mice for major radionuclides, such as ^{24}Na and ^{38}Cl, ought to be lower than the regulation limit after neutron irradiation with a BNCT system [32, 546]. The induced radioactivity and the saturated radioactivity can be expressed using $Bq \cdot g^{-1} \cdot mC^{-1}$ and $Bq \cdot g^{-1} \cdot A^{-1}$, respectively. The induced radioactivity can be therefore estimated considering the mass of the samples, and the proton (deuteron) current in each AB-BNCT system [32].

9.7.4   Determination of biological effectiveness and boron concentration ratios

RBEs and CBEs are important factors in preclinical studies, as described in Section 10.3. The N/B and T/B boron concentration ratios ought to be carefully set for preclinical tests depending on the animal and tumour model, based on pilot studies. The knowledge of pharmacokinetic parameters for each boron compound is necessary for delivered dose calculations. It is also important to estimate the ranges of uncertainties of the dose and concentration of boron compounds.

9.7.5   Radiation field size and effect on 'beam' depth direction

The range of collimator diameters for AB-BNCT is 8–25 cm [32, 38, 50, 128]. The most commonly applied collimator diameters lie in the range 12–14 cm (see Table 27 in Appendix I). Although physical doses are determined using phantoms (see Section 4.3), it is also necessary to evaluate the biological effects in an entire irradiation field and the 'out-of-field' area. This may be achieved in several ways. For example, cancer or normal cells can be set in containers to measure cell survival as described above, or mice can be distributed to measure particular biological effects in terms of endpoints such as survival and haematological status. To measure the biological effects along the beam depth direction using mice, the differences in body thickness between mice and humans ought to be considered. Annex XVIII provides extensive discussion on the experimental setup required for such studies. An example of an experimental method to overcome this difference could be to set several mice in an array and irradiate them to observe short or long term effects [544]. For short term evaluation, it may be useful to use the intestinal crypt regeneration ability of small intestine at day 3.5, which has been widely used to characterize biological effects of photon and particle beams, such as protons, carbon ions, as well as neutrons [545], along with haematological testing.

---

[20]   https://www.who.int/teams/health-product-and-policy-standards/standards-and-specifications/norms-and-standards-for-pharmaceuticals/guidelines/production

9.7.6 Preclinical tests for new boron carrier drugs

In the case of preclinical tests for new boron carrier drugs, it may be better to perform safety and efficacy evaluation using middle-sized animal models as well as small rodents. The use of middle-sized animal models is only compulsory in certain facilities. On the other hand, new experimental models that might replace middle-sized or animal models are expected to be developed due to ethical regulations. Pharmacokinetic studies ought to be carried out and administration routes and timing optimized for each boron compound. Drugs ought to be prepared under Good Manufacturing Practice and used in preclinical tests prior to clinical trials, and laboratory practice close to Good Laboratory Practice level may be requested by regulatory authorities [142–143].

9.7.7 Clinical biological studies

The determination of CBE factors for normal tissues for each boron carrier drug is important for expanding the application of BNCT to various types of cancers. The CBE factors for muscle, intestine, kidney, and liver have yet to be determined [3]. It will also be useful to carry out clinical biological studies during clinical trials and treatment with AB-BNCT systems. Comprehensive omics studies, including transcriptome, proteome, and metabolome studies will be beneficial to find various types of therapeutic and predictive biomarkers in acute response and long-term effects. These data will be the basis for application to different cancer types, multiple-port irradiation, and new boron carrier drugs.

9.7.8 For future development of neutron capture therapy

Two types of preclinical studies are needed:

- Specific preclinical tests performed in pre-determined limited conditions in preparation for clinical trials to prove safety and efficacy;
- A wide scope of biological investigations, complementary to preclinical tests and clinical trials that employ various experimental systems for individual AB-BNCT facilities. These are aimed at producing knowledge that will contribute to the advancement of BNCT.

Much progress in AB-BNCT/AB-NCT can be expected in the near future, including approval of various types of AB-BNCT systems, new cancer type applications, new boron carrier drug approvals, and other NCTs such as Gd-BNCT and NCT for treatment of other diseases such as rheumatoid arthritis [547]. Biological studies of AB-BNCT combined with precise dose reporting ought to support these developments in AB-BNCT/AB-NCT.

9.8 COMBINED THERAPIES

Combined therapies may be the answer to the treatment of heterogeneous tumours that are refractory to single treatment modalities. The fact that it is possible to achieve a high dose gradient between tumour and normal tissue with BNCT would allow for reirradiation for BNCT at full dose [8, 548] and for combined therapies that involve BNCT without exceeding normal tissue radiotolerance. Different approaches have been used to explore the benefit of combining the advantages of different therapeutic modalities. Results are scarce to date, but examples of the approaches under consideration are outlined below:

- It has been demonstrated that BNCT + intensity-modulated radiation therapy may improve treatment homogeneity and conformity, as well as possible local tumour control without increasing dose levels to normal tissues, especially for tumour volumes > 100 cm^3 [549], and this approach has been used by others [237];

- In one study, newly diagnosed glioblastoma patients were treated with BNCT and fractionated external beam XRT [218] to diminish the possibility of tumour recurrence, based on experimental animal data showing that a significant therapeutic gain could be obtained when BNCT was combined with an X ray boost. Prior to this study, BNCT had not been followed clinically by a photon boost until the time of tumour progression. It has been reported that an X ray boost after BNCT could significantly enhance survival time in an experimental brain tumour model [550]. BNCT followed by intensity modulated chemoradiotherapy has been used as a primary treatment of large head and neck cancer with intracranial involvement [551]. Four patients with inoperable tumours were treated with BNCT, followed by chemotherapy and photon irradiation [415]. A comparison of BNCT to reirradiation with conventional radiotherapy in a randomized study is warranted and it has been proposed that potential approaches to improve BNCT efficacy would include administration of BNCT with systemic cancer therapy or in sequence with conventional radiotherapy [408]. In patients with newly diagnosed GBMs, BNCT mediated by BPA and BSH resulted in favourable responses with/without an XRT boost, particularly in high risk groups [552]. Similarly improved survival by combining BNCT with a photon boost has been reported [553]. A multicentre, phase-II Japanese clinical trial to evaluate BNCT in combination with temozolomide and an X ray irradiation therapy boost for newly diagnosed GBM has been completed [552];

- The combination of BNCT and immunotherapy would enhance therapeutic efficacy without enhancing toxicity. Radioimmunotherapy provides a systemic approach whereas the effect of classical therapy is circumscribed to the tumour. Even when given alone, radiation can act as an immunostimulator by activating antigen presenting cells and by increasing T-cell infiltration. However, radiation can also induce suppression of the immune system as a result of irradiation of the draining lymph nodes and effector cell inhibition. Particle therapy would be more effective than X rays in combination with immunotherapy. The physical advantages of particles over X rays reduce damage to immune cells that are necessary for an effective immune response. Furthermore, densely ionizing radiation would have biological advantages in terms of cell death pathways and release of cytokine mediators of inflammation that contribute to an immune response [554–555]. Within this context, BNCT, based on high LET particle therapy, would be a good partner for immunotherapy. Since BNCT would spare normal cells, and more specifically immune effector cells at the site of the tumour, BNCT and systemic immunotherapy would be mutually complementary and potentially synergistic [383, 415]. Recent advances in immunotherapeutic approaches (e.g., Ref. [556]) to treat metastatic melanoma in combination with BNCT of the primary tumour looks a promising approach for this malignancy, which has a high propensity to metastasize. BNCT for the treatment of extramammary Paget's disease, combined with anti-PD1 immunotherapy may improve the therapeutic outcome significantly [415]. In addition, the combination of immunotherapy and BNCT might enhance the abscopal effect of treatment [472]. More recently, immunomodulatory effects of BNCT that contribute to tumour growth inhibition have been described [473];

- Polylactic and polyglycolic acid nanoparticles have been used for delivery of an amphiphilic gadolinium complex and a boron-curcumin complex (RbCur) simultaneously into tumour cells to perform boron and gadolinium neutron capture therapy combined with the antiproliferative effects of curcumin [348]. The synergic action of neutron treatment and curcumin cytotoxicity resulted in a significant therapeutic improvement;

- A combination of ^{252}Cf brachytherapy, BNCT and an intracavitary moderator balloon catheter applied to brain tumour and infiltrations has been proposed [557]. This would represent a new modality to selectively combat brain tumour infiltrations and metastasis;

- BNCT followed by Bevacizumab treatments for radiation necrosis or symptomatic pseudoprogression improved the clinical symptoms and might prolong the survival of recurrent malignant glioma patients. Oedema in brain necrosis would be caused by overexpression of vascular endothelial growth factor in reactive astrocytes. Bevacizumab, an anti-vascular endothelial growth factor antibody, would serve to treat and/or prevent this unwanted effect of BNCT [558–560]. A clinical trial for head and neck cancer combining BNCT with the EGFR-targeting monoclonal antibody Cetuximab [415] has been carried out [408]. The Food and Drug Administration has approved Cetuximab for use in the treatment of recurrent EGFR (+) SCCs of the head and neck. Because these tumours strongly express EGFR, an EGFR targeting agent, such as boronated Cetuximab, was used in combination with BPA for BNCT of EGFR-positive tumours with encouraging results [561]. The pharmacology is discussed in Section 7.5;

- The synthesis and biological evaluation of new BSH-conjugated chlorin derivatives as agents for both photodynamic therapy and BNCT of cancer has been reported [562]. A carboranyl-containing chlorin of high boron content has been used for this combined therapy approach [563]. BNCT and photodynamic therapy could one day be combined to treat malignant tumours, if a boronated porphyrin can be successfully developed and validated for both these purposes [564];

- Although BNCT-enhanced fast neutron therapy has attracted research interest, it has not been subjected to formal clinical trials to date. As a small proportion of the neutrons in fast-neutron therapy will thermalize within the irradiation volume, it may be possible to create an incremental absorbed dose from neutron capture selectively in the target volume. This incremental dose may in certain cases improve the tumour control probability [565];

- Neutron capture enhanced particle therapy has been proposed, which may enhance the radiation dose selectively to a tumour during proton and carbon ion therapy by capturing thermal neutrons produced inside the treatment volume during irradiation [566]. The $p + ^{11}B \rightarrow 3\alpha$ reaction, which generates high-LET α particles with a clinical proton beam, has been exploited [567]. To maximize the reaction rate, BSH was used with natural boron content rather than the ^{10}B-enriched BSH employed as a carrier for BNCT. This strategy would combine proton therapy's ballistic precision with the higher RBE of α particles and would lead to the possibility of significantly boosting the direct proton dose in proton therapy via introduction of ^{11}B into the target tissue;

- The thermal neutron component induced during a traditional high-energy radiotherapy treatment could produce a localized BNCT effect in a patient previously infused with BPA [568]. A localized therapeutic dose enhancement would be achieved, corresponding to 4% or more of photon dose, depending on tumour features. This BNCT additional dose could act as a localized radiosensitizer and improve the therapeutic outcome of radiotherapy.

9.9 FUTURE PROSPECTS

Applied and translational studies are pivotal to the advancement of BNCT. In particular, studies employing AB-BNCT are still scarce and very necessary to understand potential differences and similarities between the radiobiology involved in reactor and accelerator based BNCT.

The development of standards in BNCT radiobiology is an unmet challenge that ought to be addressed, despite the many unavoidable limitations that exist to establish standards that are informative within the context of biological variability. Approaches to design radiobiological standards/tests/systems for potential use in comparative studies and in screening for potential therapeutic success ought to be explored within the context of potential benefits and constraints.

A database of radiobiological data hosted by the International Society of Neutron Capture Therapy[21] and available to the community would be a valuable contribution.

---

[21] https://isnct.net/

10 METHODS AND MODELS OF DOSE CALCULATION

Boron neutron capture therapy is characterized by the interaction of a mixed radiation field with biological tissues. Thermal neutrons are captured by ^{10}B, producing an α particle and a Li ion (Fig. 15), and by other elements in tissue, among which the most relevant is ^{14}N, producing a proton (Table 7). These charged particles have moderate-to-high ionization density values and short ranges in tissue (Table 20). Epithermal neutrons thermalize in tissue mainly through elastic scattering from hydrogen (moderation), each losing on average half of its energy per collision and producing a recoil proton that deposits dose. Finally, neutrons are captured by ^{1}H in a reaction that produces low-LET radiation: a 2.224 MeV γ ray. Another source of low-LET radiation is the γ rays present in the beam or produced in neutron interactions with surrounding materials: this structural component is unavoidable but kept as low as possible during design of the beam and collimator system (see Section 4). Dosimetry is thus characterized by the calculation of the absorbed dose due to the charged particles heavier than electrons (depositing all their energy locally) and the electrons set in motion by the sparsely ionizing photons. With sophisticated transport calculation techniques, the different contributions are tracked separately.

TABLE 20. PROPERTIES OF CHARGED PARTICLES RELEVANT TO DOSE AND DOSE DISTRIBUTION PRODUCED BY NEUTRON CAPTURE REACTIONS [569–570]

	^{10}B(n,α)^{7}Li		^{14}N(n,p)^{14}C
	α	^{7}Li	p
LET* (keV/μm)	164	151	44
Range in tissue (μm)	9	5	11

* Averaged over track intersection segments within a cell

Biological effects are the result of the action of ionizing radiation in living systems. These effects are directly related to absorbed dose. Therefore, a deep understanding of the spatial scale and geometry of the problem and of the methods to correctly calculate dose is essential to evaluate and optimize the clinical use of BNCT. Absorbed doses can be calculated using different strategies according to the physical situation, for example, assuming charged particle or electronic equilibrium (see Section 10.1.1.4), which makes calculation more straightforward. In preclinical models such as cell cultures or small animals, or in the case of calculation of dose to skin for patients, the equilibrium hypothesis may not be correct. It is thus necessary to apply more detailed simulations. Section 10.2 deals with this issue: the need to set up the correct absorbed dose calculation in different scenarios. Two approaches have been described: macroscopic (Section 10.2.1) and microscopic (Section 10.2.2), separating the issue of dose calculation to different spatial scales and starting from different perspectives: the whole sample/tissue/patient or the single cell.

Each of the BNCT radiation components has different biological effects: high and low LET components produce different ionization densities. The high-LET radiation (Table 20), densely ionizing, directly damages the DNA. Low-LET radiation, sparsely ionizing, mainly causes indirect damage by forming free radicals. The fact that there is not a unique relationship between absorbed dose and induced biological effects, prompts the need for translation of BNCT doses into a reference radiation dose capable of predicting clinical effects. To this end, the clinical experience with photon therapy is used as a reference. Radiobiological experiments with cell cultures or animals irradiated with BNCT, with neutrons only, and with photons, provide the fundamental information for models which aim to translate the BNCT absorbed dose into the dose of the reference radiation producing the same effect. With a BNCT dose expressed in photon equivalent units, i.e., with a photon isoeffective dose, medical doctors can prescribe doses and predict the outcome of the therapy according to the clinical experience gained with photon radiotherapy. Different strategies conceived to translate BNCT dose into photon equivalent units are described in Section 10.3.4, highlighting the range of validity of the traditional and modern models and the equivalent dose unit.

To deliver a safe and effective BNCT treatment, it is necessary to calculate absorbed dose in the most exact and precise way and to know how to relate this physical quantity to its effects in tumour and in normal tissues. The key ingredients are correct dose calculation, representative radiobiological data, and reliable models to translate mixed-field absorbed dose into photon-equivalent units. Advice presented in Sections 10.2–10.3 is thus particularly important: incorrect assumptions in dose calculation and incorrect models may propagate significant errors in the determination of isoeffective dose in patients, leading to a bias in evaluating the relationship between clinical outcome and calculated dose.

10.1 GENERAL CONCEPTS

The following subsections outline some of the fundamental concepts and inputs that are required for dose calculation.

10.1.1 KERMA and absorbed dose

10.1.1.1 KERMA

KERMA is an acronym for Kinetic Energy Released per unit MAss, defined as the sum of the initial kinetic energies of all the charged particles liberated by uncharged ionizing radiation (photons and neutrons) in a volume of matter, divided by its mass (Gy) [571].

10.1.1.2 Absorbed dose

'Absorbed dose' refers to the energy imparted by ionizing radiation per unit mass of irradiated material measured by the SI unit gray (Gy) [571].

10.1.1.3 Equilibrium states of radiation

General definitions and solutions for absorbed dose require complete knowledge of the radiation fields involved at all points. This requirement, however, can be relaxed by assuming the existence of equilibrium conditions for the radiation fields. Under these hypotheses, detailed knowledge of spatial, directional or energy dependence can be omitted, depending on the type of equilibrium that is assumed. An equilibrium state is defined as one in which the balance between the radiant energy of the incoming and outgoing particles in a given volume is zero.

10.1.1.4 Charged particle equilibrium

Given the importance of this case in dosimetry, a partial equilibrium condition is defined, known as charged particle equilibrium (CPE). This means that the balance of the radiant energy of the incoming and outgoing charged particles, in an infinitesimal volume, is zero [571]. CPE is of particular importance when a material medium is irradiated by an external beam of uncharged particles (photons, neutrons).

10.1.1.5   KERMA and dose in boron neutron capture therapy

The knowledge of the mean energy imparted to matter within a certain volume, i.e., the absorbed dose, is critical for all radiobiological calculations that seek to understand the relationship to the observed effect. Simplifications like those described above need to be carefully considered and reported regarding their adoption, especially in BNCT. In BNCT, absorbed dose components are typically referred to by common names; e.g., boron dose, thermal neutron dose, fast neutron dose, and photon dose (see Table 7 and later discussion in Sections 10.3.3.1). Structural or induced photons deserve special attention since the KERMA approximation is not applicable at air–solid interfaces in the first millimetres (the electron build-up layer). On the other hand, Bremsstrahlung photons, generated from electron interactions, deposit their energy elsewhere. However, this contribution is unimportant for light ions and low energy electrons in low Z materials (i.e., living tissue), and it is usually already considered in the mass absorption coefficients. As mentioned above, in the case of interfaces such as skin, assessments based on KERMA are not reliable and a detailed dose calculation with full particle transport needs to be considered. Section 10.2 explores these considerations more deeply for typical situation where BNCT absorbed dose needs to be calculated.

Other factors that require consideration for the calculation of the absorbed dose are the accuracy of the cross-section data used in Monte Carlo simulations, and the spatial scale of interest. These issues are discussed briefly in the next section.

10.1.2 Nuclear and atomic data

Reliable data libraries to be used in the Monte Carlo transport are necessary to estimate the components of the mixed field responsible of dose deposition in tissues. For neutron transport, the quantity required for estimating the role of each element is the isotopic macroscopic cross section, defined as:

$$\Sigma_i = n_i \sigma_i = \frac{\varrho N_A x_i}{A_i} \sigma_i \tag{16}$$

where ϱ is the material density, N_A is Avogadro's number, x_i the mass proportion of the isotope, A_i its atomic mass, and σ_i the total interaction cross section. This gives the probability per unit length that a neutron of energy E interacts with a nucleus of the isotope i. These values are dependent on the neutron energy. For tissues in particular, macroscopic cross sections depend on the material composition. Table 21 gives the probability of interaction for a representative tissue for BNCT (adult brain with the composition defined by the ICRU Report 46 [572]). It shows that the probability is highest for ^1H, followed by ^{16}O, with much lower probabilities ^{12}C, ^{14}N, and negligibly low values for the minor elements and isotopes (illustrated in Fig. 60). The following interactions may take place with some of the nuclei in tissue:

- Elastic scattering (including the moderating effect of light nuclei);
- Inelastic scattering (if the neutron has enough energy to excite the target nucleus);
- Neutron capture.

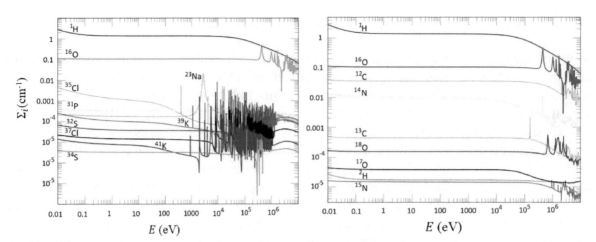

FIG. 60. Macroscopic cross sections for the components of brain tissue as a function of neutron energy. Data are taken from cross sections of the database ENDF/B-VIII.0[22] (courtesy of I. Porras, University of Granada).

TABLE 21. ELEMENTAL AND ISOTOPIC COMPOSITION OF ADULT BRAIN TISSUE (DENSITY 1.04 g/cm^3)

	A_i (g/mol)	Element mass fraction	Isotopic fraction	x_i	$\dfrac{x_i}{A_i}$ (mol/g)	$\dfrac{\varrho N_A x_i}{A_i}$ (at/(b·cm))
^{1}H	1.008	0.107	0.99985	1.070×10^{-1}	1.062×10^{-1}	6.648×10^{-2}
^{2}H	2.014	0.107	0.00015	1.605×10^{-5}	7.969×10^{-6}	4.991×10^{-6}
^{12}C	12.000	0.145	0.98900	1.434×10^{-1}	1.195×10^{-2}	7.484×10^{-3}
^{13}C	13.003	0.145	0.01100	1.595×10^{-3}	1.227×10^{-4}	7.682×10^{-5}
^{14}N	14.003	0.022	0.99634	2.192×10^{-2}	1.565×10^{-3}	9.803×10^{-4}
^{15}N	15.000	0.022	0.00366	8.052×10^{-5}	5.368×10^{-6}	3.362×10^{-6}
^{16}O	15.995	0.712	0.99762	7.103×10^{-1}	4.441×10^{-2}	2.781×10^{-2}
^{17}O	16.999	0.712	0.00038	2.706×10^{-4}	1.592×10^{-5}	9.968×10^{-6}
^{18}O	17.999	0.712	0.00200	1.424×10^{-3}	7.911×10^{-5}	4.955×10^{-5}
^{23}Na	22.990	0.002	1.00000	2.000×10^{-3}	8.700×10^{-5}	5.448×10^{-5}
^{31}P	30.974	0.004	1.00000	4.000×10^{-3}	1.291×10^{-4}	8.088×10^{-5}
^{32}S	31.972	0.002	0.95020	1.900×10^{-3}	5.944×10^{-5}	3.723×10^{-5}
^{33}S	32.971	0.002	0.00750	1.500×10^{-5}	4.549×10^{-7}	2.849×10^{-7}
^{34}S	33.968	0.002	0.04210	8.420×10^{-5}	2.479×10^{-6}	1.552×10^{-6}
^{36}S	35.967	0.002	0.00020	4.000×10^{-7}	1.112×10^{-8}	6.965×10^{-9}
^{35}Cl	34.969	0.003	0.75770	2.273×10^{-3}	6.500×10^{-5}	4.071×10^{-5}
^{37}Cl	36.966	0.003	0.24230	7.269×10^{-4}	1.966×10^{-5}	1.232×10^{-5}
^{39}K	38.964	0.003	0.93260	2.798×10^{-3}	7.181×10^{-5}	4.497×10^{-5}
^{40}K	39.964	0.003	0.00012	3.600×10^{-7}	9.008×10^{-9}	5.642×10^{-9}
^{41}K	40.962	0.003	0.06230	1.869×10^{-4}	4.563×10^{-6}	2.858×10^{-6}

The probability of each is given by the ratio between the cross section of the process and the total interaction cross section (the sum of all). With ^{1}H, only elastic scattering and (n,γ) neutron capture reactions are possible. In the latter case, a 2.224 MeV γ ray is produced. The interaction of neutrons in tissue are well described by Monte Carlo codes. All evaluated data bases show a perfect agreement in case of hydrogen: ENDF, since the version ENDF/B-V.2 (1994) to the most recent ENDF/B-VIII.0 (2018)[23] [573], JENDL data, from JENDL-3.2 (1994)[24] to JENDL-4.0 (2012) [574] and JEFF data, from 3.0 to 3.3 (2017)[25]. The agreement between all the evaluations and experimental data is also good. The capture process involving epithermal

---

[22] ENDF: https://www-nds.iaea.org/exfor/endf.htm

[23] ENDF/B-VIII.0: https://www.nndc.bnl.gov/endf/b8.0/index.html

[24] JENDL-3.2: https://inis.iaea.org/collection/NCLCollectionStore/_Public/26/038/26038454.pdf

[25] JEFF-3.3: https://www.oecd-nea.org/dbdata/jeff/jeff33/

neutrons is about four orders of magnitude less probable than elastic scattering; therefore, neutrons lose energy primarily by elastic collisions with hydrogen. When thermalized, they will still produce collisions until they are captured (by H, N or other minor elements as Cl) or escape from the body.

With respect to oxygen, the agreement is very good between different data bases (and the experimental values, except for energies above the first resonance). The dominating process is elastic scattering, and capture processes are much less important than for hydrogen. However, as can be seen in Fig. 61 the (n,α) process may play some role if neutrons have energy greater than 2.35 MeV. As the cross section is lower than 1 barn, this capture will not produce a significant perturbation of the neutron flux, but it may contribute to the dose. Some discrepancies are found between different data sets. In cases where the neutron spectrum has a high energy tail, the contribution of this process to the dose ought to be at least estimated.

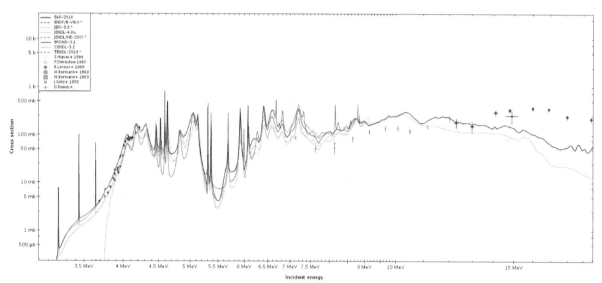

FIG. 61. Variation between cross section datasets for the $^{16}O(n,α)$ reaction. Data prepared from the JANIS database[26] (courtesy of I. Porras, University of Granada).

During the thermalization process, as the neutron energy falls the capture process (for which the cross sections usually present a $1/v$ behaviour) becomes increasingly important. The most likely capture process is radiative capture for most of the elements in organic tissues (except for ^{14}N and ^{10}B). The probability per unit length of a neutron producing secondary photons when interacting with nucleus i is given by:

$$\Sigma_{i,\gamma} = n_i\, \sigma_{i,\gamma} = \frac{\varrho N_A x_i}{A_i}\, \sigma_{i,\gamma} \qquad (17)$$

where $\sigma_{i,\gamma}$ is the (n,γ) cross section. The energy transferred to secondary photons per unit length can be evaluated from the product $\Sigma_{i,\gamma}Q$, where Q is the reaction Q-value (Section 2.1.2). These are evaluated for thermal neutrons (for which capture reactions are most likely) in Table 22 for the most relevant elements.

Although neutron capture by ^1H is the most important contributor to secondary photons, the role of ^{35}Cl ought to be considered. The impact of the ^{35}Cl(n,γ) reaction has been shown to contribute to the dose delivered in normal brain at levels up to 12% of the total normal tissue dose (when ^{10}B is absent) [394]. Data on this cross section present some uncertainties, as there

[26] JANIS: https://www.oecd-nea.org/jcms/pl_39910/janis

are discrepancies between experimental data and evaluations and the $1/v$ behaviour is not confirmed by experimental data. For this reason, a measurement of this reaction was proposed at the neutron time of flight (n_TOF) facility at CERN, and the results are currently under analysis [575].

TABLE 22. QUANTITIES TO CALCULATE THE MACROSCOPIC CROSS SECTION OF RADIATIVE CAPTURE FOR THE PRINCIPAL ELEMENTS IN TISSUES

i	$\dfrac{\varrho N_A x_i}{A_i}$ (at/(b·cm))	$\sigma_{i,\gamma}(0.025\ \text{eV})$ (b)	$\Sigma_{i,\gamma}(0.025\ \text{eV})$ (cm^{-1})	Q (MeV)	$\Sigma_{i,\gamma}Q$ (MeV/cm)	Percentage
^{1}H	6.64825×10^{-2}	0.3326000	2.21121×10^{-2}	2.224	4.9177×10^{-2}	73.844
^{12}C	7.48440×10^{-3}	0.0038600	2.88900×10^{-5}	4.946	1.4289×10^{-4}	0.214
^{14}N	9.8035×10^{-4}	0.0749913	7.35180×10^{-5}	10.833	7.9642×10^{-4}	1.196
^{16}O	2.78123×10^{-2}	0.0001700	4.72810×10^{-6}	4.143	1.9589×10^{-5}	0.029
^{23}Na	5.44841×10^{-5}	0.5280000	2.87680×10^{-5}	6.960	2.0022×10^{-4}	0.300
^{31}P	8.08798×10^{-5}	0.1693610	1.36980×10^{-5}	7.935	1.0869×10^{-4}	0.163
^{32}S	3.72261×10^{-5}	0.5282150	1.96630×10^{-5}	8.641	1.6991×10^{-4}	0.255
^{35}Cl	4.07109×10^{-5}	43.6122000	1.77549×10^{-3}	8.580	1.5234×10^{-2}	22.875
^{39}K	4.49708×10^{-5}	2.1274200	9.56720×10^{-5}	7.800	7.4624×10^{-4}	1.120

The (n,γ) reaction with ^{14}N lacks experimental data in the EXFOR database[27] (only three points which do not match the evaluations except one at thermal energy). Although this process accounts just for 1.2% of the total energy released by photons, a confirmation of the $1/v$ behaviour would be desirable by a dedicated measurement.

Different evaluated data sets are in fair agreement for the ^{14}N(n,p) reaction, except near the region of resonances, where JENDL-4.0 presents some differences with respect to ENDF/B-VIII and JEFF-3.3. However, in this region this reaction has less importance with respect to the total KERMA than in the thermal region. The great importance of this reaction for low-energy neutrons and the mentioned discrepancies between the experimental data and evaluations (Fig. 62) motivated a new measurement at the n_TOF facility[28] at CERN, Switzerland, which is currently under analysis [576]. The reduction of this uncertainty will improve the accuracy of the neutron dose determination in normal tissues.

In the same experiment, the cross section of ^{35}Cl(n,p) was also measured, which may be relevant at certain energies because of its strong resonances. The first, at 398 keV, makes this process dominate the brain KERMA factor, although its effect in a continuous epithermal beam is expected to be small as the resonance is very narrow. Nevertheless, it is interesting to measure these resonances as there is some discrepancy between evaluations and experimental data.

Finally, the boron neutron capture reaction (Fig. 15) is the main one responsible for dose deposition in BNCT, especially in tumour when ^{10}B is preferentially concentrated within cancer cells. This is a well-known reaction for which all cross section datasets coincide, in good agreement with experimental data. It is considered a standard for neutron cross section measurements.

---

[27] EXFOR: https://www-nds.iaea.org/exfor/
[28] https://home.cern/science/experiments/n_tof

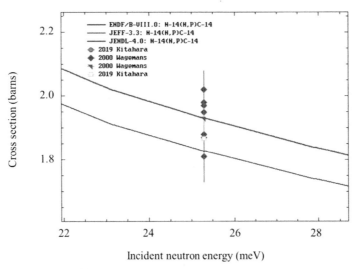

FIG. 62. Cross section data for the $^{14}N(n,p)$ reaction near thermal neutron energies showing the variation among three databases and four experimental measurements (courtesy of I. Porras, University of Granada).

Even if discrepancies are found in processes whose contribution to the total absorbed dose is low, the use of different nuclear data may result in different calculated absorbed doses, especially in the fast energy range. For this reason, it is good practice to take this aspect into consideration, when precise dose calculation is needed. These considerations show that the choice of the nuclear data may play a role in the result of dosimetry and the evaluation of the data table to be used is a very important issue. The same holds for the choice of the Monte Carlo code, because the numerical treatment (tables, interpolations, cut-offs, score calculation) and the physical models implemented can lead to differences between results of the same simulated system. For this reason, it is meaningful to benchmark different codes and to gain deeper insights into the differences between the model and the physical reality according to the simulation tool selected.

10.1.3 Macroscopic and microscopic scales

The calculation of the boron dose, D_B, calls for additional considerations regarding the ranges of the emitted α and ^7Li particles because they are comparable to the cell size (Table 20). This fact, together with the nature of the boron compounds and their cellular localization, requires distinction to be made between spatial scales of interest.

It has been shown that a large variation of boron microdistribution in tumour and normal cells is possible, either extracellular, intracellular, uniform, attached to the cell membrane, located in the cell cytoplasm, etc. and all these situations can coexist in different tissues (and change in time) during irradiation. Therefore, KERMA factors as a surrogate quantity for dose are not always applicable since a uniform distribution assumption is needed at a particular length-scale. If the scale of interest is orders of magnitude greater than the cell size (e.g., clinical treatment volume, organs at risk, etc., see Section 12) and CPE conditions are fulfilled in a given volume of interest, then an average boron concentration suffices for the calculation and KERMA factors can be used. If, however, the scale of interest is comparable to the particle ranges (e.g., monolayer of cells in culture or tumour cells scattered within normal tissue) then a detailed dose calculation (and eventually a microdosimetric approach, see Section 10.2.2), might be necessary.

The same can be said regarding nitrogen (or thermal neutron) dose, D_N. Although not as relevant as boron dose, D_B, nitrogen may be non-uniformly distributed in multiple situations (e.g., cell cultures, tumour cells interspersed in normal tissue with higher nitrogen content than normal tissues, etc.). In such situations involve the same considerations regarding the scale of interest.

10.2 METHODS FOR CALCULATING ABSORBED DOSES

The following sections discuss the methods for calculating doses over different length scales. In particular, the dose calculation is intended to be simulated with Monte Carlo transport codes, for which one of the assumptions is that the particle density in the system is sufficiently low to neglect interactions between them. In some cases, dose calculation could be performed using other tools such as analytical codes, which are not considered here.

10.2.1 Macroscopic calculation

10.2.1.1 Dose calculation in preclinical models

Radiobiological models are used to study BNCT effects and to compare them with those of conventional photon therapy. The dose–effect relationship in these models is used to convert absorbed dose into photon-equivalent units, as explained below in Section 10.3. In any such strategy, the robustness of the translation of BNCT dose into suitable units for predicting the clinical outcome depends on the validity of the dose–effect curves that are generated.

The preclinical models typically used are monolayer cell cultures (mainly tumour cells) and small animals (rats, mice). In Section 9, the types of models and the biological assays are described in detail. Here, the focus is on dose calculation to construct the dose–effect curves.

One typical radiobiological experiment involves the assessment of cell survival as a function of dose, employing monolayer cell cultures with thickness of around 10 µm. Typically, these are tumour cells: survival is taken as an indicator of the capacity of BNCT to kill a tumour, with the assumption that tumour control directly depends on the number of inactivated cells. In these experiments, cells are cultured as adherent in flasks to a certain point, treated with boron, and then irradiated in a thermal neutron 'beam' (or field). Dose–survival curves are built for photon, neutron-only, and neutron + boron irradiations. In BNCT, separate dose components are considered because they have different spatial distributions and different biological consequences. Absorbed dose needs to be calculated with Monte Carlo transport codes, and the components that contribute, as summarized in Table 7, are:

- High-LET charged particles generated by neutron capture by boron (when present): D_B;
- Intermediate LET charged particles produced by neutron capture by nitrogen and by neutron scattering on hydrogen: D_N and D_H;
- Low-LET photons coming from the irradiation beam and environment as well as from neutron capture by hydrogen: D_γ.

As explained above, the products of the neutron interactions have different LETs and ranges, and equilibrium may or may not be fulfilled in any specific case.

If equilibrium holds, dose is equal to KERMA. In this case, calculation is more straightforward, because KERMA is obtained by neutron/photon fluence multiplied by fluence-to-KERMA conversion factors which are specified in the input file for Monte Carlo simulation. In this situation, the first piece of advice is to choose suitable fluence-to-KERMA conversion factors tables. Reference [577] highlights the importance of including tables comprising factors for low energy neutrons. In fact, if the lower limit of the KERMA-energy table is too high, fluence of lower energy will be converted into KERMA using the lowest available factor. This can cause incorrect estimation of the thermal component of the dose, which is very important in the case of BNCT, because the interactions of neutrons with nitrogen and boron have larger cross sections at lower energies. In Ref. [577], conversion tables derived from ICRU Report 63 [578], which extended those from ICRU Report 46 [572] to lower energies, are used and the difference in dose calculations are shown.

Another way of calculating absorbed dose in case of equilibrium is to calculate the reaction rate of the event of interest (for example, neutron capture by nitrogen) and multiply it by the Q value of the reaction, with the assumption that all the energy of the produced proton and of the recoil nucleus is deposited locally. In both cases, only the primary radiation is transported, and the simulation can reach statistical convergence in a short time.

The accuracy of the equilibrium assumption has been tested for short ranged, high LET, charged particles in monolayer cell cultures and found to be inaccurate. In fact, for a cell layer of the order of 10 μm, KERMA overestimates dose by about 10% for the boron component and by about 40% for nitrogen [579]. Thus, the calculation needs to include transport of all the secondary particles. This involves more sophisticated computational strategies. In fact, generating and transporting the secondary particles generated from the primary neutron source (i.e., simulating all the reactions occurring in the volume of interest), would be highly inefficient, and statistical convergence may be too difficult to obtain. Moreover, as described above, cross sections of charged particle production need to be well validated. Hence, simulation might be divided into two steps: (i) characterizing the primary neutron/γ field around the volume of interest, and (ii) using this field to generate charged particles. For high-LET components, one strategy is to sample α particles, ^{7}Li nuclei, and protons in the monolayer of cells, transport them and calculate their dose contributions. If nitrogen and boron are also present elsewhere (for example in the culture medium), reaction products need to be generated there too. To be precise, ^{14}C also needs to be sampled and transported, but, given its very short range, that part of the Q value transported by the recoiling carbon nucleus can be assumed to be deposited locally. The total dose deposited then needs to be normalized by the reaction rate of each previously calculated interaction.

For the low-LET component due to photons, electron transport is always necessary, unless the experiment is specially designed to avoid this requirement. Electrons are generated in cells but also in the surrounding materials, depending on the photon spectrum present in the facility. To define the portion of the geometry which contributes to electron dose in cells, it is necessary to establish the furthest range of the generated electrons. As electron transport is quite expensive in terms of computational time, a useful strategy is to produce a track-by-track neutron/photon source at a sufficient distance from the cells to consider all the electrons that can reach the volume of interest. The track-by-track approach reproduces the local primary field allowing a shorter calculation time, because it limits the need for fully detailed dose calculations to a reduced portion of the geometry.

Dose due to α and Li ions depends critically on the boron concentration present in cells during irradiation. In any simulation it is best to specify a representative boron concentration in the material describing the cells, as its presence may influence the neutron transport. However, D_{B} ought to be calculated by normalizing the results using the actual concentration taken up by cells at the moment of irradiation (i.e., measured in cells cultivated in the same way and on the same day as those used for the survival assessment). This is particularly important when cells are irradiated without boron in the culture medium. In fact, the dose–survival curve will change if D_B is not evaluated with a representative ^{10}B-concentration, as in most cases total BNCT dose is dominated by the boron component.

Some types of cells are cultivated in suspension and irradiated in vials instead of in flasks. In this case, during irradiation they form a pellet in the bottom of the holder, constituting a thicker volume than a monolayer adherent cell layer (some tens of μm versus about 10 μm). The geometry of the pellet depends on the quantity of cells present in the vial and in each case the possibility to use a KERMA approach for dose needs to be verified. Electronic equilibrium depends on the quantity of culture medium surrounding the pellet, but a detailed transport of electrons is probably necessary.

Other in-vitro models may allow the calculation of KERMA instead of dose. For example, in tissue samples cultivated in vitro, typically with larger thicknesses, the high-LET charged component could be absorbed entirely within the volume of interest. The same holds for spheroids or other 3D-cell constructs. It is thus important to verify the validity of the equilibrium hypothesis in order to choose the most adequate and efficient computational strategy on a case-by-case basis.

For small animal models, the same best practice holds true: to verify whether the hypothesis of equilibrium holds in each specific experiment. If dose needs to be calculated inside entire internal organs, high-LET charged particle equilibrium exists, because the linear dimensions of the organs are several orders of magnitude longer than the particle range. However, sometimes dose needs to be calculated in thin parts of the animal, i.e., small superficial nodules, thin layers of skin, or exposed tissues. In this case, part of the energy of secondary particles may escape the volume of interest, requiring detailed transport calculations. For γ dose, transient electronic equilibrium may hold in the inner organs, but skin, superficial tumours, or structures close to the interface of different materials may be critical and require detailed simulation.

Charged particles are normally transported using a condensed history approach in Monte Carlo simulations [580]. When the volume of interest is very small, e.g., in cells or thin exposed tissues, it is important to set the value of the energy step into which the condensed history is divided such that a sufficient number of steps is calculated within the cells or tissues.

As explained in more detail below, transport of thermal neutrons in matter can be affected by the molecular binding of atoms and by their motion. When dose components due to low energy neutrons need to be computed, it is important to use cross sections considering these effects (often described as 'thermal treatment'). In calculations related to radiobiological models, this effect is especially relevant because they are irradiated in thermal neutron fields (or with epithermal neutron beams with an interposed moderator). It is therefore important to use the specific cross sections, as described in Ref. [577].

To summarize, it is always necessary to verify the condition of equilibrium in the radiobiological experiment:

- For monolayer adherent cell cultures, detailed transport of all the secondary particles is necessary to avoid overestimation;
- For suspensions, tissues cultivated in-vitro, 3D cell constructs and spheroids, the validity of assuming the CPE condition needs to be verified. Electronic equilibrium does not exist in typical irradiation set-ups;

- For small animals, CPE is likely to be satisfied, as well as transient electronic equilibrium in the inner organs. However, for superficial small nodules and thin layers of tissues exposed to neutrons, small volumes close to the interface between different materials may be critical for KERMA/dose calculation;
- Boron concentration needs to be known as precisely as possible, considering the biological variability that causes different boron uptake in different experiments even when the administration protocol is the same. While for animal models this is more difficult, for cells, boron measurement ought to be performed in cultures prepared on the same day and in the same way as those irradiated for survival experiments;
- For low energy neutrons, use cross sections that consider the specific molecular binding and motion of hydrogen;
- Specific computational strategies are needed to ensure efficient and statistically reliable simulation, for correct cross section selection (i.e., thermal treatment for low-energy neutrons), and to ensure correct transport of charged particles by condensed history.

As noted above, the same radiobiological models exposed to neutrons are also exposed to a reference photon dose for comparison. It is important to verify that this part of dose evaluation is also correct by verifying that the nominal dose imparted by the irradiation system corresponds to the true absorbed dose. Sometimes, in fact, equilibrium may hold in the calibration set-up, while it may be lacking in the cells/animal irradiation configuration due to geometrical constraints. In this case, a simulation is necessary to compute a normalization factor between the nominal and the true absorbed dose.

10.2.1.2   Dose calculation in patients

This section focuses on the simulation of absorbed dose in in-patient models; most of the suggestions for in-patient dose calculations are the same as those described above for radiobiological models.

For in-patient dosimetry, a model for Monte Carlo transport calculations is obtained from the medical images of the patient. Ideally, the most precise model is obtained by converting the minimal geometrical unit of the medical image into a voxel whose material is inferred from the same study. Dose would thus be calculated in these small volumes, and Monte Carlo methods could be used to perform detailed simulation of dose absorbed in each voxel. However, since equilibrium is surely satisfied in most of the patient volume, calculation has always been performed to obtain KERMA, using superimposed meshes for scoring neutron/photon fluence and coupling them to fluence-to-KERMA conversion tables [577].

This approach has some limitations, for example in skin, where secondary charged particles may escape the tissue and deposit their energy elsewhere, without being replaced by others coming from external volumes. Moreover, it may not hold in small voxels at the interface of materials with very different composition/density such as soft/lung tissues. Skin is an important organ in BNCT, both for superficial and deep-seated tumours, because it is sometimes the organ that limits irradiation time. Especially when using BPA as the boron carrier, skin takes up a higher boron concentration than other normal tissues do. Thus, the tolerance dose may be reached in skin before other organs. For this reason, a possible solution is to dedicate part of the calculation to a detailed dose deposition simulation in those regions of the patient geometry where equilibrium cannot be assumed [581]. This is relevant when the region is the one limiting the irradiation time or is a critical organ at risk (see Section 12.5).

As explained in Ref. [6], for thermal neutron transport in tissues, it is better to use cross sections that consider thermal treatment. This simulates the interaction of low-energy neutrons with the target nucleus considering both its motion and its chemical binding. In the case of biological tissues, this is especially relevant for hydrogen. For BNCT, the effect of the molecular binding for low energy neutrons is often considered by using the cross section for water molecules at 300 K. However, more accurate results can be obtained using cross sections that consider the binding and dynamics of hydrogen in specific molecules within each irradiated tissue [582–583].

In patients, even more importantly than in animal models, the presence of boron in tissues needs to be modelled in the simulation. The change in low energy neutron fluence due to interaction with ^{10}B (self-shielding) affects the calculation of the boron, nitrogen, and hydrogen dose components. Thus, it is better to include a representative ^{10}B-concentration in the description of materials (see Section 11.1.1 concerning thermal neutron flux depression due to boron). More refined results will be possible in the future when the true boron distribution can be determined (see Sections 8 and 9.3) and included in the patient model for the Treatment Planning Systems (TPS).

As for the case of radiobiological models, there follows a list of good practice for absorbed dose in human voxelized patient models:

- Be aware that KERMA may overestimate the dose in skin or in small regions close to interfaces between different materials/densities when these regions constitute a limiting tissue for irradiation (Section 10.2.1.1);
- For accurate neutron transport, describe the materials adding the expected boron concentration (within 20–30%) in irradiated tissues;
- Use thermal treatment to consider molecular binding and the motion of target atoms (hydrogen);
- Use proper fluence-to-KERMA conversion factors, accounting for low-energy neutrons (Section 10.2.1.1);
- Especially when the neutron spectrum has a fast component, consider minor reactions due to capture by Cl and O (Section 10.1.2);
- For accurate dose delivery, calculate boron dose using information on measured boron concentration in blood during the treatment to ensure that the boron dose component is known as precisely as possible.

10.2.1.3 Further considerations

The possibility of obtaining statistically robust results in very small volumes usually requires the use of non-analogue Monte Carlo techniques. This refers to statistical bias that is artificially introduced into the physical probability density functions that describe the interaction of particles, with the goal of improving effectiveness, i.e., to obtain a result with low variance in shorter calculation times, without affecting the correctness of the score. These techniques are embedded in the codes, and statistical convergence is affected in this case by complex effects, not only by the number of particles transported. A user has to pay attention and check the convergence of the results, in particular, to check if the variance associated with the result is a reliable indicator of sufficient statistics. Some codes, such as MCNP, provide several tests to check that the score calculated is statistically reliable.

Another consideration is the necessity to use adequately validated models. Monte Carlo calculations are only reliable if they can reproduce experimental results. In the case of mixed-field dosimetry, validation is particularly difficult because the measurement of absorbed dose in biological systems is challenging with respect to the set-up and for the detectors to be employed. However, it is possible to validate models in relevant parts of the calculations, for example, in their ability to adequately calculate the neutron flux and the photon dose present in the irradiation position. Thus, it is possible to prepare a model of the irradiation facility and to calculate flux/dose for simple situations. It is also possible to employ detectors (such as activation wires or foils or other small systems such as self-powered neutron detectors) applied to cell flasks or to the skin of small animals to evaluate the reproducibility of the measurements by the computational model. Such detectors can be also applied to the skin of patients during irradiation as a monitor of fluence/dose. The experimental results can be used to compute normalization factors to adjust the Monte Carlo results.

10.2.2 Microdosimetric calculations

Due to the short ranges of protons, α particles, and ^{7}Li ions produced by neutron reactions, the absorbed doses are heterogeneously distributed on the microscopic scale in BNCT radiation fields. This heterogeneity influences the biological effectiveness of this therapy, and many dosimetric studies have been devoted to its evaluation. Those studies are designated as 'microdosimetry' [29] and usually distinguished from conventional dosimetry because the stochastic nature of the energy depositions needs to be considered. Thus, the absorbed doses in microscopic sites are generally expressed by their frequency and dose probability density functions, $f(z)$ and $d(z)$, respectively, where z denotes the specific energy defined in ICRU Report 36 [584]. It has the unit of J/kg (or Gy) as for absorbed dose, but it is a stochastic quantity. The principles and application of microdosimetry to BNCT are summarized in Ref. [585].

Assuming that the scale of interest is comparable to the cell size (i.e., the microscopic scale), it is important to verify whether a microdosimetric approach is needed to better estimate the actual dose imparted to cells [584]. Here, it is necessary to consider: (a) the ionization density of a given BNCT dose component and (b) the charged particle production rate of that dose component per unit volume, examined at microscopic (i.e., μm) scales. For electrons set in motion by photons (sparsely ionizing radiation), the number of events needed to deposit 1 Gy of absorbed dose in a cell is sufficiently large (more than 1000 per Gy) to preclude the need of an approach that considers the stochastic aspects at the cellular level. For ^{10}B neutron capture reactions (high ionization density), however, the mean number of reactions needed to produce 1 Gy is of the order of 3 per cell, considering a spherical cell of about 10 μm in diameter [569]. Considering nitrogen neutron capture reactions (intermediate ionization density), although less significant than boron reactions, the mean number of protons and ^{14}C recoils that contribute to 1 Gy of absorbed dose in a cell is about 14, again considering a cell diameter of about 10 μm. Since Poisson statistics govern the behaviour of random particle traversals [586], these mean numbers are equal to the variance of the number of hits. Thus, cell-to-cell variability is very important, including the possibility that certain cells receive very few events (or none) even at therapeutic doses.

In BNCT radiation fields, both intra- and intercellular heterogeneities in ^{10}B-distribution significantly influence the specific energy in a cell nucleus, z_n, which can be regarded as a

---

[29] Microdosimetry is the branch of radiation biophysics that studies the spatial, temporal and spectral stochastic aspects of energy deposition by radiation in microscopic structures.

better index for expressing the biological effectiveness in comparison to the equilibrium absorbed dose or KERMA. The intracellular heterogeneity changes the mean value, $\bar{z}_n$, which becomes larger when ^{10}B atoms are accumulated closer to the cell nucleus.

10.2.2.1 *Analytical calculations*

Analytical approaches usually require assumptions regarding the description of the intersection (and energy deposited) between the particle track and the site, the latter typically described as a simple geometric body. Usual shapes include spheres and spheroids to simulate cells, cylinders for blood vessels, slabs for layered tissues, or experimental radiobiological set-ups, etc. Hypotheses regarding rectilinear paths and continuous slowing-down approximations are usually adequate to accurately calculate the average dose deposited under non-equilibrium conditions. In terms of boron localization, uniform, bi-valued and surface source distributions can be included in the models, including regular lattices to compute crossfire effects [570, 587–588].

One important advantage of analytical descriptions (vs. Monte Carlo calculations) is the possibility of obtaining continuous functions that describe the dependence of dose on site size, shape, and microdistribution in a simple manner. Although restricted to simplified descriptions, analytical approaches can provide a complete portrayal of the full state space of a multivariable problem. Deposited energy spectra, dependent on-site geometry and boron localization, can be straightforwardly obtained, including time-dependent calculations (particularly important for considering the kinetics of a boron compound during irradiation).

Mixed analytical and numerical approaches can be used to add some degree of complexity, especially regarding the actual shape and size of tumour and normal cells, by utilizing the concepts of stereology and integral geometry [589]. Stereology deals with obtaining higher-dimensional information from the analysis of lower-dimensional random sections of the sample in question. Thus, in terms of chord-length distributions (CLD), the actual population-averaged geometry of the sensitive site can be considered. Moreover, if stereological analysis is performed on a histological sample that contains a heterogeneous cell population, the CLD will be sensitive to the composition of that sample. It is important to consider the appropriate randomness to correctly describe the full process [590]. Briefly, the concept of randomness in radiation biophysics lies in the realm of geometric probabilities (the field of mathematics that studies the application of probabilities to random geometric sets) and allows determination of the distribution of outcomes of the intersection between two geometric objects under a given random process that rules their intersection. This random process, in radiation biophysics, is composed of a 'site' (e.g., the cell) and a 'probe' (the track), where the outcome (e.g., track length, energy deposition, DNA breaks, etc.) will be influenced by the 'sampling process' (the irradiation), which will depend, among other things, on the track source distribution. If particle tracks are uniformly and isotropically produced, then μ-randomness is the appropriate assumption to perform the analytical calculation. However, internal sources or surface distributed sources are described by different kinds of randomness ('I-randomness' and 'S-randomness') and these need to be used if the origins of the reactions are inside the volume or located on its boundaries. CLDs under different types of randomness are very different in shape, mean chord-length value, and variance, and therefore, when coupled with energy deposition along the track, lead to very different average microscopic doses [591].

10.2.2.2 Monte Carlo calculations

Monte Carlo simulation is also a powerful tool to determine the doses on microscopic length-scales, though it generally requires longer computational times compared with analytical calculations. Historically, two types of Monte Carlo codes have been developed for radiation transport: one type simulates atomic interaction event-by-event, while the other employs the continuous slowing down approximation or condensed history method with reduced computation times. The former is called a 'track-structure code' and the latter a 'general purpose particle transport simulation code'.

Track-structure codes can determine the doses even over nanometre scales, and thus, have the potential to reveal the complex nature of DNA damage induced by BNCT. However, no track-structure code that is applicable to the BNCT radiation fields is available so far because of the lack of evaluated cross section data for atomic interactions induced by α particles and ^{7}Li ions. General purpose particle transport simulation codes are widely used for cellular scale dosimetry in addition to the macroscopic dose calculations described in Section 10.2.1.2. They can also be used for confirming the accuracy of analytical calculations, which ought to agree with the corresponding Monte Carlo simulations of a representative set of points in the state space of the problem. Such comparison provides a good starting-point from which to pursue more complex descriptions.

10.2.2.3 Microscopic dose correction factors

KERMA factors for calculating boron absorbed dose depend on a single value of the boron concentration at a given point in tissue and assume that the average energy per reaction is deposited at that point. Therefore, in situations where incomplete energy balance between particle hits exists, like cells exposed to different or non-uniform boron concentrations, the actual average absorbed dose in a cell depends on a space-dependent boron concentration function, or in other words, on a scalar field of boron reactions. Superposition of sources is usually a suitable method for decomposing the reaction field into approximately constant and region-delimited boron distribution primary fields.

The simplest case of a bi-valued boron distribution leads to the definitions of intra and extra-site distributions. Any bi-valued source distribution can be expressed as the superposition of a uniform and an intra-site distribution, either positive or negative [585]. In this case, it can be demonstrated that a factor that depends on the reaction-rate ratio and site size and shape can be employed to 'correct' the boron KERMA (calculated assuming a uniform boron distribution).

It is, therefore, possible to obtain the actual average boron dose imparted to cells, and also other microdosimetric quantities depending on boron localization, through use of bi-valued microdistributions. These correction factors, referred to as 'Microscopic Dose Correction Factors' can be used to understand the dependence of the effectiveness of BNCT on the boron microdistribution, size and shape of the cell and, in general, to correct every parameter that accompanies the boron dose term in any expression, like in the case of a boron compound's CBE factor or the radiobiological parameters of photon isoeffective dose–effect relationships [585].

Empirical observations [244] suggest a relationship between the slope of BNCT survival curves and sensitive site characteristics, obtained by stereological analysis of histological sections of different tumours, expressed as the nucleocytoplasmic cell ratio and cell size. These empirical observations corroborate the fact that the biological effectiveness of the boron reaction is not only related to the radiation sensitivity of a given tumour cell line but also on the microdistribution of sources (Section 9.4). The latter arises from the pharmacokinetics of two different boron compounds which are known to have diverse mechanisms for uptake and localization. These observations are in line with the aforementioned approach, establishing the need to apply geometry and source dependent form factors to the KERMA, in order to calculate the correct dose in cells [566].

10.2.2.4 Summary

The following points are important considerations:

- The specific localization of boron atoms in tumour and normal cells. If the distribution is not uniform, it is not possible to use KERMA as a surrogate quantity for calculating the average dose deposited in the cells, since the problem is no longer 'scale-independent';
- In the case of non-uniform distributions of boron or nitrogen reactions, methods for estimating the actual average dose imparted to the cells are needed. These methods need to consider the lack of CPE conditions due to the incomplete balance of energy transferred in and out of a sensitive microscopic site;
- In the case of boron and nitrogen reactions, the small number of events needed to impart therapeutic doses has to be considered, since they determine the variance of the specific energy, and thus the biological effect. Thus, a microdosimetric approach is required.

10.3 TRANSLATING BNCT DOSES INTO A REFERENCE RADIATION DOSE

The secondary charged particles that contribute to the total absorbed dose have very different LETs. These particles produce ionization patterns in living cells that may result in different cellular injury for the same absorbed dose. Since the same total absorbed dose delivered with different relative contributions of the main components may lead to a different outcome, a method is required for translating BNCT doses into a reference radiation dose capable of quantifying clinical effects.

The reference radiation dose needs to also enable the specialist to:

- Apply the clinical experience gained with the reference therapy to optimize BNCT;
- Compare protocols and results from different institutions applying BNCT;
- Have a simple and practical language to prescribe doses in BNCT;
- Facilitate combined radiation treatments.

10.3.1 Reference radiation

Photon beam therapy has long been considered the reference radiation therapy modality. A large body of experience has been built up worldwide over about 70 years with MV photons. Most clinical radiotherapy uses photon beams in the range 4–18 MV. Less than 1% of the patients worldwide are treated with protons or other hadrons, although the number is increasing with the number of new facilities [592]. These reasons justify remaining with photon beam therapy as the reference radiation treatment modality of choice, and the benefit of any new or unconventional techniques such as BNCT ought to be evaluated in relation to photon radiotherapy.

Dose delivery schemes with photon radiotherapy and BNCT are generally different. While BNCT administers the full dose in one or two applications, traditional photon treatments require dose splitting in multiple fractions of 1.5 to 3 Gy administered daily over a period of 3–7 weeks.

To avoid making unnecessary conversions between the selected photon reference protocol and BNCT, the first choice for comparison would be that the treatment conditions are equal. In this regard, clinical data for single fraction photon dose schemes are preferable. Second choice, if single fraction radiotherapy data are not available, is hypofractionated radiation therapy, since large doses are delivered in a few fractions. The conversion of hypofractionated radiation doses to single doses can be performed using the linear–quadratic model (see Section 10.3.3.2). In that case, the absorbed dose in tumour per fraction, the so-called fraction size, as well as the α/β ratio of the reference therapy needs to be explicitly mentioned [593]. However, since there is evidence that the resulting single doses may underestimate the effects of hypofractionated radiation, special attention needs to be paid to the results of conversion.

10.3.2 Relative and compound biological effectiveness

Biological effects are the result of the action of ionizing radiation in living systems. These effects are directly related to absorbed dose. However, since they also depend on other factors, such as dose delivery scheme, dose rate, radiation quality, biological system and end points, there is no unique relationship between absorbed dose and induced biological effects.

At equal absorbed doses, radiations of different quality produce different levels of biological and clinical effects. The differences in effectiveness are related to differences in energy deposition along the particle tracks and through subcellular structures [594]. The concept of Relative Biological Effectiveness (RBE) was introduced to quantify differences in biological effectiveness of different radiation qualities. The RBE is a ratio between two absorbed doses with two radiation qualities, one of which is the 'reference radiation', that both produce the same effect in a given biological system under identical conditions [594]. It is a clear, well-defined radiobiological concept and possesses two notable characteristics:

- It is unnecessary to provide a numerical expression for quantifying the effect, i.e., qualitative descriptions on an arbitrary scale can be used to determine the ratio of doses that produce the effect (e.g., lens opacity, skin erythema, or motor skill performance);
- Once the intermediate and complex mechanisms that arise from radiation-quality dependent lesions have ended, the observed effect is independent of the nature of those different mechanisms. Such complexities just produce equal radiation-quality independent modifications, with differences only in the kinetics of lesion formation [586].

An RBE value has an associated experimental uncertainty as it is the result of an experiment. An RBE value depends on the biological system studied as well as the type and level of effect. Therefore, the dose and experimental conditions in which a given RBE value was determined needs to be specified [595]. The RBE of a given radiation quality varies markedly with dose, biological system, and effect. Thus, it cannot be considered as a single number.

In BNCT, the Compound Biological Effectiveness (CBE) is employed to emphasize the observed dependence of the relative biological effectiveness of the α and ^{7}Li particles altogether, i.e., the 'boron reaction' effectiveness, with the spatial location of the carrier compound. Compound biological effectiveness factors are measured experimentally, in vivo or in vitro, for different boron compounds, specific tissues or cell types, and for diverse biological end points [378, 435, 596]. Usually, KERMA has been used to compute the ratio between the reference and the boron dose that results in the same effect. Important differences in CBE factors are usually found depending on the compound used. These differences can be explained by decomposing the CBE factor into the product of three quantities:

- A 'form factor', which depends on the compound microdistribution and target geometry;
- The ratio of survival slopes, which accounts for any differences in intrinsic radiosensitivity between the ballistic efficiency of α and ^{7}Li particles for the particular compound microdistribution and a uniform microdistribution;
- A reference boron RBE obtained under CPE conditions or, equivalently, under a uniform production of boron neutron capture reactions.

Compound biological effectiveness factors have already been considered in practical treatment planning of BNCT. For example, the CBE values of BPA for cells in culture are greater than the corresponding values for BSH, owing to cell permeability of BPA (see Section 7.4). In contrast, the intercellular heterogeneity changes its probability density, $f(z_n)$ (see Section 10.2.2); the variance of $f(z_n)$ becomes larger in cases where ^{10}B tends to be accumulated in certain cells, and a microdosimetric approach is needed. In addition, uptake of ^{10}B-compounds is dependent on the cell cycle, and both BSH and BPA are taken up at higher rates at the G_2/M than at the G_0/G_1 phase [597]. A larger variance of $f(z_n)$ results in lower biological effectiveness particularly at higher doses because of the overkill effect [389].

10.3.3 The standard model based on single biological effectiveness values

The term 'photon-equivalent dose' was largely used in the clinical application of BNCT to indicate the dose of a reference photon radiation that is estimated to produce the same biological effect as the combination of the different absorbed doses administered with this therapy. A variety of units to indicate the administered dose in photon equivalent units have been used by the BNCT community such as Gy(W), Gy-RBE, Gy-Eq and Gy_w.

In radiation therapy, the term 'isoeffective dose' or D_{IsoE} is introduced to imply a comparison with a dose delivered under reference conditions [594]. The term 'photon isoeffective dose' ought to be used to describe a photon dose that would produce the same effects on a given biological system as the doses delivered under BNCT treatment conditions.

There is an obvious risk of confusion as the unit Gy is used for two quantities (absorbed dose and isoeffective dose, D_{IsoE}). This potential confusion could be harmful (and potentially lethal) for the patient [594]. Following the International System of Units (SI) convention, the number of grays has to be given along with the specific name of the quantity. No additional modifier (subscript, asterisk, etc.) is allowed in the name of the unit to specify the quantity that is involved [598]. Historical forms to indicate photon-equivalent units can be still found throughout this publication as they have been traditionally employed. However, considering the difficulty related to the fact that both absorbed dose and isoeffective dose have the same unit (Gy), the ICRU and the IAEA, together with particle therapy communities, agreed to introduce the unit 'Gy (IsoE)' for the isoeffective dose, D_{IsoE}. In this way, the risk of confusion in clinical practice between the quantities that could be harmful for patients, is avoided or at least reduced. The space between 'Gy' and '(IsoE)' is necessary to comply with the requirements of the International System of Units [598].

10.3.3.1 Description of the standard model

The standard model for calculating the photon dose that would produce the same effects on a given biological system as the dose delivered under BNCT treatment conditions is given by a weighted sum of the primary absorbed dose components:

$$D_{\text{IsoE}} = w_1 D_1 + w_2 D_2 + w_3 D_3 + w_4 D_4 \tag{18}$$

where the components $D_1 = D_B$, $D_2 = D_N$, $D_3 = D_H$ and $D_4 = D_\gamma$, as defined in Table 7.

The standard model converts each term of the total absorbed dose to a photon isoeffective dose term, multiplying the absorbed doses by fixed weighting factors, w_i. The w_i values used in expression (18) are the RBE and CBE of the different radiation qualities determined from radiobiological experiments (Section 9.2) for a given biological system, endpoint, and level of effect.

The biological system and associated effect most considered in BNCT for the determination of these weighting factors are in vitro cell cultures and fraction of surviving cells. Other systems and effects such as radiation damage to normal tissue and tumour control on in vivo models have been less considered.

10.3.3.2 Validity and limitations of the model

Both RBE and CBE depend upon the biological system and the type and level of effect. Therefore, the unrestricted use of single RBE and CBE factors in expression (18) determined for a given level of effect in a particular system will generally lead to inaccurate photon isoeffective dose estimates.

Figure 63 shows the limitations of the standard model. The fraction of surviving cells S for the photon reference radiation 'R' and radiation 'A' shown in Fig. 63a can be described, using the linear–quadratic model, by

$$S_R = \exp[-(\alpha_R D + \beta_R D^2)], \quad S_A = \exp[-(\alpha_A D)] \tag{19}$$

with α_R and α_A are the linear coefficients of both radiations, and β_R the quadratic coefficient of the photon reference radiation, R.

The RBE for 1% survival level is 3 (Fig. 63a). Therefore, according to the standard model, the photon reference dose D_{IsoE} that produces the same survival as the dose D delivered with radiation 'A' is

$$D_{\text{IsoE}} = 3D \tag{20}$$

However, to produce the same survival fraction as the dose D, the D_{IsoE} needs to satisfy

$$\exp[-(\alpha_R D_{\text{IsoE}} + \beta_R D_{\text{IsoE}}^2)] = \exp[-(\alpha_A D)] \tag{21}$$

Thus, the correct expression for D_{IsoE} is

$$D_{\text{IsoE}} = \frac{-\alpha_R + \sqrt{\alpha_R^2 + 4\beta_R \alpha_A D}}{2\beta_R} \tag{22}$$

Figure 63(b) compares D_{IsoE} given by Eqs (20) and (22). For absorbed doses lower than $D_x = 4$ Gy, the standard model depicted by the straight line underestimates the correct D_{IsoE} value. Conversely, for absorbed doses above $D_x = 4$ Gy the standard model overestimates D_{IsoE}. Note that, while the low dose region is associated with absorbed doses typically delivered to normal tissues, the high dose region is associated with values administered to the tumours. This fact requires special attention because the general behaviour of the standard model is not conservative; the resulting D_{IsoE} values could be potentially harmful (or insufficient) for the patients.

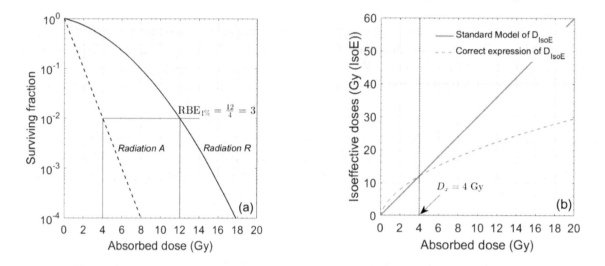

FIG. 63 (a) Fraction of surviving cells for a photon reference radiation R and an example radiation A. (b) Comparison of the standard model (Eq. (20)) and the correct expression of D_{IsoE}, (Eq. (22)) as a function of the total absorbed dose. $D_x = 4$ Gy represents the absorbed dose where the standard model and correct expressed for D_{IsoE} cross (courtesy of S. González, CNEA).

The range in which the standard model gives a reasonable approximation is very narrow:

(a) In the neighbourhood of D_x;
(b) In the dose interval from 0 Gy to a few grays, when the initial slope of Eq. (22) is similar to the slope of the standard model. In this case, both curves are similar up to a certain low dose value.

The only condition under which the standard model and Eq. (22) match is through neglection of the quadratic term in the survival model for the reference photon radiation (Eq. (19)). This approach is discouraged, however, since an important fraction of cell death due to the accumulation of sublethal damage (SLD) is typical of low LET radiation and hence the quadratic term is crucial.

10.3.4 Models for calculating photon isoeffective doses

The following sections describe the models developed in the last decade to determine photon isoeffective doses in BNCT. These models are based on widely known and well-accepted theories, and progressively incorporate degrees of sophistication according to the dose–response relationship and the phenomena considered. Specific models with parameters derived from in vitro and in vivo radiobiological data prepared for prospective use in the clinic under medical judgment are listed at the end of this section.

10.3.4.1 Definition and general considerations

Following recommendations introduced in Ref. [594], a publication co-sponsored by IAEA and ICRU, the term 'photon isoeffective dose' with the unit 'Gy (IsoE)' needs to be adopted to describe the photon dose that would produce the same effects on a given biological system as the doses delivered under BNCT treatment conditions. Where the Gy (IsoE) unit is used, some explanation needs to also be provided for the applied photon isoeffective dose model (either constant RBE model or more complex dose models).

The concept of photon isoeffective dose requires mathematical expressions which quantify the effect of interest under both photon and BNCT treatment conditions. For example, consider that a radiation specialist wants to determine the D_{IsoE} that, delivered in a single fraction scheme, would produce the same tumour control as the combination of the main absorbed doses in BNCT. Then, she/he will need to have models that describe the 'tumour control' effect after single-fraction photon radiotherapy and BNCT.

Both macroscopic and microscopic approaches were developed for estimating photon isoeffective dose. The macroscopic approach here assumes that calculations are independent of the spatial scale. In contrast, the microscopic approach is based on absorbed doses on cellular and sub-cellular scales. Information on the boron microdistribution (see the techniques in Sections 8 and discussion in Sections 9.3–9.4, and 9.6.1) is indispensable in the latter approach.

Phenomena that deserve special attention when describing the biological effects, particularly in the case of BNCT are:

(a) Synergism between different radiations;
(b) Sublethal damage repair.

10.3.4.2 Photon isoeffective doses and biological endpoints

The photon isoeffective dose model is intimately linked to the biological effect of interest and, consequently, to the dose–response relationship.

Models based on dose–response curves for cell survival

Such models can be based on formalisms over two different length scales, macroscopic and microscopic:

Macroscopic formalism

The simplest model is based on dose–response curves for cell survival and neglects sublethal lesion repair (for both photon and BNCT treatment conditions) and synergism between radiations (in the case of BNCT). It is derived by replacing radiation 'A' in Section 10.3.3.2 by the radiations D_1, D_2, D_3 and D_4 delivered in BNCT (Eq. (18)). Then, if the fraction of surviving cells for BNCT is

$$S_{BNCT} = \exp[-(\alpha_1 D_1 + \alpha_2 D_2 + \alpha_3 D_3 + \alpha_4 D_4 + \beta_4 D_4{}^2)], \tag{23}$$

with α_i, β_i ($i = 1,...,4$) the coefficients of the model, the correct expression for D_{IsoE} is

$$D_{IsoE} = \frac{-\alpha_R + \sqrt{\alpha_R^2 + 4\beta_R(\alpha_1 D_1 + \alpha_2 D_2 + \alpha_3 D_3 + \alpha_4 D_4 + \beta_4 D_4{}^2)}}{2\beta_R} \tag{24}$$

Equation (23) shows that only the γ dose component of the BNCT field is allowed a quadratic term, thus permitting accumulation and combination of sublethal damage for the low LET component exclusively.

When combining high-LET radiations with low-LET radiations simultaneously, synergistic effects may appear [599]. This means that sublethal lesions produced by one radiation (e.g., by the α particles of the BNCT reaction) can interact with the sublethal lesions produced by any other radiation (e.g., by the 583 keV protons from neutron capture by nitrogen) to form lethal damage. The yield of sublethal lesions per unit dose for each radiation component i is accounted for by $\sqrt{\beta_i}$. Then, the term $\beta_4 D_4{}^2$ in S_{BNCT} (Eq. 23) is replaced by

$$\beta_1 D_1{}^2 + \beta_2 D_2{}^2 + \beta_3 D_3{}^2 + \beta_4 D_4{}^2 + 2\sum_{i<j}\sqrt{\beta_i\beta_j}D_iD_j \tag{25}$$

and the same needs to be done in expression (24) for D_{IsoE}.

Typical treatment times in BNCT vary between 30 to 90 minutes, values that are greater than or at least comparable to the typical repair half-times of sublethal damage for the fast repair component (in the range of 5–30 min). The reduction in the probability of SLD interaction when repair mechanisms are present is accounted for by the generalized Lea–Catcheside time factors, G, that modify the quadratic terms of Eq. (25). Analytical expressions for these factors for simultaneous delivery of high and low-LET radiations are provided in Refs [392, 600].

In the case of normal tissues exposed during a BNCT treatment, more than 50% of the total absorbed dose may be due to the low-LET component. Hence, for these tissues it may be of particular importance to consider these reduction factors in the calculation of D_{IsoE}.

Microscopic formalism

In contrast to the macroscopic approach, the difference between the survival probability of each cell, S_C, and survival fraction of a certain cell group, S_G, needs to be considered in the microscopic approach [28], as expressed below:

$$S_G(D) = \int_0^\infty S_C(z_C)f_C(z_C, D)\mathrm{d}z_C \tag{26}$$

where D is the absorbed dose in the cell group and $f_C(z_C, D)$ is the probability density of the specific energy in each cell or cell nucleus, z_C, for mean absorbed dose D. Note that the specific energy is the stochastic quantity introduced to express the heterogeneity of the absorbed doses in microscopic sites.

For conventional radiotherapy such as XRT, the variance of $f_C(z_C, D)$ is negligibly small, i.e., the approximation of $f_C(z_C, D) = \delta(z_C, D)$ is well established, where δ is the Dirac delta function. In that case, $S_G(D) = S_C(D)$. In contrast, $f_C(z_C, D)$ needs to be carefully evaluated in the case of BNCT because of the large dose heterogeneity as described in Section 10.2.2 Information on the intercellular boron distribution is indispensable in the evaluation. The intracellular boron distribution (Sections 7, 9.3–9.4, and 9.6.1) is also important in solving Eq. (26) because it significantly influences the survival probability of each cell, $S_C(z_C)$, by changing the microscopic dose correction factor.

Once $f_C(z_C, D)$ and $S_C(z_C)$ are evaluated, $S_C(D)$ can be easily determined by solving Eq. (26) numerically. Then, the photon-isoeffective dose, D_{IsoE}, can be obtained from

$$D_{\text{IsoE}} = \frac{-\alpha_R + \sqrt{\alpha_R^2 + 4\beta_R \ln[S_G(D)]}}{2\beta_R} \tag{27}$$

If the fractionated size, X, is adopted as the reference treatment, Eq. (27) can be replaced with

$$D_{\text{IsoE}} = \frac{-\ln[S_G(D)]}{\alpha_R + \beta_R X} \tag{28}$$

An advantage of the microscopic model over the macroscopic one is that it can roughly predict the biological effectiveness of newly developed boron compounds if information on their intra- and intercellular heterogeneity is available.

Models based on dose–response curves for tumour control and normal tissue toxicity

In addition to models of cancer in vitro, there are several in vivo models that have been extensively used in BNCT, as they better describe the interactions between tissues and organs, as well as the biological pathways involved, which includes the vascular system and its pivotal role in tumour growth.

These models can be used to obtain the response of dose-limiting normal tissue as well as the dose–response data for local tumour control. As physiological complexity is taken into account, with some restrictions, these results can be extrapolated to clinical scenarios. In addition, since these responses are closely linked to those evaluated in the clinic, their use in determining the isoeffective dose is strongly suggested.

The determination of the photon isoeffective dose in BNCT with Eq. (29) requires suitable mathematical expressions to describe the probability of effect E for both the reference photon radiation R and for BNCT.

$$E_R(D_{\text{IsoE}}) = E_{\text{BNCT}}(D_1, D_2, D_3, D_4) \tag{29}$$

A general Tumour Control Probability (TCP) model can be written as

$$\text{TCP} = \exp[-(c_1 v^{c_2} S)] \tag{30}$$

where v represents the tumour volume (in cm^3), c_1 and c_2 the parameters that modulate its effect on local control probability and S the fraction of surviving cells.

When S for the reference radiation and BNCT are expressed by the simple linear–quadratic models of Eqs. (19) and (23) the value of D_{IsoE} that solves

$$\text{TCP}_R(D_{\text{IsoE}}) = \text{TCP}_{\text{BNCT}}(D_1, D_2, D_3, D_4) \tag{31}$$

coincides with Eq. (24) (whenever, for both radiation treatments, the target tumour cell population for the same probability of local tumour control is equal). Moreover, this result holds for models in which the TCP is simply a function of the fraction of surviving cells.

The general expression for the photon isoeffective dose $D_{\text{IsoE},}$ when both synergism and repair mechanisms are considered in Eq. (30), is

$$D_{\text{IsoE}} = \frac{-\alpha_R + \sqrt{\alpha_R^2 + 4\beta_R\left(\ln\left(\frac{c_1^*}{c_1}v^{c_2^*-c_2}\right) + \sum_{i=1}^{4}\left(\alpha_i D_i + G_i \beta_i D_i^2\right) + 2\sum_{i<j} G_{ij}\sqrt{\beta_i \beta_j}D_i D_j\right)}}{2\beta_R}$$

$$(32)$$

with c_1, c_2 and c_1^*, c_2^* being the parameters of the volume for the reference and BNCT, respectively, and G the time factors for a simultaneous mixed irradiation reported in Refs [392, 600]. If the target cells are the same for both radiation treatments, $c_1 = c_1^*$ and $c_2 = c_2^*$, and the logarithmic tumour volume term disappears.

The most widely used Normal Tissue Complication Probability (NTCP) model in clinical radiobiology is the Lyman model [601]. The model for a single-fraction photon reference dose D is given by

$$NTCP_R = \frac{1}{\sqrt{2\pi}} \int_{-\infty}^{s'} \exp\left(-\frac{t^2}{2}\right) dt \qquad (33)$$

where

$$s' = \frac{D - TD_{50}}{m \cdot TD_{50}} \qquad (34)$$

and where m is the NTCP vs. dose curve slope and TD_{50}, the tolerance dose for a complication probability of 0.5 (median toxic dose).

An NTCP model for BNCT based on Lyman's model is proposed in Ref. [385], obtained by replacing D in expression (34) by D_{IsoE} given in Eq. (24) (or for the more general case, in Eq. (32) but without the logarithmic term). Thus, the value of D_{IsoE} that solves

$$NTCP_R(D_{\text{IsoE}}) = NTCP_{BNCT}(D_1, D_2, D_3, D_4) \qquad (35)$$

coincides with Eqs (24) or (32) (without the logarithm).

10.3.5 Determination of radiobiological parameters

10.3.5.1 Coefficients from dose–response curves for cell survival.

Several groups have made cell survival measurements under various conditions to try to evaluate fixed RBE factors for different cell lines. They generally consist of determining the dose response to the:

- Photon reference radiation;
- BNCT beam only (n + γ);
- BNCT beam in the presence of the boron carrier [1].

In Section 10.2.1, details on how to determine the absorbed dose in preclinical experiments are provided. Here, a methodology is suggested to obtain a suitable set of parameters from typical cell survival experiments carried out in BNCT.

The simplest model of D_{IsoE} given by Eq. (24) involves a total of seven radiobiological parameters, i.e., the

- Two coefficients α_R, β_R of the photon reference radiation, R;
- Four linear coefficients α_i of the contributions in BNCT;
- One quadratic coefficient β_4 for the γ component of BNCT.

Based on the argument that similar biological responses are obtained when a system is exposed to radiations with comparable lineal energy spectra, the number of free model parameters can be reduced as follows:

- For photons, $\alpha_4 = \alpha_R$ and $\beta_4 = \beta_R$. The lineal energy spectra of the reference photons (e.g., from ^{60}Co, ~1 MeV) and photons in the beam (mostly ~2 MeV) are almost the same;
- For neutrons, $\alpha_2 = \alpha_3$. Fast neutrons in BNCT beams generally have $E \lesssim 1$ MeV, where elastic recoils with ^{1}H nuclei dominate the charged particle slowing down spectrum (feeding through to the value of α_3), with energies comparable to those of protons produced from thermal neutron capture by nitrogen (feeding through to the value of α_2).

Equation (19) is first used to fit the photon data to obtain the coefficients α_R and β_R. Considering the estimated coefficients and previous considerations, the survival model from Eq. (23) is used to determine the remaining free parameters, α_1 and α_2.

The fitting procedure can be carried out sequentially or simultaneously. In the sequential approach, the alpha coefficient of the neutron component ($\alpha_2 = \alpha_3$) is determined first from the fitting of the data for BNCT beam only. Then, the obtained result is used to calculate the alpha coefficient of the boron component, α_1, by fitting the (BNCT + boron compound) dose–response data. The dose–response data from irradiation with a BNCT beam in the presence of the boron compound contains information on the action of the products that result from all neutron interactions, in particular, from neutron interactions with nitrogen and hydrogen present in tissues. In the simultaneous approach, a fitting procedure involving a minimization of both BNCT beam only and (BNCT + boron compound) cell survival data can be carried out to fully exploit all the experimental information.

The general procedure and considerations to determine the parameters of the photon isoeffective dose model when synergism and SLD repair are considered are the same. However, the following points need to be considered:

- The number of free model parameters increases due to the inclusion of the quadratic coefficients for the high LET components. In this case, simultaneous minimization can be quite challenging;
- Time reduction factors, G, depend on the SLD repair kinetic model and the characteristic repair times of the cell line under study. A bi-exponential modelling characteristic slow and fast repair times that are independent of LET (i.e., of the component of the BNCT field) can be assumed for the calculation of G factors. In this case, fast and slow repair times for the specific cell line and the proportions of the SLD repaired by the two kinetics for both low and high LET radiations need to be known.

10.3.5.2 Coefficients from dose–response curves for tumour control and normal tissue toxicity

Models based on dose–response curves for tumour control can give, as previously shown, formulae for the isoeffective dose which coincide with those coming from survival curves. However, the probabilistic nature of these models means that determination of the radiobiological parameters requires a different approach. The main difference is that, instead of dose–survival curves, the input information is dose–tumour control data:

– 1, if the tumour was controlled (i.e., showed complete response);
– 0, otherwise.

There are two approaches for parameter determination:

(a) The first is based in maximum likelihood estimation and consists of finding the values of the parameters that maximize the likelihood function. This function involves the TCP model, tumour doses, volumes, and responses. From the dose and volume of each tumour, the probability that the tumour is controlled (or not) is computed as a function of the parameters. Then, the likelihood function is the product of the probabilities of all observed responses. Like models based on survival curves, one can make different simplifications or reductions in the number of parameters. Also, the maximization can be carried out for all parameters simultaneously or sequentially, with the simultaneous optimization being rather challenging;
(b) The second is to perform curve fitting, just as described in the case of dose–survival curves but considering the TCP model: the functions are given by Eq. (30) with the corresponding expression of S for the reference radiation 'R' and for BNCT, and the data to be fitted consists of dose and volume of each tumour and 0 or 1 according to the tumour response. In this way, the method searches for parameters that give TCPs close to 1 for tumours which were controlled and close to 0 for those which were not. Again, reduction of the number of parameters and sequential/simultaneous optimization can be performed.

Approaches for parameter determination when the input is dose–normal tissue radiotoxicity data are the same as those for tumour control. In this case, responses are:

– 1, if the normal tissue reaches or exceeds the radiotoxic effect considered clinically relevant;
– 0, otherwise.

The determination of model coefficients involves NTCP models, normal tissue doses, responses, and eventually irradiated volumes or areas.

10.3.6 Range of validity and advice

The macroscopic approach for derivation of the photon isoeffective doses in BNCT is based on well-known grounds and can be easily applied to any tissue and effect so long as dose–response data and numerical expressions for quantifying the effect of interest are available for both the photon reference radiation, 'R', and BNCT.

The models can be easily entered into any mathematical analysis tool or treatment planning system and, as they are fed with the absorbed doses calculated during treatment planning, their application in the clinic is straightforward. The models presented in Section 10.3.4 can be used to compute photon isoeffective doses for both uniform and non-uniform absorbed dose distribution scenarios. In the non-uniform dose case, a photon isoeffective dose is obtained for each point of the volume of interest. An equivalent uniform dose for this volume (or its corresponding TCP/NTCP) can be determined following different strategies (see, e.g., Refs [602–603]).

Parameters of the photon isoeffective dose model based on dose–response data for cell survival have been derived already for different cell lines. The models for D_{IsoE} calculation that are ready for use in the clinic are for gliosarcoma [385, 393], metastatic melanoma [385], osteosarcoma [604], and SCC [605].

In addition, parameters derived from dose–response data for tumour control and normal tissue radiotoxicity were also reported for some tumours and tissues. In particular, the models that are ready to use in the clinic are for SCC tumours [385], and for the two dose-limiting normal tissues: mucosal membrane (endpoint: mucositis G3 or higher [385]) and normal brain (endpoint: incidence of myelopathy [393]).

A link to a photon isoeffective dose calculator is provided for those who wish to explore some existing models before developing their own tools.[30]

As for RBE, photon isoeffective dose models depend upon the biological system and the type of effect. The available models cover most of the main targets and dose-limiting normal tissues involved in BNCT. However, models for other important normal tissues, such as skin, are still needed. The applicability of any of the existing models to calculate photon isoeffective doses in these cases needs to be left to medical professional judgment.

The introduction of the concept of microdosimetry in the treatment planning of BNCT is desirable for more precise estimation of its biological effectiveness. From the viewpoint of microdosimetry, development of ^{10}B-compounds that are accumulated closer to cell nucleus and homogeneously distributed in all tumour cells is the key issue because cell-nucleus specific energies with higher mean values and lower variances could result in higher therapeutic effect. In principle, the biological effectiveness of ^{10}B-compounds can be estimated from their intra- and intercellular heterogeneity and knowing the size of the target cells and its nucleus. Therefore, measurements of ^{10}B-distributions in cellular and subcellular scales are strongly needed. This requires further improvement of high-resolution imaging devices such as α cameras and track-etch detectors (see Sections 8.2.2–8.2.3).

---

[30] Photon isoeffective dose calculator, https://bnct.com.ar/calculator.html

11 PRESCRIBING AND TREATMENT PLANNING

Treatment planning for BNCT requires specialized software because it differs from conventional radiotherapy planning. Although several utilities included in the treatment planning systems (TPS) for conventional radiotherapy can be applied to BNCT treatment planning, some BNCT-specific utilities are missing. Thus, several TPSs have been developed over the years specifically for BNCT use, and recently several new ones have been developed to fulfil the needs of AB-BNCT. In general, an AB-BNCT system needs to be approved as a medical device and such approval is also needed for medical device software like a TPS. Ideally, a modern BNCT TPS may also handle conventional radiotherapy treatment planning so that plans from both modalities can be combined, because BNCT can be applied as re-irradiation or be combined with conventional radiotherapy. As a recent milestone, BNCT capability has been added into a commercial conventional TPS.

11.1 GENERAL ISSUES

This section describes issues such as patient and 'beam' models, as well as QA around dose calculation and the definition of the MU

11.1.1 Patient model

Dose calculation for BNCT requires modelling nuclear interactions of neutrons within the patient (see Section 10). For that reason, electron densities of tissues cannot be utilized as typically done in conventional radiotherapy; instead, elemental compositions of the tissues need to be modelled. Typically, tabulated values for the composition of each human tissue (see Table 21) are applied, based on references such as ICRU reports [572]. The regions of interest (ROI), i.e., organs, are first contoured based on medical images (manually, semiautomatically, or automatically, depending on the tools available in the TPS). Each ROI is either modelled to contain a single homogenous tissue type and density or information from a medical image of the patient (usually an X ray CT image) is used to define unique density and tissue composition voxel by voxel [606]. Hydrogen and nitrogen concentrations and densities give the main contribution to the patient dose arising chiefly from interactions with neutrons of different energies: the dose due to nitrogen is also referred to as the 'thermal neutron dose' and that due to hydrogen as the 'fast neutron dose', respectively (see Table 7 and Section 10.3.3.1). The ^{10}B-concentration assumed in the treatment plan ought not to differ by more than about 20–30% from that during irradiation, because the presence of ^{10}B depresses the neutron flux by 0.5% per ppm ^{10}B and influences the dose distributions [1, 607]. The biological dose depends on the organ, and it is thus strongly suggested that RBE factors, or appropriate biological model parameters, for each ROI are considered by the TPS.

The patient model can be a voxel-by-voxel model as defined by the resolution of the medical image, or a coarser model of larger voxels can be taken. If larger voxels are used, material in each voxel is either assigned based on the dominant material or a combination of each material within the voxel. This is the typical situation, as very long calculation times may be required if model resolution is high (e.g., pixel-by-pixel calculation). Either uniformly sized pixels/voxels are applied throughout the patient model, or the voxel size can be adapted based on geometry. The program MultiCell has been developed to support BNCT planning using medical image processing including image segmentation and patient voxelization so that the patient model consists of several voxel sizes [608–609]. This way, the number of voxels in the patient model is reduced, which speeds up dose calculation.

11.1.2 Neutron 'beam' model

To create an accurate neutron 'beam' model, the entire neutron generating system (i.e., the neutron source and the surrounding BSA) has to have been modelled first. The neutron field is measured at a surface close to the beam aperture (see Section 4.2.2) and used as the neutron source for clinical dose calculation. In this case, part of the BSA needs to be included in the dose calculation for treatment planning to account for backscattering from the patient to the BSA and back to the patient. The 'beam' model may be one of two alternatives:

(a) A list of particles, including the direction, energy, and location information for each neutron and photon within the beam. To be accurate, all the particles need to be included for every conceivable problem, so the list needs to be large enough that a single simulation will create statistically accurate dose calculations for every patient case that may be handled. This method may be the most accurate way to define the 'beam' if a large enough number of neutrons and photons is included (e.g., 50–100 million) to provide a clinically acceptable statistical uncertainty of around 1% on the dose. The problems with using a large list of particles as a 'beam' model are:
    (i) Large file sizes;
    (ii) Running the particles from the list is time consuming;
(b) A planar particle distribution model, defined as a distribution of particles at the source plane that describes their energy and direction. Use of the planar particle distribution as a 'beam' model requires some averaging of the particle information, and is, therefore, more accurate if the beam model is symmetric, as desired. The advantages of using the particle distribution are:
    (i) Rapid source particle initiation;
    (ii) The number of initial particles can be chosen based on the problem.

There is no need to include particles other than neutrons and photons in the source model.

11.1.3 Dose calculation

Dose calculation for BNCT requires solving the neutron transport equation, which takes into account neutron and photon scattering and absorption within the tissues in the patient model. So far, only Monte Carlo based algorithms have been used in clinical dose calculation, mainly because of the complexity of the geometry. Monte Carlo calculation is time consuming because the solution is found based on repeated random sampling, which after a sufficient number of samples converges to the correct solution. Speeding up the calculation requires either high computing power or the utilization of variance reduction methods, which are prone to errors. The Monte Carlo algorithms applied in BNCT have either been developed by the TPS developer [610] or an existing multipurpose Monte Carlo software such as MCNP (as used in NCTplan, Multicell, THORplan, JCDS), PHITS (as used in TsukubaPlan, JCDS, NeuCure Dose Engine, COMPASS), or Geant4 (as used in the dose engine used by Neutron Therapeutics) has been utilized. The TPS called NeuManta, being developed by NeuBoron, can use its own dose engine, COMPASS, or run with MCNP or PHITS (see Section 11.2).

The most accurate way to calculate dose is to model each neutron and photon interaction and sum the energies deposited by each reaction, including secondary particle transport. Since such a method may be too slow in practice, BNCT dose is commonly calculated via the KERMA approximation from neutron and photon fluxes applying flux-to-KERMA conversion factors (see Section 10.1.1.5). The flux-to-KERMA conversion factors can be generated from cross sections (see Table 22), or published data [572, 578, 611] can be utilized for each element separately or for each tissue within the patient model. Available flux-to-KERMA conversion tables sometimes lack information for certain energy regions or some elements (see Section 10.1.2). Thus, achieving a complete data set requires interpolations/extrapolations and/or generating new conversion factors. During neutron transport, neutron and photon fluxes are tallied in the voxel mesh which is superimposed over the patient model. Choice of voxel size and assigned composition for calculation are discussed in Section 11.1.1.

An alternative numerical method to solve the neutron transport equation is the deterministic discrete ordinate method, which requires discretizing the model in space (structured meshes), in energy (multi group flux and cross sections), and in angular direction (particle transport only considered along discrete directions). A deterministic method can be convenient and accurate when undertaken with optimally chosen parameters, but inadequate angular directions, mesh structure or coarse energy grouping can result in poor solutions. Deterministic methods are widely used in reactor core calculations and have been applied in BNCT dosimetry [612–613]. Some deterministic methods have also been proposed for BNCT treatment planning [613–615].

11.1.4 Quality assurance and calculation accuracy

Quality assurance (QA) of the treatment planning is essential to ensure accurate dose delivery to the patient. Commissioning of the TPS is one of the most important parts of the QA for a radiotherapy program [616–617], and the commissioning data of the AB-BNCT system can be used as references for periodic QA checks. Reference values of the facility for various quantities are determined by repeated irradiation tests during the commissioning period.

An important step in commissioning is to evaluate the 'beam' characteristics in air (Section 4.2) and in phantom (Section 4.3). When commissioning the TPS, tests to acquire data to normalize the output of the TPS to measured values are required. To normalize the thermal neutron fluence output by the TPS, the distribution of thermal neutron fluence measured within suitable phantoms can be used. It is important to set criteria for thermal neutron fluence or reaction rate of activation foils at the surface, peak, and therapeutic depth in the phantom. Similarly, the γ ray dose rate distribution measured within suitable phantoms can be used to calibrate the γ ray source model used by the TPS. The Monte Carlo code MCNP[31] has been previously suggested as a benchmark for BNCT dose calculations because it can accurately calculate BNCT dose distributions [1].

Traditionally, nearly head sized plastic or water phantoms (diameter ~20 cm) of several shapes (parallelepiped, cylindrical, cubic, ellipsoid, or anthropomorphic) have been used in the QA of TPSs for BNCT. Daily QA and calibration of the MU is typically performed at the 'reference depth' in a small phantom. In BNCT performed with epithermal neutrons, the reference depth is usually located at the depth of the thermal neutron maximum (typically ~2 cm). In addition, during commissioning of a TPS, the calculated depth and radial dose profiles need to be compared in a sufficient size of water phantom, so that each dose component can be defined down to the 5% isodose level to verify the 'beam' model. The phantom needs to be large enough so that the full scattering condition is reached at each measurement point for each dose component. It is strongly suggested that the treatment planning dose calculations are verified in an inhomogeneous anthropomorphic geometry [118]. It is also important to verify that the TPS correctly accounts for the presence of nitrogen in tissues and its effects on neutron transport, which can be achieved using a tissue-equivalent phantom material containing nitrogen, such as A-150 plastic [618].

---

[31] MCNP: https://mcnp.lanl.gov/

Ideally, QA guidelines would be defined for BNCT based on those defined for conventional radiotherapy, where applicable. No such clear guidance for BNCT treatment planning and dose calculation QA has been defined. However, Ref. [107] suggests a program and schedule for QA measurements for use in BNCT.

The accuracy of dose calculation is limited by the uncertainty of the nuclear interaction cross sections, applied KERMA factors (derived from the cross sections), statistical error of the Monte Carlo calculation, and reference dose measurement accuracy. Uncertainty in the ENDF-VII cross sections for the interactions (see Section 10.1.2) that give rise to the four main dose components (discussed in Section 10.3.3.1) are typically < 1% [572]. The statistical uncertainty of the Monte Carlo calculation in clinically important parts of the patient model needs to be as low as reasonably possible, so that the total uncertainty of the dose delivery is not affected.

Acceptance criteria for the commissioning of the dose calculations depend on the uncertainty of the applied reference dosimetry methods, and typical values are given in Section 4.7. Several uncertainty analyses have concluded that a reasonable uncertainty in total dose delivery is 7.0% (1σ), excluding the uncertainty in the boron concentration in tissues [619–622].

11.1.5 Monitor unit definition

Each treatment plan provides monitor units (MUs) for each field which is delivered during the treatment. The IEC requires that the radiation field incident on the patient is monitored with two independent measurement channels [623]. However, clear guidance on MU definition for BNCT has not been created. As examples:

- At reactors in 'the West', following guidance for conventional radiotherapy equipment, the neutron 'beam' was typically monitored with fission chambers, and MUs were defined based on neutron counts (Table   TABLE 9). In Japan, fission chambers were not used at reactors for regulatory reasons. They may also be difficult to use in a hospital, where other types of neutron detectors may be considered (see Table   TABLE 9);
- In some AB-BNCT centres, MUs have been defined based on proton beam monitor counts (see discussion in Section 4.6), which may work if the neutron yield per proton current remains consistent over the long term or if the relationships between the neutron and photon yields and the charged particle current have been sufficiently well evaluated over the lifetime of the target (see Ref. [58] and Annex VII for Li targets).

Typically, MUs are calibrated against activation measurements at the reference depth in a phantom (Section 4.6). However, there is no consensus as to what the calibration quantity has to be. In some centres, MUs are defined in relation to Au foil activation while in others they may be defined on the basis of thermal neutron fluence or on weighted dose (now isoeffective dose) to healthy brain [621, 624]. It is necessary to confirm that the thermal neutron fluence or reaction rate of the activation foils per MU does not deviate from the reference value determined during commissioning.

11.2 AVAILABLE TREATMENT PLANNING SYSTEMS

Several new TPSs have been developed for BNCT and the systems currently available or under development for BNCT and their major parameters are listed in Table 23. The four older systems developed specifically for BNCT use and applied clinically are Simulation Environment for Radiotherapy Applications (SERA) [610], NCTPlan [625], Japan Atomic

Energy Agency Computational Dosimetry System (JCDS) [626] and THORplan [627, 606].[32] These four are non-commercial systems, developed by small teams with expertise in nuclear engineering at universities and research institutes, for reactor based BNCT. None of them have been approved as a medical device, and neither have the reactors themselves.

TABLE 23. TREATMENT PLANNING SYSTEMS DEVELOPED FOR BNCT

	TPS	Developers	MC code	User site
Research Reactors	NCTPlan	Harvard-MIT-CNEA	MCNP	MITR (MIT), RA-6 (CNEA)
	SERA	INEEL/MSU	seraMC	KUR (Kyoto) FiR 1 (Helsinki University Hospital) Studsvik (Sweden) Petten (Netherlands)
	JCDS	JAEA	MCNP/PHITS	JRR-4 (JAEA)
	THORplan	Tsing Hua Univ.	MCNP	THOR, Hsinchu City
Accelerators	Tsukuba Plan	Univ. of Tsukuba	PHITS	iBNCT, University of Tsukuba Hospital
	Dose Cure Engine	SHI	PHITS	Southern Tohoku BNCT Research Center, Kansai BNCT Medical College
	NeuCure Dose Engine	RaySearch Laboratories in collaboration with SHI, NT, and TAE LS.	PHITS	Southern Tohoku BNCT Research Center, Kansai Medical BNCT Center
			A GEANT4 based dose engine by NT	Helsinki University Hospital, Shonan Kamakura General Hospital
			A dose engine by TAE LS	
	DM-BTPS	Dawon Medax	seraMC	DM-BNCT, Dawon Medax, Seoul
	NeuManta	Neuboron	COMPASS (dose engine) with support for PHITS and MCNP	Xiamen BNCT Center, Xiamen
	To be decided	CICS and NCC	PHITS	National Cancer Center, Tokyo

Note: SHI = Sumitomo Heavy Industries; NT = Neutron Therapeutics Inc; INEEL = Idaho National Engineering and Environmental Laboratory; JAEA = Japan Atomic Energy Agency; TAE LS = TAE Life Sciences; MSU = Montana State University; CNEA = Comisión Nacional de Energía Atómica, Argentina, CICS = Cancer Intelligence Care Systems, Inc.

While any of the first four TPSs can be applied to AB-BNCT, for accelerators to be approved as medical devices for BNCT requires development of new TPSs that themselves are approved as medical devices. Three BNCT accelerator vendors, Sumitomo Heavy Industry Co., Neutron Therapeutics Inc, and TAE Life Science, are collaborating with RaySearch Laboratories to add BNCT dose calculation to the RayStation system, which is commercially available for conventional radiotherapy. In addition, a TPS is under development for AB-BNCT at Tsukuba University, Japan, and another is being developed as part of the initiatives on-going in Xiamen, China (see Annex III).

---

[32] The precursors of the NCT_Plan and SERA were called MacNCTPlan and BNCT_Rtpe with minor differences from the more recent versions.

11.3 PRESCRIBING

Tumour dose in each cell from BNCT is uncertain at present because of the heterogeneity of BPA uptake within tumours (Section 9.4) and the dose needed for desired clinical outcome is unknown. Therefore, BNCT is typically prescribed based on the maximum tolerable dose to healthy tissues. Since a limited amount of toxicity data about BNCT is available from clinical experience, the tolerable doses have often been estimated based on radiobiological studies (Section 9) and conventional radiotherapy data. The toxicity profiles of brain tissue and mucosal membrane are best known because the majority of BNCT treatments have been delivered to patients with brain and head and neck cancer, mainly using BPA as the boron carrier. When neutron 'beams' with large components of fast or thermal neutrons are applied in clinical BNCT, the treatment is often prescribed based on the maximum dose to skin, because skin receives the highest dose due to high proton recoil dose (D_H, Table 7), its radiosensitivity, and its comparatively high BPA uptake [378]. When an epithermal neutron 'beam' with low contamination by thermal and fast neutrons is used, sensitive, healthy tissues deeper in the patient might be of concern. When BNCT is administered as a re-irradiation, the doses from the first line radiotherapy need to be considered when BNCT doses are prescribed.

BNCT is often prescribed for brain cancer based on maximum and average dose to normal brain tissue and to sensitive structures within or near the brain, like the lens, eyes, and optic nerves. When BNCT was applied as first-line radiotherapy for newly diagnosed brain cancer, using the units then current in the field, maximal prescribed doses to normal brain were 8–15 Gy (W)[33] and average radiation doses to normal brain were 2–6 Gy (W) without grade 3 or 4 toxicity (see Refs [160, 163, 628–632] and Table 24). When BNCT has been applied as re-irradiation of the brain, maximal prescribed doses to normal brain were 6–14 Gy (W) and average doses to normal brain were 2–6 Gy (W) [167, 408, 632].

Mucosal membranes are sensitive to BPA-mediated BNCT because of their high BPA uptake and the general radiosensitivity of all rapidly dividing cells. Furthermore, when using epithermal neutron 'beams' for head and neck cancer, the highest neutron flux is often achieved at the location of the oral mucosal membrane, where the tumour often grows. The mucosal membrane around the cancer, sometimes called precancerous tissue, is found to be even more sensitive to BPA-mediated BNCT [633]. Therefore, the dose prescription to the mucosal membrane must be carefully considered for such cases. The maximal prescribed dose to mucosal membranes has been reported as 10–12 Gy (W)[33] [222, 237] or absorbed dose of 5–6 Gy [167, 408].

The calculated dose depends on the values of several variables, such as the nitrogen and hydrogen concentrations assumed for the sensitive organ, which CBE and RBE factors (or photon isoeffective dose model) parameters are used (Section 10), and the T/B ratio assumed. CBE and RBE factors (see Section 9.2) are selected as fixed values based on the experimental conditions and may differ for different biological endpoints. In principle, the T/B ratios can be defined for each patient if an ^{18}F-FBPA PET study has been performed before the treatment (see Sections 7.6.3.2 and 8.1.2). However, the T/B ratio is known to vary with time and the patient's condition. In addition, the administration methods for BPA and ^{18}F-FBPA are different (see Section 7.4.3). Accordingly, the true value of the T/B ratio during BNCT might differ from that estimated based on ^{18}F-FBPA PET. Furthermore, the tumour tissue includes heterogeneous tumour cells, stromal and vascular cells, and other normal cells, so that the boron concentrations within each tumour cell vary from the estimated boron concentration value based on a

---

[33] Here 'Gy (W)' refers to a specific weighting scheme applied to the physical component doses and what now would be termed as isoeffective dose, Gy (IsoE). Refer to Section 10.3.4.

macroscopic ^{18}F-FBPA PET image. At present, there is no method of precisely predicting boron concentrations for BNCT (see Section 8) or the absorbed doses of each tumour cell. When a BNCT dose is prescribed, all these details need to be taken into consideration. If different parameters are used, the prescribed doses are not equivalent.

In the future, when absorbed dose to each cancer cell can be calculated accurately taking into account tumour heterogeneity and boron uptake, and when a significant number of patient datasets have been collected over time, therapeutically optimal tumour doses will become apparent, and BNCT will be prescribed accordingly. Given that BNCT is administered in various ways (different BPA infusion doses and times, and irradiation timings), the doses may not provide the same end point. More needs to be learnt about boron dose kinetics within the tumour during and after BPA administration.

12 DOSE AND VOLUME SPECIFICATION FOR REPORTING

The first IAEA publication on BNCT [1], published in 2001, was written in accordance with ICRU Report 50 [634], the guideline for the standardization of radiotherapy at that time. In the intervening two decades, the method of treatment planning for BNCT has changed, and radiotherapy in general has developed. This section contains information that may help to harmonize reporting in BNCT across different centres. Annex XIX describes practice and concepts from China.

12.1 THE NEEDS FOR HARMONIZATION OF DOSE REPORTING

The ICRU has published several reports to promote harmonization of methods for the reporting of radiotherapy applications. In 1993, ICRU Report 50 defined several volumes used to report treatment planning and processes for tumours and normal tissues [634]. Later, with the availability of 4D imaging and treatment, more detailed volume dosimetry was required, and ICRU Report 62 [635] was published in 1999 as a revised version of ICRU Report 50. With the widespread use of intensity-modulated radiation therapy, ICRU Report 83 was published in 2010 to deal with complex treatment plans [636].

In Japan, an AB-BNCT system was approved by regulators for specific indications in March 2020, and clinical practice commenced under public medical insurance in June 2020. Two centres, the Southern Tohoku BNCT Research Center and the Kansai BNCT Medical Center, (Annex X) are currently in clinical operation. Furthermore, since November 2019, the National Cancer Center Hospital has been undergoing a clinical trial using another AB-BNCT technology (Table 5). The number of centres is likely to grow, and AB-BNCT systems based on a variety of technologies will be developed around the world. This situation calls for greater harmonization of dose reporting for BNCT.

12.2 REPORTING CLINICAL DATA

Three types of data can be identified when reporting an oncology treatment (radiotherapy, in general, or BNCT treatments):

- Patient-related data;
- Radiotherapy data;
- Data specific to BNCT.

Ideally, complete, reliable oncological together with other clinical data are reported for any oncological treatment. Patient-related data include name, age, tumour extent and grading according to the TNM Classification of Malignant Tumours [637] and pathological diagnosis. Biological sex may be included to determine the elemental composition of the body for BNCT dose calculation. Height and weight are essential measures as they relate to the amount of boron drug to be administered and to the safe use of AB-BNCT.

12.3 ONCOLOGICAL CONCEPTS: GROSS TUMOUR AND CLINICAL TARGET VOLUMES

Gross tumour volume (GTV) and clinical target volume (CTV) were defined in ICRU Report 50 [634] and shown schematically in Fig. 64. The GTV and CTV are general oncological concepts originating from conventional radiotherapy and are commonly adopted for use in BNCT. The GTV is the "gross palpable or visible/clinically demonstrable location and extent of the malignant growth" [634]. The GTV may appear in different sizes and shapes depending on the examination technique used for evaluation. Therefore, it is preferable that the technique used for evaluation be reported. "The clinical target volume (CTV) is the tissue volume that contains a demonstrable GTV and/or sub-clinical microscopic malignant disease, which has to be eliminated. This volume thus has to be treated adequately in order to achieve the aim of therapy, cure or palliation" [634].

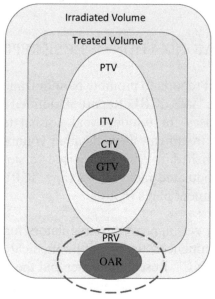

FIG. 64. General ICRU concepts of target volumes and organs at risk in common use in radiotherapy.

12.4 SPATIAL CONCEPTS: PLANNING AND INTERNAL TARGET VOLUMES

In conventional radiotherapy, the planning target volume (PTV), defined in ICRU Report 50 [634], is selected when choosing beam sizes and arrangements to ensure that the CTV receives the prescribed dose. The PTV compensates for physiological changes in position, shape, and size of the CTV (e.g., due to breathing), along with inaccuracies in the set-up and beam delivery. In ICRU Report 62 [635], the physiological changes in the CTV were defined as the internal target volume (ITV) (Fig. 64), which includes the CTV and an internal margin.

Both PTV and ITV are geometrical concepts, intended to put a spatial safety margin around the CTV for treatment planning purposes to ensure that the prescribed dose is adequately delivered to the CTV, taking into account internal and external geometric variations of this volume. In the context of BNCT, these geometrical concepts may need to be carefully considered, as suggested in Annex XIX. Geometrical safety margins do not guarantee delivering the prescribed dose to the CTV in BNCT, because the change of dose to the CTV by geometrical deviation is not compensated by adding the margin. The geometrical deviation of the CTV would fundamentally change the conditions of the neutron transport calculations. In BNCT, a Monte Carlo code is used for neutron transport calculations to generate multiple particle events, and the final distribution is obtained as the accumulation of these events. In order to improve efficiency of calculation, the patient 3D model is converted into a 'coarse'[34] voxel model, and the absorbed dose for each averaged voxel model is calculated (Sections 11.1.1 and 11.1.3). Those voxels at the boundaries of the CTV may be modelled as being filled with tumour or another tissue, with the result that the dose to the CTV may not be accurately described.

---

[34] At Southern Tohoku BNCT Research Center and at Kansai BNCT Medical Center, 'coarse' voxelization is typically defined as 5 mm × 5 mm × 5 mm, and final dose calculation is done with 2 mm × 2 mm × 2 mm voxels.

12.5 ORGANS AT RISK AND ORGANS TO BE PRESCRIBED

Organs at risk (OAR) are organs defined as normal tissues that are sensitive to BNCT or at risk of serious adverse effects from it (Fig. 64). Due to uncertainties in RBE values and in the boron concentration in tumours, dose prescriptions based on the dose constraints for normal tissue are often adopted, and the 'organs to be prescribed' are often specified as the most BNCT-sensitive OARs in the region to be irradiated. Additionally, because BNCT has also been applied to re-irradiation, OARs have to include tissues that may exceed their maximum tolerable dose even if they have low sensitivity for BNCT.

Currently, for most cases, the organs to be prescribed are normal brain tissue for brain irradiation and mucosal tissue for head and neck irradiation. However, depending on the location of the tumour and the condition of the patient, the dose limiting organ may be different (see Section 11.3).[35] In the case of a brain tumour, normal brain tissue is easily identified. However, for organs such as mucous membranes of the oral cavity and pharynx, because the contouring can vary greatly depending on the observer, identification of those mucosa has to be performed carefully. Interobserver errors for the contouring, as well as internal and external geometric variations of this volume, may be taken into account by employing a 'planning organ at risk volume', abbreviated as the PRV (Fig. 64). However, as is the case for the PTV, in the context of BNCT, this geometrical concept may need to be carefully considered, as suggested in Annex XIX.

12.6 REPORTING DOSE

Recommendations related to the recording and reporting of doses have changed over time. The previous IAEA publication [1] suggested dosimetry at the reference point, based on ICRU Report 29 [638]. After the introduction of 3D conformal radiotherapy and intensity modulated radiotherapy, 3D dose distribution needed to be reported. With the spread of CT simulators and treatment planning devices, the 3D dose distribution is easily obtained, and volume based dosimetry is now strongly suggested because it has less uncertainty in absorbed dose assessment. According to ICRU Report 83 [636], there are three levels of dosimetry:

- Level 1: Basic techniques. This level is the minimum standard required in all centres. Dose at the reference point, maximum dose in the PTV, and minimum dose are required to be reported in conventional radiotherapy with X rays. Such a reference point is meaningless for BNCT (see Section 12.6.2);
- Level 2: Advanced techniques. Level 2 includes items required for accurate exchange of information among institutions, such as 3D dose distribution, dose–volume histogram, and other volume dosimetry items, as well as the processes used to determine target volume and OARs. It seems suitable that the near-minimal (D98%), near-maximal (D2%), and median (D50%) doses of GTVs and CTVs be reported for BNCT, in accordance with the recommendations of ICRU Report 83 [636];
- Level 3: Developmental techniques. Level 3 is an evaluation item for which the reporting format has not yet been determined, and which may become Level 2 in the future, depending on the development of new treatments, technologies and equipment.

---

[35] Depending on the condition of the patient (such as pre- or post-surgery/external beam radiation therapy) and the location of the tumour, the dose limiting organ may vary; e.g., at Kansai BNCT Medical Center, for head and neck cancer, the dose to the skin, the eye, and brain are taken into consideration as the organs to be prescribed.

Information concerning dose configuration on the standard phantom specific to each AB-BNCT system ought to be reported as Level 1 information. While some authors have suggested that items up to Level 2 need to be reported clinically [635], there is no generally accepted and widely adopted standard for dose reporting in BNCT.

12.6.1 Dose components

At any point in the irradiated tissues, one can identify four components contributing to the absorbed dose (Table 7). Since there is no unique relationship between total absorbed dose and induced biological effects, it is necessary to apply a model to estimate the effect of the integrated dose from these dose components. Section 10.3 presents different models introduced in BNCT for this purpose, highlighting their range of validity and limitations.

BNCT is a radiation treatment delivered under non-reference conditions. Thus, different dose values ought to be reported:

(a) Absorbed doses D_B, D_N, D_H and D_γ and the total absorbed dose D, in Gy;
(b) Photon isoeffective dose, D_{IsoE}, in Gy (IsoE).[36]

When reporting the photon isoeffective dose, additional relevant information can be added in parentheses when considered useful. For example, if a tumour receives 30 Gy (IsoE) computed with Eq. (24) and coefficients of each radiation $\alpha_R = 0.5$; $\beta_R = 0.03$; $\alpha_1 (= \alpha_B) = 0.5$, etc., then the description could be:

"Tumour received 30 Gy (IsoE; $\alpha_R = 0.5$; $\beta_R = 0.03$; $\alpha_B = 0.5$, …)."

Or, if a tumour receives 30 Gy (IsoE) computed with constant RBE and CBE factors, using Eq. (18), the description could be

"Tumour received 30 Gy (IsoE; $RBE_H = 3.2$, $RBE_N = 3.2$, $CBE = 3.8$)."

At present, BNCT doses are reported as the total weighted dose calculated from the sum of each dose component (as above). However, it is not clear whether different dose component compositions result in different antitumour effects or different normal tissue reactions. To examine this, a large volume of information in the form of 3D dose distributions for each dose component would need to be reported. However, such suggestions are based on developing methods that go far beyond common RBE-weighted dosimetry and are not essential to clinical practice at present. Such reports would fall under Level 3.

12.6.2 Reference point(s) and volumes for prescription and reporting

Specific reference points currently used for conventional photon beam therapy are not suitable for BNCT. Because of the poor dose uniformity in BNCT, a volumetric dose prescription needs to be considered. Thus, it may be preferable that the dose constraints are determined by volumetric dose parameters. However, there have as yet been no clear rules for volumetric dose constraints based on scientific evidence and point dose prescriptions based on the maximum organ dose or prescriptions based on the facility's own volumetric dose constraints are tentatively used.

---

[36] The '(IsoE)' after 'Gy' denotes the photon isoeffective dose D_{IsoE}, not absorbed dose. A space between 'Gy' and '(IsoE)' is needed to comply with SI notation.

12.6.3 Boron-10 concentration and concentration ratios

The method by which boron concentration in tumour tissue is evaluated needs to be reported. The measurement conditions as well as the time elapsed since the start of boron administration to the measurement need to be reported. Methods of boron concentration determination used at AB-BNCT facilities include blood concentration measurements using ICP-OES or ICP-MS which are then multiplied by the T/B ratio (see Section 8) to calculate tumour concentration. The time when the blood was collected, whether whole blood or serum was used, and details of the method used all need to be reported for ICP methods. As discussed in Sections 7.8–7.9 and 8.1.2, in the case of BNCT using BPA, the distribution of BPA in the body may be assessed beforehand using ^{18}F-FBPA PET. The quantitative nature of standardized uptake values (SUVs) derived from ^{18}F-FBPA PET depends not only on the resolution of the PET device but also on the tumour conditions such as cell density or infiltration, and size of tumour. Therefore, when introducing T/B or T/N based on SUV parameters derived from ^{18}F-FBPA PET, it is preferable that the PET-CT devices used, the imaging conditions, and tumour conditions all be reported as supplementary information to the quantitative properties. On the other hand, if a value derived independently of ^{18}F-FBPA PET is used, that value and the method used to determine it need to be reported.

12.6.4 Tissue element composition and biological effectiveness

The elemental composition of each tissue used in the transport calculation is often based on data taken from ICRU Report 46 [572] (see Section 10.1.2). However, the choice of data set may differ among facilities depending on the patient's age and sex, and, therefore, needs to be reported. In addition, the data set of RBE values is considered a Level 1 reportable item because institution-specific values may be adopted for each organ, taking the accumulated data of clinical cases into account.

12.6.5 Equipment configuration and treatment planning process

In the case of an AB-BNCT system, the parameters related to the treatment device including the accelerating voltage and charge (time-integrated proton or deuteron current) delivered by the accelerator, and the size (diameter and height) of the collimator of the irradiation system need to be reported. For treatment planning, the computational processes adopted, including the voxelization size and composition adopted in the voxels for the transport calculations (multiple tissue averaging, tissue replacement based on the occupied volume fraction), Monte Carlo code used in the computational process, the cross-section data used by the Monte Carlo code, the calculated MU, the parameters related to the statistical accuracy adopted in the Monte Carlo calculation, the parameters related to the smoothing of the neutron flux distribution, and the voxel size adopted to apply the RBE to the neutron flux distribution also need to be reported.

12.6.6 Patient position and positioning error

Uncertainties in X ray radiation therapy due to set-up errors are compensated by providing spatial safety margins. However, spatial safety margins cannot guarantee delivery of the prescribed dose to the tumour in BNCT (Sections 12.4–12.5). Errors in patient positioning fundamentally alter the transport calculation conditions for BNCT. In addition, because of the lack of dose uniformity in BNCT, deviation in position brings a substantial change to RBE weighted doses. Therefore, positional uncertainties need to be reduced as much as possible by robust immobilization of the patient. For this reason, it is necessary to report the patient treatment position adopted (sitting or supine position) as an indicator of potential positional deviation during treatment.

12.7 ADDITIONAL POINTS OF NOTE

Harmonized reporting methods that are neither excessive nor insufficient are needed to exchange treatment information among BNCT facilities. However, as stressed in various ICRU reports, the concepts of target volumes, selection of OAR, dosimetry and reference points etc. have changed with time. It has become more important to depict their uncertainties and lessen the causes of those uncertainties for BNCT. Commercially available TPSs do not yet have sufficient reporting functions for all the reporting information outlined above. This needs to be improved in the future. By adhering to the proposals outlined above, the variation of reporting within and across institutions can be reduced, which will lead to standardization of this treatment.

13 CLINICAL TRIAL DESIGN AND PROCEDURES

This section gives an overview of clinical trials that have examined BNCT for various cancer types. Some of the earliest trials are described in detail in Annex XII.

13.1 BACKGROUND OF MALIGNANT GLIOMAS INCLUDING GLIOBLASTOMA

A high-grade glioma with predominantly glial differentiation is called a malignant glioma (MG). The latest WHO classification of tumours [639] of the central nervous system incorporates not only morphological but also molecular components of diagnosis into MGs:

- The most important molecular marker for MGs is immunoglobulin heavy chain (IDH) [640]. IDH-wild type glioblastoma is the most common and most malignant astrocytic glioma, accounting for 90% of all glioblastomas. Its synonym, 'IDH-wild type primary glioblastoma', indicates that this glioblastoma usually arises without a recognizable lower-grade precursor lesion. The tumour diffusely infiltrates adjacent and distant brain structures. IDH-mutant, generally secondary, glioblastoma has a better prognosis than IDH-wild type glioblastoma;
- The other molecular marker that may affect patient prognosis is O6-methylguanine-DNA-methyltransferase (MGMT). The methylation status of the MGMT promotor of the tumour influences the sensitivity of the tumour to alkylating agents, such as temozolomide. Tumours with methylated MGMT promotor status show better responses to alkylating agents than do tumours with unmethylated promoter status. The annual incidence is about 3–4 cases per 100,000 population in the USA [641], whereas the incidence is relatively low in eastern Asia, such as in Japan and Korea [642].

For MG, surgery and postoperative chemo-radiotherapy are the standard of care, and the standard chemotherapeutic agent used for MG is temozolomide alone [643], but in some cases, it is used in combination with bevacizumab [644]. The usual initial treatment is fractionated external beam X ray radiation, but for recurrence there are limited or no treatment options. Thus, external beam XRT with concomitant temozolomide administration is generally used for newly diagnosed glioblastoma. However, this treatment regimen shows median overall survival (mOS) of only 14.5 months after diagnosis. Overall, temozolomide prolongs the mOS by only 2.5 months compared to XRT alone after surgery.

For recurrent MGs, especially for recurrent glioblastoma, no standard of care has been established. Since radiotherapy is usually applied as a first line treatment, its use at the time of recurrence is normally discouraged. In these situations, chemotherapy is looked to as a treatment for relapse. Many drugs have been tested, including bevacizumab and dose-dense temozolomide. Moreover, a new medical device, the NovoTTF-100A system, has recently been developed, but to date nothing has prolonged the overall survival of patients with recurrence. The 1-year survival rate is 34.5% for patients receiving bevacizumab, 20% for Novo-TTF, and < 30% for stereotactic radiotherapy [645–647].

13.2 BACKGROUND OF HIGH-GRADE MENINGIOMA

Meningiomas are mostly benign, slow-growing neoplasms that most likely derive from the meningothelial cells of the arachnoid layer. Meningiomas generally produce neurological signs and symptoms due to their compression of adjacent structures. Headache and seizures are common but non-specific presentations.

Most meningiomas correspond histologically to WHO grade I (benign). Certain histological subtypes or meningiomas with less favourable clinical outcomes correspond to WHO grades II and III. The lifetime risk of developing meningioma is approximately 1%. At 36% of all brain tumours, it is the most commonly reported brain tumour in the USA. WHO grade II and III meningiomas are termed high-grade meningiomas (HGMs), and ca. 20–25% and 1–6% of meningiomas fall into these two categories, respectively [648–650]. The five-year recurrence rate of HGM has been reported as 78–84% [651] and the median survival time is 6.89 years; late mortality occurs in 69% of cases due to recurrence after initial surgery [652]. HGMs that recur after radiotherapy have a particularly poor prognosis: mOS and the median progression-free survival values are 24.6 and 5.2 months, respectively [653]. No standard treatment for HGM has been established [654]. Therefore, effective treatments HGMs, especially for recurrent and refractory ones, need to be developed.

13.3   BNCT FOR MALIGNANT GLIOMAS

Initially, clinical interest and application of BNCT was focused on high-grade gliomas [659–661], chiefly glioblastomas (GBM) [155, 655]. Clinical application of BNCT was initiated in USA using Brookhaven National Laboratory in the 1950s and is described in Annex XII and in Refs [656–658].

13.3.1 Newly diagnosed glioblastomas

The results of BNCT theoretically depend on the product of boron concentration in tumour tissue or tumour cells and the fluence of neutrons that reaches them. The IAEA TECDOC 1223 published in 2001 [1] reported the results of several clinical studies using thermal neutron beams derived from nuclear reactors [659–664]. In addition, from 1960 to 1961 a clinical trial was conducted by Massachusetts General Hospital using the Massachusetts Institute of Technology research reactor. Seventeen cases of MGs were treated with BNCT in this study, but the clinical results were disappointing: the mOS after BNCT was only 87 days (see Annex XII). Since remedied, the major deficiencies at that time include the use of:

- Thermal neutron beams. While the neutron capture reaction is more probable with neutrons in the thermal energy range (Fig. 16), many competing scattering and absorption reactions increase in cross section in this neutron energy range, so that the beam becomes strongly depleted on the way to the tumour. Epithermal neutrons more readily penetrate the body, and the tissues thermally moderate the neutrons at depth within the body. Therefore, after the 1990's, epithermal beams were used in many reactors around the world for BNCT;
- Non-optimized boron compounds with limited specificity.

The BNCT clinical trial results for GBM (chiefly, newly diagnosed and partly recurrent cases) are summarized in Table 24, with data taken from review articles [128, 161]. (Reference [665] also compares BNCT to conventional treatments for newly diagnosed GBM.) From the 1990's, epithermal beams became available in the USA, Germany, Finland, Sweden, Czech Republic, Taiwan, China, and Japan for brain tumours. As boron carriers, only BPA and BSH were examined (see Section 7). There were several modifications in each clinical study and case numbers were very limited, and therefore systematic review (meta-analysis) for this field was difficult.

However, some trends can be deduced from Table 24.

- The American studies of the 1990's showed marked improvement in clinical results compared to those from previous studies conducted using thermal 'beams';
- The European studies showed almost the same level of response as the American studies [643];
- In some of the Japanese trials, investigators used BSH and BPA simultaneously. These compounds accumulate in different fashions to different subpopulations in glioma cells and tissues (see Section 9.6.1). Therefore, a combination of both compounds might compensate each other's weak points. Japanese researchers also applied additional XRT, which seemed to improve the clinical results of BNCT for newly diagnosed GBM with mOS of up to 23.5 to 27.1 months. These results seem to be good in comparison to large-scale clinical trials for newly diagnosed GBM [643, 644, 666]. Follow-on XRT might make up for any deficiency of ^{10}B atoms in tumour cells compared to that intended from the boron compound prescription. It could also enhance the delivered dose, especially for deeply seated tumours, where the neutron fluence has been attenuated. Prospective multi-centre clinical trials of BNCT and additional XRT with the combination of temozolomide for newly diagnosed GBM have been completed (Osaka-TRIBRAIN 0902, NCT00974987) and the data are now in preparation for publication [667]. In this prospective study, the mOS and 2-year survival rate are 21.1 months and 45.5 %, respectively. The clinical regimen for newly diagnosed GBM used in this study has been published elsewhere [217–218].

13.3.2 Recurrent glioblastomas

Initially, BNCT was mainly applied to recurrent MGs: marked tumour shrinkage was recognized in neuroimages in the initial series of trials [217, 668]. In 8 out of 12 cases, the majority of contrast enhanced lesions had disappeared during follow-up [217, 552]. The survival data of some clinical studies of BNCT for recurrent MGs are also summarized in Table 24: the mOS for BNCT alone for recurrent GBM was 7–10.8 months. For newly diagnosed GBM, some large-scale clinical trials with chemo-radiotherapy have been reported as mentioned above [128, 643–644], but only a few reports have been published for recurrent GBM. Therefore, it is difficult to judge whether the mOS achieved by BNCT for recurrent GBM is good or not. In 2007, an article concerning recursive portioning analysis applied to recurrent gliomas was published [669]. In subsequent work, BNCT results for recurrent gliomas were analyzed based on their recursive portioning analysis classification and it was found that BNCT prolonged the survival of recurrent MG cases in every recursive portioning analysis class and, moreover, BNCT achieved excellent prolongation especially for the cases in poor classes [670]; the clinical regimen used for the recurrent MGs has been published [632].

The biggest shortcoming in BNCT for recurrent MGs is brain radiation necrosis (BRN). Almost all cases of recurrent MGs have already received nearly 60 Gy XRT as an initial radiotherapy prior to second radiation treatment at the recurrence. Even with the tumour-selective particle radiation of BNCT, BRN often occurs after BNCT in recurrent glioma cases, which may cause severe brain oedema leading to severe neurological deficits that can sometimes be life threatening. Bevacizumab, an anti-VEGF antibody, is a powerful weapon to treat BRN [671–672] and is also useful in BRN after BNCT [559, 673–674]. In Fig. 65, a representative case of recurrent GBM treated by BNCT in Osaka Medical College, followed by successful treatment of BRN with bevacizumab is shown.

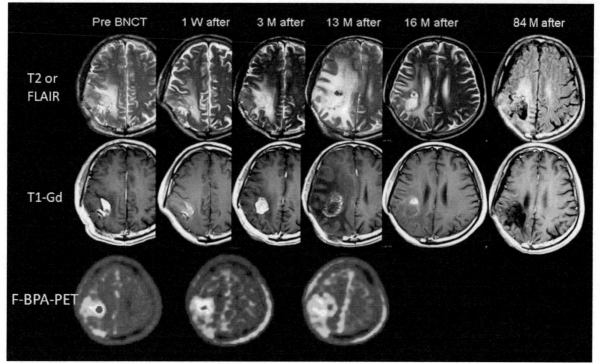

FIG. 65. A representative case of recurrent GBM treated by BNCT with successive bevacizumab experienced by Osaka Medical College. The ^{18}F-FBPA-PET image indicated marked tracer uptake by the right parietal region with a 3.8 lesion/normal (L/N) ratio, indicating that the lesion was a GBM that had recurred after standard chemoradiotherapy. The patient was given BNCT. Periodic MRIs showed that both the enhanced lesion and perifocal oedema gradually enlarged, whereas uptake of ^{18}F-FBPA-PET gradually decreased. Thirteen months after BNCT, the L/N ratio was 2.3, suggesting the lesion was BRN. The patient was given four biweekly treatments of intravenous bevacizumab (5 mg/kg), and MRI showed marked improvement in the perifocal oedema and left hemiparesis. After 84 months following BNCT, the patient was doing well, without progression of tumour or recurrence of BRN (courtesy of S. Miyatake, Osaka Medical College).

A pilot clinical study of administration of bevacizumab just after BNCT for recurrent MGs was undertaken [558], which demonstrated the prevention of BRN and potential clinical benefits.

13.4 BNCT FOR HIGH-GRADE MENINGIOMA

As stated in Section 13.2, no standard treatments have been established for refractory and recurrent HGMs. Since 2005, reactor based BNCT was applied to 44 recurrent HGMs [675–677]. A representative case is shown in Fig. 66. In all cases, good shrinkage of the mass was observed as an initial response to BNCT. However, many cases were lost due to systemic metastasis and distant intracranial recurrence outside of the neutron irradiation field [675]. The clinical regimen used for reactor based BNCT for HGM has been published [675]. Now, a randomized trial of AB-BNCT for recurrent HGMs is being performed, comparing a group that receives BNCT treatment and another in best supportive care, where the estimation of progression free survival is the primary endpoint.

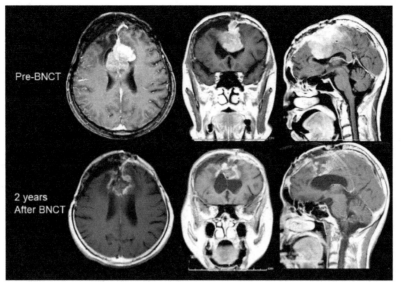

FIG. 66. A representative case of recurrent HGM. The patient had recurrent anaplastic meningioma, undertook three operations, stereotactic radiosurgery and XRT. The lesion recurred and reactor based BNCT was administered. Two years after BNCT, the lesion had shrunk vigorously without neurological deterioration (courtesy of S. Miyatake, Osaka Medical College).

13.5 FROM REACTOR TO ACCELERATOR

Up until 2012, all clinical BNCT treatments were performed using neutrons generated by a research reactor. Not only has BNCT demonstrated potency against MGs and HGMs, it has also shown promising effects for recurrent and refractory head and neck cancers [165–166, 408, 551, 678–679]. Despite this, BNCT is not yet a standard treatment modality for many cancers. The operation and maintenance of nuclear reactors for clinical purposes was a major impediment. More than 10 reactors were used for clinical BNCT activities in the USA, Europe, Argentina, and Asia; however, currently only two reactors (one in Taiwan, China and one in Japan) still perform clinical BNCT. After the severe accidents at the Fukushima-Daiichi nuclear power plant in 2011, oversight and regulatory controls of nuclear reactors have increased. Therefore, compact accelerator based neutron sources for clinical BNCT are being strongly considered around the world. The two main obstacles that had to be overcome were the target cooling and stability of the beam intensity.

A cyclotron-based accelerator neutron source (1 mA proton beam at 30 MeV with a Be target) was approved by the Japanese Ministry of Health, Labor and Welfare in March 2020 (Annex XVI). Phase I and II clinical trials for recurrent MGs using this system have been undertaken since 2012 [680]. These were followed by a clinical trial for head and neck cancers, and subsequently this system was approved for clinical use with head and neck cancers in Japan.[37] The GMP-grade BPA medicine manufactured by Stella Pharma Corporation received simultaneous approval for use (Annex XVI).[38] Currently, an investigator-initiated, randomized, controlled clinical trial for recurrent and refractory high-grade meningioma is being performed using this system.

---

[37] SHI announcement of medical device approval: https://www.shi.co.jp/english/info/2019/6kgpsq0000002ji0.html

[38] STELLA PHARMA announcement of marketing and manufacturing approval: https://stella-pharma.co.jp/cp-bin/wordpress5/wp-content/uploads/2020/03/Press-release_Steboronine-approvalENG.pdf

TABLE 24. CLINICAL TRIALS INVESTIGATING GLIOBLASTOMAS WITH BNCT

Medical institution	Treatment dates	Tumour type & No. of patients	Boron compound (treatment[a])	Clinical outcome[b]	Refs
Brookhaven National Laboratory, Upton, NY, USA	1994–1999	nGBM 53	BPA 250–330 mg/kg in 2 h	mOS: 12.8 mos.	[163, 681–683]
Beth Israel Deaconess Medical Center, Harvard Medical School, Boston, USA	1996–1999	nGBM 20	BPA 250–350 mg/kg in 1.5 h	mOS: 11.1 mos.	[611, 628, 682]
	2002–2003	nGBM 6	BPA 14 g/m^2 in 1.5 h	NA	[684]
Universitätsklinikum Essen, Essen, Germany	1997–2002	nGBM 26	BSA 100 mg/kg in 1.7 h	mOS: 10.4–13.2 mos.	[685]
Helsinki University Central Hospital, Helsinki, Finland	1999–2001	nGBM 30	BPA 290–500 mg/kg in 2 h	mOS: 11.0–21.9 mos.	[629]
	2001–2008	rGBM 20	BPA 290–450 mg/kg in 2 h	mOS: 7 mos (post BNCT)	[417]
Faculty Hospital of Charles University, Prague, Czech Republic	2000–2002	nGBM 5	BSH 100 mg/kg in 1 h	NA	[161]
Nyköping Hospital, Nyköping, Sweden	2001–2003	nGBM 29	BPA 900 mg/kg in 6 h	mOS: 17.7 mos.	[161, 179, 483, 631, 686]
	2001–2005	rGBM 12	BPA 900 mg/kg in 6 h	mOS: 8.7 mos (post BNCT)	[687]
University of Tsukuba, Tsukuba City, Ibaraki, Japan	1999–2002	nGBM 5	BSH 100 mg/kg in 1–1.5 h (IO-BNCT)	mOS: 23.2 mos.	[688]
	1998–2007	nGBM 7	BSH 5 g/body in 1 h (IO-BNCT)	mOS: 23.3 mos.	[553]
	1998–2007	nGBM 8	BSH 5 g/body in 1 h + BPA or 250 mg/kg in 1 h (BNCT + XRT)	mOS: 27.1 mos.	[553]
University of Tokushima, Tokushima, Japan	1998–2000	nGBM 6	BSH 64.9-178.6 mg/kg (IO-BNCT)	mOS: 15.5 mos.	[689–691]
	2001–2004	nGBM 11	BSH 64.9-178.6 mg/kg (IO-BNCT)	mOS: 19.5 mos.	[689–691]
	2005–2008	nGBM 6	BSH 100 mg/kg & BPA 250 mg/kg in 1h (BNCT + XRT)	mOS: 26.2 mos.	[689–691]
Osaka Medical College, Takatsuki, Japan	2002–2003	nGBM 10	BSH 5 g/body + BPA 250 mg/kg in 1h	mOS: 14.5 mos.	[217–218]
	2003–2006	nGBM 11	BSH 5 g/body in 1h + BPA 700 mg/kg in 3h (BNCT + XRT)	mOS: 23.5 mos.	[218]
	2002–2007	rGBM 19	BSH 5 g/body + BPA 250 mg/kg in 1h or BSH 5 g/body + BPA 700 mg/kg in 3h.	mOS: 10.8 mos.	[632]
	2010–2013	nGBM 32	BSH 5 g/body in 1h + BPA 500 mg/kg in 3h (BNCT + XRT+TMZ)	mOS: 21.1 mos. (2 y OS: 45.5%)	Manuscript in preparation
	2013–2018	rGBM 10	BPA 500 mg/kg in 3 h (BNCT+Bev)	mOS: 12 mos.	[558], unpublished data

Note: n: newly-diagnosed; r: recurrent; GBM: glioblastoma; mOS: median overall survival; mos: months; NA: not available; IO-BNCT: intraoperative BNCT; XRT: X ray treatment; Bev: bevacizumab; TMZ: temozolomide.

[a] The treatment in parenthesis was not only external beam BNCT.

[b] For some cases, a range of survival is given to summarize survival reported for different cohorts.

13.6 BACKGROUND OCCURRENCES OF HEAD AND NECK CANCERS

13.6.1 Squamous cell carcinoma

Worldwide, there are more than 700,000 occurrences and 350,000 deaths each year from head and neck squamous cell carcinoma (SCC) cancer [692]. Most such patients are diagnosed with locally advanced disease (Stages III and IV). Multimodality treatments including surgery, radiotherapy and chemotherapy have been applied to advanced head and neck cancer. Unfortunately, 25–60% of patients with a locally advanced disease experienced develop a recurrence [693]. Patients with recurrent or metastatic head and neck SCC cancer have poor prognosis with a median overall survival of less than one year [694].

13.6.2 Non-squamous cell carcinoma

Non-squamous cell carcinoma of the head and neck is rare; e.g., adenoid cystic carcinoma of the head and neck accounts for only 1% of head and neck cancers [695]. Multimodality treatment involving surgery, radiation therapy and /or chemotherapy has been applied to non-squamous cell carcinoma (nSCC) of the head and neck, as it is generally resistant to photon therapy and chemotherapy alone. Reference [696] compares the safety and efficacy of BNCT for patients with this condition that have been previously irradiated to that of Pt-based chemotherapy.

13.7 BNCT CLINICAL STUDIES FOR HEAD AND NECK CANCER

In 2001, a patient with recurrent parotid gland tumour who had received standard therapies including surgery, radiotherapy and chemotherapy was treated at the Kyoto University Research Reactor Institute. This was the world's first attempt [165] to use BNCT for this purpose. Locoregional control was achieved for 7 years, although the patient subsequently died due to recurrence. This prompted clinical trials of BNCT for head and neck cancer in several centres in Japan, Finland, and Taiwan, China. In the clinical trials conducted in Japan and Finland, the prescribed doses were set to those delivered to mucosa (see discussions in Sections 11.3 and 12.5). In the clinical trial conducted in Taiwan, China, the prescribed dose was set to that for the tumour. Table 25 summarizes the BNCT clinical trials for head and neck cancers [222, 237, 408]. Annex XX provides a clinical case report for recurrent head and neck cancer treated by an AB-BNCT system.

13.8 SKIN CANCER

Skin cancer — the abnormal growth of skin cells — is caused by ultraviolet light damaging the DNA in skin cells. The main source of ultraviolet light is sunlight, but it can occur in areas not frequently exposed. The three major types of skin cancer are SCC, basal cell carcinoma, and melanoma. Surgery is the main treatment for skin cancer, although it may depend on the stage of the disease, the location of the tumour and complications. With early diagnosis and appropriate management, the overall five-year survival rate for most skin cancers is 95%. However, malignant melanoma is associated with higher morbidity and mortality and causes 65% of all skin cancer deaths.

13.8.1 Cutaneous malignant melanoma

Melanomas are malignant tumours arising from the melanin-producing cells called melanocytes. Cutaneous melanoma is the most common type. Melanoma accounts for less than 5% of all skin cancers; however, it causes the greatest number of skin cancer–related deaths worldwide. Patients with metastatic melanoma have a long-term survival rate of only 5%, but early detection of CMm with surgical excision is often curative. Early detection of is the best means of reducing mortality.

TABLE 25. BNCT CLINICAL TRIALS FOR RECURRENT OR UNTREATED HEAD AND NECK CANCER

Institution	Treatment dates	Tumour type, No. patients	Boron carrier and administration	Outcome	Refs
			Research reactors		
Kyoto University Research Reactor Institute, Japan	2001–2007	rH&N, 49 urH&N, 13	BSH+BPA (13 cases); BPA (72 cases); 250 mg/kg (5 cases) 500 mg/kg: 200 mg/kg/h ×2 h + 100 mg/kg/h×1h during irradiation (67 cases)	PR: 28%, CR: 29% MeST: 10.1 mos. 2y OS : 24.2%	[222]
Helsinki University Hospital, Finland	2003–2012	79	BPA 350–400 mg/kg in 2h before irradiation	PR: 32% [601], CR: 36% 2y LRPFS: 38% 2y OS: 21%	[408]
Taipei, Veterans General Hospital, Taiwan, China	2010–2013	SCC: 11; nSCC: 6	BPA 450 mg/kg (180 mg·kg^{-1}·h^{-1} × 2h; 90 mg·kg^{-1}·h^{-1} × 0.5h during irradiation)	PR: 35%, CR: 35% 2y LRPFS: 28% 2y OS: 47%	[237]
			Accelerators		
Southern Tohoku General Hospital, Japan	2016–2019	SCC: 13; nSCC: 8	BPA 500 mg/kg (200 mg·kg^{-1}·h^{-1} × 2h; 100 mg·kg^{-1}·h^{-1} × 1h during irradiation)	PR: 48%, CR: 24% 2y LRPFS: 56% 2y OS: 85.3%	[51]

Note: Tumour types: H&N: head and neck cancer; SCC: squamous cell carcinoma; nSCC: non-squamous cell carcinoma; with prefixes: r: recurrent; ur: unresectable; Responses: PR: partial response; CR: complete response; 2y OS: 2-year overall survival; MeST: median survival time; LRPFS Locoregional progression-free survival.

13.8.1.1   New directions in treatment

The current treatment for cutaneous malignant melanoma (CMm) is wide surgical excision of the lesion, reconstruction with a skin flap or graft, or treatment with neoadjuvant chemoimmunotherapy [697]. Wide surgical excision is a highly invasive procedure and may lead to a variety of functional, aesthetic and psychological problems that reduce quality of life. Other treatment modalities, including topical chemotherapy, immunotherapy, and hadron radiotherapy have been used for local control [698]. Recent advances in immunotherapeutic approaches [556, 699] to treat metastatic melanoma represent a breakthrough in treating this malignancy, which has a high propensity to metastasize. Examples include:

- A Phase III clinical trial has shown that high-dose interferon-alpha can produce a significant increase in overall survival;
- Checkpoints help keep T cells from attacking normal cells in the body. As shown in a recent clinical trial, cytotoxic T lymphocyte antigen-4 blockade with ipilimumab has produced a significant improvement in overall clinical survival. Approved in 2011 by the U.S. Food and Drug Administration stage IV melanoma patients, ipilimumab was the first checkpoint blockade therapy that shrank many tumours significantly and extended life;
- Phase III trials have examined targeting of the programmed cell-death-1 ligand with multiple programmed cell-death-1 monoclonal antibodies. These have also resulted in impressive clinical responses. In 2014, the U.S. Food and Drug Administration approved two additional checkpoint blockade drugs for use in patients with stage IV melanoma.

13.8.1.2 Boron neutron capture therapy

In 1973, an experimental study of BNCT for CMm was initiated at Kobe University, Japan. They first proposed using ^{10}B-chlorpromazine as a boron carrier to target the specific melanin synthesis activity of melanoma cells [700]. The use of BPA was later re-evaluated [701], as melanoma cells selectively absorb it and it exhibits a high tumour to normal skin concentration ratio. The first clinical study was initiated in 1987 by this group for BNCT of CMm using BPA [162, 702–707], and the radiation dose and local response of CMm were reported for cases treated from 1987 to 2001 [708]. Since then, several other groups [582, 611, 709–713] have initiated BNCT with their own protocols and primary end points. But the number of melanoma patients treated with BNCT is very small, as surgical excision is considered to be the most effective and curative therapy, and melanoma is considered to have a high tolerance to irradiation. Therefore, there is no literature on long-term local response and survival of melanoma patients after therapy with BNCT. Table 26 gives a summary of the clinical trials for CMm.

TABLE 26. CLINICAL REPORTS OF BNCT FOR CUTANEOUS MELANOMA

Country Research group	No. patients	Treatment dates	Neutron source	Clinical outcome	Refs
Japan Kobe University group	22	1987–2001	reactor	15/22: CR	[162, 708]
USA Harvard/MIT group	4	1994–1996	reactor	1/4: CR, 3/4: PR	[611, 709]
Europe Petten/Essen group	4	2002–	reactor	4/4: PR or SD	[710–711]
Argentina, CNEA RA-6 reactor	7[a]	2003–2007	reactor	52/88: CR, 9/88: PR	[519]
Japan Kawasaki Medical School	8	2003–2014	reactor	6/8: CR	[712]
China	1	2014–	reactor	1/1: CR	[713]
Japan National Cancer Center Hospital	On-going	2019–	accelerator		

Note: CR: complete response, PR: partial response, SD: stable disease

[a] The patients in the Argentinian study presented 88 nodular lesions among them in the period 2003–2007. In the middle of 2015, the clinical trial was restarted, and a new patient was treated with a new configuration of the RA-6 reactor. The results of this treatment showed consistency with the previous findings [714].

Both BNCT and carbon ion radiotherapies provide high LET radiation doses to the target volume. However, as BNCT is administered to relatively large areas, a wide margin is allowed, as BPA selectively accumulates in melanoma cells, which are subsequently killed by the boron neutron capture reaction (Fig. 15) without causing significant damage to neighbouring normal tissue. BNCT can deposit a large, cellular-level dose gradient between melanoma and surrounding normal skin, whereas carbon ion radiotherapy deposits dose uniformly within the target volume. This feature of BNCT is particularly useful for treating melanoma as histological involvement typically extends beyond the gross visible lesion. The large safety margin around the visible area is important in avoiding local recurrence both following surgery and radiotherapy.

The safety and efficacy of BNCT for treatment of cutaneous melanoma has not been tested in a randomized clinical trial. However, outside of clinical trials, patients with CMm have received BNCT as an alternative therapy and as a therapeutic application for specific cases and their responses have shown excellent local tumour control. The few case reports available indicate that BNCT may be promising for early stage CMm when wide local excision is not feasible.

13.8.2 Other skin malignancies

The clinical results of BNCT for other skin malignancies have been reported [383]. Although only four patients were treated, the clinical results obtained in treating patients with mucosal melanoma and extramammary Paget's disease of the genital region have been impressive. Extramammary Paget's disease is a slowly growing, rare cutaneous adenocarcinoma of apocrine gland-bearing skin. The vulva is the most common site involved in women; scrotal and penile skin in men; and perineal and perianal in both men and women. Extramammary Paget's disease constitutes 1–5% of all vulvar malignancies and has a peak incidence at age 65. Papers on reactor and accelerator based BNCT have reported results of treatments on sarcomas [78, 715].

13.9 REGULATORY ASPECTS OF CLINICAL TRIALS

From the regulatory standpoint, two sets of regulations are relevant to the development of AB-BNCT in Japan. The situation would be similar in other countries. One set consists of the regulations concerning ionizing radiation and the other of medical device regulations:

(a) The system composed of accelerator, target, and treatment facility has to comply with regulations concerning ionizing radiation. Suitable controlled areas need to be defined (i.e., 'zoning', see Section 5.1.2), in which the radiation hazards to different classes of personnel are well defined, so that their annual radiation doses are within the administrative and regulatory limits (see Section 6.4). Radioactive material produced by neutron activation has to be adequately shielded or disposed of. This reduces workplace exposure to staff and non-therapeutic doses to the patients under treatment. In order to commission an AB-BNCT system, preclinical biological and physical experiments are required including irradiation of water phantoms, foils, and small animals etc. The accelerator and neutron generating system will also have to satisfy the electrical as well as radiological safety inspections with qualifying documents issued from the relevant authorities (see Annexes XV–XVI);

(b) An AB-BNCT system which will be used in medicine has to obtain permission of the U.S. Food and Drug Agency, or similar agencies in other countries. In Japan, this is the Pharmaceutical and Medical Device Agency (PMDA). Such permission is a prerequisite for the clinical application of AB-BNCT in humans. If it is not obtained, the therapy remains an experimental trial and cannot be widely applied in routine clinical oncology. To obtain permission, the safety and efficacy of the accelerator and boron compound have to be demonstrated at least in phase I and phase II trials. BNCT is a very unique treatment modality using thermal as well as epithermal neutrons combined with one or more ^{10}B-compounds (Section 7). Such compounds are treated in the same way as any other stable pharmaceutical: the pharmaceutical company has to deliver all the non-clinical pharmaceutical experimental data to the regulatory agency and extensive critical review processes follow (see Annex XVII).

Uniquely in BNCT, both pharmaceutical and medical device reviews are necessary. After approval, a first-in-human clinical trial can be launched under supervision of the relevant regulatory agency:

(a) In a phase I trial, candidate patients with a specific cancer undergo BNCT, and the safety of the procedure is demonstrated. Phase I trials are dose-increasing trials, in which the dose of therapeutics is stepwise increased up to the maximum tolerable dose. In the case of BNCT, either radiation dose or pharmaceutical dose can be increased. In phase I trials conducted in Japan, the BPA dose was fixed at 500 mg/kg as experimental BNCT previously conducted on humans at a research reactor had confirmed that this pharmaceutical dose was both safe and sufficient for the boron neutron capture reaction in tumours. Therefore, the neutron dose to patients was increased stepwise to determine the maximum tolerable radiation dose;[39]

(b) After determining the maximum tolerable doses of neutrons and of BPA, a phase II trial demonstrating efficacy of BNCT will be launched. If statistically significant and clinically meaningful results are obtained, permission in accordance with the pharmaceutical and medical device regulations will be given to the AB-BNCT system encompassing the accelerator and BPA. After permission is received, the system can be applied clinically as a treatment modality;

(c) However, the establishment of a firm role of BNCT in oncology has to await the results of phase III trials, which compare the efficacy and morbidity of BNCT versus conventional therapy in a randomized fashion.

Another issue concerning regulatory permission is predictive assay by PET. Although immunohistochemical examination of the LAT-1 amino acid transporter can disclose whether the tumour will accumulate BPA (see Sections 7.6.4 and 9.6.1), the quantitative amount of BPA accumulated in the tumour and surrounding normal tissue cannot be obtained by this measurement. However, by using ^{18}F-FBPA, BPA concentrations in the tumour as well as the normal tissues can be calculated (see Sections 7.6.3 and 8.1.2). Temporal changes of BPA accumulation in the gross tumour tissue may also be measured (see Section 10.2). If ^{18}F-FBPA PET imaging shows no accumulation of ^{18}F-FBPA in the tumour, the gross tumour concentration of ^{10}B is low. However, caution needs to be used when considering eligibility for BNCT treatment based on ^{18}F-FBPA PET images. PET imaging using ^{18}F-FBPA is affected by the cellularity of the gross tissue so that boron concentrations determined using this method may not directly reflect the ^{10}B intracellular concentration of the tumour cells. However, ^{18}F-FBPA PET examination may be a useful imaging modality, and its availability could promote BPA-BNCT by visualizing gross tumour uptake of BPA in patients exhibiting high ^{18}F-FBPA uptake. Its clinical usefulness requires evidence from future clinical studies. A review by PMDA of AB-BNCT applied to head and neck cancer has been published [716].

---

[39] The alternative would be to fix the prescribed radiation dose and increase the dose of BPA to determine its maximum tolerable pharmaceutical dose.

Appendix I.

FAST NEUTRON DOSE CONTRIBUTION

This Appendix is focussed on the reference values suggested in Section 3.3, Table 6 and specifically the fast neutron fluence per unit epithermal neutron fluence, $D_H/\int\phi_{epi}(t)\cdot dt$. The reference value suggested in this publication is 7×10^{-13} Gy·cm^2, whereas in Ref. [1] a reference value of 2×10^{-13} Gy·cm^2 was suggested. The reasons for this change are described below.

First, it is important to consider the values for this parameter which have been achieved by a variety of clinical BNCT facilities based on reactors, along with the achieved or anticipated values for selected modern accelerator systems (Table 27).

TABLE 27. FAST NEUTRON DOSE CONTRIBUTIONS AT BNCT FACILITIES

Facility	$D_H/\int\phi_{epi}(t)\cdot dt$ $\times 10^{-13}$ Gy·cm^2	Reference / comments	Beam diameter (cm)	Approx. No. of patients treated (to May 2022)
Research reactor neutron sources				
FiR-1, Finland	3.1	Calculated from Ref. [119]	14	250
JRR-4, Japan	3.3	Ref. [717]	12	112
KUR, Japan	6.2	Ref. [717]	12	>500
R2-0, Sweden	5.6	Ref. [718]	14 × 10 (rectangular)	~60
	7.9	Calculated from Ref. [119]		
MIT-FCB, USA	1.55	Calculated from Ref. [119]	12	?
HFR-Petten, Netherlands	12.1	Ref. [719]	12	~30
THOR, Taiwan, China	2.8	Ref. [719]	14	~250
BMRR, USA	2.6	Ref. [719]	12	~53
Accelerator neutron sources				
NeuCure, Japan	6.6	Ref. [720]	12	>100
nuBeam, Neutron Therapeutics	1.83	MCNP simulated value	14	Preparing
CICS-1, Japan	4.7	Best current estimate	24	10
iBNCT001, Japan	3.8	Refs [721–722]	12	Preparing

It is clear from Table 27 that it is possible to achieve the reference value for $D_H/\int\phi_{epi}(t)\cdot dt$ of 2×10^{-13} Gy·cm^2 for the fast neutron dose component suggested in IAEA TECDOC-1223 [1] but many successful clinical programs were delivered and continue to be delivered with beams that had higher values. In particular, the beams at KUR and R2-0 had two of the higher values of the widely used reactor beams, while the THOR and FiR-1 beams had two of the lower values. It is also to be noted that there is some variability in the values reported for some beams, most likely due to small variations in the conditions between publications (e.g., the field size) or perhaps because of the inherent variability associated with delivery on a design calculation along with real uncertainties of subsequent measurement validation (see Section 4.7). Hence, the precise values cannot be relied upon to better than a variation of perhaps ±10%.

When considering patient outcomes, in general terms, a higher fast neutron dose component would be expected to show most clearly as an increase in skin reactions and other signs of superficial damage, where it is one of several contributing components of total dose. Depending on the site of treatment, these may be reported in a variety of ways such as incidence and grade of alopecia (hair loss), dermatitis (skin reactions – possibly scored as grades of erythema) possibly epistaxis (bleeding from the nose), otitis externa (redness and swelling on the ear canal), and perhaps other measures in addition. To identify in clinical data an effect from a higher than acceptable fast neutron dose component it might be expected that these indices of toxicity would be more prevalent and more severe than expected for this patient group. If patients are found to experience a wide range of toxicities at both superficial and deeper sites, then that is more likely to be associated with the overall delivered dose (see Section 10) rather than the effect of the fast neutron dose component. Clinical trials reported at four different reactor based facilities have reported the following effects:

- FiR-1, Finland: Low grade alopecia was the most common side effect occurring in 82% of the patients treated for brain tumours [408]. The incidence of grade 1–2 radiation dermatitis in the 30 patients with recurrent head and neck cancer was ~7% (2/30 patients) [408];
- KUR, Japan: In excess of 500 patients were treated, with the most comprehensive publication being that of Ref. [222] reporting on a cohort of 62 patients with cancer in the head and neck, including 87 BNCT procedures. Toxicities at grades 3 and 4 were observed and of those that might be specifically related to dose to the skin / superficial tissues there are reports of radiation dermatitis in two patients (3.2%) and ulceration in two patients (3.2%). Such observations are by no means unusual in a patient population of this kind;
- R2-0, Sweden: Over 50 patients were treated, with the most comprehensive publication being that of Ref. [631] describing 30 patients treated for high grade glioma. These patients were prescribed perhaps the highest dose of BPA of any study (900 mg ^{10}B per kg body mass over 6 hours) so it would be reasonable to anticipate comparatively high levels of ^{10}B in skin which may cloud any impact on skin toxicity of the higher fast neutron component of this beam. Reference [603] reports 'adverse events' in a table where there are 13 adverse events in 8 patients in skin and/or mucosa. The authors note explicitly in the text that "Nine grades 3–4 adverse events were reported in 4 of the 29 patients (epileptic seizures, haematuria, thrombosis, erythema) in close proximity to the BNCT procedure." One of these (erythema) might specifically be related to the skin / superficial dose but the reporting does not suggest an unusual prevalence in this patient cohort;
- THOR, Taiwan, China: All 34 patients reported by Ref. [723] treated for brain tumours experienced hair loss. Of the 17 patients reported by Ref. [724] treated for recurrent head and neck cancers (now with the continuous boron infusion technique and treatment delivered over 2 fractions), 14 (82%) of these experienced grade 1–2 radiation dermatitis.

In order to reliably determine a difference in skin toxicity between different facilities would require a large and well controlled clinical study, which has not been performed. These differences include characteristics of the neutron 'beams' applied, the clinical protocols used, the method of administration of the boron drug, and the number of irradiations performed.

Sumitomo's C-BENS NeuCure system has a fast neutron dose of 6.6×10^{-13} Gy·cm^2 comparable to that used in several reactor based facilities, although substantially higher than that originally suggested in Ref. [1]. Nevertheless, the level of adverse events experienced by patients appears to be tolerable for a medical device used to deliver life-saving treatment. There are now published clinical data, including detailed reporting of toxicities where this system has been used to treat patients with recurrent tumours in the head and neck [51] and patients with recurrent brain tumours [680], and in both cases the 'continuous infusion' technique (see Annex XIII) for BPA administration was used:

- The study reported by Ref. [51] delivered a prescribed weighted dose of 12 GyE[40] to the oral mucosa of 21 patients with head and neck cancers which had recurred. Weighted doses to skin are not explicitly reported but are likely to be in the 5–6 GyE[40] range. Reference [51] presents the toxicities experienced by patients in detail. Only one patient (5% of the total reported) experienced a skin toxicity above grade 2. This was a grade 3 dermatitis which would not be unusual in a similar patient population treated with X ray radiotherapy;
- In the study of Ref. [680], 27 patients with malignant glioma which had recurred were treated. In order to allow careful examination of toxicity, the target tissue for dose prescription was the skin, where a weighted dose of 8.5 GyE[40] was prescribed. Reference [680] reports that a large proportion of patients suffered low grade alopecia but none at grades 3 or 4, and no other reactions typical of superficial tissues are reported.

From the viewpoint of in-phantom (or in-patient) dose distribution, the results of the computational evaluation by Ref. [543] can be considered. This study demonstrates that for well optimised accelerator neutron beams, increases in fast neutron contribution from 2×10^{-13} to 6×10^{-13} and even 1×10^{-12} Gy·cm^2 cause little change in dose characteristics at depth. It is, however, noted that the peak tumour dose tends to decrease slightly as the fast neutron dose increases. Overall, this study suggests strongly that an increase in the fast neutron contribution from 2×10^{-13} to 7×10^{-13} Gy·cm^2, for a well-designed BSA, would have negligible impact on the dose in the deeper regions of the patient. The critical factor in the work of Ref. [543] is that the BSA has been well designed in each case (i.e., for each value of fast neutron dose component). In principle, as the fast neutron contribution increases in the beam, the dose rate to the skin may increase, and thus, the irradiation time may have to be shortened due to restrictions on the treatment protocol. This would then cause a reduction in the therapeutic dose delivered to a tumour at depth in the patient. Therefore, in designing a BSA, it is essential to consider the dose distribution and to avoid any decrease in the therapeutic dose to the tumour due to a large fast neutron contribution. Similarly, increasing the moderation effectiveness of the BSA to reduce the fast neutron component may increase the BSA's size and absorption (Section 3), reducing the overall therapeutic flux. Therefore, the optimization of the non-therapeutic fast neutron component in a system design involves many factors including the effects on therapeutic dose, therapy time, technology choice, and cost.

Finally, it is perhaps worth noting that studies with cell cultures indicate that the biological effect of some accelerator beams is likely to be lower than for some reactor beams. Of direct relevance to this discussion is the data of Ref. [388] which compared the biological effect (and neutron spectrum) of the beam from the R2-0 research reactor, Sweden, and the experimental Li(p,n) accelerator beam in Birmingham, UK [624]. In the cell system studied (V79 cells), the fast neutron dose component of the accelerator beam had a lower RBE by a perhaps 20% (evaluated at 2 cm depth). This may provide a degree of protection for some accelerator neutron beams.

---

[40] Now considered as IsoE with the constant RBE model (see Section 10.3.3).

OUT-OF-FIELD LEAKAGE IN LIGHT ION BEAM
MEDICAL ELECTRICAL EQUIPMENT

For conventional radiation therapy, international guidelines and standards for out-of-field leakage have been specified at the International Electrotechnical Commission (IEC). The recent Draft Proposal of the Light Ion Working Party, Amendment to IEC 60601-2-64:2014 (Light Ion Safety) [725] specified target limits for leakage dose separately for neutrons. In addition, AAPM Task Group 158 has published recommendations for reporting the leakage dose in external beam radiation therapy [98]. Since IEC provides an equipment safety standard, only the leakage dose coming from the equipment not from the patient (or patient simulating phantom) is considered. Thus, the out-of-field leakage measurements are performed in air without a phantom present.

AAPM Task Group 158 ('Measurement and calculation of doses outside the treated volume from external-beam radiation therapy') has gathered published data of the out-of-field doses in external photon, electron, proton, and carbon ion therapy, as well as brachytherapy [98]. Out-of-field doses have been reported in units of dose equivalents or ambient dose equivalent (mSv/Gy) as a fraction of the maximum dose on the central axis or as a fraction of the dose delivered to the target. Out-of-field doses are defined sometimes in air and sometimes in a phantom. An in-air measurement enables separation of the external out-of-field dose from unavoidable scattering dose in the patient, which is not of interest since the aim is to give guidelines for system manufacturers to design a well shielded system. Measurement or calculation of physical dose is recommended because it is well standardized, easily measured and precise, unlike biological models. Neutron doses associated with high-energy photon, proton, and carbon ion therapy have been most commonly measured in air as it is considered the most straightforward condition to analyze out-of-field leakage dose.

For light ion beam therapy, the IEC safety standard [725] requires evaluation of the out-of-field leakage dose from all radiation types at two lateral regions in air in the patient plane, from 150 mm to 500 mm and from 500 mm to 2000 mm from the field edge. The irradiation field size is defined by the 50% isodose level. According to the standard, the maximum ratio of the total leakage absorbed dose from all radiation types in the patient plane at the specified lateral distances and the dose delivered at the equipment reference point (ERP) on the central axis of the field (at a depth coincident with the centre-of-modulation depth for 60 mm of range modulation) is not to exceed 0.5% and 0.1% at respective regions. Separately, the IEC standard requires evaluation of the absorbed dose from neutron radiation at distances lateral to the beam axis from 150 mm to 2000 mm beyond the irradiated field, which is not to exceed 0.08% of the dose delivered at the ERP on the central axis of the field (at a depth coincident with the centre-of-modulation depth for 60 mm of range modulation).

The Light Ion Working Party has proposed an amendment to the IEC safety standard [725], which defines changes in the neutron specific leakage dose unit and limits. The goal is to harmonize the dose unit with ICRU, ICRP, and AAPM and make it traceable to a primary standard. According to the proposed recommendation, the neutron dose is to be defined in units of ambient dose equivalent [H*(10)], which is defined as "the dose equivalent at a depth of 10 mm in a 30-cm diameter sphere (the ICRU sphere) when exposed to an isotropic field of neutrons". The advantage of this unit is that ambient dose equivalent is a measurable quantity and, practically, can be calculated from the neutron spectrum by using energy specific conversion coefficients [726–727]. Ambient dose equivalent describes an in-air assessment of dose equivalent and overestimates the dose equivalent (even at the surface of the patient). According to the proposed IEC amendment, the estimated maximum ambient dose equivalent from neutrons is not to exceed 0.5% (Sv/Gy) of the dose that would be deliverer at the ERP on the central axis of the field. The estimated value of the leakage dose is to be delivered from measurements, calculations or their combination. Ambient dose equivalent in air is the recommended standard measure of out-of-field neutron doses in light-ion therapy by IEC amendment and in photon therapy by AAPM.

Appendix III.

OUT-OF-FIELD LEAKAGE PARAMETERS FOR BNCT

If international guidelines and standards for the light ion beam therapy technique are followed and adjusted for BNCT, the following consideration could be applied in the out-of-field leakage definition. In BNCT, the location of the patient plane varies based on the tumour location. On average, the surface of the patient's body is located approximately at 5 cm distance from the beam exit plane. Thus, in BNCT, an appropriate plane for the out-of-field leakage dose definition is the patient plane at 5 cm distance in front of the beam exit. In conventional radiotherapy, the field edge is defined as the lateral distance of the 50% isodose from the field axis defined in a phantom. Field edge is defined at the skin surface for source to surface distance treatments, and at the axis depth (typically 10 cm) for source-to-axis distances in isocentric treatments. In BNCT, the field edge location has not been defined so far. As low energy neutrons scatter strongly in the patient, the useful radiation field in BNCT can be wider than that of other radiotherapy modalities. In brain cancer BNCT, the maximum tumour dose at the beam axis applying typical RBE and CBE factors and assuming T/N ratio of 3.5 can be from 60 to 70 Gy (IsoE)[41], when maximum dose delivered to normal brain is fixed to 10 to 12 Gy (IsoE). The minimum desired tumour dose for head and neck cancer is thought to be approximately "20 Gy-Eq" [728], and following this rationale, the field edge in BNCT could be defined based on the lateral location of the 20 Gy-Eq (or the equivalent dose as expressed in Gy (IsoE); see Section 10.3.3) tumour isodose at the depth of the maximum thermal neutron fluence. This depth corresponds to the tumour dose maximum and can be directly measured with activation detectors. The location of the field edge may be defined in a large enough water phantom, minimum of 50–60 cm cube (depending on the beam size), to allow full back scattering at the field edge.

Following the rationale of the IEC standard, when reporting the out-of-field leakage dose in BNCT, the prescribed dose could be defined as the tumour dose at the depth of the maximum in the thermal neutron fluence along the beam axis in the large water phantom placed in front of the beam so that the depth of the maximum in the thermal neutron flux coincides with the plane where the leakage dose is defined (5 cm from the beam exit). For example, if the thermal neutron flux maximum is located at 2.5 cm depth, the phantom surface-to-beam exit plane distance would be 2.5 cm. The total tumour dose could be calculated according to Eq. (33). The common parameters for calculating the prescribed (tumour) dose are defined in Table 28 chosen based on the values applied with the NeuCure system, as approved by the Japanese regulatory agency, PMDA [729].

Following the amendment (March 25, 2019 Light Ion Working Party) to IEC 60601-2-64 [725], the out-of-field leakage dose could be reported as the ambient dose equivalent for neutrons in the lateral region from 150 mm to 2000 mm and for the total dose including dose due to neutrons and photons in two regions from 150 mm to 500 mm and from 500 mm to 2000 mm from the field edge. The leakage doses are determined by calculations and measurements in air. Ambient dose equivalent can be defined from the calculated neutron and photon fluxes with the fluence-to-ambient dose equivalent conversion factors from Ref. [726] for neutrons and from ICRU 47 [727] for photons. For out-of-field leakage dose measurements, the IEC standard recommends using a detector measuring directly in Sv. Currently, such a detector is not available for BNCT beams. Instead, activation foils and wires can be applied for neutron fluence measurements and the ionization chambers for photon and total dose measurements at the out-of-field locations.

---

[41] Formerly Gy (W) or Gy-Eq.

TABLE 28. PARAMETERS FOR OUT-OF-FIELD LEAKAGE DEFINITION IN BNCT

Parameter	Proposed definition
Field edge	Lateral 30% tumour isodose at the thermal neutron maximum depth in a large water phantom (50–60 cm cube, depending on the field size)
Where to define out-of-field leakage	In air, at the 'patient plane', 5 cm distance from the beam exit, at distances lateral to the neutron field axis from 1500 mm to 2000 mm beyond the treatment field edge
Treatment dose at the neutron field axis (the denominator to calculate the leakage dose ratio)	Maximum tumour dose in the large water phantom placed so that the maximum dose is located at the 5 cm distance from the neutron field exit plane
T/B ratio in tumour dose definition	3.5 [729]
Tissue definition	ICRU 33 [571] and ICRU 46 [572]: average soft tissue, adult (2.6% of ^{14}N, 10.1% of ^{1}H)
Nitrogen dose RBE	2.9 [729]
Hydrogen dose RBE	2.4 [729]
Photon dose RBE	1.0 [729]
Tumour dose CBE	4 [729]
Blood boron concentration	25 ppm [729]
Collimator size	The reference collimator size at the facility
Detectors for leakage dose measurement in air	Neutron flux: Au and/or Mn activation foils or wires Photon dose: TLDs or neutron-insensitive ionization chamber
Fluence-to-ambient dose equivalent conversion factors	Neutrons: Ref. [726] Photons: ICRU 47 [727]

Before reasonable target values for maximum allowed out-of-field leakage dose in BNCT can be specified, the ratio of the leakage dose and prescribed dose needs to be evaluated for existing BNCT facilities using common parameters listed in Table 28. After evaluating the achievable leakage doses, a common guideline can be established for BNCT systems in the future. Furthermore, the target values can be consented by IEC. Appropriate methods to measure out-of-field leakage in BNCT need to be developed, and further international standardization and unification of these methods established.

In addition to the out-of-field leakage due to the non-primary radiation defined here, the out-of-field dose including the scattering from the primary neutron field in the patient, might be interesting to compare between different BNCT systems. The out-of-field scattering from the primary neutron field depends on the energy and directionality of the field entering the patient, and thus may vary based on the initial neutron energy spectrum. The total out-of-field dose including the in-patient scattering could be defined in an anthropomorphic phantom, probably including the tissue boron concentration in the evaluation. The specific parameters for such evaluation will be defined in the future.

Appendix IV.

INDUCTIVELY COUPLED PLASMA ANALYTICAL METHODS IN USE AROUND THE WORLD.

Table 29 provides a summary of some of the ICP methods and operating parameters used for boron concentration determination in BNCT.

TABLE 29. METHODS TO MEASURE BORON IN BLOOD BY ICP-OES (or ICP-AES) AND ICP-MS DURING TREATMENTS AND BIODISTRIBUTIONS IN BNCT

Country	Japan			Finland	Argentina	Taiwan	China	
Centre	OMPU KBMC[a]	Southern Tohoku	National Cancer Center	Helsinki	CNEA[b]	Taiwan	Xiamen[c]	Beijing
Method's name	–	Whole-blood [B] measurement by ICP-AES	–	–	Determination of total boron in whole blood by ICP-OES	Patient blood boron measurement by ICP-AES	Determination of ^{10}B in plasma by ICP-MS	–
	Agilent 5110 VDV ICP-OES, Agilent Technologies	ICPE-9000, SHIMADZU	SPS-3520 DD, Hitachi High-Tech Science	Perkin-Elmer [d]3200 DV ICP-AES. [e]ICP-MS-VG Plasma Quad 2+ (for back-up)	ICP-OES (Perkin-Elmer 3100 AV)	ICP-OES	Agilent 7900 ICP-MS	Thermo iCAP 6300 DV ICP-OES
Sample	Whole blood with heparin	Whole blood with heparin	Whole blood with heparin	Whole blood with heparin	Whole blood with EDTA	Whole blood with heparin	Whole blood with EDTA	–
Sample pre-treatment	Whole blood 0.5 ml diluted by surfactant (2.5% triton-X100 0.5 ml), acid (1% of HNO$_3$ 8 ml), and internal standard 1 ml.	Whole blood 0.5 ml diluted by surfactant (2.5% triton-X100 0.5 ml), acid (HNO$_3$ 8 ml), and internal standard 1ml.	Whole blood 0.5 ml diluted by surfactant (2.5% triton-X100), acid (HNO$_3$), and Internal Standard	[d]Tricholoacetic acid [e]Wet-ashing by microwave digestion (HNO$_3$ + H$_2$O$_2$)	Dilution by surfactant (triton)	Dilution by surfactant and acid (triton+HNO$_3$)	Dilution by surfactant and acid (0.1% TritonX-100 +1% HNO$_3$)	Dilution by surfactant (triton)
RF generator	27 MHz	–	27 MHz	–	40 MHz (axial view)	–	27.12 MHz	–

189

TABLE 29. METHODS TO MEASURE BORON IN BLOOD BY ICP-OES (or ICP-AES) AND ICP-MS DURING TREATMENTS AND BIODISTRIBUTIONS IN BNCT (cont.)

Country	Japan			Finland	Argentina	China		
Centre	OMPU KBMC[a]	Southern Tohoku	National Cancer Center	Helsinki	CNEA[b]	Taiwan	Xiamen[c]	Beijing
Detector	CCD	CCD	CCD	–	Segmented array CCD	–	EM	–
Instrument quality control	Yes. B at 208.959 nm Y at 371.030 nm ratio for wave peak area: < ±3.0% SD	Yes. B at 208.959 nm Y at 371.030 nm ratio for wave peak area: less than ±3.0% SD	Yes. B at 208.959 nm Y at 371.030 nm ratio for wave peak area: <±3.0% SD	–	Yes (sensitivity)	Yes (Wave peak and pattern of the B atom detection are normal, detection value of deionized water <1200 cps).	Yes (sensitivity of ^{10}B is above 10 Mcps/ppm, BEC is 1.5 ppb and LOD is lower than 6 ppt)	–
Polychromator/ resolution	Echelle	Echelle mounting 0.21 nm/mm (at 200 nm)/ 0.68 nm/mm (at 600 nm)	–	–	Echelle	–	–	–
Torch Injector	Quartz	–	Quartz	–	Alumina	–	Quartz	–
Spray chamber	Cyclonic chamber	–	Cyclonic chamber	Scott chamber	Scott chamber	–	Scott chamber	–
Nebulizer	C-spray Nebulizer	Glass expansion Conikal U-Series Nebulizer #046-00092-20	Glass expansion Conikal U-Series	[d]Cross flux [e]V-groove	Cross flux	–	Quartz concentric nebulizer	–
Autosampler	–	ASC-6100, SHIMADZU	–	Gilson autosampler	Model AS 90, Tray Type B	–	SPS-4 autosampler	–
RF power (kW)	1.2	1.6	1.2	[e]1.3 [e]1.35	1.3	–	1.55	–
Plasma view	Axial	Axial	Radial	Radial	Axial	–	–	–

TABLE 29. METHODS TO MEASURE BORON IN BLOOD BY ICP-OES (or ICP-AES) AND ICP-MS DURING TREATMENTS AND BIODISTRIBUTIONS IN BNCT (cont.)

Country	Japan			Finland	Argentina	Taiwan	China	
Centre	OMPU KBMC[a]	Southern Tohoku	National Cancer Center	Helsinki	CNEA[b]	Taiwan	Xiamen[c]	Beijing
Plasma Ar flow rate [l/min]	12	10	15	15	15	–	15	–
Auxiliary Ar flow rate [l/min]	1.0	0.6	0.5	0.5	0.5	–	0–1.0	–
Nebulizer Ar flow rate [l/min]	–	0.7	0.4	0.81	0.75	–	0–1.3	–
Purge flow	–	Low	–	–	Normal	–	0.1–0.5 ml/min	–
Resolution	–	–	–	–	Normal	–	0.3–1.0 amu	–
Auto integration	–	–	–	–	10 s [min]–20 s [max]	–	–	–
Read time [s]/replicate	–	–	–	1	–	–	3	–
Read delay time [s]	–	50	–	–	25	–	30	–
Source equilibration decay	–	–	–	–	25 s	–	–	–
Sample flow rate [ml/min]	–	–	2	1	1	2	0.2	–
Flush time [s]	–	25	–	–	25	–	30	–
Wash time [s], rate	–	10	20	30	40 at 3 ml/min	–	45 at 0.5 ml/min	–
Extra time [s]	–	–	–	–	40 s at 3 ml/min	–	45 at 0.5 ml/min	–
Replicates intra sample	–	–	–	–	[d]3 [e]2	–	3	–
Analytical wavelength [nm]	B: 208.959	B: 208.959	B: 208.959	B: 249.772	B: 249.677. Alternative: B: 203.889	B: 249.773	not applicable	–

TABLE 29. METHODS TO MEASURE BORON IN BLOOD BY ICP-OES (or ICP-AES) AND ICP-MS DURING TREATMENTS AND BIODISTRIBUTIONS IN BNCT (cont.)

Country	Japan			Finland	Argentina	Taiwan	China	
Centre	OMPU KBMC[a]	Southern Tohoku	National Cancer Center	Helsinki	CNEA[b]	Taiwan	Xiamen[c]	Beijing
Control wavelength [nm]	–	–	–	–	B: 249.772; B: 208.957; Y: 360.073	–	–	–
Mass-to-charge ratio (m/z)	–	–	–	–	–	–	10 for $(^{10}B)^+$	–
Internal standard emission lines [nm]	–	Y: 371.030	–	Be: 234.861	Y: 371.029; Y: 324.227; Sr: 232.235.	–	^6Li or ^{40}Sc	Y
Blanks and QCs between samples	–	–	–	–	Yes. Every 6 samples.	–	Yes. Every 5 samples.	–
Background correction	–	–	–	–	2 points, manual selection	–	–	–
Read algorithm	Peak area	Peak area	Peak area	–	Peak area	–	Peak area	–
Number of pixels per peak	–	4	–	–	5	–	20 ($\geq$3)	–
Calibration	Linear	Linear	Linear	–	Linear	–	Linear ($R\geq$0.999)	–
Points of calibration curve [ppb]	0, 500, 1000, 2000, 5000 (optional 10000)	0, 500, 1000, 2000, 3000, 5000, 7000	0, 25, 50, 100, 500, 1000, 2500, (optional 5000)	0, 1000, 2000, 5000	0, 50, 100, 200, 500, 1000	5, 10, 25, 50, 100	1, 5, 10, 20, 50	0.2–5.0 µg/g
B detection limit	0.2 ppb	–	–	–	20 ppb	–	6 ppt	–
Patient's first blood sample	Before boron infusion	1h after boron infusion	Before boron infusion	Before boron infusion	Before boron infusion	Before boron infusion	Before boron infusion	–
Other samples	1 h after boron infusion; 2 h after boron infusion; after	2 h after boron infusion; 5 min after neutron irradiation	1 h after boron infusion; 2 h after boron infusion; less than 30 min	Every 20 min during the BPA–fructose infusion and	Every 15 min during boron infusion, immediately after irradiation	Other time point is specified in BNCT treated protocol	Other time point is specified in BNCT treated protocol	–

TABLE 29. METHODS TO MEASURE BORON IN BLOOD BY ICP-OES (or ICP-AES) AND ICP-MS DURING TREATMENTS AND BIODISTRIBUTIONS IN BNCT (cont.)

Country	Japan			Finland	Argentina	China		
Centre	OMPU KBMC[a]	Southern Tohoku	National Cancer Center	Helsinki	CNEA[b]	Taiwan	Xiamen[c]	Beijing
	neutron irradiation		after neutron irradiation	3 h after the end of infusion	and at 12 and 24 h from the initial			
Measurement time [s]	10	10	–	10 (pulse counting mode)	–	–	3	–
Dwell time	–	–	0.1 s	0.32 ms	–	–	50 ms	–

Note: BEC: Background equivalent concentration; LOD: Limit of detection

[a] OMPU KBMC: Osaka Medical and Pharmaceutical University; Kansai BNCT Medical Center.
[b] Advantages: Quick, robust and reproducible method. Accepts 20× dissolved solids.
Disadvantages: Memory effect. Not being able to use the most sensitive line (249.772 nm). B line at 249.772 nm is interfered with by Fe, present in high amounts in whole blood.
[c] Advantage: Wide linear range and no interference.
[d] ICP-AES method.
[e] ICP-MS method.

REFERENCES

[1] INTERNATIONAL ATOMIC ENERGY AGENCY, Current Status of Neutron Capture Therapy, TECDOC 1223, IAEA, Vienna (2001).

[2] ZHANG, Z., LIU, T., A review of the development of In-Hospital Neutron Irradiator-1 and boron neutron capture therapy clinical research on malignant melanoma, Ther. Radiol. Oncol. **2** (2018) 49.

[3] ONO, K., Prospects for the new era of boron neutron capture therapy and subjects for the future, Ther. Radiol. Oncol. **2** (2018) 40.

[4] SUMITOMO HEAVY INDUSTRIES, Japanese BNCT facilities have started clinical treatments under the national health insurance system with NeuCure™ BNCT system and NeuCure™ Dose Engine provided by Sumitomo Heavy Industries, Ltd (2020),
https://www.shi.co.jp/english/info/2020/6kgpsq0000002p30.html.

[5] CHRISTENSEN, C.J., et al., Free-neutron beta-decay half-life, Phys. Rev. D **5** (1972) 628.

[6] INTERNATIONAL ATOMIC ENERGY AGENCY, Compact Accelerator based Neutron Sources, IAEA-TECDOC-1981, IAEA, Vienna (2021).

[7] KIYANAGI, Y., SAKURAI, Y., KUMADA, H., TANAKA, H., Status of the accelerator based BNCT projects worldwide, AIP Conf. Proc. 21601 (2019) 050012.

[8] KREINER, A.J., et al., Present status of accelerator-based BNCT, Rep. Pract. Oncol. Radiother. **21** 2 (2016) 95–101.

[9] CARTELLI, D.E., et al., Status of low-energy accelerator-based BNCT worldwide and in Argentina, Appl. Radiat. Isot. **166** (2020) 109315.

[10] KIYANAGI, Y., Accelerator-based neutron source for boron neutron capture therapy, Ther. Radiol. Oncol. **2** (2018) 55.

[11] KONONOV, V.N., et al., Accelerator-based fast neutron sources for neutron therapy, Nucl. Instrum. Methods A **5641** (2006) 525–531.

[12] CAPOULAT, M.E., et al., Neutron spectrometry of the ^{9}Be(d (1.45 MeV), n)^{10}B reaction for accelerator-based BNCT, Nucl. Inst. Methods B **445** (2019) 57–62. CAPOULAT, M.E., KREINER, A.J., A ^{13}C(d,n)-based epithermal neutron source for boron neutron capture therapy, Phys. Med. **33** (2017) 106–113.

[13] HAWKESWORTH, M.R., Neutron radiography equipment and methods, At. Energy Rev. **15** 2 (1977) 169–220.

[14] COLONNA, N., et al., Measurements of low-energy (d,n) reactions for BNCT, Phys. Med. **26** (1999) 793–798.

[15]   HASHIMOTO, Y., HIRAGA, Y., KIYANAGI, Y., Effects of proton energy on optimal moderator system and neutron-induced radioactivity of compact accelerator-driven ^{9}Be(p,n) neutron sources for BNCT, Phys. Procedia **60** (2014) 332–340.

[16]   LEE, C.L., et al., A Monte Carlo dosimetry based evaluation of the ^{7}Li(p,n)^{7}Be reaction near threshold for accelerator boron neutron capture therapy, Phys. Med. **27** 1 (2000) 192–202.

[17]   ALLEN, D.A., BEYNON, T.D., A design study for an accelerator-based epithermal neutron beam for BNCT, Phys. Med. Biol. **40** (1995) 807–821.

[18]   CAPOULAT, M.E., MINSKY, D.M., KREINER, A.J., Computational assessment of deep-seated tumour treatment capability of the ^{9}Be(d,n)^{10}B reaction for accelerator-based boron neutron capture therapy, Phys. Med. **30** (2014) 133–146.

[19]   KAMATA, S., Measurements of differential thick target neutron yields – Be(p,xn) reactions induced by 11 MeV protons, Japan Atomic Energy Society Spring Meeting (2006) K42.

[20]   WATERS, L.S., et al., The MCNPX Monte Carlo Radiation Transport Code, AIP Conf. Proc. **896** (2007) 81–90.

[21]   JONES, D.T.L., BARTLE, C.M., Neutrons from the 2 MeV deuteron bombardment of thick ^{7}Li targets, Nucl. Inst. Methods **118** 2 (1974) 525–529.

[22]   GUZEK, J., TAPPER, U.A.S., MCMURRAY, W.R., WATTERSON, J.I.W., Characterization of the ^{9}Be(d,n)^{10}B reaction as a source of neutrons employing commercially available Radio frequency quadrupole (RFQ) (Proc. 5th Int. Conf. Crete, 1996) SPIE (1997) 509–512.

[23]   DURISI, E., et al., Design and simulation of an optimized e-linac based neutron source for BNCT research, Appl. Radiat. Isot. **106** (2015) 63–67.

[24]   HIRAGA, F., Monte Carlo simulation-based design of an electron-linear accelerator-based neutron source for boron neutron capture therapy, Appl. Radiat. Isot. **162** (2020) 109203.

[25]   LEUNG, K.N., New compact neutron generator system for multiple applications, Nucl. Technol. **206** 10 (2020) 1607–1614.

[26]   MOSS, C.E, et al., Survey of Neutron Generators for Active Interrogation, Rep. LA-UR-17-23592, Los Alamos National Laboratory, NM, USA (2017).

[27]   SZTEJNBERG GONÇALVES-CARRALVES, M.L., MILLER, M.E., Neutron flux assessment of a neutron irradiation facility based on inertial electrostatic confinement fusion, Appl. Radiat. Isot. **106** (2015) 95–100.

[28]   TANAKA, H., et al., Characteristics comparison between a cyclotron-based neutron source and KUR-HWNIF for boron neutron capture therapy, Nucl. Instrum. Methods B **267** 11 (2009) 1970–1977.

[29]   KATO, T., Design and construction of an accelerator-based boron neutron capture therapy (AB-BNCT) facility with multiple treatment rooms at the Southern Tohoku BNCT Research Center, Appl. Radiat. Isot. **156** (2020) 108961.

[30]   KANSAI BNCT MEDICAL CENTER, https://www.omp.ac.jp/en/kbmc.html.

[31]   KUMADA, H., et al., Development of beryllium-based neutron target system with three-layer structure for accelerator-based neutron source for boron neutron capture therapy, Appl. Radiat. Isot. **106** (2015) 78–83.

[32]   NAKAMURA, S., et al., Evaluation of radioactivity in the bodies of mice induced by neutron exposure from an epithermal neutron source of an accelerator-based boron neutron capture therapy system, Proc. Jpn. Acad. Ser. B **93** 10 (2017) 821–831.

[33]   HALFON, S., et al., Demonstration of a high-intensity neutron source based on a liquid-lithium target for accelerator based boron neutron capture therapy, Appl. Radiat. Isot. **106** (2015) 57–62.

[34]   SUAREZ ANZORENA, S., GAGETTI, L., DEL GROSSO, M.F., KREINER, A.J., Characterization of Be deposits on Mo, W and Cu substrates, to implement in a neutron production target for AB-BNCT, Procedia Mater. Sci. **8** (2015) 471–477.

[35]   GAGETTI, L., et al., Proton irradiation of beryllium deposits on different candidate materials to be used as a neutron production target for accelerator-based BNCT, Nucl. Inst. Methods A **874** (2017) 28–34.

[36]   NAKAMURA, S., et al., Characterization of the relationship between neutron production and thermal load on a target material in an accelerator-based boron neutron capture therapy system employing a solid-state Li target, PLOS ONE **14** 11 (2019) e0225587.

[37]   NAKAMURA, S., et al., Dependence of neutrons generated by ^{7}Li(p,n) reaction on Li thickness under free-air condition in accelerator-based boron neutron capture therapy system employing solid-state Li target, Phys. Med. **58** (2019) 121–130.

[38]   PORRA, L., et al., Accelerator-based boron neutron capture therapy facility at the Helsinki University Hospital, Acta Oncol. **61** 2 (2022) 269273.

[39]   YOSHIHASHI, Y., et al., High heat removal technique for a lithium neutron generation target used for an accelerator-driven BNCT system, JINST **16** (2021) P04016.

[40]   ASTRELIN, V.T., et al., Blistering of the selected materials irradiated by intense 200 keV proton beam, J. Nucl. Mater. **396** 1 (2010) 43–48.

[41]   KURIHARA, Y., KOBAYASHI, H., MATSUMOTO, H., YOSHIOKA, M., Neutron target research and development for BNCT: Direct observation of proton induced blistering using light-polarization and reflectivity changes, J. Radioanal. Nucl. Chem. **305** (2015) 935–942.

[42]   YAMAGATA, Y., et al., Development of a neutron generating target for compact neutron sources using low energy proton beams, J. Radioanal. Nucl. Chem. **305** (2015) 787–794.

[43]   BYKOV, T., et al., In situ study of the blistering effect of copper with a thin lithium layer on the neutron yield in the $^7Li(p,n)^7Be$ reaction, Nucl. Instrum. Methods B **481** (2020) 62–81.

[44]   ALEYNIK, V., et al., BINP accelerator based epithermal neutron source, Appl. Radiat. Isot. **69** 12 (2011) 1635–1638.

[45]   TASKAEV, S., BINP accelerator based neutron source (Proc. Int. Workshop Legnaro, 2014), INFN (2014).

[46]   KREINER, A.J., et al., A tandem-ESQ for accelerator-based boron neutron capture therapy, Nucl. Instrum. Methods B **261** (2007) 751–754.

[47]   KAPCHINSKII, I.M., TEPLYAKOV, V.A., A linear ion accelerator with spatially uniform hard focusing, Prib. Tekh. Eksp. **2** (1970) 19–22.

[48]   WEISS, M., Radio-frequency Quadrupole, Rep. CERN-PS/87-51, CERN Accelerator School, Aarhus (1986).

[49]   MITSUMOTO, T., et al., Cyclotron-based neutron source for BNCT, AIP Conf. Proc. **1525** (2013) 319–322.

[50]   TANAKA, H., et al., Experimental verification of beam characteristics for cyclotron-based epithermal neutron source (C-BENS), Appl. Radiat. Isot. **69** 12 (2011) 1642–1645.

[51]   HIROSE, K., et al., Boron neutron capture therapy using cyclotron-based epithermal neutron source and borofalan (^{10}B) for recurrent or locally advanced head and neck cancer (JHN002): An open-label phase II trial, Radiother. Oncol. **155** (2021) 182–187.

[52]   INTERNATIONAL ATOMIC ENERGY AGENCY (2021) Interactive Map of Accelerators, https://nucleus.iaea.org/sites/accelerators/Pages/default.aspx

[53]   NCT NEWS, The accelerator-driven BNCT facility at Southern Tohoku General Hospital: The world's first BNCT hospital, Japanese Society of Neutron Capture Therapy Issue 1 May 2016, http://jsnct.jp/e/pdf/02_nctletter.pdf.

[54]   KIM, S.S., A-BNCT Project. Dawonsys (2018), https://indico.ibs.re.kr/event/191/material/slides/52.pdf.

[55]   KUMADA, H., et al., Beam performance of the iBNCT as a compact linac-based BNCT neutron source developed by University of Tsukuba (Proc. 25th Int. Conf. Grapevine, TX, 2018) AIP Conf. Proc. **2160** (2019) 050013.

[56]   PISENT, E., FAGOTTI, E., COLAUTTI, P., MUNES project: an intense Multidisciplinary Neutron Source for BNCT based on a high intensity RFQ accelerator. (Proc. 16th Int. Cong. Geneva) INFN/LNL (2014) 66.

198

[57] CHINA SPALLATION NEUTRON SOURCE, China builds first accelerator-based facility for boron neutron capture therapy experiments, http://english.ihep.cas.cn/csns/doc/3958.html.

[58] NAKAMURA, S., et al., Neutron flux evaluation model provided in the accelerator-based boron neutron capture therapy system employing a solid-state lithium target, Sci. Rep. **11** (2021) 8090.

[59] BIKCHURINA, M., et al., The measurement of the neutron yield of the $^7Li(p,n)^7Be$ reaction in lithium targets, Biology **10** (2021) 824.

[60] TAE LIFE SCIENCES, Alphabeam System (2021), https://taelifesciences.com/alphabeam-neutron-system/.

[61] NEUBORON, boron neutron capture therapy (2021), https://en.neuboron.com/bnct

[62] URITANI, A., et al., Design of beam shaping assembly for an accelerator-driven BNCT system in Nagoya University (Proc. Int. Conf. Nara, 2017), JPS Conf. Proc. **22** (2018) 011002.

[63] WATANABE, K., et al., First experimental verification of the neutron field of Nagoya University accelerator-driven neutron source for boron neutron capture therapy, Appl. Radiat. Isot. **168** (2021) 109553.

[64] UNIVERSITY OF BIRMINGHAM, High Flux Accelerator-Driven Neutron Facility (2020), https://www.nnuf.ac.uk/high-flux-accelerator-driven-neutron-facility, https://physicsworld.com/a/boron-neutron-capture-therapy-is-back-on-the-agenda/.

[65] FREEMAN, T., PHYSICS WORLD, Boron neutron capture therapy is back on the agenda (2020), https://physicsworld.com/a/boron-neutron-capture-therapy-is-back-on-the-agenda/.

[66] NEUTRON THERAPEUTICS, INC., Neutron Therapeutics installs Europe's first accelerator-based boron neutron capture therapy (BNCT) platform – On track for first cancer patient treatment in 2019 (2019), http://www.neutrontherapeutics.com/news/pr-041819/.

[67] NEUTRON THERAPEUTICS, INC., Neutron Therapeutics to install its BNCT cancer therapy system into Japan's largest private hospital chain, http://www.neutrontherapeutics.com/news/pr-071619/.

[68] BUSINESS WIRE, Neutron Therapeutics and University Hospital of Brussels Announce Their Collaboration to Offer Cancer Patients a Revolutionary New Treatment Method at Institut Jules Bordet,https://www.businesswire.com/news/home/20220623005093/en/Neutron-Therapeutics-and-University-Hospital-of-Brussels-Announce-Their-Collaboration-to-Offer-Cancer-Patients-a-Revolutionary-New-Treatment-Method-at-Institut-Jules-Bordet.

[69]   PORRAS, I., et al., BNCT research activities at the Granada group and the project NeMeSis: Neutrons for Medicine and Sciences, towards an accelerator-based facility for new BNCT therapies, medical isotope production and other scientific neutron applications, Appl. Radiat. Isot. **163** (2020) 109247.

[70]   TORRES-SÁNCHEZ, P., et al., Optimal beam shaping assembly for a low energy accelerator-based neutron source for boron neutron capture therapy, Sci. Rep. **11** (2021) 7576.

[71]   ALMENAS, K., LEE, R., "Neutron Diffusion, Basic Concepts", Nuclear Engineering: An Introduction, Springer Science & Business Media (2012) Ch4.

[72]   CARPENTER, J.M., LOONG, C.-K., Elements of Slow-Neutron Scattering: Basics, Techniques, and Applications, Cambridge University Press (2015).

[73]   HASHIMOTO, Y., HIRAGA, F., KIYANAGI, Y., Optimal moderator materials at various proton energies considering photon dose rate after irradiation for an accelerator-driven ^{9}Be(p,n) boron neutron capture therapy neutron source, Appl. Radiat. Isot. **106** (2015) 88–91.

[74]   KUMADA, H., KISHI, T., HORI, N., YAMAMOTO, K., TORII, Y., "Development of the Multi-Leaf Collimator for Neutron Capture Therapy", Research and development in Neutron Capture Therapy (Proc. Int. Congress Essen, 2002) Essen (2002) 115–119.

[75]   HU, N., et al., Improvement in the neutron beam collimation for application in boron neutron capture therapy of the head and neck region, Sci. Rep. **12** (2022) 1377812.

[76]   SAKURAI, Y., TAKATA, T., TANAKA, H., SUZUKI, M., Simulation for improved collimation system of gamma-ray telescope system for boron neutron capture therapy at Kyoto University Reactor, Appl. Radiat. Isot. **165** (2020) 109256.

[77]   YU, H.-T., LIU, Y.-W.H., LIN T.-Y., WANG, L.-W., BNCT treatment planning of recurrent head-and-neck cancer using THORplan, Appl. Radiat. Isot. **69** 12 (2011) 1907–1910.

[78]   IGAKI, H., et al., Scalp angiosarcoma treated with linear accelerator-based boron neutron capture therapy: A report of two patients, Clin. Transl. Radiat. Oncol. **33** (2022) 128–133.

[79]   YUE, A.T., et al., Precision determination of absolute neutron flux, Metrologia **55** 4 (2018) 460.

[80]   INTERNATIONAL ATOMIC ENERGY AGENCY, Neutron Monitoring for Radiation Protection, IAEA Safety Report Series No. 115, IAEA, Vienna (in preparation).

[81]   MIRZAJANI, N., et al., Application of a Bonner sphere spectrometer for the determination of the angular neutron energy spectrum of an accelerator-based BNCT facility, Appl. Radiat. Isot. **88** (2014) 216–220.

[82]   LICEA, A., et al., "The measurement of neutron source terms using the nested neutron spectrometer", Modern Neutron Detection: Proceedings of a Technical Meeting, IAEA-TECDOC-1935, IAEA, Vienna (2020) 183–189.

[83]   MASUDA, A., et al., Neutron spectral fluence measurements using a Bonner sphere spectrometer in the development of the iBNCT accelerator-based neutron source, Appl. Radiat. Isot. **127** (2017) 47–51.

[84]   KOSLOWSKY, M., "Spectral unfolding: A mathematical perspective", Modern Neutron Detection: Proceedings of a Technical Meeting, IAEA-TECDOC-1935, IAEA, Vienna (2020) 97–115.

[85]   RADULOVIC, V., KOS., B., TRKOV, A., SNOJ, L., "Neutron spectrum adjustment and cross-section validation activities at the Jozef Stefan Institute", Modern Neutron Detection: Proceedings of a Technical Meeting, IAEA-TECDOC-1935, IAEA, Vienna (2020) 117–130.

[86]   THOMAS, D.J., "Use of a priori information in Bonner Sphere unfolding", Modern Neutron Detection: Proceedings of a Technical Meeting, IAEA-TECDOC-1935, IAEA, Vienna (2020) 131–136.

[87]   VEGA-CARRILLO, H.R., et al., "Neutron spectrometry around linacs using passive bonner sphere spectroscopy in planetary mode", Modern Neutron Detection: Proceedings of a Technical Meeting, IAEA-TECDOC-1935, IAEA, Vienna (2020) 191–200.

[88]   STAMELATOS, I.E., et al., "Novel neutron activation detector for fusion", Modern Neutron Detection: Proceedings of a Technical Meeting, IAEA-TECDOC-1935, IAEA, Vienna (2020) 301–312.

[89]   KWON, S., et al., "Activation foil candidates for intense d-Li neutron measurement up to 60 MeV", Modern Neutron Detection: Proceedings of a Technical Meeting, IAEA-TECDOC-1935, IAEA, Vienna (2020) 337–343.

[90]   GOMEZ-ROS, J.M., et al., The directional neutron spectrometer CYSP: Further developments for measuring low intensity fields, Radiat. Meas. **106** (2017) 580–584.

[91]   BEDOGNI, R., POLA, A., GOMEZ-ROS, J.M., "Neutron spectrometers from thermal energies to GeV based on single moderators", Modern Neutron Detection: Proceedings of a Technical Meeting, IAEA-TECDOC-1935, IAEA, Vienna (2020) 177–181.

[92]   BEDOGNI, R., et al., NCT-WES: A new single moderator directional neutron spectrometer for neutron capture therapy. Experimental validation. Eur. Phys. Lett. **34** 4 (2021) 42001.

[93]   GRYZIŃSKI, M.A., MACIAK, M., WIELGOSZ, M., Summary of recent BNCT Polish programme and future plans, Appl. Radiat. Isot. **106** (2015) 10–17.

[94] GRYZIŃSKI, M., ZIELCZYŃSKI, M., GOLNIK, N., JAKUBOWSK, E., Improvement of construction of recombination chambers for mixed radiation dosimetry at work places, https://www.irpa.net/members/TS2d.1.pdf.

[95] HU, N., et al., Microdosimetric quantities of an accelerator-based neutron source used for boron neutron capture therapy measured using a gas-filled proportional counter, J. Radiat. Res. **61** 2 (2020) 214–220.

[96] MAIRE, D., et al., Neutron energy reconstruction and fluence determination at 27 keV with the LNE-IRSN-MIMAC μ-TPC recoil detector, IEEE Trans. Nucl. Sci. **63** 3 (2016) 1934–1941.

[97] SAUZET, N., et al., Fast neutron spectroscopy from 1 MeV up to 15 MeV with MIMAC-FASTn, a mobile and directional fast neutron spectrometer, Nucl. Instrum. Methods A **965** (2020) 163799.

[98] KRY, S.F., et al., AAPM TG 158: Measurement and calculation of doses outside the treated volume from external-beam radiation therapy, Med. Phys. **44** (2017) e391–e429.

[99] TANAKA, K., et al., Measurement of spatial distribution of neutrons and gamma rays for BNCT using multi-imaging plate system, Appl. Radiat. Isot. **106** (2015) 125–128.

[100] HSIAO, M.C., JIANG, S.H., In-phantom neutron dose measurement using Gafchromic film dosimeter for QA of BNCT beams, Appl. Radiat. Isot. **143** (2019) 79–86.

[101] KUMADA, H., et al., Evaluation of the characteristics of the neutron beam of a linac-based neutron source for boron neutron capture therapy, Appl. Radiat. Isot. **165** (2020) 109246.

[102] SANTOS, D., SAUZET, N., GUILLAUDIN, O., MURAZ, J.F., Neutron spectroscopy from 1 to 15 MeV with Mimac-FastN, a mobile and directional fast neutron spectrometer and an active phantom for BNCT and PFBT (Proc. 8th Int. Meeting Paris, 2019), EPJ Web Conf. **231** (2020) 05003.

[103] TSUKAMOTO, T., et al., A phantom experiment for the evaluation of whole body exposure during BNCT using cyclotron-based epithermal neutron source (C-BENS), Appl. Radiat. Isot. **69** 12 (2011) 1830–1833.

[104] MILLER, M.E., MARIANI, L.E., SZTEJNBERG GONÇALVES-CARRALVES, M.L., SKUMANIC, M., THORP, S.I., Implantable self-powered detector for on-line determination of neutron flux in patients during NCT treatment, Appl. Radiat. Isot. **61** 5 (2004) 1033–1037.

[105] GADAN, M.A., CRAWLEY, V., THORP, S., MILLER, M., Preliminary liver dose estimation in the new facility for biomedical applications at the RA-3 reactor, Appl. Radiat. Isot. **67** 7–8 (2009) S206–S209.

[106]   MILLER, M.E., et al., Rhodium self-powered neutron detector as a suitable on-line thermal neutron flux monitor in BNCT treatments, Med. Phys. **38** (2011) 6502–6512.

[107]   JÄRVINEN, H., VOORBRAAK, W., (Eds), Recommendations for the Dosimetry of Boron Neutron Capture Therapy, NRG Report 21425/03.55339/C, Petten, The Netherlands (2003).

[108]   SAKURAI, Y., KOBAYASHI, T., Characteristics of the KUR Heavy Water Neutron Irradiation Facility as a neutron irradiation field with variable energy spectra, Nucl. Instrum. Methods A **453** 3 (2000) 569–596.

[109]   BINNS, P.J., et al., An international dosimetry exchange for boron neutron capture therapy, Part I: Absorbed dose measurements, Phys. Med. **32** 12 (2005) 3729–3736.

[110]   JEON, S.-J., CHOI, J., LEE, K.Y., YI, J.G., "Physical Dosimetry of Dawonmedax BNCT", paper presented at IAEA Technical Meeting on Technical Meeting on Best Practices in Boron Neutron Capture Therapy, Vienna, Austria, March 14–18, 2022.

[111]   KAMADA, S., BABA, M., ITOGA, T., UNNO, Y., Measurement of energy-angular neutron distribution for ^{7}Li, ^{9}Be(p,xn), reaction at E_p = 70 MeV and 11 MeV, J. Korean Phys. Soc. **59** 2 (2011) 1676–1680.

[112]   LIU, Y.H., CHEN, W.L., JIANG, S.H., Improvement of activation-detector-based spectrum unfolding technique for BNCT, IEEE Trans. Nucl. Sci. **60** 2 (2013) 783–791.

[113]   AUTERINEN I., SEREN, T., KOSUNEN, A., SAVOLAINEN, S., Measurement of free beam neutron spectra at eight BNCT facilities worldwide, Appl. Radiat. Isot. **61** 5 (2004) 1021–1026.

[114]   MAREK, M., VIERERBL, L., Bonner sphere spectrometer for characterization of BNCT beam, Appl. Radiat. Isot. **69** 12 (2011) 1918–1920.

[115]   MORO, D., et al., BNCT dosimetry performed with a mini twin tissue-equivalent proportional counters (TEPC), Appl. Radiat. Isot. **67** Suppl. 7–8 (2009) S171–S174.

[116]   LIN, H.X., CHEN, W.L., LIU, Y.H., SHEU, R.J., Characteristics and application of spherical-type activation detectors in neutron spectrum measurements at a boron neutron capture therapy (BNCT) facility, Nucl. Instrum. Methods A **811** (2016) 94–99.

[117]   LIU, Y.H., TSAI, P.E., LIU, H.M., JIANG, S.H., The angular and spatial distributions of the thermal neutron source description of the THOR BNCT beam, Radiat. Meas. **45** 10 (2010) 1432–1435.

[118]   KOIVUNORO, H., et al., Validation of dose planning calculations for boron neutron capture therapy using cylindrical and anthropomorphic phantoms, Phys. Med. Biol. **55** 12 (2010) 3515–3533.

[119]   BINNS, P.J., RILEY, K.J., HARLING, O.K., epithermal neutron beams for clinical studies of boron neutron capture therapy: A dosimetric comparison of seven beams, Radiat. Res. **164** (2005) 212–220.

[120] ASCHAN, C., TOIVONEN, M., SAVOLAINEN, S., AUTERINEN, I., Epithermal neutron beam dosimetry with thermoluminescence dosemeters for boron neutron capture therapy, Radiat. Prot. Dosim. **81** 1 (1999) 47–56.

[121] KOSUNEN, A., et al., twin ionisation chambers for dose determinations in phantom in an epithermal neutron beam, Radiat. Prot. Dosim. **81** 3 (1999) 187–194.

[122] UUSI-SIMOLA, J., et al., Dosimetric comparison at FiR 1 using microdosimetry, ionisation chambers and computer simulation, Appl. Radiat. Isot. **61** 5 (2004) 845–848.

[123] INTERNATIONAL COMMISSION ON RADIATION UNITS AND MEASUREMENTS, Clinical neutron Dosimetry-Part: I: Determination of absorbed dose in a patient treated by external beams of fast neutrons, ICRU Report 45 (1989).

[124] TANNER, V., et al., On-line neutron beam monitoring of the Finnish BNCT facility, Nucl. Instrum. Methods A **422** 1–3 (1999) 101–105.

[125] AUTERINEN, I., et al., Quality assurance procedures for the neutron beam monitors at the FiR 1 BNCT facility, Appl. Radiat. Isot. **61** 5 (2004) 1015–1019.

[126] ISHIKAWA, M., et al., Early clinical experience utilizing scintillator with optical fiber (SOF) detector in clinical boron neutron capture therapy: Its issues and solutions, Radiat. Oncol. J. **11** (2016) 105.

[127] MALOUFF, T.D., et al., Boron neutron capture therapy: A review of clinical applications, Front. Oncol. **11** (2021) 601820.

[128] SUZUKI, M., Boron neutron capture therapy (BNCT): A unique role in radiotherapy with a view to entering the accelerator-based BNCT era, Int. J. Clin. Oncol. **25** 1 (2020) 43–50.

[129] INTERNATIONAL ATOMIC ENERGY AGENCY, Specific Considerations and Guidance for the Establishment of Ionizing Radiation Facilities, IAEA Radiation Technology Series 7, IAEA, Vienna (2023).

[130] INTERNATIONAL ATOMIC ENERGY AGENCY, Setting Up a Radiotherapy Programme: Clinical, Medical Physics, Radiation Protection and Safety Aspects, Non serial Publication STI/PUB/1296, IAEA, Vienna (2008).

[131] INTERNATIONAL ATOMIC ENERGY AGENCY, Regulatory Control of the Safety of Ion Radiotherapy Facilities, IAEA-TECDOC-1891, IAEA, Vienna (2020).

[132] INTERNATIONAL ATOMIC ENERGY AGENCY, Monitoring for Compliance with Exemption and Clearance Levels, Report Series No. 67, IAEA, Vienna (2012).

[133] INTERNATIONAL ATOMIC ENERGY AGENCY, Application of the Concepts of Exclusion, Exemption and Clearance: Safety Guide. Safety Standards Series No. RS-G-1.7, IAEA, Vienna (2004).

[134]   INTERNATIONAL ATOMIC ENERGY AGENCY, Decommissioning of Particle Accelerators, Nuclear Energy Series No. NW-T-2.9, IAEA, Vienna (2020).

[135]   CANCER INTELLIGENCE CARE SYSTEMS INC., CICS (2020), https://www.cics.jp/page/english.html.

[136]   KUMADA, H., et al., Project for the development of the linac based NCT facility in University of Tsukuba, Appl. Radiat. Isot. **88** (2014) 211–215.

[137]   SUMITOMO HEAVY INDUSTRIES, LTD., Sumitomo Heavy Industries, Ltd. (2020), https://www.shi.co.jp/english/index.html.

[138]   INTERNATIONAL ATOMIC ENERGY AGENCY, "Cyclotron facility design", Chapter 3, Cyclotron Produced Radionuclides: Guidelines for Setting Up a Facility, Technical Report Series 471, IAEA, Vienna (2009).

[139]   CARROLL, L.R., Predicting Long-Lived, Neutron-Induced Activation of Concrete in a Cyclotron Vault, Carroll-Ramsey Associates, http://www.carroll-ramsey.com/docs/bibliography_and_selected_articles/2000%20Predicting%20long-lived%20activation.pdf.

[140]   EUROPEAN COMMISSION, Evaluation of The Radiological and Economic Consequences of Decommissioning Particle Accelerators, EUR 19151 (1999).

[141]   INTERNATIONAL ATOMIC ENERGY AGENCY, Radiation Protection in the Design of Radiotherapy Facilities, IAEA Safety Reports Series No. 47, IAEA, Vienna (2006).

[142]   WORLD HEALTH ORGANIZATION, Handbook: Good Laboratory Practice. Quality practices for regulated non-clinical research and development. Second Edition, WHO reference number: TDR/PRD/GLP/01.2, WHO, Geneva (2010).

[143]   ORGANISATION FOR ECONOMIC CO-OPERATION AND DEVELOPMENT, OECD Principles on Good Laboratory Practice, OECD Series on Principles of Good Laboratory Practice and Compliance Monitoring Number 1, env/mc/chem(98)17, OECD, Paris (1998).

[144]   INTERNATIONAL ATOMIC ENERGY AGENCY, Planning a Clinical PET Centre, IAEA Human Health Series No. 11, IAEA, Vienna (2010).

[145]   INTERNATIONAL ATOMIC ENERGY AGENCY, Staffing in Radiotherapy: An Activity Based Approach, IAEA Human Health Reports No. 13, IAEA, Vienna (2015).

[146]   INTERNATIONAL ATOMIC ENERGY AGENCY, Radiation Protection and Safety in Medical Uses of Ionizing Radiation, IAEA Safety Standards Series No. SSG-46, IAEA, Vienna (2018).

[147]   AMERICAN SOCIETY FOR RADIATION ONCOLOGY, Safety is No Accident: A Framework for Quality Oncology Care (2019).

[148]   INTERNATIONAL COMMISSION ON RADIOLOGICAL PROTECTION, The 2007 Recommendations of the International Commission on Radiological Protection, ICRP Publication 103, Ann. ICRP **37** (2–4) (2007).

[149]   INTERNATIONAL ATOMIC ENERGY AGENCY, Radiation Protection and Safety of Radiation Sources: International Basic Safety Standards, General Safety Requirements Part 3, No. GSR Part 3, IAEA, Vienna (2014).

[150]   CHAN, C.-H., LIU, H.-M., CHEN, Y.-W., CHANG, S.-L., TSAI, H.-Y., Activation analysis of patients and establishment of release criteria following boron neutron capture therapy at Tsing Hua Open-Pool Reactor, Radiat. Phys. Chem. **198** (2022) 110226.

[151]   HUANG, C.-K., LIU, C.-T., JIANG, S.-H., Neutron activation survey on patients following BNCT clinical trials at THOR, Appl. Radiat. Isot. **174** (2021) 109747.

[152]   WITTIG, A., et al., Neutron activation of patients following boron neutron capture therapy of brain tumors at the high flux reactor (HFR) Petten (EORTC trials 11961 and 11011). Strahlenther. Onkol. **181** (2005) 774–782.

[153]   INTERNATIONAL COMMISSION ON RADIOLOGICAL PROTECTION, ICRP Statement on Tissue Reactions / Early and Late Effects of Radiation in Normal Tissues and Organs – Threshold Doses for Tissue Reactions in a Radiation Protection Context. ICRP Publication 118, Ann. ICRP **41**(1/2) (2012).

[154]   SOLOWAY, A.H., et al., The chemistry of neutron capture therapy, Chem. Rev. **98** 4 (1998) 1515–1562.

[155]   BARTH, R.F., CODERRE, J.A., VICENTE, M.G., BLUE, T.E., Boron neutron capture therapy of cancer: Current status and future prospects, Clin. Cancer Res. **11** 11 (2005) 3987–4002.

[156]   TANAKA, H., et al., Development of a simple and rapid method of precisely identifying the position of ^{10}B atoms in tissue: An improvement in standard alpha autoradiography, J. Radiat. Res. **55** 2 (2014) 373–380.

[157]   SWEET, W.H., SOLOWAY, A.H., WRIGHT, R.L., Evaluation of boron compounds for use in neutron capture therapy of brain tumors. II. Studies in man, J. Pharmacol. Exp. Ther. **137** 2 (1962) 263.

[158]   SOLOWAY, A.H., HATANAKA, H., DAVIS, M.A., Penetration of brain and brain tumor. VII. tumor-binding sulfhydryl boron compounds 1,2, J. Med. Chem. **10** 4 (1967) 714–717.

[159]   HATANAKA, H., A revised boron-neutron capture therapy for malignant brain tumors, J. Neurol. **209** 2 (1975) 81–94.

[160]   YAMAMOTO, T., NAKAI, K., MATSUMURA, A., Boron neutron capture therapy for glioblastoma, Cancer Lett. **262** 2 (2008) 143–152.

[161]   BARTH, R.F., et al., Current status of boron neutron capture therapy of high-grade gliomas and recurrent head and neck cancer, Radiat. Oncol. **7** (2012) 146–166.

[162]   MISHIMA, Y., et al., Treatment of malignant melanoma by single thermal neutron capture therapy with melanoma-seeking ^{10}B-compound, Lancet **334** 8659 (1989) 388–389.

[163]   CHANANA, A.D., et al., Boron neutron capture therapy for glioblastoma multiforme: Interim results from the phase I/II dose-escalation studies, Neurosurgery **44** 6 (1999) 1182–1192, discussion 1192–1193.

[164]   ZONTA, A., et al., Extra-corporeal liver BNCT for the treatment of diffuse metastases: What was learned and what is still to be learned, Appl. Radiat. Isot. **67** (2009) 67–75.

[165]   KATO, I., et al., Effectiveness of BNCT for recurrent head and neck malignancies, Appl. Radiat. Isot. **61** 5 (2004) 1069–1073.

[166]   AIHARA, T., et al., First clinical case of boron neutron capture therapy for head and neck malignancies using ^{18}F-BPA PET, Head Neck **28** 9 (2006) 850–855.

[167]   KANKAANRANTA, L., et al., Boron neutron capture therapy in the treatment of locally recurred head and neck cancer, Int. J. Radiat. Oncol. Biol. Phys. **69** 2 (2007) 475–482.

[168]   YOSHINO, K., et al., "Examination of stability of *p*-, *m*-, *o*-boronophenylalanine in blood with high performance liquid chromatography", Cancer Neutron Capture Therapy (MISHIMA, Y., Ed.), Springer, Boston, MA.

[169]   HIRATSUKA, J., et al., Biodistribution of boron concentration on melanoma-bearing hamsters after administration of *p*-, *m*-, *o*-boronophenylalanine, Jpn. J. Cancer Res. **91** 4 (2000) 446–450.

[170]   CODERRE, J.A., et al., Selective targeting of boronophenylalanine to melanoma in BALB/c mice for neutron capture therapy, Cancer Res. **47** 23 (1987) 6377–6383.

[171]   CODERRE, J.A., et al., Selective delivery of boron by the melanin precursor analogue *p*-boronophenylalanine to tumors other than melanoma, Cancer Res. **50** 1 (1990) 138–141.

[172]   DETTA, A., CRUICKSHANK, G.S., L-amino acid transporter-1 and boronophenylalanine-based boron neutron capture therapy of human brain tumors, Cancer Res. **69** 5 (2009) 2126–2132.

[173]   WONGTHAI, P., et al., Boronophenylalanine, a boron delivery agent for boron neutron capture therapy, is transported by $ATB^{0,+}$, LAT1 and LAT2, Cancer Sci. **106** 3 (2015) 279–286.

[174]   MORI, Y., SUZUKI, A., YOSHINO, K., KAKIHANA, H., Complex formation of *p*-boronophenylalanine with some monosaccharides, Pigm. Cell Res. **2** 4 (1989) 273–277.

[175]   SHULL, B.K., et al., Fructose and related carbohydrates: Chemical and ^{13}C NMR evidence for the β-D-fructofuranose 2,3,6-(*p*-phenylalanylorthoboronate) structure, J. Pharm. Sci. **89** 2 (2000) 215–222.

[176]   HEIKKINEN, S., SAVOLAINEN, S., MELKKO, P., In vitro studies on stability of L-*p*-boronophenylalanine–fructose complex (BPA-F), J. Radiat. Res. **52** (2011) 360–364.

[177]   JALILIAN, A.R., et al., Potential theranostic boron neutron capture therapy agents as multimodal radiopharmaceuticals, Cancer Biother. Radiopharm. **37** 5 (2022) 342–354.

[178]   KOGANEI, H., et al., Development of high boron content liposomes and their promising antitumour effect for neutron capture therapy of cancers, Bioconjug. Chem. **24** (2013) 124–132.

[179]   CAPALA, J., et al., Boronated epidermal growth factor as a potential targeting agent for boron neutron capture therapy of brain tumors, Bioconjug. Chem. **7** (1996) 7–15.

[180]   WU, G., et al., Site-specific conjugation of boron containing dendrimers to anti-EGF receptor monoclonal antibody cetuximab (IMC-C225) and its evaluation as a potential delivery agent for neutron capture therapy, Bioconjug. Chem. **15** (2004) 185–194.

[181]   RONDINA, A., et al., A Boron Delivery Antibody (BDA) with boronated specific residues: new perspectives in boron neutron capture therapy from an in silico investigation, Cells **10** (2021) 3225.

[182]   CARPENTER, G., COHEN, S., Epidermal growth factor, Annu. Rev. Biochem. **48** (1979) 193–216.

[183]   GULLICK, W.J., et al., Expression of epidermal growth factor receptors on human cervical, ovarian, and vulval carcinomas, Cancer Res. **46** (1986) 285–292.

[184]   INTERNATIONAL ATOMIC ENERGY AGENCY, Atlas of Non-FDG PET–CT in Diagnostic Oncology, IAEA Human Health Series No. 38, IAEA, Vienna (2021).

[185]   IMAHORI, Y., et al., "A basic concept for a PET-BNCT system". Cancer Neutron Capture Therapy, (MISHIMA, Y., Ed.), New York, Plenum Press (1996) 691–696.

[186]   MISHIMA, Y., et al., In vivo diagnosis of human malignant melanoma with positron emission tomography using specific melanoma-seeking ^{18}F-DOPA analogue, J. Neuro-Oncol. **33** (1997) 163–169.

[187]   KABALKA, G.W., et al., Evaluation of fluorine-18-BPA–fructose for boron neutron capture treatment planning, J. Nucl. Med. **38** (1997) 1762–1767.

[188]   IMAHORI, Y., et al., Positron emission tomography-based boron neutron capture therapy using boronophenylalanine for high-grade gliomas: Part I, Clin. Cancer Res. **4** (1998) 1825–1832.

[189]   ISHIWATA, K., et al., Synthesis and radiation dosimetry of 4-borono-2-[18]fluoro-d,l-phenylalanine: A target compound for PET and boron neutron capture therapy, Appl. Radiat. Isot. **42** (1991) 325328.

[190] BERGMAN, J., SOLIN, O., Fluorine-18-labeled fluorine gas for synthesis of tracer molecules, Nucl. Med. Biol. **24** (1997) 677–683.

[191] MAIRINGER, S., Automated electrophilic radiosynthesis of [18F]FBPA using a modified nucleophilic GE TRACERlab FXFDG, Appl. Radiat. Isot. **104** (2015) 124–127.

[192] INTERNATIONAL ATOMIC ENERGY AGENCY, Cyclotron Produced Radionuclides: Operation and Maintenance of Gas and Liquid Target. IAEA Radioisotopes and Radiopharmaceuticals Series No. 4, IAEA, Vienna (2012).

[193] INTERNATIONAL ATOMIC ENERGY AGENCY, Production and Quality Control of Fluorine-18 Labelled Radiopharmaceuticals, IAEA TECDOC No. 1968, IAEA, Vienna (2021).

[194] INTERNATIONAL SOCIETY FOR NEUTRON CAPTURE THERAPY (2021) Newsletter #8, August 2021, https://isnct.net/news-2/newsletters/

[195] HE, J., et al., Nucleophilic radiosynthesis of boron neutron capture therapy-oriented PET probe [18F] FBPA using aryldiboron precursors, Chem. Commun. **57** (2021) 8593–8596.

[196] WATANABE, T., et al., Comparison of the pharmacokinetics between l-BPA and l-FBPA using the same administration dose and protocol: A validation study for the theranostic approach using [18F]-l-FBPA positron emission tomography in boron neutron capture therapy, BMC Cancer **16** (2016) 859.

[197] ISHIWATA, K., et al., A unique in vivo assessment of 4-[10B]borono-l-phenylalanine in tumour tissues for boron neutron capture therapy of malignant melanoma using positron emission tomography and 4-borono-2-[18F]fluoro-l-phenylalanine, Melanoma Res. **2** (1992) 171–179.

[198] WANG, H.E., et al., Evaluation of 4-borono-2-18F-fluoro-l-phenylalaninefructose as a probe for boron neutron capture therapy in a glioma bearing rat model, J. Nucl. Med. **45** (2004) 302–308.

[199] HANAOKA, K., et al., FBPA PET in boron neutron capture therapy for cancer: Prediction of 10B-concentration in the tumor and normal tissue in a rat xenograft model, Eur. J. Nucl. Med. Mol. Imaging Res. **4** (2014) 70.

[200] CHANDRA, S., Imaging of fluorine and boron from fluorinated boronophenylalanine in the same cell at organelle resolution by correlative ion microscopy and confocal laser scanning microscopy, Clin. Cancer Res. **8** (2002) 2675–2683.

[201] KUBOTA, R., et al., Cellular accumulation of 18F-labelled boronophenylalanine depending on DNA synthesis and melanin incorporation: a double-tracer microautoradiographic study of B16 melanomas in vivo, Br. J. Cancer **67** (1993) 701–705.

[202] WITTIG, A., SAUERWEIN, W.A., CODERRE, J.A., Mechanisms of transport of *p*-borono-phenylalanine through the cell membrane in vitro, Radiat. Res. **153** 2 (2000) 173–180.

[203] SHIMOSEGAWA, E., et al., Assessment of [10]B-concentration in boron neutron capture therapy: Potential of image-guided therapy using [18]FBPA PET, Ann. Nucl. Med. **30** (2016) 749–755.

[204] YOSHIMOTO, M., et al., Predominant contribution of L-type amino acid transporter to 4-borono-2-^{18}F-fluoro-phenylalanine uptake in human glioblastoma cells, Nucl. Med. Biol. **40** (2013) 625–629.

[205] YOSHIMOTO, M., et al., Non-invasive estimation of [10]B-4-boronol-phenylalanine derived boron concentration in tumors by PET using 4-borono-2-^{18}F-fluoro-phenylalanine, Cancer Sci. **109** (2018) 1617–1626.

[206] WATABE, T., et al., ^{18}F-FBPA as a tumor-specific probe of L-type amino acid transporter 1 (LAT1): A comparison study with ^{18}F-FDG and ^{11}C-methionine PET, Eur. J. Nucl. Med. Mol. Imaging **44** (2017) 321–331.

[207] ISOHASHI, K., et al., Comparison of the image-derived radioactivity and blood-sample radioactivity for estimating the clinical indicators of the efficacy of boron neutron capture therapy (BNCT): 4-borono-2-^{18}F-fluorophenylalanine (FBPA) PET study, Eur. J. Nucl. Med. Mol. Imaging Res. **6** (2016) 75.

[208] KONO, Y., et al., Radiation absorbed dose estimates for ^{18}F-BPA PET, Acta Radiol. **58** (2017) 1094–1101.

[209] ISHIWATA, K., et al., 4-Borono-2-[^{18}F]fluoro-d,l-phenylalanine as a target-compound for boron neutron capture therapy: tumor imaging potential with positron emission tomography, Nucl. Med. Biol. **18** (1991) 745–751.

[210] GRUNEWALD, C., et al., On the applicability of [^{18}F]FBPA to predict l-BPA concentration after amino acid preloading in HuH-7 liver tumor model and the implication for liver boron neutron capture therapy, Nucl. Med. Biol. **44** (2017) 83–89.

[211] SAKATA, M., et al., Direct comparison of radiation dosimetry of six PET tracers using human whole-body imaging and murine biodistribution studies, Ann. Nucl. Med. **27** (2013) 285–296.

[212] ROMANOV, V., et al., Evaluation of the total distribution volume of ^{18}F-FBPA in normal tissues of healthy volunteers by non-compartmental kinetic modelling, Ann. Nucl. Med. **34** 3 (2020) 155–162.

[213] WATABE, T., et al., Practical calculation method to estimate the absolute boron concentration in tissues using ^{18}F-FBPA PET, Ann. Nucl. Med. **31** (2017) 481–485.

[214] TAKAHASHI, Y., IMAHORI, Y., MINEURA, K., Prognostic and therapeutic indicator of fluoroboronophenylalanine positron emission tomography in patients with gliomas, Clin. Cancer Res. **9** (2003) 5888–5895.

[215]   BESHR, R., et al., Preliminary feasibility study on differential diagnosis between radiation-induced cerebral necrosis and recurrent brain tumor by means of [18F]fluoro-borono-phenylalanine PET/CT, Ann. Nucl. Med. **32** (2018) 702–708.

[216]   IMAHORI, Y., et al., Fluorine-18-labeled fluoroboronophenylalanine PET in patients with glioma, J. Nucl. Med. **39** (1998) 325–333.

[217]   MIYATAKE, S., et al., Modified boron neutron capture therapy for malignant gliomas performed using epithermal neutron and two boron compounds with different accumulation mechanisms: An efficacy study based on findings on neuroimages, J. Neurosurg. **103** (2005) 1000–1009.

[218]   KAWABATA, S., et al., Boron neutron capture therapy for newly diagnosed glioblastoma, J. Radiat. Res. **50** (2009) 51–60.

[219]   MIYASHITA, M., et al., Evaluation of fluoride-labeled boronophenylalanine-PET imaging for the study of radiation effects in patients with glioblastomas, J. Neuro-Oncol. **89** 2 (2008) 239–246.

[220]   KABALKA, G.W., et al., The use of positron emission tomography to develop boron neutron capture therapy treatment plans for metastatic malignant melanoma, J. Neuro-Oncol. **62** (2003) 187–195

[221]   ARIYOSHI, Y., et al., Boron neuron capture therapy using epithermal neutrons for recurrent cancer in the oral cavity and cervical lymph node metastasis, Oncol. Rep. **18** (2007) 861–866.

[222]   SUZUKI, M., et al., Boron neutron capture therapy outcomes for advanced or recurrent head and neck cancer, J. Radiat. Res. **55** 1 (2014) 146–153.

[223]   AIHARA, T., et al., Boron neutron capture therapy for advanced salivary gland carcinoma in head and neck, Int. J. Clin. Oncol. **19** 3 (2014) 437–444.

[224]   ISHIWATA, K., 4–Borono–2–18F–fluoro–L–phenylalanine PET for boron neutron capture therapy–oriented diagnosis: Overview of a quarter century of research, Ann. Nucl. Med. **33** (2019) 223–236.

[225]   MORITA, T., et al., Dynamic changes in 18F-borono-L-phenylalanine uptake in unresectable, advanced, or recurrent squamous cell carcinoma of the head and neck and malignant melanoma during boron neutron capture therapy patient selection, Radiat. Oncol. **13** 1 (2018) 4.

[226]   LAAKSO, J., et al., Atomic emission method for total boron in blood during neutron-capture therapy, Clin. Chem. **47** (2001) 1796–1803.

[227]   CODERRE, J.A., et al., Biodistribution of boronophenylalanine in patients with glioblastoma multiforme: Boron concentration correlates with tumor cellularity, Radiat. Res. **149** (1998) 163–170.

[228]   CARDOSO, J., et al., Boron biodistribution study in colorectal liver metastases patients in Argentina, Appl. Radiat. Isot. **67** Suppl. 7–8 (2009) S76–S79.

[229]   LIBERMAN, S.J., et al., Biodistribution studies of boronophenylalanine–fructose in melanoma and brain tumor patients in Argentina, Appl. Radiat. Isot. **61** (2004) 1095–1100.

[230]   KIGER, W., et al., Pharamacokinetic modeling for boronophenylalanine–fructose mediated neutron capture therapy: ^{10}B-concentration predictions and dosimetric consequences, J. Neuro-Oncol. **62** (2003) 171–186.

[231]   INAMOTO, H., et al., "A new image reconstruction technique with limited view-angle projection data for BNCT-SPECT", M-08-149, Virtual 2020 IEEE Nuclear Science Symposium and Medical Imaging Conference (2020 IEEE NSS-MIC).

[232]   LEE, T., et al, Monitoring the distribution of prompt gamma rays in boron neutron capture therapy using a multiple-scattering Compton camera: A Monte Carlo simulation study, Nucl. Instrum. Methods Phys. Res. Sect. A **798** (2015) 135–139.

[233]   MENÉNDEZ, P.R., et al., BNCT for skin melanoma in extremities: Updated Argentine clinical results, Appl. Radiat. Isot. **67** Suppl. 7–8 (2009) S50–S53.

[234]   ZHANG, Z., et al., Biodistribution studies of boronophenylalanine in different types of skin melanoma, Appl. Radiat. Isot. **163** (2020) 109215.

[235]   LINKO, S., et al., Boron detection from blood samples by ICP-AES and ICP-MS during boron neutron capture therapy, Scand. J. Clin. Lab. Invest. **68** (2008) 696–702.

[236]   TRIATMOKO, I.M., SUTJIPTO, Assessment of analytical instrumentation for boron measurement in BNCT system, Indones. J. Phys. Nucl. Appl. **2** (2017) 20–33.

[237]   WANG, L.W., LIU, Y.H., CHOU, F.I., JIANG, S.H., Clinical trials for treating recurrent head and neck cancer with boron neutron capture therapy using the Tsing-Hua Open Pool Reactor, Cancer Commun. (London) **38** (2018) 37.

[238]   THOMAS, R., Practical Guide to ICP-MS, Marcel Dekker Inc., New York, USA, 2014.

[239]   DUTTA, A., JINDAL, D., PATIL, R.K., PATIL, H.C., Analytical techniques for elemental analysis: a review, Eur. J. Pharm. Med. Res. **7** 10 (2020) 267–272.

[240]   WILSCHEFSKI, S.C., BAXTER, M.R., Inductively coupled plasma mass spectrometry: Introduction to analytical aspects, Clin. Biochem. Rev. **40** (2019) 115–133.

[241]   CUBADDA, F., Inductively coupled plasma-mass spectrometry for the determination of elements and elemental species in food: A review, J. AOAC Int. **87** (2004) 173–204.

[242]   HOFFMAN, E.J., HUANG, S.C., PHELPS, M.E., Quantitation in positron emission computed tomography: 1. Effect of object size, J. Comput. Assist. Tomogr. **3** (1979) 299–308.

[243] MATSUNAGA, K., et al., Quantitative pulmonary blood flow measurement using ^{15}O-H$_2$O PET with and without tissue fraction correction: A comparison study, Eur. J. Nucl. Med. Mol. Imaging Res. **7** 1 (2017) 102.

[244] ONO, K., et al., Proposal for determining absolute biological effectiveness of boron neutron capture therapy – The effect of ^{10}B(n,α)^{7}Li Dose can be predicted from the nucleocytoplasmic ratio or the cell size, J. Radiat. Res. **60** 1 (2019) 29–36.

[245] MOSES, W.W., Fundamental limits of spatial resolution in PET, Nucl. Instrum. Methods A. **648** Suppl. 1 (2011) S236–S240.

[246] YANO, F., et al., Diagnostic accuracy of positron emission mammography with ^{18}F-fluorodeoxyglucose in breast cancer tumor less than 20 mm in size. Asia Oceania J. Nucl. Med. Biol. **7** (2019) 13–21.

[247] WITTIG, A., et al., Boron analysis and boron imaging in biological materials for boron neutron capture therapy (BNCT), Crit. Rev. Oncol. Hematol. **68** (2008) 66–90.

[248] LINDSTROM, R.M., REVAY, Z., Prompt gamma neutron activation analysis (PGAA): Recent developments and applications, J. Radioanal. Nucl. Chem. **314** (2017) 843–885.

[249] RILEY, K.J., HARLING, O.K., An improved prompt gamma neutron activation analysis facility using a focused diffracted neutron beam, Nucl. Instrum. Methods B **143** (1998) 414–421.

[250] BORTOLUSSI, S., ALTIERI, S., Boron concentration measurement in biological tissues by charged particle spectrometry, Radiat. Environ. Biophys. **52** (2013) 493–503.

[251] GABEL, D., et al., Quantitative neutron capture radiography for studying the biodistribution of tumor-seeking boron-containing compounds, Cancer Res. **47** (1987) 5451–5454.

[252] MURATA, I., et al., Design of SPECT for BNCT to measure local boron dose with GAGG scintillator, Appl. Radiat. Isot. **181** (2022) 110056.

[253] ALDOSSARI, S., MCMAHON, G., LOCKYER, N.P., MOORE, K.L., Microdistribution and quantification of the boron neutron capture therapy drug BPA in primary cell cultures of human glioblastoma tumour by NanoSIMS, Analyst. **144** (2019) 6214–6224.

[254] BUSSER, B., et al., Elemental imaging using laser-induced breakdown spectroscopy: A new and promising approach for biological and medical applications, Coord. Chem. Rev. **358** (2018) 70–79.

[255] TIMONEN, M., et al., ^{1}H MRS phantom studies of BNCT ^{10}B-carrier, BPA-F using STEAM and PRESS MRS sequences: Detection limit and quantification, Spectroscopy **18** (2004) 133–142.

[256] SHIBATA, S., et al., "Design of collimator for T/N-SPECT for BNCT", PS1 B 01, 18th International Congress on Neutron Capture Therapy, Oct. 28–Nov. 2, 1998, Taipei (1998).

[257] KOBAYASHI, T., SAKURAI, Y., ISHIKAWA, M., A noninvasive dose estimation system for clinical BNCT based on PG-SPECT—Conceptual study and fundamental experiments using HPGe and CdTe semiconductor detectors, Med. Phys. **27** (2001) 2124–2132.

[258] NAKAMURA, S., MUKAI, T., MAMABE, M., MURATA, I., Precise numerical simulation of gamma-ray pulse height spectrum measured with a CdTe detector designed for BNCT-SPECT, Prog. Nucl. Sci. Technol. **3** (2012) 52–55.

[259] HALES, B., et al., Predicted performance of a PG-SPECT system using CZT primary detectors and secondary Compton-suppression anti-coincidence detectors under near-clinical settings for boron neutron capture therapy, Nucl. Instrum. Methods A **875** (2017) 51–56.

[260] ISHIKAWA, M., KOBAYASHI, T., SAKURAI, Y., KANDA, K., Optimization technique for a prompt gamma-ray SPECT collimator system, J. Radiat. Res. **42** (2001) 387–400.

[261] MINSKY, D.M., et al., First tomographic image of neutron capture rate in a BNCT facility, Appl. Radiat. Isot. **69** (2011) 1858–1861.

[262] MINAMI, K., et al., "Design of a proto-type array detector using GAGG scintillator for BNCT-SPECT", P8-3, 9th Young Researchers' BNCT Meeting (YBNCT 9), Nov. 13–15, 2017, Uji Obaku Plaza, Kyoto, Japan (2017).

[263] ROSENSCHÖLD, P.M., et al., Prompt gamma tomography during BNCT – A feasibility study, J. Instrum. **1** (2006) P05003.

[264] VERBAKEL, W.F.A.R., SAUERWEIN, W., HIDEGHETY, K., STECHER-RASMUSSEN, F., Boron concentrations in brain during boron neutron capture therapy: In vivo measurements from the Phase I Trial EORTC 11961 Using a gamma-ray telescope, Int. J. Radiat. Oncol. Biol. Phys. **55** (2003) 743–756.

[265] FURUKAWA, Y., KOYAMA, M., YUKI, M., Determination of boron content in several mediums by prompt gamma ray analysis, Radioisotopes **16** (1967) 711.

[266] KOBAYASHI, T., KANDA, K., Microanalysis system of ppm order B-10 concentrations in tissue for neutron capture therapy by prompt gamma-ray spectrometry, Nucl. Instrum. Methods **204** (1983) 525–531.

[267] SZTEJNBERG GONÇALVES-CARRALVES, M., et al., Development of prompt gamma neutron activation analysis facility for ^{10}B measurements at RA-3: Design stage, Appl. Radiat. Isot. **69** (2011) 1928–1931.

[268] VALERO, M., ROGULICH, L., THORP, S.I., MILLER, M.E., SZTEJNBERG, M., "Improvements in design of prompt gamma facility for BNCT at RA-3: On the verge of the desired detection limit," 19th International Congress on Neutron Capture Therapy, Granada, Spain (2021).

[269] CARPTENTER, J. M., LOONG, C.-K., "Chapter 3 Scattering theory: Nuclear". Elements of Slow-Neutron Scattering: Basics, Techniques and Applications. Materials Research Society, Cambridge University Press, Cambridge, UK (2015).

[270] RAAIJMAKERS, C.P.J., et al., Monitoring of blood-^{10}B-concentration for boron neutron capture therapy using prompt gamma-ray analysis, Acta Oncol. **34** (1995) 517–523.

[271] KAZUHIKO, A.O.K.P., KOBAYASHI T., KANDA, K., Phantom experiment and analysis for in vivo measurement of boron-10 concentrations in melanoma for boron neutron capture therapy, J. Nucl. Sci. Technol. **21** (1984) 647–656.

[272] HONDA, C., et al., In situ detection of cutaneous melanoma by prompt gamma-ray spectrometry using melanoma-seeking "B-DOPA" analogue, J. Dermatol. Sci. **1** (1990) 23–32.

[273] MATSUMOTO, T., et al., Phantom experiment and calculation for in vivo boron analysis by prompt gamma ray spectroscopy, Phys. Med. Bid. **36** (1991) 329–338.

[274] VERBAKELA, W.F.A.R., STECHER-RASMUSSEN, T.F., A γ-ray telescope for on-line measurements of low boron concentrations in a head phantom for BNCT, Nucl. Instrum. Methods A **394** (1997) 163–172.

[275] CHIARAVIGLIO, D., et al., Evaluation of selective boron absorption in liver tumors, Strahlenther. Onkol. **165** (1989) 170–172.

[276] ROVEDA, L., et al., How to study boron biodistribution in liver metastases from colorectal cancer, J. Chemother. **16** (2004) 15–18.

[277] STELLA, S., et al., Measurement of α particle energy loss in biological tissue below 2 MeV, Nucl. Instrum. Methods B **267** (2009) 2938–2943.

[278] BORTOLUSSI, S., et al., Boron uptake measurements in a rat model for boron neutron capture therapy of lung tumours, Appl. Radiat. Isot. **69** (2011) 394–398.

[279] PROVENZANO, L., et al., Measuring the stopping power of α particles in compact bone for BNCT, J. Phys. Confer. Ser. **583** (2015) 012047–012050.

[280] PROVENZANO, L., et al., Charged particle spectrometry to measure ^{10}B-concentration in bone, Radiat. Environ. Biophys. **58** (2019) 237–245.

[281] BORTOLUSSI, S., et al., Boron concentration measurements by alpha spectrometry and quantitative neutron autoradiography in cells and tissues treated with different boronated formulations and administration protocols, Appl. Radiat. Isot. **88** (2014) 94–98.

[282] LAMART, S., et al., Actinide bioimaging in tissues: Comparison of emulsion and solid track autoradiography techniques with the iQID camera, PLOS ONE **12** (2017) e0186370.

[283] FAIRCHILD, R.G., et al., Neutron autoradiographic determination of boron-10 concentration distribution in mammalian tissue, Radiat. Res. **36** (1968) 87–97.

[284]   GABEL, D., HOCKE, I., ELSEN, W., Determination of sub-ppm amounts of boron in solutions by means of solid state track detectors, Phys. Med. Biol. **28** (1983) 1453–1457.

[285]   ABE, M., et al., Boron distribution analysis by alpha-autoradiography, J. Nucl. Med. **27** (1986) 677–684.

[286]   FAIRCHILD, R.G., et al., Microanalytical techniques for boron analysis using the ^{10}B(n, α)^{7}Li reaction, Med. Phys. **13** (1986) 50–56.

[287]   ALTIERI, S., et al., Neutron autoradiography imaging of selective boron uptake in human metastatic tumours, Appl. Radiat. Isot. **66** (2008) 1850–1855.

[288]   SCHÜTZ, C., et al., Boron determination in liver tissue by combining quantitative neutron capture radiography (QNCR) and histological analysis for BNCT treatment planning at the TRIGA Mainz, Radiat. Res. **176** (2011) 388–396.

[289]   DURRANI, S.A., BULL, R.K., International Series in Natural Philosophy (TER HAAR, D., Ed), Pergamon Press, Oxford (1987).

[290]   GUO, S.L., et al., "Solid-state nuclear track detectors", Handbook of Radioactivity Analysis: Volume 1: Radiation Physics and Detectors (ANNUNZIATA, M.F.L., Ed), Academic Press (2020).

[291]   YANAGIE, H., et al., Neutron capture autoradiographic determination of ^{10}B-distributions and concentrations in biological samples for boron neutron capture therapy, Nucl. Instrum. Methods A **424** (1999) 122–128.

[292]   FLEISCHER, R.L., PRICE, P., WALKER, R.M., Nuclear Tracks in Solids, University of California Press, Berkeley (1975).

[293]   PINELLI, T., et al., "Neutron radiography: a methodology to study the boron distribution in samples including tumor and normal hepatic tissues", INIS-XA-08N1446, Paper presented at the World TRIGA Users' Conference, Pavia (Italy), 16–20 June 2002.

[294]   SUZUKI, M., et al., Intra-arterial administration of sodium borocaptate (BSH)/lipiodol emulsion delivers B-10 to liver tumors highly selectively for boron neutron capture therapy: Experimental studies in the rat liver model, Int. J. Radiat. Oncol. Biol. Phys. **59** (2004) 260–266.

[295]   ALTIERI, S., et al., Boron absorption imaging in rat lung colon adenocarcinoma metastases, J. Phys. Confer. Ser. **41** (2006) 123–126.

[296]   GARABALINO, M.A., et al., Electroporation optimizes the uptake of boron-10 by tumor for boron neutron capture therapy (BNCT) mediated by GB-10: A boron biodistribution study in the hamster cheek pouch oral cancer model, Radiat. Environ. Biophys. **58** (2019) 455–467.

[297]   LIN, Y.C., et al., Macro- and microdistributions of boron drug for boron neutron capture therapy in an animal model, Anticancer Res. **32** (2012) 2657–2664.

[298] VIDAL, C., et al., Optical density analysis in autoradiographic images from BNCT protocols, Radiat. Meas. **119** (2018) 121–124.

[299] PORTU, A., et al., Measurement of ^{10}B-concentration through autoradiography images in polycarbonate nuclear track detectors, Radiat. Meas. **46** (2011) 1154–1159.

[300] SAINT MARTIN, G., PORTU, A., SANTA CRUZ, G.A., BERNAOLA, O.A., Stochastic simulation of track density in nuclear track detectors for ^{10}B measurements in autoradiography, Nucl. Instrum. Methods B **269** (2011) 2781–2785.

[301] POSTUMA, I., et al., An improved neutron autoradiography set-up for ^{10}B-concentration measurements in biological samples, Rep. Pract. Oncol. Radiother. **21** (2016) 123–128.

[302] CEBERG, C.P., et al., Neutron capture imaging of ^{10}B in tissue specimens, Radiother. Oncol. **26** (1993) 139–146.

[303] PORTU, A., et al., Reference systems for the determination of ^{10}B through autoradiography images: Application to a melanoma experimental model, Appl. Radiat. Isot. **69** (2011) 1698–1701.

[304] PIGNOL, J.P., et al., Neutron capture radiography applied to the investigation of boron-10 biodistribution in animals: Improvements in techniques of imaging and quantitative analysis, Nucl. Instrum. Methods B **94** (1994) 516–522.

[305] PROVENZANO, L., et al., Extending neutron autoradiography technique for boron concentration measurements in hard tissues, Appl. Radiat. Isot. **137** (2018) 62–67.

[306] PORTU, A., et al., Neutron autoradiography to study boron compound microdistribution in an oral cancer model, Int. J. Radiat. Biol. **91** (2015) 329–335.

[307] SCHÜTZ, C.L., et al., Intercomparison of inductively coupled plasma mass spectrometry, quantitative neutron capture radiography, and prompt gamma activation analysis for the determination of boron in biological samples, Anal. Bioanal. Chem. **404** (2012) 1887–1895.

[308] PORTU, A., et al., Inter-comparison of boron concentration measurements at INFN-University of Pavia (Italy) and CNEA (Argentina), Appl. Radiat. Isot. **105** (2015) 35–39.

[309] CANSOLINO, L., et al., Comparative study of the radiobiological effects induced on adherent vs suspended cells by BNCT, neutrons and gamma rays treatments, Appl. Radiat. Isot. **106** (2015) 226–232.

[310] POSTUMA, I., et al., Colocalization of tracks from boron neutron capture reactions and images of isolated cells, Appl. Radiat. Isot. **167** (2020) 109353.

[311] PORTU, A., et al., Simultaneous observation of cells and nuclear tracks from the boron neutron capture reaction by UV-C sensitization of polycarbonate, Microsc. Microanal. **21** (2015) 796–804.

[312]   YANAGIE, H., et al., Accumulation of boron compounds to tumor with polyethylene-glycol binding liposome by using neutron capture autoradiography, Appl. Radiat. Isot. **61** (2004) 639–646.

[313]   SOLARES, G.R., ZAMENHOF, R.G., A novel approach to the microdosimetry of neutron capture therapy. Part I. High-resolution quantitative autoradiography applied to microdosimetry in neutron capture therapy, Radiat. Res. **144** (1995) 50–58.

[314]   GADAN, M.A., et al., Set-up and calibration of a method to measure ^{10}B-concentration in biological samples by neutron autoradiography, Nucl. Instrum. Methods B **274** (2012) 51–56.

[315]   ESPECTOR N.M., PORTU, A., SANTA CRUZ, G.A., SAINT MARTIN, G., Evaporation process in histological tissue sections for neutron autoradiography, Radiat. Environ. Biophys. **57** (2018) 153–162.

[316]   PORTU, A.M., et al., Qualitative autoradiography with polycarbonate foils enables histological and track analyses on the same section, Biotech. Histochem. **88** (2013) 217–221.

[317]   LAURENT-PETTERSSON, M., DELPECH, B., THELLIER, M., The mapping of natural boron in histological sections of mouse tissues by the use of neutron-capture radiography, Histochem. J. **24** (1992) 939–950.

[318]   ESPAIN, M.S., et al., Neutron autoradiography to study the microdistribution of boron in the lung, Appl. Radiat. Isot. **165** (2020) 109331.

[319]   TANAKA, H., et al., Evaluation of thermal neutron irradiation field using a cyclotron-based neutron source for alpha autoradiography, Appl. Radiat. Isot. **88** (2014) 153–156.

[320]   KIGER, W.S. III, MICCA P.L., MORRIS, G.M., CODERRE, J.A., Boron microquantification in oral mucosa and skin following administration of a neutron capture therapy agent, Radiat. Prot. Dosim. **99** (2002) 409–412.

[321]   AMEMIYA, K., et al., High-resolution nuclear track mapping in detailed cellular histology using CR-39 with the contact microscopy technique, Radiat. Meas. **40** (2005) 283–288.

[322]   KONISHI, T., et al., A new method for the simultaneous detection of mammalian cells and ion tracks on a surface of CR-39, J. Radiat. Res. **48** (2007) 255–261.

[323]   PORTU, A.M., et al., Experimental set up for the irradiation of biological samples and nuclear track detectors with UV C, Rep. Pract. Oncol. Radiother. **21** (2016) 129–134.

[324]   GADAN, M.A., et al., Neutron autoradiography combined with UV-C sensitization: toward the intracellular localization of boron, Microsc. Microanal. **25** (2019) 1331–1340.

[325]   SAINT MARTIN, G., et al., UV-C radiation effect on nuclear tracks of different ions in polycarbonate, Radiat. Phys. Chem. **173** (2020) 108936.

[326] CIANI, L., et al., Rational design of gold nanoparticles functionalized with carboranes for application in boron neutron capture therapy, Int. J. Pharm. **458** (2013) 340–346.

[327] PIETRANGELI, D., et al., Water-soluble carboranyl-phthalocyanines for BNCT. Synthesis, characterization, and in vitro tests of the Zn (II)-nido-carboranyl-hexylthiophthalocyanine, Dalton Trans. **44** (2015) 11021–11028.

[328] NAR, I., et al., A phthalocyanine-ortho-carborane conjugate for boron neutron capture therapy: Synthesis, physicochemical properties, and in vitro tests, ChemPlusChem **84** (2019) 345–351.

[329] ROSSINI, A.E., et al., Assessment of biological effectiveness of boron neutron capture therapy in primary and metastatic melanoma cell lines, Int. J. Radiat. Biol. **91** (2015) 81–89.

[330] FERRARI, C., et al., Selective uptake of p-boronophenylalanine by osteosarcoma cells for boron neutron capture therapy, Appl. Radiat. Isot. **67** Suppl. 7–8 (2009) S341–S344.

[331] MOLINARI, A.J., et al., Assessing advantages of sequential boron neutron capture therapy (BNCT) in an oral cancer model with normalized blood vessels, Acta Oncol. **54** (2015) 99–106.

[332] BORTOLUSSI, S., et al., Neutron flux and gamma dose measurement in the BNCT irradiation facility at the TRIGA reactor of the University of Pavia, Nucl. Instrum. Methods B **414** (2018) 113–120.

[333] OGURA, K., YANAGIE, H., ERIGUCHI, M., MATSUMOTO, T., KOBAYASHI, H., Application of CR-39 for alpha-autoradiography and in vivo detection of ^{10}B accumulations in tumor bearing mice, Radiat. Meas. **31** (1999) 389–394.

[334] LU, X.Q., KIGER, W.S., III, Application of a novel microdosimetry analysis and its radiobiological implication for high-LET radiation, Radiat. Res. **171** (2009) 646–656.

[335] LEVITT, M.H., Spin Dynamics: Basics of Nuclear Magnetic Resonance, Wiley, second edition (2008).

[336] DEAGOSTINO, A., et al., Insights into the use of gadolinium and gadolinium/boron-based agents in imaging-guided neutron capture therapy applications, Future Med. Chem. **8** (2016) 899–917.

[337] McROBBIE, D.W., et al., MRI from Picture to Proton. Second edition, Cambridge University Press, Cambridge, UK (2010).

[338] FORD, T.C., CREWTHER, D.P., A comprehensive review of the ^{1}H-MRS metabolite spectrum in autism spectrum disorder, Front. Mol. Neurosci. **9** (2016) 14.

[339] BENDEL, P., MARGALIT, R., SALOMON, Y., Optimized ^{1}H MRS and MRSI methods for the in vivo detection of boronophenylalanine, Magn. Reson. Med. **53** (2005) 1166–117.

[340] ZUO, C.S., et al., Proton nuclear magnetic resonance measurement of *p*-boronophenylalanine (BPA): A therapeutic agent for boron neutron capture therapy, Med. Phys. **26** (1999) 1230–1236.

[341] TIMONEN, M., et al., ^{1}H MRS studies in the Finnish boron neutron capture therapy project: Detection of [10]B-carrier, L-*p*-boronophenylalanine–fructose, Eur. J. Radiol. **56** (2005) 154–159.

[342] TIMONEN, M., Proton magnetic resonance spectroscopy of a boron neutron capture therapy [10]B-carrier, L-*p*-boronophenylalanine–fructose complex, Academic thesis, University of Helsinki, Report Series in Physics, HU-P-D171 (2010).

[343] WAHSNER, J., et al., Chemistry of MRI contrast agents: Current challenges and new frontiers, Chem. Rev. **119** (2019) 957–1057.

[344] TERRENO, E., et al., Advances in metal-based probes for MR molecular imaging applications, Curr. Med. Chem. **17** (2010) 3684–3700.

[345] TATHAM, A.T., NAKAMURA, H., WIENER, E.C., YAMAMOTO, Y., Relaxation properties of a dual-labeled probe for MRI and neutron capture therapy, Magn. Reason. Med. **42** (1999) 32–36.

[346] NAKAMURA, H., et al., In vivo evaluation of carborane gadolinium-DTPA complex as an MR imaging boron carrier, Chem. Pharm. Bull. **48** (2000) 1034–1038.

[347] BONORA, M., et al., ^{1}H and [10]B NMR and MRI investigation of boron- and gadolinium-boron compounds in boron neutron capture therapy, Appl. Radiat. Isot. **69** (2011) 1702–1705.

[348] ALBERTI, D., et al., Theranostic nanoparticles loaded with imaging probes and rubrocurcumin for combined cancer therapy by folate receptor targeting, ChemMedChem. **12** (2017) 502–509.

[349] BAN, H.S., NAKAMURA, H., Boron-based drug design, Chem. Rec. **15** (2015) 616–635.

[350] ICTEN, O., et al., Gadolinium borate and iron oxide bioconjugates: Nanocomposites of next generation with multifunctional applications, Mater. Sci. Eng. C **92** (2018) 317–328.

[351] ALBERTI, D., et al., A theranostic approach based on the use of a dual boron/Gd agent to improve the efficacy of boron neutron capture therapy in the lung cancer treatment, Nanomed. Nanotechnol. Biol. Med. **11** (2015) 741–750.

[352] GENINATTI-CRICH, S., et al., MRI-guided neutron capture therapy by use of a dual gadolinium/boron agent targeted at tumour cells through upregulated low-density lipoprotein transporters, Chem. Eur. J. **17** (2011) 8479–8486.

[353] ALBERTI, D., et al., An innovative therapeutic approach for malignant mesothelioma treatment based on the use of Gd/boron multimodal probes for MRI guided BNCT, J. Control. Release **280** (2018) 31–38.

[354] BARTUSIK, D., AEBISHER, D., ^{19}F applications in drug development and imaging-a review, Biomed. Pharmacother. **68** (2014) 813–817.

[355] PORCARI, P., et al., In vivo ^{19}F MR imaging and spectroscopy for the BNCT optimization, Appl. Radiat. Isot. **67** (2009) 6979.

[356] BENDEL, P., KOUDINOVA, N., SALOMON, Y., In vivo imaging of the neutron capture therapy agent BSH in mice using ^{10}B MRI, Magn. Reson. Med. **46** (2001) 13–17.

[357] SMITH, D.R., et al., quantitative imaging and microlocalization of boron-10 in brain tumors and infiltrating tumor cells by SIMS ion microscopy: Relevance to neutron capture therapy, Cancer Res. **61** (2001) 8179–8187.

[358] HOPPE, P., COHEN, S., MEIBOM, A., NanoSIMS: Technical aspects and applications in cosmochemistry and biological geochemistry, Geostand. Geoanal. Res. **37** (2013) 111–154.

[359] ARLINGHAUS, H.F., SPAAR, M.T., SWITZER, R.C., KABALKA, G.W., Imaging of boron in tissue at cellular level for boron neutron capture therapy, Anal. Chem. **69** (1997) 3169–3176.

[360] ARLINGHAUS, H.F., et al., Mass spectrometric characterization of elements and molecules in cell cultures and tissues, Appl. Surf. Sci. **252** (2006) 69416948.

[361] MICHEL, J., SAUERWEIN, W., WITTIG, A., BALOSSIER, G., ZIEROLD, K., Subcellular localization of boron in cultured melanoma cells by electron energy-loss spectroscopy of freeze-dried cryosections, J. Microsc. **210** (2003) 25–34.

[362] LEAPMAN, R.D., et al., Three dimensional distribution of elements in biological samples by energy filtered electron tomography, Ultramicroscopy **100** (2004) 115–125.

[363] SANCEY, L., et al., Laser spectrometry for multi-elemental imaging of biological tissues, Sci. Rep. **4** (2014) 6065.

[364] SANCEY, L., et al., Laser-induced breakdown spectroscopy: a new approach for nanoparticle's mapping and quantification in organ tissue, J. Vis. Exp. **18** (2014) e51353.

[365] MOTTO-ROS, V., et al., Mapping of native inorganic elements and injected nanoparticles in a biological organ with laser-induced plasma, Appl. Phys. Lett. **101** (2012) 223702.

[366] MOTTO-ROS, V., et al., Imagerie LIBS : Aux portes de la clinique, Photoniques **103** (2020) 34–37.

[367] BUSSER, B., et al., Characterization of foreign materials in paraffin-embedded pathological specimens using in situ multi-elemental imaging with laser spectroscopy, Mod. Pathol. **31** (2018) 378–384.

[368] MONCAYO, S., et al., Multi-elemental imaging of paraffin-embedded human samples by laser-induced breakdown spectroscopy, Spectrochim. Acta Part B **133** (2017) 40–44.

[369] DETAPPE, A., et al., Advanced multimodal nanoparticles delay tumor progression with clinical radiation therapy, J. Control. Release **238** (2016) 103–113.

[370] GIMENEZ, Y., et al., 3D imaging of nanoparticle distribution in biological tissue by laser-induced breakdown spectroscopy, Sci. Rep. **6** (2016) 29936.

[371] KUNJACHAN, S., et al., Nanoparticle mediated tumor vascular disruption: A novel strategy in radiation therapy, Nano Lett. **15** (2015) 7488–7496.

[372] MOUSSARON, A.S., et al., Ultrasmall nanoplatforms as calcium-responsive contrast agents for magnetic resonance imaging, Small **11** (2015) 4900–4909.

[373] SANCEY, L., et al., Long-term in vivo clearance of gadolinium-based AGuIX nanoparticles and their biocompatibility after systemic injection, ACS Nano **9** (2015) 2477–2488.

[374] LEPRINCE, M., et al., Elemental imaging using laser-induced breakdown spectroscopy: latest medical applications, Med. Sci. (Paris) **35** (2019) 682–688.

[375] BUSSER, B., BONNETERRE, V., SANCEY, L., MOTTO-ROS, V., LIBS imaging is entering the clinic as a new diagnostic tool, Spectroscopy **35** (2020) 17–19.

[376] LOCHER, G.L., Biological effects and therapeutic possibilities of neutrons, Am. J. Roentgenol. Radium Ther. **36** (1936) 1–13.

[377] HOPEWELL, J.W., MORRIS, G.M., SCHWINT, A., CODERRE, J.A., The radiobiological principles of boron neutron capture therapy: A critical review, Appl. Radiat. Isot. **69** 12 (2011) 1756–1759.

[378] CODERRE, J.A., MORRIS, G.M., The radiation biology of boron neutron capture therapy, Radiat. Res. **151** (1999) 1–18.

[379] FUTAMURA, G., et al., A case of radiation-induced osteosarcoma treated effectively by boron neutron capture therapy, Radiat. Oncol. **9** (2014) 237.

[380] POZZI, E.C.C., et al., Boron neutron capture therapy (BNCT) for liver metastasis: Therapeutic efficacy in an experimental model, Radiat. Environ. Biophys. **51** 3 (2012) 331–339.

[381] MONTI HUGHES, A., et al., Histamine reduces boron neutron capture therapy-induced mucositis in an oral precancer model, Oral Dis. **21** 6 (2015) 770.

[382] HAAPANIEMI, A., et al., boron neutron capture therapy in the treatment of recurrent laryngeal cancer, Int. J. Radiat. Oncol. Biol. Phys. **95** 1 (2016) 404–410.

[383] HIRATSUKA, J., et al., Boron neutron capture therapy for vulvar melanoma and genital extramammary Paget's disease with curative responses, Cancer Commun. **38** (2018) 38.

[384] SANTA CRUZ, G.A., Microdosimetry: Principles and applications, Rep. Pract. Oncol. Radiother. **21** 2 (2016) 135–139.

[385] GONZÁLEZ, S.J., et al., Photon iso-effective dose for cancer treatment with mixed field radiation based on dose-response assessment from human and an animal model: Clinical application to boron neutron capture therapy for head and neck cancer, Phys. Med. Biol. **62** 20 (2017) 7938–7958.

[386] HOPEWELL, J.W., et al., Boron neutron capture therapy for newly diagnosed glioblastoma multiforme: An assessment of clinical potential, Appl. Radiat. Isot. **69** 12 (2011) 1737–1740.

[387] MCNALLY, N.J., DE RONDE, J., FOLKARD, M., Interaction between X-ray and α particle damage in v79 cells, Int. J. Radiat. Biol. **53** 6 (1988) 917–920.

[388] MASON, A.J., et al., Interaction between the biological effects of high- and low-LET radiation dose components in a mixed field exposure, Int. J. Radiat. Biol. **87** 12 (2011) 1162–1172.

[389] SATO, T., MASUNAGA, S., KUMADA, H., HAMADA, N., Microdosimetric modeling of biological effectiveness for boron neutron capture therapy considering intra- and intercellular heterogeneity in ^{10}B-distribution, Sci. Rep. **8** 1 (2018) 988.

[390] SATO, T., MASUNAGA, S., KUMADA, H., HAMADA, N., Depth distributions of RBE-weighted dose and photon-isoeffective dose for boron neutron capture therapy, Radiat. Prot. Dosimetry **183** 1–2 (2019) 247–250.

[391] ONO, K., An analysis of the structure of the compound biological effectiveness factor, J. Radiat. Res. **57** (2016) i83–i89.

[392] GONZÁLEZ, S., SANTA CRUZ, G., The photon-isoeffective dose in boron neutron capture therapy, Radiat. Res. **178** (2012) 609–621.

[393] PEDROSA-RIVERA, M., et al., A simple approximation for the evaluation of the photon isoeffective dose in boron neutron capture therapy based on dose-independent weighting factors, Appl. Radiat. Isot. **157** (2020) 109018.

[394] PORRAS, I., et al., "Perspectives on neutron capture therapy of cancer" (Proc. 15th Int. Confer. on Nuclear Reaction Mechanisms, Varenna, July 11–15, 2018, Italy) CERN, Geneva (2019) 295–304.

[395] PROVENZANO, L., et al., The essential role of radiobiological figures of merit for the assessment and comparison of beam performances in boron neutron capture therapy, Phys. Medica **67** (2019) 9–19.

[396] POSTUMA, I., et al., "Design of a neutron beam for BNCT treatment of deep-seated tumours and evaluation of its therapeutic potential and suitability", Technical Meeting on Advances in Boron Neutron Capture Therapy 27–31 July 2020, IAEA, Vienna (2020).

[397] LUDERER, M.J., DE LA PUENTE, P., AZAB, A.K., Advancements in tumor targeting strategies for boron neutron capture therapy, Pharm. Res. **32** 9 (2015) 2824–2836.

[398]   GARABALINO, M.A., et al., Boron neutron capture therapy (BNCT) for the treatment of liver metastases: Biodistribution studies of boron compounds in an experimental model, Radiat. Environ. Biophys. **50** 1 (2011) 199.

[399]   MARUSYK, A., ALMENDRO, V., POLYAK, K., Intra-tumour heterogeneity: A looking glass for cancer? Nat. Rev. Cancer **12** 5 (2012) 323–334.

[400]   BHATIA, S., FRANGIONI, J.V., HOFFMAN, R.M., IAFRATE, A.J., POLYAK, K., The challenges posed by cancer heterogeneity, Nat. Biotechnol. **30** 7 (2012) 604–610.

[401]   MASUNAGA, S., et al., An attempt to improve the therapeutic effect of boron neutron capture therapy using commonly employed ^{10}B-carriers based on analytical studies on the correlation among quiescent tumor cell characteristics, tumor heterogeneity and cancer stemness, J. Radiat. Res. **61** 6 (2020) 876–885.

[402]   VAUPEL, P., Tumor microenvironmental physiology and its implications for radiation oncology, Semin. Radiat. Oncol. **14** 3 (2004) 198–206.

[403]   SCHWINT, A.E., TRIVILLIN, V.A., "Close-to-ideal" tumor boron targeting for boron neutron capture therapy is possible with "less-than-ideal" boron carriers approved for use in humans, Ther. Delivery **6** 3 (2015) 269–272.

[404]   DAGROSA, M.A., et al., Experimental application of boron neutron capture therapy to undifferentiated thyroid carcinoma, Int. J. Radiat. Oncol. Biol. Phys. **57** 4 (2003) 1084–1092.

[405]   TRIVILLIN, V.A., et al., Therapeutic success of boron neutron capture therapy (BNCT) mediated by a chemically non-selective boron agent in an experimental model of oral cancer: A new paradigm in BNCT radiobiology, Radiat. Res. **166** 2 (2006) 387–396.

[406]   ZONTA, A., et al., Clinical lessons from the first applications of BNCT on unresectable liver metastases, J. Phys. Conf. Ser. **41** 1 (2006) 484.

[407]   MATSUMURA, A., et al., Current practices and future directions of therapeutic strategy in glioblastoma: Survival benefit and indication of BNCT, Appl. Radiat. Isot. **67** Suppl. 7–8 (2009) S12–S14.

[408]   KANKAANRANTA, L., et al., Boron neutron capture therapy in the treatment of locally recurred head and neck cancer: Final analysis of a phase I/II trial, Int. J. Radiat. Oncol. Biol. Phys. **82** (2012) e67–e75.

[409]   BARTH, R.F., From the laboratory to the clinic: How translational studies in animals have led to clinical advances in boron neutron capture therapy, Appl. Radiat. Isot. **106** (2015) 22–28.

[410]   BARTH, R.F., Boron neutron capture therapy at the crossroads: Challenges and opportunities, Appl. Radiat. Isot. **67** Suppl. 7–8 (2009) S3–S6.

[411]   BARTH, R.F., MI, P., YANG, W., Boron delivery agents for neutron capture therapy of cancer, Cancer Commun. **38** 1 (2018) 1–5.

[412] PAUL, S.M., et al., How to improve R&D productivity: The pharmaceutical industry's grand challenge, Nat. Rev. Drug Discovery **9** 3 (2010) 203–214.

[413] CRUICKSHANK, G.S., et al., A cancer research UK pharmacokinetic study of BPA-mannitol in patients with high grade glioma to optimise uptake parameters for clinical trials of BNCT, Appl. Radiat. Isot. **67** Suppl. 7–8 (2009) S31–S33.

[414] MOSS, R.L., Critical review, with an optimistic outlook, on boron neutron capture therapy (BNCT), Appl. Radiat. Isot. **88** (2014) 2–11.

[415] BARTH, R.F., ZHANG, Z., LIU, T., A realistic appraisal of boron neutron capture therapy as a cancer treatment modality, Cancer Commun. **38** 1 (2018) 1–7.

[416] KREIMANN, E.L., et al., The hamster cheek pouch as a model of oral cancer for boron neutron capture therapy studies: Selective delivery of boron by boronophenylalanine, Cancer Res. **61** 24 (2001) 8775.

[417] KANKAANRANTA, L., et al., L-boronophenylalanine-mediated boron neutron capture therapy for malignant glioma progressing after external beam radiation therapy: A Phase I study, Int. J. Radiat. Oncol. Biol. Phys. **80** (2011) 369–376.

[418] KREIMANN, E.L., et al., Boron neutron capture therapy for the treatment of oral cancer in the hamster cheek pouch model, Cancer Res. **61** 24 (2001) 8638.

[419] KAWABATA, S., et al., Survival benefit from boron neutron capture therapy for the newly diagnosed glioblastoma patients, Appl. Radiat. Isot. **67** Suppl. 7–8 (2009) S15–S18.

[420] MASUNAGA, S., et al., Evaluation of apoptosis and micronucleation induced by reactor neutron beams with two different cadmium ratios in total and quiescent cell populations within solid tumors, Int. J. Radiat. Oncol. Biol. Phys. **51** 3 (2001) 828–839.

[421] MOLINARI, A.J., et al., Tumor blood vessel "normalization" improves the therapeutic efficacy of boron neutron capture therapy (BNCT) in experimental oral cancer, Radiat. Res. **177** 1 (2012) 59–68.

[422] MOLINARI, A.J., et al., Blood vessel normalization in the hamster oral cancer model for experimental cancer therapy studies, Anticancer Res. **32** 7 (2012) 2703.

[423] AKAGAWA, Y., KAGEJI, T., MIZOBUCHI, Y., KUMADA, H., NAKAGAWA, Y., Clinical results of BNCT for malignant brain tumors in children, Appl. Radiat. Isot. **67** Suppl. 7–8 (2009) S27–S30.

[424] ONO, K., KINASHI, Y., SUZUKI, M., TAKAGAKI, M., MASUNAGA, S., The combined effect of electroporation and borocaptate in boron neutron capture therapy for murine solid tumors, Jpn. J. Cancer Res. **91** 8 (2000) 853–858.

[425] GABEL, D., et al., Pharmacokinetics of $Na_2B_{12}H_{11}SH$ (BSH) in patients with malignant brain tumours as prerequisite for a Phase I clinical trial of boron neutron capture, Acta Neurochir. **139** 7 (1997) 606–612.

[426] BAUER, W.F., BRADSHAW, K.M., RICHARDS, T.L., "Interaction between boron containing compounds and serum albumin observed by nuclear magnetic resonance", Progress in Neutron Capture Therapy for Cancer (ALLEN, B.J., MOOORE, D.E., HARRINGTON, B.V., Eds) Springer Science+Business Media, LLC (1992), p 339–343.

[427] PATEL, H., SEDGWICK, E., "BPA & BSH Accumulation in Experimental Tumors", Proc. Ninth International Symposium on Neutron Capture Therapy for Cancer, Osaka, Japan (2000) 59–60.

[428] KATO, I., et al., Effectiveness of boron neutron capture therapy for recurrent head and neck malignancies, Appl. Radiat. Isot. **67** Suppl. 7–8 (2009) S37–S42.

[429] HEBER, E., et al., Biodistribution of GB-10 ($Na_2{}^{10}B_{10}H_{10}$) compound for boron neutron capture therapy (BNCT) in an experimental model of oral cancer in the hamster cheek pouch, Arch. Oral Biol. **49** 4 (2004) 313–324.

[430] STELZER, K.J., et al., "Boron neutron capture-enhanced fast neutron therapy (BNC/FNT) for non-small cell lung cancer in canine patients", Frontiers in Neutron Capture Therapy, Vol. 1 (HAWTHORNE, M.F., SHELLY, K., WIERSEMA, R.J., Eds) Springer Science+Business Media, LLC (2001) 735–739.

[431] SWEET, W.H., SOLOWAY, A.H., BROWNELL, G.L., Boron-slow neutron capture therapy of gliomas, Acta Radiol. Ther. Phys. Biol. **1** 2 (1963) 114.

[432] DIAZ, A., STELZER, K., LARAMORE, G., WIERSEMA, R., "Pharmacology studies of $Na_2{}^{10}B_{10}H_{10}$ (GB-10) in human tumor patients", Research and Development in Neutron Capture Therapy. (Proc. 10th International Congress on Neutron Capture Therapy, Essen, Germany Sept. 8–13, 2002) (2002) 993.

[433] TRIVILLIN, V.A., et al., Translational boron neutron capture therapy (BNCT) studies for the treatment of tumors in lung, Int. J. Radiat. Biol. **95** 5 (2019) 646–654.

[434] MORRIS, G.M., et al., Boron microlocalizatian in oral mucosal tissue: Implications for boron neutron capture therapy, Br. J. Cancer **82** 11 (2000) 1764–1771.

[435] FUKUDA, H., Response of normal tissues to boron neutron capture therapy (BNCT) with ^{10}B-borocaptate sodium (BSH) and ^{10}B-paraboronophenylalanine (BPA), Cells **10** (2021) 2883.

[436] OKAMOTO, E., YAMAMOTO, T., NAKAI, K., FUMIYO, Y., MATSUMURA, A., Detection of DNA double-strand breaks in boron neutron capture reaction, Appl. Radiat. Isot. **106** (2015) 185–188.

[437] KONDO, N., et al., DNA damage induced by boron neutron capture therapy is partially repaired by DNA ligase IV, Radiat. Environ. Biophys. **55** 1 (2016) 89–94.

[438] JAMSRANJAV, E., et al., DNA strand breaks induced by fast and thermal neutrons from Yayoi research reactor in the presence and absence of boric acid, Radiat. Res. **191** 5 (2019) 483–489.

[439] MASUTANI, E.M., KINOSHITA, C.K., TANAKA, T.T., ELLISON, A.K.D., YOZA, B.A., Increasing thermal stability of gelatin by UV-induced cross-linking with glucose, Int. J. Biomater. **2014** (2014) 979636.

[440] MASUTANI, M., et al., Histological and biochemical analysis of DNA damage after BNCT in rat model, Appl. Radiat. Isot. **88** (2014) 104–108.

[441] FAIÃO-FLORES, F., COELHO, P.R.P., ARRUDA-NETO, J.D.T., MARIA-ENGLER, S.S., MARIA, D.A., Cell cycle arrest, extracellular matrix changes and intrinsic apoptosis in human melanoma cells are induced by boron neutron capture therapy, Toxicol. Vitr. **27** 4 (2013) 1196–1204.

[442] CHANDRA, S., AHMAD, T., BARTH, R.F., KABALKA, G.W., Quantitative evaluation of boron neutron capture therapy (BNCT) drugs for boron delivery and retention at subcellular-scale resolution in human glioblastoma cells with imaging secondary ion mass spectrometry (SIMS), J. Microsc. **254** 3 (2014) 146–156.

[443] MALISZEWSKA-OLEJNICZAK, K., KANIOWSKI, D., ARASZKIEWICZ, M., TYMIŃSKA, K., KORGUL, A., Molecular mechanisms of specific cellular DNA damage response and repair induced by the mixed radiation field during boron neutron capture therapy, Front. Oncol. **11** (2021) 676575.

[444] KONDO, N., et al., Cerebrospinal fluid dissemination of high-grade gliomas following boron neutron capture therapy occurs more frequently in the small cell subtype of IDH1R132H mutation-negative glioblastoma, J. Neuro-Oncol. **133** 1 (2017) 107–118.

[445] KONDO, N., SUZUKI, M., ONO, K., "Radiation biology on BNCT – What we have known and don't know", Technical Meeting on Advances in Boron Neutron Capture Therapy 27–31 July 2020, IAEA, Vienna (2020).

[446] FAIÃO-FLORES, F., et al., Apoptosis through Bcl-2/Bax and cleaved caspase up-regulation in melanoma treated by boron neutron capture therapy, PLOS ONE **8** 3 (2013) e59639.

[447] FAIÃO-FLORES, F., COELHO, P.R.P., ARRUDA-NETO, J., MARIA, D.A., Boron neutron capture therapy induces cell cycle arrest and DNA fragmentation in murine melanoma cells, Appl. Radiat. Isot. **69** 12 (2011) 1741–1744.

[448] PERONA, M., et al., In vitro studies of cellular response to DNA damage induced by boron neutron capture therapy, Appl. Radiat. Isot. **69** 12 (2011) 1732–1736.

[449] SUN, T., et al., Boron neutron capture therapy induces cell cycle arrest and cell apoptosis of glioma stem/progenitor cells in vitro, Radiat. Oncol. **8** 1 (2013) 1–8.

[450] FUJITA, Y., et al., Role of p53 mutation in the effect of boron neutron capture therapy on oral squamous cell carcinoma, Radiat. Oncol. **4** 1 (2009) 63.

[451] EKI, K., KINASHI, Y., TAKAHASHI, S., Influence of p53 status on the effects of boron neutron capture therapy in glioblastoma, Anticancer Res. **35** 1 (2015) 169–174.

[452]   WANG, P., et al., Boron neutron capture therapy induces apoptosis of glioma cells through Bcl-2/Bax, BMC Cancer **10** (2010) 661.

[453]   MASUNAGA, S., et al., Radiobiological characteristics of solid tumours depending on the p53 status of the tumour cells, with emphasis on the response of intratumour quiescent cells, Eur. J. Cancer **38** 5 (2002) 718–727.

[454]   SCHMID, T.E., et al., The effectiveness of the high-LET radiations from the boron neutron capture $[^{10}B(n,\alpha)^7Li]$ reaction determined for induction of chromosome aberrations and apoptosis in lymphocytes of human blood samples, Radiat. Environ. Biophys. **54** 1 (2015) 91–102.

[455]   AROMANDO, R.F., et al., Insight into the mechanisms underlying tumor response to boron neutron capture therapy in the hamster cheek pouch oral cancer model, J. Oral Pathol. Med. **38** 5 (2009) 448–454.

[456]   KAMIDA, A., et al., Effects of boron neutron capture therapy on human oral squamous cell carcinoma in a nude mouse model, Int. J. Radiat. Biol. **82** 1 (2006) 21–29.

[457]   RODRIGUEZ, C., et al., In vitro studies of DNA damage and repair mechanisms induced by BNCT in a poorly differentiated thyroid carcinoma cell line, Radiat. Environ. Biophys. **57** 2 (2018) 143–152.

[458]   SANTA CRUZ, G.A., "Microdosimetry and BNCT", Technical Meeting on Advances in Boron Neutron Capture Therapy 27–31 July 2020, IAEA, Vienna (2020).

[459]   SATO, A., et al., Proteomic analysis of cellular response induced by boron neutron capture reaction in human squamous cell carcinoma SAS cells, Appl. Radiat. Isot. **106** (2015) 213–219.

[460]   FERRARI, E., et al., Urinary proteomics profiles are useful for detection of cancer biomarkers and changes induced by therapeutic procedures, Molecules **24** 4 (2019) 794.

[461]   MAURI, P., et al., Proteomics of bronchial biopsies: Galectin-3 as a predictive biomarker of airway remodelling modulation in omalizumab-treated severe asthma patients, Immunol. Lett. **162** 1 (2014) 2–10.

[462]   MAURI, P.L., et al., "Clinical trial investigations on BNCT by molecular and network medicine", Technical Meeting on Advances in Boron Neutron Capture Therapy 27–31 July 2020, IAEA, Vienna (2020).

[463]   SAUERWEIN, W.A.G., et al., Theranostics in boron neutron capture therapy, Life **11** 4 (2021) 330.

[464]   MOLE, R.H., Whole body irradiation; radiobiology or medicine? Br. J. Radiol. **26** 305 (1953) 234–241.

[465]   OKUMA, K., YAMASHITA, H., NIIBE, Y., HAYAKAWA, K., NAKAGAWA, K., Abscopal effect of radiation on lung metastases of hepatocellular carcinoma: A case report, J. Med. Case Rep. **5** (2011) 111.

[466] GOLDEN, E.B., DEMARIA, S., SCHIFF, P.B., CHACHOUA, A., FORMENTI, S.C., An abscopal response to radiation and ipilimumab in a patient with metastatic non-small cell lung cancer, Cancer Immunol. Res. **1** 6 (2013) 365–372.

[467] DEMARIA, S., et al., Ionizing radiation inhibition of distant untreated tumors (abscopal effect) is immune mediated, Int. J. Radiat. Oncol. Biol. Phys. **58** 3 (2004) 862–870.

[468] VATNER, R.E., COOPER, B.T., VANPOUILLE-BOX, C., DEMARIA, S., FORMENTI, S.C., Combinations of immunotherapy and radiation in cancer therapy, Front. Oncol. **4** (2014) 325.

[469] PARK, B., YEE, C., LEE, K.M., The effect of radiation on the immune response to cancers, Int. J. Mol. Sci. **15** 1 (2014) 927–943.

[470] LOCK, M., MUINUDDIN, A., KOCHA, W.I., DINNIWELL, R., RODRIGUES, G., D'SOUZA, D., Abscopal effects: case report and emerging opportunities, Cureus **7** 10 (2015) e344.

[471] TRIVILLIN, V.A., et al., Evaluation of local, regional and abscopal effects of boron neutron capture therapy (BNCT) combined with immunotherapy in an ectopic colon cancer model, Br. J. Radiol. **94** (2021) 20210593.

[472] TRIVILLIN, V.A., et al., "BNCT Combined with Immunotherapy: Evaluation of Local, Immunologic and Cytotoxic Effects in an Ectopic Colon Cancer Model", PTCOG58 (The Particle Therapy Co-Operative Group), Manchester, England (2019).

[473] KHAN, A.A., MAITZ, C., QUANYU, C., HAWTHORNE, F., BNCT induced immunomodulatory effects contribute to mammary tumor inhibition, PLOS ONE **14** 9 (2019) e0222022.

[474] ANDOH, T., et al., Boron neutron capture therapy (BNCT) as a new approach for clear cell sarcoma (CCS) treatment: Trial using a lung metastasis model of CCS, Appl. Radiat. Isot. **106** (2015) 195.

[475] HUGHES, A.M., et al., Boron neutron capture therapy (BNCT) inhibits tumor development from precancerous tissue: An experimental study that supports a potential new application of BNCT, Appl. Radiat. Isot. **67** Suppl. 7–8 (2009) S313–S317.

[476] ONO, K., et al., The combined effect of boronophenylalanine and borocaptate in boron neutron capture therapy for SCCVII tumors in mice, Int. J. Radiat. Oncol. Biol. Phys. **43** 2 (1999) 431–436.

[477] HEBER, E.M., et al., Homogeneous boron targeting of heterogeneous tumors for boron neutron capture therapy (BNCT): Chemical analyses in the hamster cheek pouch oral cancer model, Arch. Oral Biol. **51** 10 (2006) 922–929.

[478] HUGHES, A.M., et al., Boron neutron capture therapy for oral precancer: Proof of principle in an experimental animal model, Oral Dis. **19** 8 (2013) 789–795.

[479]   BARTH, R.F., et al., Boron neutron capture therapy of brain tumors: Enhanced survival and cure following blood-brain barrier disruption and intracarotid injection of sodium borocaptate and boronophenylalanine, Int. J. Radiat. Oncol. Biol. Phys. **47** 1 (2000) 209–218.

[480]   FUTAMURA, G., et al., Evaluation of a novel sodium borocaptate-containing unnatural amino acid as a boron delivery agent for neutron capture therapy of the F98 rat glioma, Radiat. Oncol. **12** 1 (2017) 26.

[481]   AZZALIN, A., et al., A new pathway promotes adaptation of human glioblastoma cells to glucose starvation, Cells **9** 5 (2020) 1249.

[482]   JOEL, D.D., CODERRE, J.A., MICCA, P.L., NAWROCKY, M.M., Effect of dose and infusion time on the delivery of p-boronophenylalanine for neutron capture therapy, J. Neuro-Oncol. **41** 3 (1999) 213–221.

[483]   SKOLD, K., et al., Boron neutron capture therapy for glioblastoma multiforme: Advantage of prolonged infusion of BPA-f, Acta Neurol. Scand. **122** (2010) 58–62.

[484]   ONO, K., et al., "Neutron irradiation under continuous BPA injection for solving the problem of heterogeneous distribution of BPA", Proc. 12th International Conference on Neutron Capture Therapy, Takamatsu, Japan (2006) 27–30.

[485]   CAPUANI, S., PORCARI, P., FASANO, F., CAMPANELLA, R., MARAVIGLIA, B., ^{10}B-editing ^{1}H-detection and ^{19}F MRI strategies to optimize boron neutron capture therapy, Magn. Reson. Imaging **26** 7 (2008) 987–993.

[486]   WINGELHOFER, B., et al., Preloading with L-BPA, L-tyrosine and L-DOPA enhances the uptake of [^{18}F]FBPA in human and mouse tumour cell lines, Appl. Radiat. Isot. **118** (2016) 67–72.

[487]   WATANABE, T., et al., L-phenylalanine preloading reduces the ^{10}B(n,α)^{7}Li dose to the normal brain by inhibiting the uptake of boronophenylalanine in boron neutron capture therapy for brain tumours, Cancer Lett. **370** 1 (2016) 27–32.

[488]   SATO, E., et al., Intracellular boron accumulation in CHO-K1 cells using amino acid transport control, Appl. Radiat. Isot. **88** (2014) 99–103.

[489]   MASUNAGA, S., et al., Effect of bevacizumab combined with boron neutron capture therapy on local tumor response and lung metastasis, Exp. Ther. Med. **8** 1 (2014) 291–301.

[490]   LUDERER, M.J., et al., Thermal sensitive liposomes improve delivery of boronated agents for boron neutron capture therapy, Pharm. Res. **36** 10 (2019) 144.

[491]   MOLINARI, A.J., et al., "Sequential" boron neutron capture therapy (BNCT): A novel approach to BNCT for the treatment of oral cancer in the hamster cheek pouch model, Radiat. Res. **175** 4 (2011) 463–472.

[492]   OLAIZ, N., et al., "Electroporation enhances tumor control induced by GB-10-BNCT in the hamster cheek pouch oral cancer model", 17th International Congress on Neutron Capture Therapy, Columbia, Missouri, USA (2016).

[493]   WU, C.Y., et al., Pulsed-focused ultrasound enhances boron drug accumulation in a human head and neck cancer xenograft-bearing mouse model, Mol. Imaging Biol. **16** 1 (2014) 95–101.

[494]   YANG, C., LI, Y., DU, M., CHEN, Z., Recent advances in ultrasound-triggered therapy, J. Drug Targeting **27** (2019) 33–50.

[495]   BURGESS, A., SHAH, K., HOUGH, O., HYNYNEN, K., Focused ultrasound-mediated drug delivery through the blood–brain barrier, Expert Rev. Neurother. **15** (2015) 477–491.

[496]   BURGESS, A., HYNYNEN, K., Microbubble-assisted ultrasound for drug delivery in the brain and central nervous system, Adv. Exp. Med. Biol. **880** (2016) 293–308.

[497]   WOOD, A.K.W., SEHGAL, C.M., A review of low-intensity ultrasound for cancer therapy, Ultrasound Med. Biol. **41** (2015) 905–928.

[498]   YANAGIE, H., et al., Pilot clinical study of boron neutron capture therapy for recurrent hepatic cancer involving the intra-arterial injection of a ^{10}BSH-containing WOW emulsion, Appl. Radiat. Isot. **88** (2014) 32.

[499]   YANAGIE, H., et al., Selective boron delivery by intra-arterial injection of BSHWOW emulsion in hepatic cancer model for neutron capture therapy, Br. J. Radiol. **90** 1074 (2017) 2017004.

[500]   YANAGIE, H., et al., "Boron neutron capture therapy planning for recurrent hepatocellular carcinoma using boron entrapped water-in-oil-in-water emulsion", Technical Meeting on Advances in Boron Neutron Capture Therapy 27–31 July 2020, IAEA, Vienna (2020).

[501]   LIN, Y.C., CHOU, F.I., LIAO, J.W., LIU, Y.H., HWANG, J.J., The effect of low-dose gamma irradiation on the uptake of boronophenylalanine to enhance the efficacy of boron neutron capture therapy in an orthotopic oral cancer model, Radiat. Res. **195** 4 (2021) 347–354.

[502]   PERONA, M., et al., Improvement of the boron neutron capture therapy (BNCT) by the previous administration of the histone deacetylase inhibitor sodium butyrate for the treatment of thyroid carcinoma, Radiat. Environ. Biophys. **52** 3 (2013) 363–373.

[503]   MASUNAGA, S., et al., Usefulness of combination with both continuous administration of hypoxic cytotoxin and mild temperature hyperthermia in boron neutron capture therapy in terms of local tumor response and lung metastatic potential, Int. J. Radiat. Biol. **95** 12 (2019) 1708–1717.

[504]   MASUNAGA, S., et al., The usefulness of 2-nitroimidazole-sodium borocaptate-^{10}B conjugates as ^{10}B-carriers in boron neutron capture therapy, Appl. Radiat. Isot. **61** (2004) 953–958.

[505]   MASUNAGA, S., et al., Effect of oxygen pressure during incubation with a ^{10}B-carrier on ^{10}B uptake capacity of cultured p53 wild-type and mutated tumor cells: Dependency on p53 status of tumor cells and types of ^{10}B-carriers, Int. J. Radiat. Biol. **92** 4 (2016) 187–194.

[506]   WADA, Y., et al., Impact of oxygen status on ^{10}B-BPA uptake into human glioblastoma cells, referring to significance in boron neutron capture therapy, J. Radiat. Res. **59** 2 (2018) 122–128.

[507]   MASUNAGA, S., et al., Evaluation of the radiosensitivity of the oxygenated tumor cell fractions in quiescent cell populations within solid tumors, Radiat. Res. **174** 4 (2010) 459–466.

[508]   MOHYELDIN, A., GARZÓN-MUVDI, T., QUIÑONES-HINOJOSA, A., Oxygen in stem cell biology: A critical component of the stem cell niche, Cell Stem Cell **7** (2010) 150–161.

[509]   LI, Z., et al., Hypoxia-inducible factors regulate tumorigenic capacity of glioma stem cells, Cancer Cell **15** 6 (2009) 501–513.

[510]   SUN, T., et al., Targeting glioma stem cells enhances anti-tumor effect of boron neutron capture therapy, Oncotarget **7** 28 (2016) 43095.

[511]   KONDO, N., et al., Glioma stem-like cells can be targeted in boron neutron capture therapy with boronophenylalanine, Cancers (Basel) **12** 10 (2020) 3040.

[512]   MASUNAGA, S.-I., et al., Combination of the vascular targeting agent ZD6126 with boron neutron capture therapy, Int. J. Radiat. Oncol. Biol. Phys. **60** 3 (2004) 920–927.

[513]   MIYABE, J., et al., Boron delivery for boron neutron capture therapy targeting a cancer-upregulated oligopeptide transporter, J. Pharmacol. Sci. **139** 3 (2019) 215–222.

[514]   LIU, J., AI, X., ZHANG, H., ZHUO, W., MI, P., Polymeric micelles with endosome escape and redox-responsive functions for enhanced intracellular drug delivery, J. Biomed. Nanotechnol. **15** 2 (2019) 373–381.

[515]   KALOT, G., et al., Aza-BODIPY: A new vector for enhanced theranostic boron neutron capture therapy applications, Cells **9** 9 (2020) 1953.

[516]   CARPANO, M., et al., Experimental studies of boronophenylalanine (^{10}BPA) biodistribution for the individual application of boron neutron capture therapy (BNCT) for malignant melanoma treatment, Int. J. Radiat. Oncol. Biol. Phys. **93** 2 (2015) 344–352.

[517]   HUGHES, A.M., et al., Different oral cancer scenarios to personalize targeted therapy: Boron neutron capture therapy translational studies, Ther. Deliv. **10** 6 (2019) 353–362.

[518]   GENG, C., ZHANG, X., TANG, X., "Assessment of secondary cancer risks for boron neutron capture therapy", Technical Meeting on Advances in Boron Neutron Capture Therapy 27–31 July 2020, IAEA, Vienna (2020).

[519]   GONZÁLEZ, S.J., et al., First BNCT treatment of a skin melanoma in Argentina: Dosimetric analysis and clinical outcome, Appl. Radiat. Isot. **61** 5 (2004) 1101.

[520]   WITTIG, A., COLLETTE, L., MOSS, R., SAUERWEIN, W.A., Early clinical trial concept for boron neutron capture therapy: A critical assessment of the EORTC trial 11001, Appl. Radiat. Isot. **67** Suppl. 7–8 (2009) S59–S62.

[521]   YANG, C.H., et al., Autoradiographic and histopathological studies of boric acid-mediated BNCT in hepatic VX2 tumor-bearing rabbits: Specific boron retention and damage in tumor and tumor vessels, Appl. Radiat. Isot. **106** (2015) 176–180.

[522]   LIN, S.Y., et al., Therapeutic efficacy for hepatocellular carcinoma by boric acid-mediated boron neutron capture therapy in a rat model, Anticancer Res. **33** 11 (2013) 4799–4809.

[523]   SCHAFFRAN, T., et al., Hemorrhage in mouse tumors induced by dodecaborate cluster lipids intended for boron neutron capture therapy, Int. J. Nanomed. **9** 1 (2014) 3583.

[524]   ANUJA, K., et al., Prolonged inflammatory microenvironment is crucial for pro-neoplastic growth and genome instability: A detailed review, Inflammation Res. **66** (2017) 119–128.

[525]   MONTI HUGHES, A., et al., Boron neutron capture therapy (BNCT) translational studies in the hamster cheek pouch model of oral cancer at the new "B2" configuration of the RA-6 nuclear reactor, Radiat. Environ. Biophys. **56** 4 (2017) 377.

[526]   RAO, M., et al., BNCT of 3 cases of spontaneous head and neck cancer in feline patients, Appl. Radiat. Isot. **61** (2004) 947–952.

[527]   MARDYNSKII, I.S., et al., Fast neutrons in the treatment of malignant neoplasms, Vopr. Onkol. **43** 5 (1997) 515.

[528]   POZZI, E.C.C., et al., Dosimetry and radiobiology at the new RA-3 reactor boron neutron capture therapy (BNCT) facility: Application to the treatment of experimental oral cancer, Appl. Radiat. Isot. **67** Suppl. 7–8 (2009) S309–S312.

[529]   HANAHAN, D., Hallmarks of cancer: New dimensions, Cancer Discovery **12** 1 (2022) 31-46.

[530]   GUPTA, S., AHMED, M.M., A global perspective of radiation-induced signal transduction pathways in cancer therapeutics, Indian J. Exp. Biol. **42** (2004) 1153–1176.

[531]   KRAFT, S.L., et al., Borocaptate sodium: A potential boron delivery compound for boron neutron capture therapy evaluated in dogs with spontaneous intracranial tumors, Proc. Natl. Acad. Sci. USA **89** 24 (1992) 11973–11977.

[532]   MITIN, V.N., et al., Comparison of BNCT and GdNCT efficacy in treatment of canine cancer, Appl. Radiat. Isot. **67** Suppl. 7–8 (2009) S299–S301.

[533]   TAKEUCHI, A., Possible application of boron neutron capture therapy to canine osteosarcoma, Nippon juigaku zasshi [Jpn. J. Vet. Sci.] **47** 6 (1985) 869–878.

[534]   TRIVILLIN, V.A., et al., Boron neutron capture therapy (BNCT) for the treatment of spontaneous nasal planum squamous cell carcinoma in felines, radiation and environmental Biophysics **47** (2008) 147–155.

[535]   SCHWINT, A.E., et al., Clinical veterinary boron neutron capture therapy (BNCT) studies in dogs with head and neck cancer: Bridging the gap between translational and clinical studies, Biology (Basel) **9** 10 (2020) 327.

[536]   IROSE, K., et al., Updated results of a phase II study evaluating accelerator-based boron neutron capture therapy (AB-BNCT) with borofalan(^{10}B) (SPM-011) in recurrent squamous cell carcinoma (R-SCC-HN) and recurrent and locally advanced non-SCC (R/LA-nSCC-HN) of the head and neck, Ann. Oncol. **30** (2019) v460.

[537]   MIYATAKE, S.I., WANIBUCHI, M., HU, N., ONO, K., Boron neutron capture therapy for malignant brain tumors, J. Neuro-Oncol. **149** (2020) 1–11.

[538]   IGAKI, H., et al., "A clinical trial of lithium-targeted accelerator-based boron neutron capture therapy at National Cancer Center Hospital", The 10th Young Member's Boron Neutron Capture Therapy Meeting, Helsinki, Finland (2019).

[539]   NEDUNCHEZHIAN, K., ASWATH, N., THIRUPPATHY, M., THIRUGNANAMURTHY, S., Boron neutron capture therapy – A literature review, J. Clin. Diagn. Res. **10** 12 (2016) ZE01.

[540]   ONISHI, T., et al., Investigation of the neutron spectrum measurement method for dose evaluation in boron neutron capture therapy, Appl. Radiat. Isot. **140** (2018) 5.

[541]   KREINER, A.J., et al., Accelerator-based BNCT, Appl. Radiat. Isot. **88** (2014) 185.

[542]   SATO, E., et al., Radiobiological response of U251MG, CHO-K1 and V79 cell lines to accelerator-based boron neutron capture therapy, J. Radiat. Res. **59** 2 (2018) 101.

[543]   HIRAGA, F., OOIE, T., Synergistic effects of fast-neutron dose per epithermal neutron and ^{10}B-concentration on relative-biological-effectiveness dose for accelerator-based boron neutron capture therapy, Appl. Radiat. Isot. **144** (2019) 1–4.

[544]   IMAMICHI, S.Y., et al., "Evaluation of the BNCT system in National Cancer Center Hospital using cells and mice", The International Congress of Radiation Research (2019).

[545]   GUEULETTE, J., OCTAVE-PRIGNOT, M., DE COSTERA, B.M., WAMBERSIE, A., GREGOIRE, V., Intestinal crypt regeneration in mice: a biological system for quality assurance in non-conventional radiation therapy, Radiother. Oncol. **73** 2 (2004) S148.

[546] PROTTI, N., et al., Gamma residual radioactivity measurements on rats and mice irradiated in the thermal column of a TRIGA Mark II reactor for BNCT, Health Phys. **107** 6 (2014) 534.

[547] TRIVILLIN, V.A., et al., Boron neutron capture synovectomy (BNCS) as a potential therapy for rheumatoid arthritis: Radiobiological studies at RA-1 nuclear reactor in a model of antigen-induced arthritis in rabbits, Radiat. Environ. Biophys. **55** 4 (2016) 467–475.

[548] HUGHES, A.M., et al., Boron neutron capture therapy (BNCT) in an oral precancer model: Therapeutic benefits and potential toxicity of a double application of BNCT with a six-week interval, Oral Oncol. **47** 11 (2011) 1017–1022.

[549] LEE, J.C., et al., A comparison of dose distributions in gross tumor volume between boron neutron capture therapy alone and combined boron neutron capture therapy plus intensity modulation radiation therapy for head and neck cancer, PLOS ONE **14** 4 (2019) e0210626.

[550] BARTH, R.F., et al., Combination of boron neutron capture therapy and external beam radiotherapy for brain tumors, Int. J. Radiat. Oncol. Biol. Phys. **58** 1 (2004) 267–277.

[551] KANKAANRANTA, L., et al., Boron neutron capture therapy (BNCT) followed by intensity modulated chemoradiotherapy as primary treatment of large head and neck cancer with intracranial involvement, Radiother. Oncol. **99** (2011) 98–99.

[552] MIYATAKE, S., Boron neutron capture therapy for malignant brain tumors, Neurol. Med. Chir. (Tokyo) **56** (2016) 361–371.

[553] YAMAMOTO, T., et al., Boron neutron capture therapy for newly diagnosed glioblastoma, Radiother. Oncol. **91** (2009) 80–84.

[554] DURANTE, M., FORMENTI, S., Harnessing radiation to improve immunotherapy: Better with particles? Br. J. Radiol. **93** (2020) 20190224.

[555] FRIEDRICH, T., HENTHORN, N., DURANTE, M., Modeling radioimmune response— Current status and perspectives, Front. Oncol. **11** (2021) 647272.

[556] ACHKAR, T., TARHINI, A.A., The use of immunotherapy in the treatment of melanoma, J. Hematol. Oncol. **10** (2017) 88.

[557] BRANDÃO, S.F., CAMPOS, T.P.R., Intracavitary moderator balloon combined with ^{252}Cf brachytherapy and boron neutron capture therapy, improving dosimetry in brain tumour and infiltrations, Br. J. Radiol. **88** 1051 (2015) 20140829.

[558] SHIBA, H., et al., Boron neutron capture therapy combined with early successive bevacizumab treatments for recurrent malignant gliomas – a pilot study, Neurol. Med. Chir. (Tokyo) **58** (2018) 487–494.

[559] MIYATAKE, S.I., et al., Boron neutron capture therapy with bevacizumab may prolong the survival of recurrent malignant glioma patients: Four cases, Radiat. Oncol. **9** 1 (2014) 6.

[560] MIYATAKE, S.I., et al., Boron neutron capture therapy of malignant gliomas, Prog. Neurol. Surg. **32** (2018) 48–56.

[561] WU, G., et al., Molecular targeting and treatment of an epidermal growth factor receptor-positive glioma using boronated cetuximab, Clin. Cancer Res. **13** 4 (2007) 1260–1268.

[562] HIRAMATSU, R., et al., Tetrakis(p-carboranylthio-tetrafluorophenyl)chlorin (TPFC): Application for photodynamic therapy and boron neutron capture therapy, J. Pharm. Sci. **104** 3 (2015) 962–970.

[563] ASANO, R., et al., Synthesis and biological evaluation of new boron-containing chlorin derivatives as agents for both photodynamic therapy and boron neutron capture therapy of cancer, Bioorg. Med. Chem. Lett. **24** 5 (2014) 1339–1343.

[564] NAKAHARA, Y., ITO, H., MASUOKA, J., ABE, T., Boron neutron capture therapy and photodynamic therapy for high-grade meningiomas, Cancers **12** (2020) 1334.

[565] BUCHHOLZ, T.A., et al., Boron neutron capture enhanced fast neutron radiotherapy for malignant gliomas and other tumors, J. Neuro-Oncol. **33** 1–2 (1997) 171–178.

[566] SAFAVI-NAEINI, M., et al., Opportunistic dose amplification for proton and carbon ion therapy via capture of internally generated thermal neutrons, Sci. Rep. **8** 1 (2018) 16257.

[567] CIRRONE, G.A.P., et al., First experimental proof of Proton Boron Capture Therapy (PBCT) to enhance protontherapy effectiveness, Sci. Rep. **8** 1 (2018) 1141.

[568] ALIKANIOTIS, K., et al., Radiotherapy dose enhancement using BNCT in conventional LINACs high-energy treatment: Simulation and experiment, Rep. Pract. Oncol. Radiother. **21** 2 (2016) 117–122.

[569] SANTA CRUZ, G.A., ZAMENHOF, R.G., The microdosimetry of the ^{10}B reaction in boron neutron capture therapy: a new generalized theory, Radiat. Res. **162** 6 (2004) 702–710.

[570] SANTA CRUZ, G.A., PALMER, M.R., MATATAGUI, E., ZAMENHOF, R.G., A theoretical model for event statistics in microdosimetry, Part I: Uniform distribution of heavy ion tracks, Med. Phys. **28** 6 (2001) 988–996.

[571] INTERNATIONAL COMMISSION ON RADIATION UNITS AND MEASUREMENTS, Radiation Quantities and Units. International Commission on Radiation Units and Measurements, ICRU Report 33, Washington, D.C., USA (1980).

[572] INTERNATIONAL COMMISSION ON RADIATION UNITS AND MEASUREMENTS, Photon, Electron, Proton and Neutron Interaction Data for Body Tissues, International Commission on Radiation Units and Measurements, ICRU Report 46, Bethesda, MD, USA (1992).

[573]   BROWN, D.A., et al., ENDF/B-VIII.0: The 8th major release of the nuclear reaction data library with CIELO-project cross sections, new standards and thermal scattering data, Nucl. Data Sheets **148** (2018) 1–142.

[574]   SHIBATA, K., et al., JENDL-4.0: A new library for nuclear science and engineering, J. Nucl. Sci. Technol. **48** 1 (2011) 1–30.

[575]   PORRAS, I., WRIGHT, T., "Measurement of the ^{35}Cl(n,γ) cross section at n_TOF EAR1", ISOLDE and Neutron Time-of-Flight Experiments Committee, Rep. CERN-INTC-2018-010; INTC-P-541, CERN, Geneva (2018).

[576]   PRAENA, J., et al., "The ^{14}N(n,p)^{14}C and ^{35}Cl(n,p)^{35}S reactions at n_TOF-EAR2: dosimetry in BNCT and astrophysics", ISOLDE and Neutron Time-of-Flight Experiments Committee, Rep. CERN-INTC-2017-039; INTC-P-510, CERN, Geneva (2017).

[577]   GOORLEY, J.T., KIGER, W.S. III, ZAMENHOF, R.G., Reference dosimetry calculations for neutron capture therapy with comparison of analytical and voxel models, Med. Phys. **29** 2 (2002) 145–156.

[578]   INTERNATIONAL COMMISSION ON RADIATION UNITS AND MEASUREMENTS, Nuclear Data for Neutron and Proton Radiotherapy and for Radiation Protection. International Commission on Radiation Units and Measurements, ICRU Report 63, Bethesda, MD, USA (2000).

[579]   VIEGAS, A.M., et al., Detailed dosimetry calculation for in-vitro experiments and its impact on clinical BNCT, Physica Med. **89** (2021) 282–292.

[580]   NAHUM, A., Condensed-history Monte-Carlo simulation for charged particles: What can it do for us? Radiat. Environ. Biophys. **38** (1999) 163–173.

[581]   GONZÁLEZ, S.J., CARANDO, D.G., SANTA CRUZ, G.A., ZAMENHOF, R.G., Voxel model in BNCT treatment planning: performance analysis and improvements, Phys. Med. Biol. **50** (2005) 441–458.

[582]   RAMOS, R., SZTEJNBERG, M., CANTARGI, F., Bioneutronics: thermal scattering in organic tissues and its impact on BNCT dosimetry, Appl. Radiat. Isot. **104** (2015) 55–59.

[583]   RAMOS, R., CANTARGI, F., MARQUEZ DAMIAN, J., SZTEJNBERG, M., Study of thermal scattering nuclear data for organic tissues through molecular dynamic, Eur. Phys. J. Confer. **146** (2017) 13008.

[584]   INTERNATIONAL COMMISSION ON RADIATION UNITS AND MEASUREMENTS, Microdosimetry, ICRU Report 36, Bethesda, MD, USA (1983).

[585]   ZAMENHOF, R., Microdosimetry for boron neutron capture therapy: A review, J. Neuro-Oncol. **33** 1–2 (1997) 81–92.

[586]   ROSSI, H.H., ZAIDER, M., Microdosimetry and its Applications, Springer, Berlin (1996).

[587] SANTA CRUZ, G.A., PALMER, M.R., MATATAGUI, E., ZAMENHOF, R.G., A theoretical model for event statistics in microdosimetry, Part II: Nonuniform distribution of heavy ion tracks, Med. Phys. **28** 6 (2001) 997–1005.

[588] KOBAYASHI, T., KANDA, K., Analytical calculation of boron-10 dosage in cell nucleus for neutron capture therapy, Radiat. Res. **91** 1 (1982) 77–94.

[589] SANTALÓ, L.A., Integral Geometry and Geometric Probability, Encyclopedia of Mathematics and its Applications, Vol. 1, Addison-Wesley, Reading, MA (1976).

[590] COLEMAN, R., Random paths through convex bodies, J. Appl. Prob. **6** (1969) 430–441.

[591] KELLERER, A.M., Chord-length distributions and related quantities for spheroids, Radiat. Res. **98** (1984) 425–437.

[592] MOHAN, R., GROSSHANS, D., Proton therapy – Present and future, Adv. Drug Deliv. Rev. **109** (2017) 26–44.

[593] BENTZEN, S.M., et al., Bioeffect modeling and equieffective dose concepts in radiation oncology – terminology, quantities and units, Radiother. Oncol. **105** (2012) 266–268.

[594] INTERNATIONAL ATOMIC ENERGY AGENCY, Relative Biological Effectiveness in Ion Beam, Technical Reports Series 461, IAEA, Vienna (2008).

[595] WAMBERSIE, A., MENZEL, H.G., RBE in fast neutron therapy and on boron neutron capture therapy. A useful concept or a misuse? Strahlenther. Onkol. **169** (1993) 57–64.

[596] CAPALA, J., MAKAR, M.S., CODERRE, J.A., Accumulation of boron in malignant and normal cells incubated in vitro with boronophenylalanine, mercaptoborane or boric acid, Radiat. Res. **146** (1996) 554–560.

[597] YOSHIDA, F., et al., Cell cycle dependence of boron uptake from two boron compounds used for clinical neutron capture therapy, Cancer Lett. **187** (2002) 135–141.

[598] BUREAU INTERNATIONAL DES POIDS ET MESURES, The International System of Units (SI), Pavillon de Breteuil, Sèvres, France (1998).

[599] ZAIDER, M., ROSSI, H.H., The synergistic effects of different radiations, Radiat. Res. **83** (1980) 732–739.

[600] SUZUKI, S., A theoretical model for simultaneous mixed irradiation with multiple types of radiation, J. Radiat. Res. **39** (1998) 215–221.

[601] LYMAN, J.T., Complication probability as assessed from dose-volume histograms, Radiat. Res. **8** (1985) S13–S19.

[602] NIEMIERKO, A., Reporting and analyzing dose distributions: A concept of equivalent uniform dose, Med. Phys. **24** (1997) 103–110.

[603]   GONZÁLEZ, S.J., CARANDO, D.G., A general tumor control probability model for non-uniform dose distributions, Math. Med. Biol. **2** (2008) 171–184.

[604]   BORTOLUSSI, S., et al., Understanding the potentiality of accelerator based-boron neutron capture therapy for osteosarcoma: Dosimetry assessment based on the reported clinical experience, Radiat. Oncol. **12** (2017) 130.

[605]   GONZÁLEZ, S., et al., "How do photon isoeffective tumor doses derived from in-vitro BNCT studies compare to those from in-vivo cancer model data?" 18th International Conference on Neutron Capture Therapy, Taipei (2018).

[606]   LI, H.S., et al., Verification of the accuracy of BNCT treatment planning system THORplan, Appl. Radiat. Isot. **67** Suppl. 7–8 (2009) S122–S125.

[607]   ZAMENHOF, R.G., et al., "Monte Carlo based dosimetry and treatment planning for neutron capture therapy," Neutron Beam Design, Development, and Performance for Neutron Capture Therapy (HARLING, O.K., BERNARD, J.A., ZAMENHOF, R.G., Eds) Plenum Press, New York (1990).

[608]   FARIAS, R.O., GONZÁLEZ, S.J., "MultiCell model as an optimized strategy for BNCT treatment planning". (Proc. 15th Int. Congr. Neutron Capture Therapy, Tsukuba, Japan, 10–14 September 2012), Int. Soc. Neutron Capture Ther. (2012) 148–150.

[609]   FARIAS, R.O., Dosimetría y modelado computacional para irradiaciones extracorpóreas en humanos en el marco de la Terapia por Captura Neutrónica en Boro. Ph.D. Thesis, Comision Nacional de Energia Atomica, Istituto de Tecnologia, Universidad Nacional de General San Martin, Buenos Aires, Argentina (2015).

[610]   NIGG, D.W., et al., SERA – An advanced treatment planning system for neutron therapy and BNCT, Trans. Am. Nucl. Soc. **80** (1999) 6668.

[611]   PALMER, M.R., et al., Treatment planning and dosimetry for the Harvard-MIT Phase I clinical trial of cranial neutron capture therapy, Int. J. Radiat. Oncol. Biol. Phys. **53** 5 (2002) 1361–1379.

[612]   SEPPÄLÄ, T., FiR 1 Epithermal Neutron Beam Model and Dose Calculation for Treatment Planning in Neutron Capture Therapy, PhD thesis Report Series in Physics, HU-P-D103, University of Helsinki, Helsinki, Finland (2002).

[613]   NIGG, D.W., RANDOLPH, P.D., WHEELER, F.J., Demonstration of three-dimensional deterministic radiation transport theory dose distribution analysis for boron neutron capture therapy, Med. Phys. **18** (1991) 43–53.

[614]   KOTILUOTO, P., HIISMÄKI, P., SAVOLAINEN, S., Application of the new MultiTrans radiation transport code in BNCT dose planning, Med. Phys. **28** (2001) 19051910.

[615]   KOTILUOTO, P., PYYRY, J., HELMINEN, H., MultiTrans SP3 code in coupled photon–electron transport problems, Radiat. Phys. Chem. **76** 1 (2007) 914.

[616] INTERNATIONAL ATOMIC ENERGY AGENCY, Commissioning of Radiotherapy Treatment Planning Systems: Testing for Typical External Beam Treatment Techniques, TECDOC-1583, IAEA, Vienna (2008).

[617] SMILOWITZ, J.B., AAPM Medical Physics Practice Guideline 5.a.: Commissioning and QA of treatment planning dose calculations—Megavoltage photon and electron beams, J. Appl. Clin. Med. Phys. **16** (2015) 14–34.

[618] WOJNECKI, C., GREEN, S., A computational study into the use of polyacrylamide gel and A-150 plastic as brain tissue substitutes for boron neutron capture therapy, Phys. Med. Biol. **46** 5 (2001) 1399.

[619] ROGUS, R.D., HARLING, O.K., YANCH, J.C, Mixed field dosimetry of epithermal neutron beams for boron neutron capture therapy at the MITR-II research reactor, Med. Phys. **21** 10 (1994) 1611–1625.

[620] MIJNHEER, B.J., BATTERMANN, J.J., WAMBERSIE, A., What degree of accuracy is required and can be achieved in photon and neutron therapy? Radiother. Oncol. **8** (1987) 237–252.

[621] SEPPÄLÄ, T., et al., Dose planning with comparison to in vivo dosimetry for epithermal neutron irradiation of the dog brain, Med. Phys. **29** (2002) 2629–2640.

[622] MUNCK AF ROSENSCHÖLD, P.M., et al., Quality assurance of patient dosimetry in boron neutron capture therapy, Acta Oncol. **43** (2004) 404–411.

[623] INTERNATIONAL ELECTROTECHNICAL COMMISSION, Medical Electrical Equipment. Particular Requirements for the Safety of Magnetic Resonance Equipment for Medical Diagnosis. International Standard, IEC60601-2-33 (2002).

[624] CULBERTSON, C.N., et al., In-phantom characterisation studies at the Birmingham Accelerator-Generated epIthermal Neutron Source (BAGINS) BNCT facility, Appl. Radiat. Isot. **61** 5 (2004) 733738.

[625] ZAMENHOF, R.G., et al., Monte Carlo based treatment planning for boron neutron capture therapy using custom designed models automatically generated from CT data, Int. J. Radiat. Oncol. Biol. Phys. **35** (1996) 383–389.

[626] KUMADA, H., et al., Verification of the computational dosimetry system in JAERI (JCDS) for boron neutron capture therapy, Phys. Med. Biol. **49** 15 (2004) 3353–3365.

[627] LIN, T., LIU, Y.H., "Recent development of THORplan", Advances in Neutron Capture Therapy 2006 (NAKAGAWA, Y., KOBAYASHI, T., FUKUDA, H. Eds). Proc. 12th Int. Congr. Neutron Capture Ther., Japan (2006).

[628] BUSSE, P.M., et al., A critical examination of the results from the Harvard-MIT NCT program phase I clinical trial of neutron capture therapy for intracranial disease, J. Neuro-Oncol. **62** (2003) 111–121.

[629] JOENSUU, H., et al., Boron neutron capture therapy of brain tumors: Clinical trials at the Finnish facility using boronophenylalanine, J. Neuro-Oncol. **62** (2003) 123–134.

[630] STENSTAM, B.H., PELLETTIERI, L., SORTEBERG, W., REZAEI, A., SKOLD, K., BNCT for recurrent intracranial meningeal tumours – case reports, Acta Neurol. Scand. **115** 4 (2007) 243–247.

[631] HENRIKSSON, R., et al., Boron neutron capture therapy (BNCT) for glioblastoma multiforme: A phase II study evaluating a prolonged high-dose of boronophenylalanine (BPA), Radiother. Oncol. **88** 2 (2008) 183–191.

[632] MIYATAKE, S., et al., Survival benefit of boron neutron capture therapy for recurrent malignant gliomas, J. Neuro-Oncol. **91** (2009) 199–206.

[633] HEBER, E.M., et al., Therapeutic effect of boron neutron capture therapy (BNCT) on field cancerized tissue: Inhibition of DNA synthesis and lag in the development of second primary tumors in precancerous tissue around treated tumors in DMBA-induced carcinogenesis in the hamster cheek pouch oral cancer model, Arch. Oral Biol. **52** 3 (2007) 273–279.

[634] INTERNATIONAL COMMISSION ON RADIATION UNITS AND MEASUREMENTS, Prescribing, Recording and Reporting Photon Beam Therapy, International Commission on Radiation Units and Measurements, ICRU Report 50, Washington, D.C., USA (1993).

[635] INTERNATIONAL COMMISSION ON RADIATION UNITS AND MEASUREMENTS, Prescribing, Recording and Reporting Photon Beam Therapy (Supplement to ICRU Report 50), ICRU Report 62, Washington, D.C., USA (1999).

[636] INTERNATIONAL COMMISSION ON RADIATION UNITS AND MEASUREMENTS, Prescribing, Recording, and Reporting Intensity-Modulated Photon-Beam Therapy (IMRT), ICRU Report 83, Washington, D.C., USA (2001).

[637] BRIERLEY, J.D., GOSPODAROWICZ, M.K., WITTEKIND, C., (Eds), TNM Classification of Malignant Tumours, 8th Edition, John Wiley & Sons (2017).

[638] INTERNATIONAL COMMISSION ON RADIATION UNITS AND MEASUREMENTS, Dose Specification for Reporting External Beam Therapy with Photons and Electrons, ICRU Report 29, Washington, D.C., USA (1978).

[639] LOUIS, D.N., et al., The 2021 WHO classification of tumors of the central nervous system: a summary. Neuro. Oncol. **23** 8 (2021) 1231–1251.

[640] LOUIS, D.N., et al., "Glioblastoma, IDH-wildtype", WHO Classification of Tumours of the Central Nervous System, International Agency for Research on Cancer, Lyon, France (2016) 28–45.

[641] OSTROM, Q.T., et al., CBTRUS statistical report: Primary brain and central nervous system tumors diagnosed in the United States in 2007–2011, Neuro. Oncol. **16** Suppl. 4 (2014) 1–63.

[642] OSTROM, Q.T., et al., The epidemiology of glioma in adults: A "state of the science" review, Neuro. Oncol. **16** (2014) 896–913.

[643]   STUPP, R., et al., Radiotherapy plus concomitant and adjuvant temozolomide for glioblastoma, N. Engl. J. Med. **352** (2005) 987–996.

[644]   CHINOT, O.L., et al., Bevacizumab plus radiotherapy–temozolomide for newly diagnosed glioblastoma, N. Engl. J. Med. **370** (2014) 709–722.

[645]   COMBS, S.E., et al., Efficacy of fractionated stereotactic reirradiation in recurrent gliomas: long-term results in 172 patients treated in a single institution, J. Clin. Oncol. **23** (2005) 8863–8869.

[646]   NAGANE, M., et al., Phase II study of single-agent bevacizumab in Japanese patients with recurrent malignant glioma, Jpn. J. Clin. Oncol. **42** (2012) 887–895.

[647]   STUPP, R., et al., NovoTTF-100A versus physician's choice chemotherapy in recurrent glioblastoma: a randomised phase III trial of a novel treatment modality, Eur. J. Cancer **48** (2021) 2192–2202.

[648]   OLAR, A., et al., Mitotic index is an independent predictor of recurrence-free survival in meningioma, Brain Pathol. **25** (2015) 266–275.

[649]   PERRY, A., et al., Malignancy" in meningiomas: A clinicopathologic study of 116 patients, with grading implications, Cancer **85** (1999) 2046–2056.

[650]   SUGHRUE, M.E., et al., Outcome and survival following primary and repeat surgery for World Health Organization Grade III meningiomas, J. Neurosurg. **113** (2010) 202–209.

[651]   JAASKELAINEN, J., HALTIA, M., SERVO, A., Atypical and anaplastic meningiomas: Radiology, surgery, radiotherapy, and outcome, Surg. Neurol. **25** (1986) 233–242.

[652]   PALMA, L., CELLI, P., FRANCO, C., CERVONI, L., CANTORE, G., Long-term prognosis for atypical and malignant meningiomas: A study of 71 surgical cases, J. Neurosurg. **86** (1997) 793–800.

[653]   POLLOCK, B.E., STAFFORD, S.L., LINK, M.J., GARCES, Y.I., FOOTE, R.L., Stereotactic radiosurgery of World Health Organization grade II and III intracranial meningiomas: Treatment results on the basis of a 22-year experience, Cancer **118** (2012) 1048–1054.

[654]   ROGERS, L., et al., Meningiomas: Knowledge base, treatment outcomes, and uncertainties. A RANO review, J. Neurosurg. **122** (2015) 4–23.

[655]   BARTH, R.F., A critical assessment of boron neutron capture therapy: An overview, J. Neuro-Oncol. **62** (2003) 1–5.

[656]   ASBURY, A.K., OJEMANN, R.G., NIELSEN, S.L., SWEET, W.H., Neuropathologic study of fourteen cases of malignant brain tumor treated by boron-10 slow neutron capture radiation, J. Neuropathol. Exp. Neurol. **31** (1972) 278–303.

[657]   FARR, L.E., SWEET, W.H., LOCKSLEY, H.B., ROBERTSON, J.S., Neutron capture therapy of gliomas using boron, Trans. Am. Neurol. Assoc. **13** (1954) 110–113.

[658]   GOODWIN, J.T., FARR, L.E., SWEET, W.H., ROBERTSON, J.S., Pathological study of eight patients with glioblastoma multiforme treated by neutron-capture therapy using boron 10, Cancer **8** (1955) 601–615.

[659]   INTERNATIONAL ATOMIC ENERGY AGENCY, "Clinical", Current Status of Neutron Capture Therapy, TECDOC 1223, IAEA, Vienna (2001) Ch. 9.

[660]   YAMAMOTO, T., et al., "Medical set-up of boron neutron capture therapy (BNCT) for malignant glioma at the Japan research reactor (JRR)-4", Current Status of Neutron Capture Therapy, TECDOC 1223, IAEA, Vienna (2001) 233–239.

[661]   NAKAGAWA, Y., et al., "Clinical practice in BNCT to the brain", Current Status of Neutron Capture Therapy, TECDOC 1223, IAEA, Vienna (2001) 240–249.

[662]   SAUERWEIN, W., et al., "First clinical results from the EORTC Phase I Trial postoperative treatment of glioblastoma with BNCT at the Petten irradiation facility"", Current Status of Neutron Capture Therapy, TECDOC 1223, IAEA, Vienna (2001) 250–256.

[663]   DIAZ, A.Z., "The Phase I/II BNCT Trials at the Brookhaven medical research reactor: Critical considerations", Current Status of Neutron Capture Therapy, TECDOC 1223, IAEA, Vienna (2001) 257–267.

[664]   MOSS, R., et al., "The BNCT facility at the HFR Petten: Quality assurance for reactor facilities in clinical trials", Current Status of Neutron Capture Therapy, TECDOC 1223, IAEA, Vienna (2001) 268–274.

[665]   KAGEJI, T., et al., Boron neutron capture therapy (BNCT) for newly-diagnosed glioblastoma: comparison of clinical results obtained with BNCT and conventional treatment, J. Med. Invest. **61** 3.4 (2014) 254–263.

[666]   GILBERT, M.R., et al., A randomized trial of bevacizumab for newly diagnosed glioblastoma, N. Engl. J. Med. **370** (2014) 699–708.

[667]   KAWABATA, S., et al., Phase II clinical study of boron neutron capture therapy combined with X-ray radiotherapy/temozolomide in patients with newly diagnosed glioblastoma multiforme–study design and current status report, Appl. Radiat. Isot. **69** (2011) 1796–1799.

[668]   KAWABATA, S., et al., The early successful treatment of glioblastoma patients with modified boron neutron capture therapy. Report of two cases, J. Neuro-Oncol. **65** (2003) 159–165.

[669]   CARSON, K.A., GROSSMAN, S.A., FISHER, J.D., SHAW, E.G., Prognostic factors for survival in adult patients with recurrent glioma enrolled onto the new approaches to brain tumor therapy CNS consortium phase I and II clinical trials, J. Clin. Oncol. **25** (2007) 2601–2606.

[670]   MIYATAKE, S., et al., Pseudoprogression in boron neutron capture therapy for malignant gliomas and meningiomas, Neuro. Oncol. **11** (2009) 430–436.

[671]   FURUSE, M., KAWABATA, S., KUROIWA, T., MIYATAKE, S., Repeated treatments with bevacizumab for recurrent radiation necrosis in patients with malignant brain tumors: A report of 2 cases, J. Neuro-Oncol. **102** (2011) 471–475.

[672]   LEVIN, V.A., et al., Randomized double-blind placebo-controlled trial of bevacizumab therapy for radiation necrosis of the central nervous system, Int. J. Radiat. Oncol. Biol. Phys. **79** (2011) 1487–1495.

[673]   FURUSE, M., et al., A prospective multicenter single-arm clinical trial of bevacizumab for patients with surgically untreatable symptomatic brain radiation necrosis, Neuro. Oncol. Pract. **3** (2016) 272–280.

[674]   MIYATAKE, S., et al., Bevacizumab treatment of symptomatic pseudoprogression after boron neutron capture therapy for recurrent malignant gliomas. Report of 2 cases, Neuro. Oncol. **15** (2013) 650–655.

[675]   KAWABATA, S., HIRAMATSU, R., KUROIWA, T., ONO, K., MIYATAKE, S.I., Boron neutron capture therapy for recurrent high-grade meningiomas, J. Neurosurg. **119** (2013) 837–844.

[676]   MIYATAKE, S., et al., Boron neutron capture therapy for malignant tumors related to meningiomas, Neurosurgery **61** (2007) 82–90; discussion 90–91.

[677]   TAMURA, Y., et al., Boron neutron capture therapy for recurrent malignant meningioma. Case report, J. Neurosurg. **105** (2006) 898–903.

[678]   KOIVUNORO, H., et al., Boron neutron capture therapy for locally recurrent head and neck squamous cell carcinoma: An analysis of dose response and survival, Radiother. Oncol. **137** (2019) 153–158.

[679]   WANG, L.W., et al., BNCT for locally recurrent head and neck cancer: Preliminary clinical experience from a phase I/II trial at Tsing Hua Open-Pool Reactor, Appl. Radiat. Isot. **69** (2011) 1803–1806.

[680]   KAWABATA S., et al., Accelerator-based BNCT for patients with recurrent glioblastoma: a multicenter phase II study, Neuro. Oncol. Adv. **3** 1 (2021) vdab067.

[681]   CHADHA, M., et al., Boron neutron-capture therapy (BNCT) for glioblastoma multiforme (GBM) using the epithermal neutron beam at the Brookhaven National Laboratory, Int. J. Radiat. Oncol. Biol. Phys. **40** (1998) 829–834.

[682]   CODERRE, J.A., et al., Boron neutron capture therapy: Cellular targeting of high linear energy transfer radiation, Technol. Cancer Res. Treat. **2** (2003) 355–375.

[683]   DIAZ, A.Z., Assessment of the results from the phase I/II boron neutron capture therapy trials at the Brookhaven National Laboratory from a clinician's point of view, J. Neuro-Oncol. **62** (2003) 101–109.

[684]   KIGER, W.S. III, et al., Preliminary treatment planning and dosimetry for a clinical trial of neutron capture therapy using a fission converter epithermal neutron beam, Appl. Radiat. Isot. **61** (2004) 1075–1081.

[685] VOS, M.J., et al., Radiologic findings in patients treated with boron neutron capture therapy for glioblastoma multiforme within EORTC trial 11961, Int. J. Radiat. Oncol. Biol. Phys. **61** (2005) 392–399.

[686] SKOLD, K., et al., Boron neutron capture therapy for newly diagnosed glioblastoma multiforme: an assessment of clinical potential, Br. J. Radiol. **83** (2010) 596–603.

[687] PELLETTIERI, L., et al., An investigation of boron neutron capture therapy for recurrent glioblastoma multiforme, Acta Neurol. Scand. **117** (2008) 191–197.

[688] YAMAMOTO, T., et al., Current clinical results of the Tsukuba BNCT trial, Appl. Radiat. Isot. **61** (2004) 1089–1093.

[689] KAGEJI, T., MIZOBUCHI, Y., NAGAHIRO, S., NAKAGAWA, Y., KUMADA, H., Long-survivors of glioblatoma treated with boron neutron capture therapy (BNCT), Appl. Radiat. Isot. **69** (2011) 1800–1802.

[690] KAGEJI, T., MIZOBUCHI, Y., NAGAHIRO, S., NAKAGAWA, Y., KUMADA, H., Clinical results of boron neutron capture therapy (BNCT) for glioblastoma, Appl. Radiat. Isot. **69** (2011) 1823–1825.

[691] KAGEJI, T., et al., Boron neutron capture therapy using mixed epithermal and thermal neutron beams in patients with malignant glioma-correlation between radiation dose and radiation injury and clinical outcome, Int. J. Radiat. Oncol. Biol. Phys. **65** (2006) 1446–1455.

[692] BRAY, F., et al., Global cancer statistics 2018: GLOBOCAN estimates of incidence and mortality worldwide for 36 cancers in 185 countries, CA Cancer J. Clin. **68** (2018) 394–424.

[693] BOSSI, P., et al., Prognostic and predictive factors in recurrent and/or metastatic head and neck squamous cell carcinoma: A review of the literature, Crit. Rev. Oncol. Hematol. **37** (2019) 84–91.

[694] LEÓN, X., et al., A retrospective analysis of the outcome of patients with recurrent and/or metastatic squamous cell carcinoma of the head and neck refractory to a platinium-based chemotherapy, Clin. Oncol. **17** (2005) 418–424.

[695] DILLON, P.M., et al., Adenoid cystic carcinoma: A review of recent advance, molecular targets, and clinical trials, Head Neck **38** (2016) 620–627.

[696] LIM, D., QUAH, D.S., LEECH, M., MARIGNOL, L., Clinical potential of boron neutron capture therapy for locally recurrent inoperable previously irradiated head and neck cancer, Appl. Radiat. Isot. **106** (2015) 237–241.

[697] WOLCHOK, J.-D., et al., Overall survival with combined nivolumab and ipilimumab in advanced melanoma, N. Engl. J. Med. **377** (2017) 1345–1356.

[698] KARASAWA, K., et al., Clinical trial of carbon ion radiotherapy for gynecological melanoma, J. Radiat. Res. **55** (2014) 343–350.

[699]   YU, Z., SI, L., Immunotherapy of patients with metastatic melanoma, Chin. Clin. Oncol. **6** (2017) 20.

[700]   MISHIMA, Y., "Neutron capture treatment of malignant melanoma using ^{10}B-chlorpromazine compound", Pigment Cell, Vol. 1, Mechanisms in Pigmentation (MCGOVERN, V.J., RUSSELL, P., Eds), S. Karger, Basel (1973) 215–221.

[701]   YOSHINO, K., et al., "Chemical behavior of dopaborate and ^{10}B-p-boronophenylalanine", Neutron Capture Therapy (HATANAKA, H., Ed.), Nishimura, Niigata, Japan (1986) 55–60.

[702]   FUKUDA, H., et al., RBE of a thermal neutron beam and the ^{10}B(n, α)^{7}Li reaction on cultured B-16 melanoma cells, Int. J. Radiat. Biol. **51** (1987) 167–175.

[703]   YOSHINO, K., et al., Improvement of solubility of p-boronophenylalanine by complex formation with monosaccharides, Strahlenther. Onkol. **165** (1989) 127–129.

[704]   HIRATSUKA, J., KONO, M., MISHIMA, Y., RBEs of thermal neutron capture therapy and ^{10}B(n, α)^{7}Li reaction on melanoma-bearing hamsters, Pigment Cell Res. **2** (1989) 352–355.

[705]   MISHIMA, Y., et al., New thermal neutron capture therapy for malignant melanoma: Melanogenesis-seeking ^{10}B molecule–melanoma cell interaction from in vitro to first clinical trial, Pigment Cell Res. **2** 4 (1989) 226–234.

[706]   HIRATSUKA, J., et al., The relative biological effectiveness of ^{10}B-neutron capture therapy for early skin reaction in the hamster, Radiat. Res. **128** (1991)186–191.

[707]   FUKUDA, H., et al., Boron neutron capture therapy of malignant melanoma using ^{10}B-paraboronophenylalanine with special reference to evaluation of radiation dose and damage to the normal skin, Radiat. Res. **138** (1994) 435–442.

[708]   FUKUDA, H., et al., Boron neutron capture therapy (BNCT) for malignant melanoma with special reference to absorbed doses to the normal skin and tumor, Australas. Phys. Eng. Sci. Med. **26** (2003) 78–84.

[709]   BUSSE, P.-M., et al., "Clinical follow-up of patients with melanoma of the extremity treated in a phase I boron neutron capture therapy protocol", Advances in Neutron Capture Therapy, Volume I. (LARSSON, B., CRAWFORD, J., WEINREICH, R., Eds), Elsevier, Amsterdam (1997) 60–64.

[710]   WITTIG, A., et al., "Early phase I study of BNCT in metastatic malignant melanoma using the boron carrier BPA (EORTC protocol 11011)" Advances in Neutron Capture Therapy 2006 (NAKAGAWA, Y., KOBAYASHI, T., FUKUDA, H., Eds) (Proc. Int. Congr. Neutron Capture Ther. ICNCT-12) 2006, 284–287.

[711]   SAUERWEIN, W., ZURDO, A., The EORTC boron neutron capture therapy (BNCT) group: Achievements and future projects, Eur. J. Cancer **38** (2002) 31–34.

[712] HIRATSUKA, J., et al., Long-term outcome of cutaneous melanoma patients treated with boron neutron capture therapy (BNCT), J. Radiation Res. **61** 6 (2020) 945–951.

[713] YONG, Z., et al., Boron neutron capture therapy for malignant melanoma: First clinical case report in China, Chin. J. Cancer Res. **28** (2016) 634–640.

[714] SANTA CRUZ, G.A., et al., (2016) "Re-start of Clinical and Pre-Clinical BNCT Activities at the Argentine RA-6 Nuclear Reactor". 17th International Congress on Neutron Capture Therapy. Columbia, Missouri, USA (02–07 October 2016).

[715] FUJIMOTO, T., et al., BNCT for primary synovial sarcoma, Appl. Radiat. Isot. **169** (2021) 109407.

[716] KANNO, H., et al., Designation products: Boron neutron capture therapy for head and neck carcinoma, The Oncologist **26** 7 (2021) e1250–e1255.

[717] SAKURAI, Y., KOBAYASHI, Y., The medical-irradiation characteristics for neutron capture therapy at the Heavy Water Neutron Irradiation Facility of Kyoto University Research Reactor, Med. Phys. **29** 10 (2002) 2328–2337.

[718] CAPALA, J., et al., Boron neutron capture therapy for glioblastoma multiforme: Clinical studies in Sweden, J. Neuro-Oncol. **62** (2003) 135–144.

[719] SAUERWEIN, W.A.G., MOSS, R., Requirements for Boron Neutron Capture Therapy (BNCT) at a Nuclear Research Reactor, Office for Official Publications of the European Communities, EUR 23830 EN (2009).

[720] HU, N., et al., Accelerator based epithermal neutron source for clinical boron neutron capture therapy, (Proc. UCANS9, 28–31 March, 2022, RIKEN, Japan) (BAXTER, D., et al., Eds), J. Neutron Res. Spec. Issue (2023) 359–366.

[721] KUMADA, H., "Current development status of the linac-based BNCT device of the iBNCT Tsukuba project", eProceedings of the International Conference on Accelerators for Research and Development: from Good Practices Towards Socioeconomic Impact, 23–27 May 2022, IAEA, Vienna, (CHARISOPOULOS, C., Ed.) (2022) 104–108.

[722] KUMADA, H., et al., Current development status of iBNCT0001, the demonstration device of a LINAC-based neutron source for BNCT, (Proc. UCANS9, 28–31 March, 2022, RIKEN, Japan) (BAXTER, D., et al., Eds), J. Neutron Res. Spec. Issue (2023) 347–358.

[723] CHEN, Y.-W., et al., Salvage boron neutron capture therapy for malignant brain tumor patients in compliance with emergency and compassionate use: Evaluation of 34 cases in Taiwan, Biology **10** (2021) 334.

[724] WANG, L-W., et al., Fractionated boron neutron capture therapy in locally recurrent head and neck cancer: A prospective phase I/II Trial, Int. J. Radiat. Oncol. Biol. Phys. **95** 1 (2016) 396e403.

[725]   INTERNATIONAL ELECTROTECHNICAL COMMISSION, Medical electrical equipment – Part 2-64: Particular requirements for the basic safety and essential performance of light ion beam medical electrical equipment. International Standard, IEC 60601-2-64:2014 (2014).

[726]   SCHUHMACHER, H., HOLLNAGEL, R., SIEBERT, B.R.L., Sensitivity study of parameters influencing calculations of fluence to-ambient dose equivalent conversion coefficients for neutrons, Radiat. Prot. Dosim. **54** (1994) 221–225.

[727]   INTERNATIONAL COMMISSION ON RADIATION UNITS AND MEASUREMENTS, Measurement of Dose Equivalents from External Photon and Electron Radiations, International Commission on Radiation Units and Measurements, ICRU Report 47, Bethesda, MD, USA (1992).

[728]   AIHARA, T., et al., BNCT for advanced or recurrent head and neck cancer, Appl. Radiat. Isot. **88** (2014) 12–15.

[729]   PHARMACEUTICALS AND MEDICAL DEVICES AGENCY, Medical Device Evaluation Division. Pharmaceutical Safety and Environmental Health Bureau, Ministry of Health, Labour and Welfare, Report on Deliberation. February 19, 2020, https://www.pmda.go.jp/files/000237993.pdf.

Annex I.

ACCELERATOR-BASED NEUTRON SOURCE OF NAGOYA UNIVERSITY

KAZUKI TUSICHIDA[1], AKIRA URITANI[1], YOSHIAKI KIYANAGI[1], TAKEO NISITANI[1], SACHIKO YOSHIHASHI[1], SHOGO HONDA[1], KENICHI WATANABE[2]

[1]Graduate School of Engineering, Nagoya University, Nagoya 464-8601, Japan
[2]Graduate School of Engineering, Kyushu University, Fukuoka 819-0395, Japan

I-1.   INTRODUCTION

There is a worldwide trend to develop accelerator-based neutron sources for boron neutron capture therapy (BNCT), known as AB-BNCT, that can be installed in hospitals, as research reactors are being decommissioned [I-1]. An AB-BNCT system generally consists of a 'proton accelerator', 'neutron generating target', and 'beam shaping assembly' (BSA) that reduces the neutron energy to the therapeutic range.

The basic requirements for the AB-BNCT system installed in a hospital are as follows:

- Sufficient neutrons with a low impact on normal cells;
- Agreeable patient positioning during treatment;
- Low radiation exposure of medical workers;
- Low activation accelerator and facility;
- Affordable construction and operation costs.

To satisfy those requirements, an AB-BNCT System Research Laboratory for academic–industrial partnerships at Nagoya University's Graduate School of Engineering was inaugurated on November 1, 2013, to develop a neutron source equipped with an electrostatic accelerator and a sealed Li target (an AB-BNCT system).

In December 2018, the system passed a facility inspection for generating neutrons. After that, a series of physical tests confirmed that irradiated neutrons met the target values of IAEA TECDOC-1223 [I-2]. Since January 2019, with the collaboration of the Neutron Therapy Research Center, Okayama University, we have been conducting non-clinical experiments (in vitro tests) to verify the safety of the system for clinical use. To demonstrate the system's efficacy and safety as medical equipment, several in vivo experiments are being conducted.

I-2.   SPECIFICATIONS OF THE NAGOYA UNIVERSITY SYSTEM

I-2.1.   Electrostatic accelerator

The Dynamitron (Ion Beam Applications, IBA, Belgium), an electrostatic accelerator with a maximum output of 2.8 MeV, 15 mA is used (Fig. I-1). Protons are generated in an electron cyclotron resonance ion source installed in the main body of the accelerator and are accelerated in the acceleration tube after removing molecular ion components (H_2^+ and H_3^+) using a bending magnet. Since the maximum beam energy is low (2.8 MeV), activation of the accelerator and beamline is almost negligible, and maintenance and repair services can be easily performed. When the accelerator is operated at a maximum beam power of 15 mA (42 kW), the input power of the water-cooled triode vacuum tube used to supply high-voltage (HV) power is only 56 kW (efficiency: 75%), resulting in a low operating cost of electricity for the accelerator.

Dynamitron (2.8MeV, 15mA) Beamline

Target beamline

FIG. I-1. Dynamitron and the beamline.

A high-current proton beam up to 15 mA emitted from the Dynamitron is deflected by 20° by a bending magnet installed in the beamline. The proton beam is incident on a Li target positioned within a water-cooled jacket flange. To transmit a proton beam of up to 42 kW to the Li target with low loss, the shape and position of the proton beam have to be adjusted using a 'quadrupole magnet' and 'beam positioning magnet.' We have developed a compact rotating beam measurement system that can be installed in a hospital with limited space and performed precise beam control by determining the beam position and shape using beam orbit analysis software. Furthermore, to reduce the heat density in the Li target, a 2-axis scanning magnet (X-Y, 300 Hz) is used to increase the beam irradiation area (max. 8 cm × 8 cm). Then, four quadrant collimators (cf. Fig. I-1) are used to prevent the fringe of the proton beam from directly hitting the target containment vessel, made of a low-activation aluminium alloy. The scanning width and centre position are adjusted by monitoring the current incident on each collimator.

I-2.2. Sealed Li target

The Li target generates neutrons efficiently from a low-energy proton beam. The maximum neutron energy is less than 1.1 MeV. This makes moderation to the epithermal region suitable for BNCT relatively easy. However, Li is chemically active and mechanically soft. Additionally, radioactive ^7Be is produced in Li during the neutron generation reaction. As a countermeasure to this problem, we have developed a Li target sealed within a thin Ti film (sealed Li target), which enables stable operation for a long time. The thin Ti foil confines Li and ^7Be in the target. The Cu backing plate removes the heat generated during irradiation with the use of highly efficient turbulent water flow [I-2].

I-2.3. Beam shaping assembly

The beam shaping assembly (BSA) is a device to moderate fast neutrons produced by the Li target to the epithermal range (0.5 eV–10 keV) suitable for cancer treatment and to shape a neutron beam suitable for treatment. For epithermal neutrons to reach cancer cells effectively, it is necessary to adjust the patient's position such that neutrons can sufficiently irradiate cancer cells while the patient is close to the neutron outlet, and the patient needs to be restrained during irradiation.

In the development of the BSA, a special nozzle (15-cm long) attached to the neutron exit hole was developed, allowing patients to be irradiated in a comfortable position. The neutron intensity and quality at the nozzle outlet comply with the target values in the IAEA TECDOC-1223 [I-3]. Note that MgF_2 was selected as the material to moderate fast neutrons generated by the Li target, based on Monte Carlo simulations using the PHITS code and the nuclear data library of JENDL-4. The epithermal neutrons are emitted through a Bi γ-ray shielding material. A neutron reflector is placed on the inner surface of the nozzle to guide epithermal neutrons to the nozzle outlet while limiting degradation of the neutron quality (Fig. I-2) [I-4– I-5]. As our aim is to reduce to reduce the exposure of medical workers, we select low activation BSA materials.

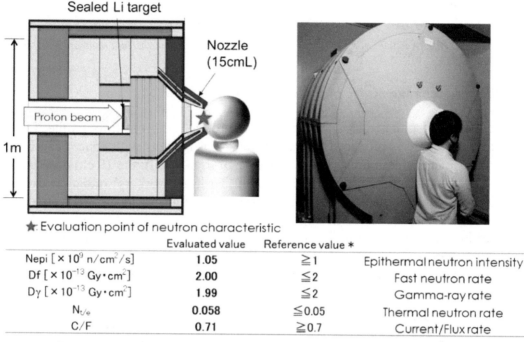

★ Evaluation point of neutron characteristic

	Evaluated value	Reference value ★	
Nepi [× 10^9 n/cm^2/s]	1.05	$\geqq 1$	Epithermal neutron intensity
Df [× 10^{-13} Gy·cm^2]	2.00	$\leqq 2$	Fast neutron rate
Dγ [× 10^{-13} Gy·cm^2]	1.99	$\leqq 2$	Gamma-ray rate
$N_{t/e}$	0.058	$\leqq 0.05$	Thermal neutron rate
C/F	0.71	$\geqq 0.7$	Current/Flux rate

★ IAEA-TECDOC-1223 "Current states of neutron capture therapy", IAEA (2001).

FIG. I-2. The BSA developed with a nozzle and its neutron characteristics.

In order to understand the characteristics of the BSA as a neutron source for BNCT, the thermal neutron flux distribution was evaluated in a water phantom (20 cm × 20 cm × 20 cm, 3-mm thick acrylic wall) by analysis and experiment as shown in Fig. I-3. Figure I-3(a) shows the neutron energy spectrum at the exit of the BSA calculated by PHITS. Figure I-3(b) shows the thermal neutron distribution measured in the water phantom and analysed with PHITS. The thermal neutron distribution in the phantom was measured using a ^{6}LiF/Eu:CaF_2 optical fibre neutron detector. This neutron detector is suitable for measuring thermal neutron distribution in water phantoms because it has a small neutron detection area (approximately 100 μm) and can measure neutrons without being affected by γ-rays. After calibrating the experimental results with the results of Au foil activation measurements, the thermal neutron distribution in the water phantom agreed well with the analysed neutron distribution [I-6].

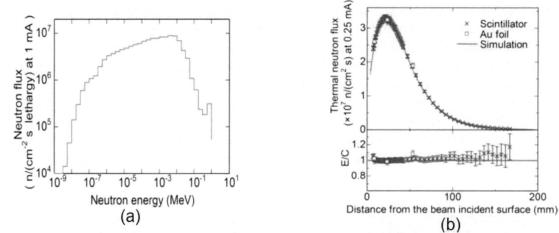

FIG. I-3. (a) Neutron energy distribution at the exit position of the BSA and (b) thermal neutron distribution in the water phantom.

I-2.4. Remote exchange device of target

When the Li target is irradiated by a proton beam to produce neutrons through the $^7Li(p,n)^7Be$ reaction, radioactive ^7Be is produced in the target. If 100 patients with cancer can be treated with the sealed Li target (assuming 100-h proton beam irradiation), the 10 cm dose equivalent near the spent target is expected to be 460 mSv/h. To limit workers' radiation exposure, a remote target exchange system, the 'robotic target exchange device,' has been developed (Fig. I-4). In this device, the sealed Li target in the BSA, along with the water-cooled jacket, is removed, put into a lead container for transportation, and transferred to the target exchange area. The aluminium flange that holds the target is removed by remote operation with a robot, and then the sealed Li target is picked up and stored in the lead storage container. Beryllium-7 contained in Li is also sealed in the target using Ti foil, enabling safe target replacement.

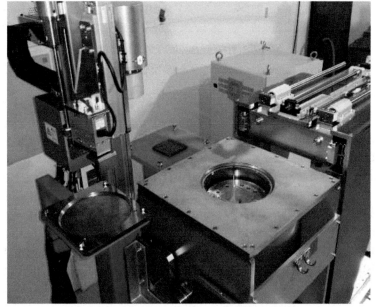

FIG. I-4. Remote target exchange device.

I-3. FUTURE DEVELOPMENT

We have developed an accelerator based neutron source from an electrostatic accelerator and a sealed Li target. In cooperation with Okayama University, we have conducted a series of non-clinical in vitro experiments to confirm the effectiveness of the developed system [I-7].

When cancer treatment is performed under one-directional irradiation using epithermal neutrons, BNCT can only be applied to cancer cells at a depth less than 5–6 cm. Conventional X ray, proton, and carbon radiotherapy systems use a crossfire technique, where radiation is delivered to deeply located cancer cells from multiple directions. Our BSA is compact and lightweight, allowing it to be mounted on a gantry system (Fig. I-5). By mounting the BSA with a nozzle on the gantry, it will be able to treat deeply located cancer cells by cross firing while the patient is in a comfortable position.

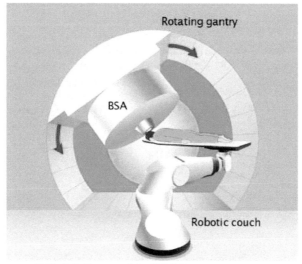

FIG. I-5. Scope of the BNCT gantry system.

REFERENCES TO ANNEX I

[I-1]   KIYANAGI, Y., Accelerator-based neutron source for boron neutron capture therapy, Ther. Radiol. Oncol. **2** (2018) 55.

[I-2]   INTERNATIONAL ATOMIC ENERGY AGENCY, Current Status of Neutron Capture Therapy, TECDOC-1223, IAEA, Vienna (2001).

[I-3]   YOSHIHASHI, S., et al., High heat removal technique for a lithium neutron generation target used for an accelerator-driven BNCT system, J. Instrum. **16** 4 (2021) P04016.

[I-4]   URITANI, A., et al., Design of beam shaping assembly for an accelerator-driven BNCT system in Nagoya University, JPS Conf. Proc. **22** (2018) P011002-1-7.

[I-5]   SATO, K., Improved design of the exit of a beam shaping assembly for an accelerator-driven BNCT system in Nagoya University, JPS Conf. Proc. **22** (2018) P011003-1-7.

[I-6]   WATANABE, K., et al., First experimental verification of the neutron field of Nagoya University accelerator-driven neutron source for boron neutron capture therapy, Appl. Radiat. Isot. **168** (2021) 109553.

[I-7]   MICHIUE, H., et al., Self-assembling A6K peptide nanotubes as a BSH delivery system for boron neutron capture therapy (BNCT), J. Controlled Release **330** (2021) 788–796.

Annex II.

NEUTRON SOURCE VITA

SERGEY TASKAEV[1,2]
[1]Budker Institute of Nuclear Physics, Novosibirsk, Russian Federation
[2]Novosibirsk State University, Novosibirsk, Russian Federation

Abstract

Compact accelerator based epithermal neutron sources are required for boron neutron capture therapy (BNCT). The Budker Institute has developed a neutron source to equip BNCT clinics. The neutron source VITA is used for BNCT development and for other topical applications.

II-1.   INTRODUCTION

The main objective for an AB-BNCT system is to develop a compact accelerator based neutron source of epithermal neutrons with minimal contributions of fast and thermal neutrons. Ideally, a neutron source for BNCT would be monoenergetic with an energy of ~10 keV. The $^7Li(p,n)^7Be$ reaction was chosen to create the narrowest epithermal neutron spectrum from a high-current, low-energy proton beam [II-1]. Such a compact accelerator-based neutron source has been created [II-2]. The source comprises a tandem accelerator of original design (the vacuum insulated tandem accelerator – VITA), solid Li target, and a beam shaping assembly. The neutron source can produce high neutron fluxes in various energy ranges, from fast to thermal, for boron neutron capture therapy and other applications.

II-2.   PROJECT

The research project is ultimately aimed at clinical application. However, as the project progressed, it became clear that the use of the facility may not be limited to only BNCT. At present, the facility provides production of protons or deuterons, the formation of neutron fluxes of almost any energy range: cold, thermal, epithermal, over-epithermal, monoenergetic or fast, the generation of 478 keV, 511 keV, or 9.17 MeV photons, as well as α particles and positrons.

Of the main results obtained at the facility, three can be singled out:
(a)   The facility became a prototype of neutron sources for BNCT clinics. The first commercial neutron source has been installed at the Xiamen Humanity Hospital in China. The next two neutron sources are being made for the National Oncological Hadron Therapy Centre in Pavia (Italy) and for the National Medical Research Centre of Oncology in Moscow (Russia);
(b)   The facility is used for the development of the BNCT technique, namely, to:
   (i)   Study the effect of neutron radiation on cell cultures and laboratory animals [II-3–II-4];
   (ii) Treat large pets with spontaneous tumours [II-5];
   (iii) Develop dosimetry tools [II-6–II-8];
   (iv) Test new boron delivery drugs [II-9–II-11].

(d) The facility is used for a number of other applications, namely:
  (i)   Measuring the cross section and the yield of nuclear reactions [II-12–II-13];
  (ii) Studying radiation blistering of metals under ion implantation [II-14–II-15];
  (iii) Radiation testing of advanced materials [II-16–II-17];
  (iv) Measuring the thickness of the Li layer [II-18], etc.

II-2.1. System

The layout of the BNCT facility is shown in Fig. II-1. The neutron producing target can be placed in five positions (A–E).

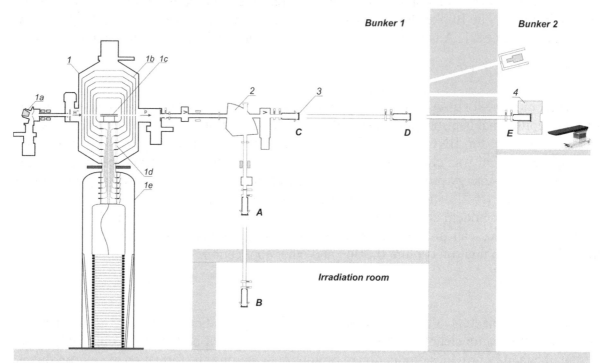

FIG. II-1. Layout of the BNCT facility: 1 – VITA (1a – negative ion source, 1b –electrodes, 1c – gas stripper, 1d – feedthrough insulator, 1e – high-voltage power supply), 2 – bending magnet, 3 – Li target, 4 – beam-shaping assembly. A, B, C, D, E – Li target placement positions.

II-2.1.1. Type of accelerator

The Budker Institute of Nuclear Physics' tandem accelerator, called the Vacuum-Insulated Tandem Accelerator (VITA), is an original design of electrostatic tandem accelerator. The term 'electrostatic' means that a static electric field is used to accelerate ions. The term 'tandem' means that the applied accelerating voltage is used twice. First, H^- ions are injected into the tandem accelerator and are accelerated by the positive potential applied to the central electrode. The electrons are subsequently stripped to create H^+ ions, which are then accelerated again by the same potential. By this scheme, a tandem accelerator reduces the required accelerating voltage by half, which simplifies the electrostatic insulation and reduces the size and cost of the system.

Unlike conventional tandem accelerators, VITA does not use accelerating tubes, but nested intermediate electrodes (1b) fixed at a feedthrough insulator (1d) are used (Fig. II-1). The ceramic parts of the feedthrough insulator are then further away from the ion beam, which has the advantage that the high-voltage strength of the accelerating gaps can be increased, given the high ion beam current.

VITA is characterized by the following features:

- The design of the accelerator provides a high voltage gradient – up to 25 kV/cm. Because of this, the input electrostatic lens of the accelerator is strong. To obtain a proton beam, it is necessary to refocus the beam of negative hydrogen ions into the accelerator inlet with high accuracy;

- The stripping of negative hydrogen ions into protons is carried out in a gas stripper, for which a sufficiently large flow of argon is supplied. The supply of additional gas flow is usually considered to be a disadvantage for a tandem accelerator. In this case, the additional gas flow became more of an advantage, since it made it possible to visualize the ion beam, and was useful when implementing a method for suppressing secondary charged particles [II-19];

- The accelerator was originally proposed for BNCT and was designed for quite specific parameters: 10 mA and 2.3 MeV. However, as development progressed, this accelerator began to be used for other applications requiring not only other values of current and energy, but even the type of ions: not only protons, but also deuterons. For this reason, the accelerator is equipped with a set of diagnostic tools that make it possible to ensure long-term stable production of protons or deuterons with energy and current varying over a wide range.

The ion beam energy at VITA can be controlled over 0.6–2.3 MeV with a stability of 0.1%. The beam current can also be controlled over a wide range (from 1 nA to 10 mA) with high current stability (0.4%). At the exit from the accelerator, the ion beam has a transverse dimension of 10 mm, an angular divergence of up to ± 1.5 mrad, and a normalized emittance of 0.2 mm mrad. Such a beam can be transported over an appreciable distance without the use of focusing elements, since its transport is not accompanied by a noticeable influence of the space charge [II-20]. The proton/deuteron beam is accompanied by a stream of neutrals, which is used to measure the efficiency of the gas stripper [II-21], and a negligibly low current of argon ions flowing from the gas stripper [II-22].

II-2.1.2.   Number of beamlines

The facility has a bending magnet that allows the ion beam to pass through in the horizontal direction or to direct it downwards. BNCT clinics equipped with such neutron sources may use rotation of the ion beam in the horizontal plane or the vertical.

II-2.1.3.   Beam energy, current

The ion beam energy can be varied from 0.6 MeV to 2.3 MeV, and the current from 1 nA to 10 mA.

II-2.1.4.   Target materials

The Li target, 10 cm in diameter, has three layers: pure Li to generate neutrons [II-13], anti-blistering material [II-14–II-15], and a Cu substrate for heat removal [II-23]. It provides a stable neutron yield for a long time with an acceptably low level of contamination of the beam transport path by the inevitably formed radioactive isotope ^7Be.

II-2.1.5. Beam shaper assembly design

A Beam Shaper Assembly (BSA) is used to convert the neutron flux into a beam of epithermal neutrons suitable for clinical applications. The BSA consists of an MgF_2 moderator, composite reflector (graphite in the front hemisphere and Pb in the back), an absorber, and a filter [II-24]. The flux of epithermal neutrons is 1.04×10^9 $cm^{-2} \cdot s^{-1}$ at a proton current of 10 mA while the contamination by thermal neutrons ($_{th}$), fast neutrons ($_{fn}$) and γ radiation ($_\gamma$) is acceptably small: $\phi_{th}/\phi_{epi} = 1/30$, $D_{fn}/\phi_{epi} = 1.25 \times 10^{-13}$ $Gy \cdot cm^2$, $D_\gamma/\phi_{epi} = 1.89 \times 10^{-13}$ $Gy \cdot cm^2$ [II-25]. Such a beam provides a dose rate of 85 RBE Gy/h in the tumour at a concentration of 52.5 ppm ^{10}B. The treatable depth is 7.52 cm and the maximum dose ratio of tumour to normal tissue is 5.38.

II-2.1.6. Real time monitor

The energy and current of the ion beam, the dose rate of neutrons and gamma radiation and many other parameters are monitored in real time.

II-2.1.7. Collimator

A collimator can be used.

II-2.1.8. Patient positioning system

An X ray surgical table is used to position the patient.

II-2.2. Facility design and layout

The neutron source is placed in two bunkers as shown in Fig. II-1. Each bunker is 10.8 m × 9.1 m and 10 m high, the wall thickness of the bunker is 1.2–1.3 m, and the wall thickness between the bunkers is 1.47 m.

The facility is equipped with dosimeters, neutron detectors, neutron flux sensors, γ-ray spectrometers, CCD-cameras, infrared cameras, scanners, atomic emission spectrometer, etc.

II-2.3. Costs and management

II-2.3.1. Financial cost

The cost of construction of the facility depends on the parameters and equipment and is about ten million Euros. Operational and treatment costs are insignificant compared to the cost of boron delivery drugs.

II-2.3.2. Staff

Currently, the facility is used for scientific research in various fields. The staff consists of a dozen permanent employees, including researchers, engineers and laboratory assistants, and a dozen Ph.D. students. For the operation of the facility in a clinic, several operators with a physical education and a couple of laboratory assistants are needed.

ACKNOWLEDGEMENTS

The study was supported mainly by the Russian Science Foundation (projects 14-32-00006, 19-72-30005).

REFERENCES TO ANNEX II

[II-1]   BAYANOV, B., et al., Accelerator based neutron source for the neutron-capture and fast neutron therapy at hospital, Nucl. Instrum. Methods Phys. Res. Sect. A **413** 2–3 (1998) 397–426.

[II-2]   TASKAEV, S., et al., Neutron source based on vacuum insulated tandem accelerator and lithium target, Biology **10** (2021) 350.

[II-3]   KANYGIN, V., et al., Dose-dependent suppression of human glioblastoma xenograft growth by accelerator-based boron neutron capture therapy with simultaneous use of two boron-containing compounds, Biology **10** (2021) 1124.

[II-4]   ZAVJALOV, E., et al., Accelerator-based boron neutron capture therapy for malignant glioma: a pilot neutron irradiation study using boron phenylalanine, sodium borocaptate and liposomal borocaptate with a heterotopic U87 glioblastoma model in SCID mice, Int. J. Radiat. Biol. **96** (2020) 868–878.

[II-5]   KANYGIN, V., et al., In vivo accelerator-based boron neutron capture therapy for spontaneous tumors in large animals: case series, Biology **11** (2022) 138.

[II-6]   BYKOV, T., et al., Evaluation of depth–dose profiles in a water phantom at the BNCT facility at BINP, JINST **16** (2021) P10016.

[II-7]   DYMOVA, M., et al., Method of measuring high-LET particles dose, Radiat. Res. **196** (2021) 192–196.

[II-8]   ZABORONOK, A., et al., Gold nanoparticles permit in situ absorbed dose evaluation in boron neutron capture therapy for malignant tumors, Pharmaceutics **13** (2021) 1490.

[II-9]   USPENSKII, S., et al., Elemental boron nanoparticles: production by ultrasonication in aqueous medium and application in boron neutron capture therapy, Dokl. Chem. **491** (2020) 45–48.

[II-10]   VOROBYEVA, M., et al., Tumor cell-specific 2′-fluoro RNA aptamer conjugated with *closo*-dodecaborate as a potential agent for boron neutron capture therapy, Int. J. Mol. Sci. **22** (2021) 7326.

[II-11]   POPOVA, T., et al., Homocystamide conjugates of human serum albumin as a platform to prepare bimodal multidrug delivery systems for boron-neutron capture therapy, Molecules **26** (2021) 6537.

[II-12]   TASKAEV, S., et al., Measurement of the $^7Li(p,p'\gamma)^7Li$ reaction cross-section and 478 keV photon yield from a thick lithium target at proton energies from 0.65 MeV to 2.225 MeV, Nucl. Instrum. Methods Phys. Res. Sect. B **502** (2021) 85–94.

[II-13]   BIKCHURINA, M., et al., The measurement of the neutron yield of the $^7Li(p,n)^7Be$ reaction in lithium targets, Biology **10** (2021) 824.

[II-14]   BADRUTDINOV, A., et al., In situ observations of blistering of a metal irradiated with 2-MeV protons, Metals **7** (2017) 558.

[II-15] BYKOV, T., et al., In situ study of the blistering effect of copper with a thin lithium layer on the neutron yield in the ^{7}Li(p,n)^{7}Be reaction, Nucl. Instrum. Methods Phys. Res. Sect. B **481** (2020) 62–81.

[II-16] SHOSHIN, A., et al., Test results of boron carbide ceramics for ITER port protection, Fusion Eng. Des. **168** (2021) 112426.

[II-17] KASATOV, D., et al., A fast-neutron source based on a vacuum-insulated tandem accelerator and a lithium target, Instrum. Exp. Tech. **63** (2020) 611–615.

[II-18] KASATOV, D., et al., Method for in situ measuring the thickness of a lithium layer, JINST **15** (2020) P10006.

[II-19] IVANOV, A., et al., Suppression of an unwanted flow of charged particles in a tandem accelerator with vacuum insulation, JINST **11** (2016) P04018.

[II-20] BYKOV, T., et al., A study of the spatial charge effect on 2-MeV proton beam transport in an accelerator-based epithermal neutron source, Tech. Phys. **66** (2021) 98–102.

[II-21] KOLESNIKOV, Y.A., KOSHKAREV, A., TASKAEV, S., SHCHUDLO, I., Diagnostics of the efficiency of a gas stripping target of a tandem accelerator with vacuum insulation, Instrum. Exp. Tech. **63** (2020) 314–318.

[II-22] KOLESNIKOV, Y.A., et al., Measuring the current of a beam of argon ions accompanying a beam of protons in a tandem accelerator with vacuum insulation, Instrum. Exp. Tech. **64** (2021) 503–507.

[II-23] BAYANOV, B., BELOV, V., KINDYUK, V., OPARIN, E., TASKAEV, S., Lithium neutron producing target for BINP accelerator-based neutron source, Appl. Radiat. Isot. **61** (2004) 817–821.

[II-24] ZAIDI, L., et al., Neutron-beam-shaping assembly for boron neutron-capture therapy, Phys. At. Nucl. **80** (2017) 60–66.

[II-25] ZAIDI, L., BELGAID, M., TASKAEV, S., KHELIFI, R., Beam shaping assembly design of ^{7}Li(p,n)^{7}Be neutron source for boron neutron capture therapy of deep-seated tumor, Appl. Radiat. Isot. **139** (2018) 316–324.

THE NEUPEX SYSTEM AND THE
XIAMEN HUMANITY HOSPITAL BNCT CENTRE

YUAN-HAO LIU, DI-YUN SHU, WEN-YU XU
BNCT Centre, Xiamen Humanity Hospital
No. 3777, Rd. Xianyue, Huli District, Xiamen City, Fujian Province, P.R. China

In May 2019, Neuboron and Xiamen Humanity Hospital (located in Xiamen, an island city and a special economic zone in Fujian Province, China), an AAA General Hospital focusing on the development of cancer, nerves, ear, nose and throat etc. launched the construction of the first Chinese BNCT Centre together. Neuboron has completed the installation of an accelerator-based boron neutron capture therapy (AB-BNCT) system 'NeuPex' in the Xiamen BNCT centre which was ready at the end of 2020. In August 2021, NeuPex has successfully delivered its first epithermal neutron beam using 4 mA of protons at 2.3 MeV. After a dedicated neutron beam characterization, a series of animal preclinical studies, and a successful treatment of a beagle with a spontaneous tumour, NeuPex and its treatment planning system (TPS) NeuMANTA were successfully used on the first-in-human clinical research, investigator initiated trial on October 9th 2022 using only 8 mA of protons at 2.3 MeV. The official clinical trial is planned in 2023 and clinical service is planned in late 2024.

III-1. INTRODUCTION

Before 2014, the development of BNCT in China was based on a reactor-based neutron source and very few cases of clinical research were performed. Limited by the lack of neutron sources suitable for hospital environments, BNCT was not well used in clinical practice. After 2014, this situation altered. A close BNCT consortium organized by Neuboron Medtech Ltd. (a.k.a. Neuboron Medical Group) has been devoted to developing AB-BNCT technology and the first AB-BNCT research centre in China, together with Nanjing University of Aeronautics and Astronautics, China Pharmaceutical University, Peking Union Medical College Hospital, Dongcheng Pharmaceutical Group Co., Ltd., the Xiamen Humanity Hospital, and other top institutions.

The first AB-BNCT developed by Neuboron was named NeuPex (Model Block-I) using an electrostatic accelerator with an advanced neutron beam shaping technology as well as other new techniques. NeuPex is known to be the most power-efficient AB-BNCT neutron source, which is attributed to its compact beam shaping assembly technology as well as its target station design. Figure III-1 shows the outside of the new facility.

The system has been successfully installed in the Xiamen Humanity Hospital of Xiamen City (Fig. III-1). NeuPex has gone through a dedicated neutron beam characterization, a series of animal preclinical studies, as well as a treatment of a spontaneous tumour in a beagle. The encouraging results has led to the first-in-human investigator-initiated trial in October 2022.

FIG. III-1. Xiamen Humanity Hospital BNCT centre.

III-2.   PROJECT

The focus of the project at Xiamen Humanity Hospital is clinical application and the system is described in the following sections.

III-2.1. System

The AB-BNCT system, installed in the Xiamen Humanity Hospital, is named NeuPex Block-I, and utilizes low energy protons at a relatively low beam power compared to other systems. NeuPex is a novel, highly integrated AB-BNCT system of medical device grade, protected by hundreds of patents globally. Figure III-2 shows the internal design of the facility with the accelerator in the far left and two horizontal treatment rooms on the right-hand side. Figure III-3 shows one of the horizontal treatment rooms with the patient positioning system.

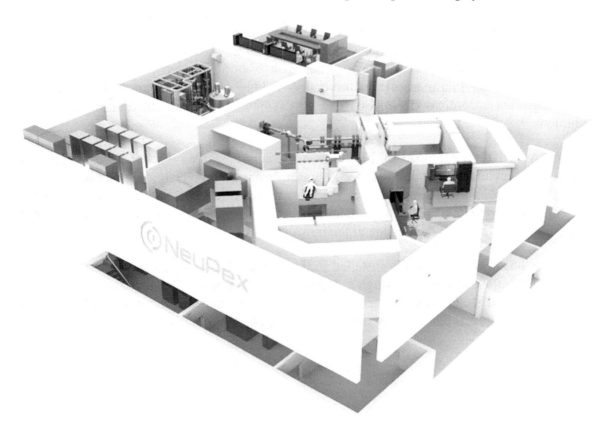

FIG. III-2. The system configuration of Neuboron NeuPex AB-BNCT system.

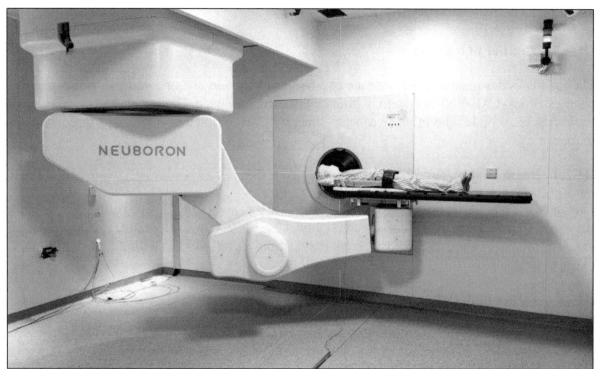

FIG. III-3. The first horizontal treatment room of Xiamen BNCT Centre.

III-2.2. Type of accelerator

NeuPex Block-I uses a tandem electrostatic accelerator to provide a low-energy proton beam at a relatively high beam current. Electrostatic accelerators are well-known as a power-efficient, stable machines. Figure III-4 shows the main vessels of the accelerator system.

FIG. III-4. Tandem electrostatic accelerator of NeuPex (a) electrostatic power vessel, (b) main accelerator vessel.

III-2.3. Number of beamlines

The facility is designed to safely operate two horizontal beams and one vertical beam using the NeuPex Block-I AB-BNCT system, within a very compact space. Currently, the first horizontal beam has been built. The other two beam lines will be completed in the next three years in sequence.

III-2.4. Beam energy, current

The energy range of the proton beam is 1.9–2.75 MeV. The nominal value of the proton beam current is 10 mA. Currently, NeuPex runs stably at 2.3 MeV, 8 mA (Nov 2022), and is expected to reach full power in Q2 2023. For clinical application, a proton beam with 2.5 MeV, 10 mA is likely to be applied.

III-2.5. Target materials

Neutron generation is based on the $^7Li(p,n)^7Be$ reaction, using a stationary, solid Li target.

III-2.6. Beam shaper assembly design

A patent-protected beam shaping assembly (BSA) with a target-embedded, bi-cone shaped moderator, using MgF_2 as the main moderation material, as well as other advanced technologies, has been developed. Neutron activation and the corresponding induced radioactive by-products have been considered during design and development. The accumulated activity of the BSA can be lower than the limit of the definition of low-level radioactive waste after a reasonable decay time. The designed advantage depth is 11 cm, and the epithermal neutron beam flux, ϕ_{epi}, exceeds 1.0×10^9 cm^{-2}·s^{-1} (with 10 mA of protons at 2.5 MeV), with a large beam aperture up to 40 cm in diameter.

III-2.7. Neutron beam characterization

The epithermal neutron flux of NeuPex is adjustable using a tuneable BSA design and currently could deliver $\phi_{epi} > 9 \times 10^8$ cm^{-2}·s^{-1}, at the BSA exit centre using a proton beam energy at 2.3 MeV and a proton beam current at 8 mA, which has fulfilled the need of clinical application. The epithermal neutron beam conversion efficiency reaches ~5×10^7 cm^{-2}·s^{-1}·kW^{-1}. The above values are measured by using Au, Cu, and Mn foils by instrumental neutron activation analysis. The measured γ ray contamination, $D_\gamma / \int \phi_{epi}(t) \cdot dt$, is ~$2.3 \times 10^{-13}$ Gy·cm^2 per unit epithermal neutron fluence. The neutron beam performance will be further optimized. The in-phantom characterization was performed by using gold foils in a $31 \times 31 \times 31$ cm^3 PMMA phantom. The calculated and measured profiles were matched well. The thermal neutron flux peak shows at ~3-cm depth and exceeds 2×10^9 cm^{-2}·s^{-1}, when the phantom is placed against the BSA exit.

III-2.8. Real time monitor

Three neutron detectors and one gamma detector are installed at the rear of the BSA, but not within the beam centre to avoid influencing the beam quality. They are installed around the BSA to obtain spatial information concerning the beam.

III-2.9. Collimator

Two types of collimators have already been designed and fabricated. One is externally attached to the BSA, and the other is embedded in the BSA. For the external collimator, a multi-layer structure is adopted with a diameter of 12 cm. Additional collimators are available using the same design and development technology.

III-2.10. Treatment planning system

NeuPex AB-BNCT system utilizes the treatment planning system (TPS) named NeuMANTA (Neuboron Multifunctional Arithmetic for Neutron Transportation Analysis) to perform its clinical irradiation as well as animal experiment evaluation. NeuMANTA is a dedicated BNCT TPS. It is DICOM-RT compatible and supports PET, CT, and MRI images as well as corresponding fusions. NeuMANTA has a standalone Monte Carlo dose engine named COMPASS (COMpact Particle Simulation System). The calculation speed of COMPASS is about ~2 fold faster than the famous MCNP6 code with the same accuracy and calculation configuration. Currently, the calculation time per head and neck cancer case is less than 10 minutes on two AMD 7950X CPUs. Figure III-5 shows two views of the interface.

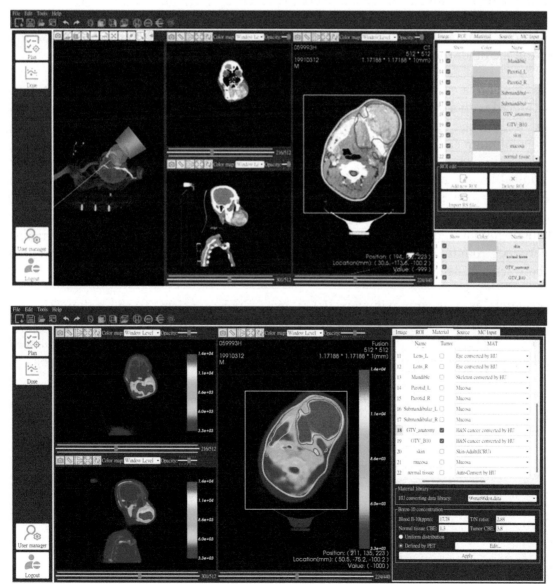

FIG. III-5. Views of the NeuMANTA treatment planning system interface.

III-2.11. Patient positioning system

The patient positioning system consists of a pair of high-precision, 6-axis robotic arms, laser positioning system, treatment couch, treatment chair, etc.; one robot arm is installed in the treatment room, and the other one is installed in a simulation room. The accuracy of the positioning system can reach sub-millimetre level. The system is designed with the capabilities of anti-collision and anti-activation.

III-3. FACILITY DESIGN AND LAYOUT

The Xiamen Humanity Hospital BNCT centre is a two-storey high, and two-storey underground building, which is dedicated to BNCT clinical and research purposes (Fig. III-6). The centre has a compact layout with a core area of approximately 800 m^2. The main shielding walls of core area use patent-protected boron-containing barite concrete as the material, which not only ensures radiation safety but also improves space utilization. The core area consists of the following rooms: 1) accelerator room, 2) high energy beam line (HEBL) room, 3) treatment rooms (three), 4) accelerator control room, 5) treatment control room, 6) auxiliary rooms, 7) maintenance hall, 8) radioactive storage room. Among them, the accelerator room, HEBL room, treatment rooms and radioactive storage room belong to the radiation-controlled area. And the other rooms of core area belong to the radiation-supervised area.

Besides the core area where NeuPex is installed, the Xiamen Humanity Hospital BNCT centre is configured with a clinical area, and a research area, to make the centre not only for BNCT cancer treatment, but also for BNCT-related new-technology development and new boron/gadolinium carrier clinical translational research.

The clinical area is approximately 700 m^2, which consists of the following rooms: 1) nurse stations for patient reception, 2) doctor's offices, 3) drug preparation and storage room, 4) transfusion rooms for patient drug infusion before BNCT irradiation, 5) observation rooms for patient monitoring and residual radiation dosage confirmation after BNCT irradiation, 6) emergency rooms, 7) simulation rooms. For the Xiamen BNCT centre, the configuration ratio between simulation room and treatment room is 1:1.

The research area is approximately 1000 m^2, which consists of the following rooms: 1) preparation rooms and personnel dressing rooms for GLP-like requirements, 2) cell laboratory, 3) biological evaluation laboratory, 4) chemical laboratory, 5) ICP-MS laboratory, 6) nuclear physics laboratory, 7) medical physics laboratory, and 8) engineering workshop.

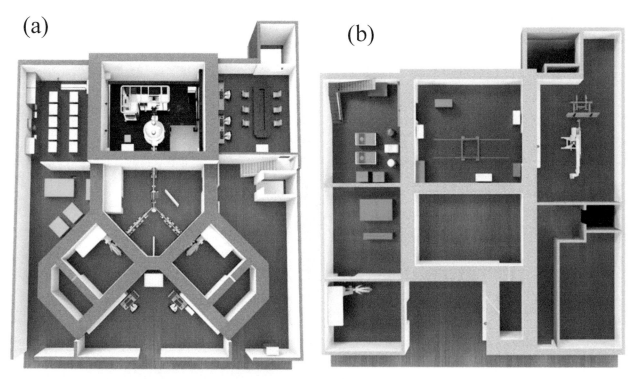

FIG. III-6. The core area layout of Xiamen Humanity Hospital BNCT centre (a) B1, (b) B2.

III-4. STAFFING AND MANAGEMENT

Table III-1 provides the list of staff expected in the facility.

TABLE III-1. NUMBERS OF STAFF AND QUALIFICATIONS

Position	Number of staff	Responsibilities
Accelerator system engineer	2–3	Operation and maintenance of accelerator and its auxiliary equipment system
Radiation safety engineer	1–2 per treatment room	Quality assurance and control of beam quality, ambient radiation measurement, personnel dose management, radioactive waste management
Medical laboratory technician	2–3	Boron concentration measurement
Medical physicist	1.5–2 per treatment room	BNCT system control, BNCT treatment planning design, patient positioning
Nurse	1.5–2 per treatment room	Boron drug infusion and patient care

III-4.1. Number of patients per year expected

With the above facility and staff, it is expected to offer more than 4000 irradiations per year, which is equal to more than 2000 patients treated.

Annex IV.

BNCT AT THE ITALIAN NATIONAL CENTER FOR ONCOLOGICAL HADRONTHERAPY

S. AGOSTEO[1,2], S. ALTIERI[1,3], F. BALLARINI[1,3], D. BETTONI[1], G.M. CALVI[4], V. CONTE[1], C. LEE[5], L. LICITRA[6], P. CIRRONE[1], A. FACOETTI[6], F. FORNERIS[3], P. FRANZOSI[4], C. FERRARI[3], M. FERRARINI[6], G. MAGRO[6], S. MOLINELLI[6], M. NECCHI[6], O. NICROSINI[1], E. ORLANDI[6], A. PEDOTTI[2], P. PEDRAZZOLI[3], I. POSTUMA[1], N. PROTTI[3], A. RETICO[1,3], S. ROSSI[6], A. SERRA[6], G. VAGO[6], G. VENCHI[6], V. VERCESI[1], G. ZANONI[3]

[1] INFN-Istituto Nazionale di Fisica Nucleare, Italy
[2] Politecnico di Milano, Italy
[3] Università degli Studi di Pavia, Italy
[4] Studio Calvi, Italy
[5] TAE Life Sciences, Foothill Ranch, California, USA
[6] Fondazione CNAO, Strada Campeggi 23, 16600 Pavia, Italy

Abstract

At the National Centre for Oncological Hadrontherapy (CNAO) in Pavia (Italy), a new project is ongoing to introduce boron neutron capture therapy (BNCT) in addition to the modern therapies with protons and carbon ions. On November 2020, TAE Life Sciences (TLS) and CNAO Foundation signed an agreement to create at CNAO a new facility equipped with a dedicated tandem accelerator to perform BNCT treatments. This collaboration aims to create, within the clinical and research environment of CNAO, the optimal conditions to pursue the goal of treating patients with this innovative modality.

Pavia has a role in the history of BNCT. The study of BNCT has been ongoing since the 1980s, taking advantage of the availability of a nuclear research reactor at the University (Laboratory of Applied Nuclear Energy, LENA). The studies started with a challenging protocol of explanted and irradiated organ and led to the clinical application of ex-situ BNCT to the liver of two patients (2001 and 2003).

To be successful, BNCT requires a multidisciplinary approach. A network of collaborations involving CNAO, INFN, University of Pavia and Polytechnic of Milan, with links to international study groups, has been created to tackle the basic topics: regulatory aspects, radiobiology, computational dosimetry and treatment planning, experimental and environmental dosimetry, boron measurement and clinical dosimetry, clinical trial procedure for BNCT, and development of new borated compounds.

IV-1. INTRODUCTION

At the National Centre for Oncological Hadrontherapy in Pavia (Italy), the modern therapies with protons and carbon ions have broadened the spectrum of tumours that can be controlled. However, there are many patients who cannot be treated due to the nature or the location of their disease. For them, BNCT represents a novel, interesting approach. BNCT is an experimental form of radiotherapy based on neutron irradiation of the tumour following perfusion with a boron drug that can concentrate boron atoms (^{10}B) in the tumour cells. The thermal neutron capture reaction on ^{10}B is highly probable and gives rise to two ionising particles: an α particle and a ^{7}Li ion. These particles lose all their energy over a distance comparable to the diameter of a cell and can cause irreparable damage to DNA when they pass through the nucleus.

To date, the clinical application of BNCT has been carried out mainly to treat brain tumours such as multiform glioblastoma, malignant meningiomas, melanoma metastases, and head and neck cancers [IV-1–IV-5]. Among all, head and neck cancers are probably the most successful targets and may represent an interesting starting point for clinical trials in the new Centre.

Pavia has a role in the history of BNCT. The study of BNCT has been ongoing since the 1980s, taking advantage of the availability of a nuclear research reactor at the University (Laboratory of Applied Nuclear Energy, LENA). The studies started with a challenging protocol of explanted and irradiated organ and led to the clinical application of ex-situ BNCT to the liver of two patients (2001 and 2003) [IV-6–IV-7]. This experience has been fundamental to continue and foster the research activities of the groups at the University of Pavia, together with INFN and other collaborating institutes.

IV-2. PROJECT

The recent development of compact accelerator-based neutron sources, replacing nuclear reactors, represents the most important innovation in the field of BNCT [IV-8], and allows the deployment and the application of this technique in a hospital environment. In November 2020, TLS and CNAO Foundation signed an agreement to create at CNAO a new facility equipped with a dedicated tandem accelerator to perform BNCT treatments. This collaboration aims to create, within the clinical and research environment of CNAO, the optimal conditions to pursue the goal of treating patients with this innovative modality. Section IV-2.1 presents the accelerator based BNCT system characteristics and Section IV 2.2 outlines the facility design and layout.

To be successful, BNCT requires a multidisciplinary approach. This approach involves physicists and engineers for the design and implementation of the technology needed to produce and exploit the neutron beams with efficacy and safety; chemists and biologists for the study and optimisation of boron bio-distribution and the analysis of radiobiological effects; medical physicists and physicians for dosimetry, preparation of treatment plans and patient management. Thus, a network of collaborations involving CNAO, INFN, University of Pavia and Polytechnic of Milan, with links to international study groups, has been created to tackle the basic topics: regulatory aspects, radiobiology, computational dosimetry and treatment planning, experimental and environmental dosimetry, boron measurement and clinical dosimetry, clinical trial procedure for BNCT, development of new borated compounds. The basic activities in these seven areas are described in Section IV-2.3.

IV-2.1. System

IV-2.1.1. Tandem accelerator

BNCT at CNAO will use the Alphabeam system from TLS, whose neutron source consists of a high current (nominally 10 mA) beam of protons at a nominal operational energy of 2.5 MeV in combination with a Li target. A compact tandem accelerator (Fig. IV-1) will produce a DC current at the required energy and intensity.

The accelerator is called a 'tandem' because it uses the same accelerating voltage twice. Negative hydrogen ions are injected into the entrance and accelerated by a positive potential gradient toward the centre of the cylindrical geometry. A charge exchanger located at the centre of the accelerator strips both electrons from the negative ions, converting them to positively charged protons. The same voltage gradient that accelerated the negative ions towards the centre now accelerates the protons away from the centre to the exit on the opposite side from the entrance. The accelerating voltage is reduced to half the nominal voltage, allowing a more compact accelerator by simplifying the electrostatic insulation requirements.

The accelerator voltage gradient is created with a series of nested electrodes and a high voltage power supply, connected by a high voltage feedthrough column. The power supply and lower portion of the feedthrough are housed in an electrostatic vessel (ELV) that extends below the tandem accelerator. Electrostatic breakdown in the accelerator and ELV is prevented using a combination of vacuum and dielectric gas (SF_6). The design of the tandem accelerator and ELV is an expansion of the Vacuum Insulated Tandem Accelerator, or VITA, created at the Budker Institute of Nuclear Physics [IV-9].

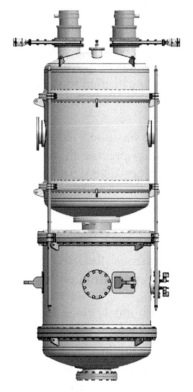

FIG. IV-1. Tandem accelerator and ELV.

IV-2.1.2.   Low energy beam line and pre-accelerator

A collection of components called the Low Energy Beam Line (LEBL) produces the negative ion source introduced into the tandem accelerator. It includes an ultra-high purity hydrogen gas source; negative hydrogen ion source; electrostatic and magnetic elements for beam steering and focusing; air, water cooling and vacuum systems; and a pre-accelerator that increases the beam energy to 80–180 keV before entering the tandem accelerator. These components are located on a single high voltage platform isolated up to 180 kV.

The negative hydrogen ion source is designed to operate up to 15 mA to provide margin for the nominal operating proton current of 10 mA. The negative ion beam uses a combination of XY position shifter, beam profiler and current monitor to maintain beam shape and directionality. It also includes a Faraday cup to support the characterization and tuning of the beam. Beam scrapers also remove any beam halo to prevent stray ions from entering the accelerator.

IV-2.1.3.   High energy beam lines

The High Energy Beam Line (HEBL) system supports the transport of the full-energy proton beam from the exit of the tandem accelerator to multiple treatment rooms. The beam line consists of aluminium beam tubes and flanges, vacuum pumps, magnets for directing and focusing the beam, diagnostic components, beam dump and target assemblies (one target per treatment room). The HEBL magnets include:

- Dipole magnets to change beam direction; e.g., 90° bending magnets to direct the beam from the accelerator to the two treatment rooms;
- XY shifter magnets to keep the beam aimed along the centre line of the vacuum chamber;
- Quadrupole magnets for beam focusing;
- Raster magnets to move the beam to points around the total surface area of the target. Rastering the beam improves thermal management of the target.

The facility configuration at CNAO requires a series of two 90° bending (dipole) magnets: a single beam splitter sending the beam either left or right from a single magnet is not possible due to neutrons back-streaming from each target to the opposite room. The beam dump is designed to accept the full beam current as needed in the event of magnet power loss and is also used during system commissioning.

Supports for beam line components are designed for structural rigidity, position adjustability and to reduce neutron activation. A highly precise alignment system is used to install all LEBL, accelerator and HEBL components in a consistent 3-D coordinate system.

IV-2.1.4.   Neutron producing targets

The BNCT target assembly consists of a layer of natural Li deposited onto a copper backing disk with cooling channels. The neutrons for BNCT treatments are produced from the endothermic $^7Li(p,n)^7Be$ reaction, with approximately 92% of natural Li as the ^7Li isotope.

The 2.5 MeV protons slow down continuously once they penetrate the Li surface, producing neutrons until they decelerate below the reaction threshold energy of 1.88 MeV. This energy loss occurs over a distance of 90 μm; additional Li thickness will increase the maximum temperature of the Li without producing more neutrons. In addition, radiative proton capture and inelastic proton scattering reactions in Li will produce more gamma rays, considered a contaminant to the clinical BNCT beam. Cooling water passing through the copper backing maintains the 90–100 μm Li layer below approximately 150 °C to prevent target melting. Rastering the proton beam over the surface of the Li target is also critical to maintaining target temperature uniformity.

The ^7Be product of the (p,n) reaction is radioactive with a 53-day half-life, and approximately 10% of its decays produce a 478-keV γ ray. As a new target is exposed to the proton beam, the quantity of ^7Be will increase in the target and create a source of γ rays in the HEBL and treatment rooms. The target assembly will be replaced periodically to keep its surface dose rate low enough for safe removal from the beam line and storage in a dedicated location for cooldown. In addition to ^7Be, other components in the target assembly will become radioactive due to the intense neutron field, including natural isotopes of copper, aluminium, and trace amounts of manganese. The components will have much shorter half-lives than ^7Be, ranging from 2.25 minutes for ^{28}Al to 12.7 hours for ^{64}Cu. To reduce overall personnel exposure in the

HEBL from activated target assembly components, the end of each beam line will be housed in a small alcove with shielded doors.

IV-2.1.5. Neutron source model

For a pencil beam of monoenergetic protons striking a Li target, a spectrum of neutron energies E_n is produced that is strongly dependent on the angle of emission θ_n [IV-10]. The relative neutron distribution is a function of the (p,n) differential cross section in the centre-of-mass frame $\partial\sigma_{pn}/\partial\Omega_{CM}$, typically tabulated as a function of the proton energy E_p and neutron emission angle in the centre-of-mass frame (θ_{CM}). However, due to the kinematic constraints of the (p,n) reaction in ^{7}Li, only one (E_n, θ_n) combination corresponds to a particular (E_p, θ_{CM}) pair, allowing the calculation of thick target neutron yields (an integral over proton energy) to collapse to a pointwise function:

$$Y(E_n, \theta_L) = K \frac{d\sigma_{pn}}{d\Omega_{CM}}(E_p, \theta_{CM}) \frac{\left|\begin{matrix} \frac{\partial E_p \partial \Omega_{CM}}{\partial E_n \; \partial \Omega_L} & \frac{\partial E_p \partial \Omega_{CM}}{\partial \Omega_L \; \partial E_n} \end{matrix}\right|}{-S(E_p)}$$

(IV-1)

where $S(E_p)$ is the stopping power of protons in natural Li, $\partial\theta_L$ and $\partial\Omega_{CM}$ are differential solid angle elements in the lab and centre-of-mass frames, and K is a constant.

This neutron yield expression does not take energy or angular straggling into account, but the relative amount of proton straggling over the 1.88–2.5 MeV range is limited (Fig. IV-2), and comparisons of these results with Monte Carlo models with straggling included result in minimal differences. The expression is easily adjusted for Li compounds with a change of the constant K and a new stopping power.

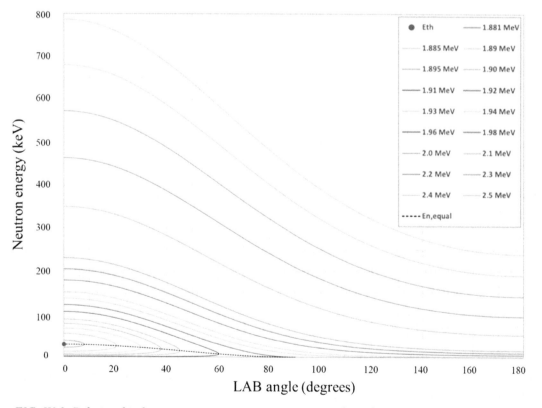

FIG. IV-2. Relationship between neutron energy, emission angle and corresponding proton energy.

IV-2.1.6. Beam shaping assembly and collimation

The clinical BNCT neutron beam would ideally consist of epithermal neutrons (from 0.5–1.0 eV up to 10–30 keV) with minimum contamination by fast neutrons, thermal neutrons, and photons [IV-11]. However, the neutrons produced from 2.5 MeV protons on Li have a maximum energy just under 800 keV. A static arrangement of carefully chosen materials called a beam shaping assembly (BSA) is required to adjust the original neutrons into a clinically useful neutron source, via judicious moderation and reflection of neutrons, as well as absorption of thermal neutrons and photons exiting the surface.

The inner core of the BSA has layers of low-*Z*, non-hydrogenous scattering material alternating with layers of compounds with a combination of high-*Z* elements and isotopes with large scattering resonances above the epithermal region and limited resonances in the epithermal, e.g., magnesium, aluminium, and fluorine. This arrangement is very efficient at slowing down neutrons a limited amount so that the large resonances can be encountered more frequently than when those materials alone are used. The outer layers of the BSA use different materials to rapidly moderate and capture neutrons that have escaped the reflector region. The total mass of the BSA exceeds three tons, primarily due to the large amount of lead used as both a reflector and photon shield.

Two interrelated guidelines exist for the design of the BSA: in-air beam parameters from the original IAEA TECDOC-1223 [IV-11], shown in Table IV-1, and dosimetry parameters, such as advantage depth and treatment ratio. It is important to evaluate BSA designs with both sets of metrics; for TAE Life Sciences' BSA, higher priority was given to dosimetric parameters when making design decisions. For this reason, not all IAEA in-air beam parameters are achieved, but this is a conscious decision that improves other metrics, such as maximum achievable dose to tumour, maximum achievable treatment depth (for a single beam) and typical treatment time.

TABLE IV-1. IN-AIR NEUTRON BEAM PARAMETERS FROM REF. [IV-11]

In-air neutron beam parameter	Reference values
Epithermal neutron flux	$\gtrsim 10^9$ cm$^{-2}\cdot$s^{-1}
Fast neutron contamination	$< 2 \times 10^{-13}$ Gy$_{eq}\cdot$cm^2
Thermal neutron contamination	$< 2 \times 10^{-13}$ Gy$_{eq}\cdot$cm^2
Ratio of thermal to epithermal neutron flux	< 0.05
Ratio of total neutron current to total neutron flux	0.7 or more

The open beam port diameter of the BSA is 25 cm, and circular collimators ranging from 8 cm to 20 cm nominal diameter are available for patient treatments. The collimators are interlocked to prevent inadvertently treating with a collimator that differs from what was planned. In additional, a movable photon shield is used to block the BSA when treatments are not being performed to reduce personnel radiation exposure. Collimator exchange is designed to maintain the clinician's distance from the BSA and to keep the movable photon shield in place.

IV-2.1.7. Patient positioning system

Patient positioning for treatment utilizes a 6-degree of freedom robot that is ceiling-mounted to a linear rail that runs perpendicular to the treatment beam centre line. The robot has been designed to support a carbon fibre treatment couch, as well as a seated treatment option in the future. The dimensions of the treatment room allow the base of the couch to stay at the far end of the rail during treatments in order to minimize neutron activation.

Collision sensors are used along with control software to prevent couch collisions with the BSA, collimators or the movable photon shield. Floor-mounted optic controls are also used to maintain millimetre-level couch positioning reproducibility.

IV-2.2. Facility design and layout

IV-2.2.1. Facility rooms layout

In CNAO, a compact neutron source based on a tandem accelerator will be installed, part of TAE Life Sciences' Alphabeam system. The system is designed to be installed in hospital environments and in CNAO will be configured in a double-room BNCT suite to meet various clinical, research and capacity needs: the first irradiation room will be reserved for patient treatments, the second one will be devoted to research activities.

The BNCT facility design balances Alphabeam system requirements with the inherent space constraints and logistical challenges of a 'building expansion' type of project. Hereafter some details will be presented concerning the main features of the facility, hosted in a new building which is being realized next to and integrated with the existing one. The new building has been designed as developed on two underground levels, two above ground and a roof that will house the technological equipment.

The logistic of the BNCT areas comply with the radiation protection requirements: this means that the rooms housing the accelerator and the treatment rooms, as well, have therefore been placed in the most extreme area of the building, with a clear separation of the access flows and internal paths for CNAO staff, patients, technicians, and goods. In the following, a detailed description of the BNCT facility will be given, with pictures representing the most important areas and parts of the overall layout.

IV-2.2.2. Underground levels

The BNCT Facility will be located on the first basement floor of the new building (floor level −4.10 m), with a functional contiguity with the existing spaces devoted to hadron therapy. The Alphabeam beamline was newly configured for CNAO to place each treatment room on either side of the high energy beamline, to fit within the spatial constraints of the expansion.

The proton accelerator (tandem) will extend vertically along two levels: for this reason, a second basement level has been designed and will be built, with a smaller plan extension than that of the other levels, as it will house only the accelerator (IAT48 in Fig. IV-3) and some ancillary technical rooms (IAT49, Assembly and Storage, as represented in Fig. IV-3).

Key elements of the BNCT system foreseen at the first underground level include the patient treatment rooms (IAT01 and IAT04 in Fig. IV-4) the accelerator beamline (IAT02 in Fig. IV-4) and some ancillary spaces. In particular, the design of the patient treatment rooms has accounted for diverse criteria to support BNCT clinical processes and potential future expansions, including possible pre-treatment patient CT imaging, surface guided imaging and medical components, such as anaesthesia equipment. The treatment room sizes were also optimized to take into account shielding requirements, primarily based on patient positioning and neutron activation requirements for the ceiling-mounted robotic couch.

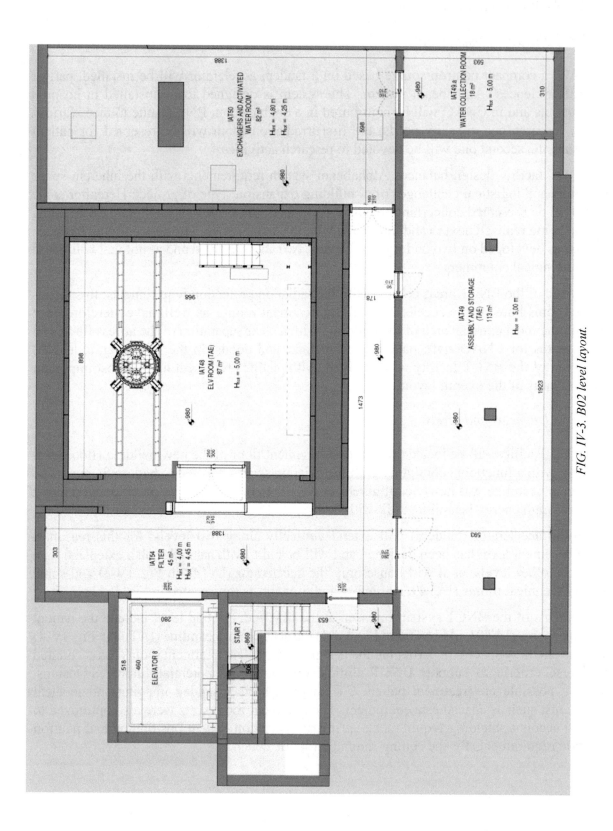

FIG. IV-3. B02 level layout.

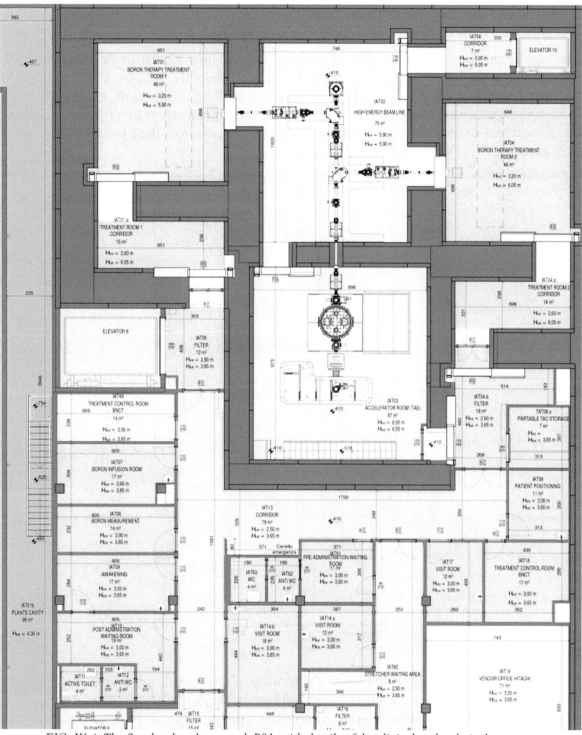

FIG. IV-4. The first level underground, B01, with details of the clinical and technical rooms.

Ground floor

The ground floor of the new building hosting the BNCT areas will be separated from the existing one. At this level the Service Control room and a Technical room reserved to BNCT will be located.

First floor

At the first level (+ 4.10 m), the new building could be divided into two parts: the southernmost one will house the technical areas for the operation of the high-technology devices, while the

277

other will host the new research laboratories (the so-called research area). The two parts will be separated by a terrace arranged as an extensive hanging garden.

The high-technology building will be accessible from the internal distribution corridors using Elevator 8 (on the bottom left corner in Fig. IV-5) and externally from the above roof garden. The target storage room (PAT07 in Fig. IV-5) will be built on this level: its capacity is chosen to be consistent with the expected patient workload, minimum target lifetime and radioactive component half-lives. This room will have a dedicated access from the Elevator 10 (on the top right corner in Fig. IV-5).

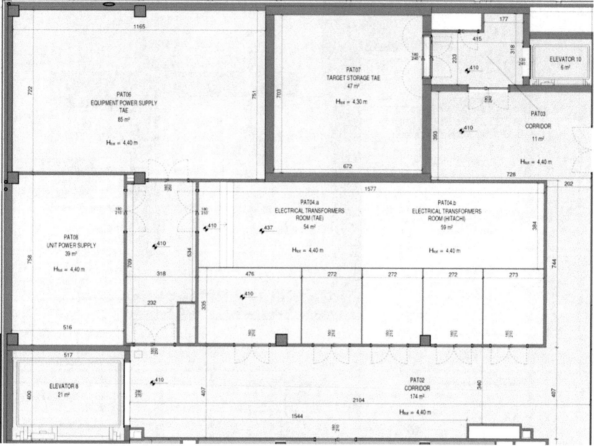

FIG. IV-5. The part of the first floor dedicated to BNCT (PAT07, PAT06, PAT04a).

Rooftop

The plant's equipment will be housed on the rooftop of the high-tech building (L02 level). A dedicated space for hosting the transformers and heat pumps will be built there. An expansion project foresees the installation of some photovoltaic panels: they will be housed partly on the rooftop and partly on a dedicated metal structure. The rooftop will be accessible both with the Elevator 8 (on the bottom left corner in Fig. IV-5) and through the free-to-air path connecting the roof of the new research laboratories with the existing building.

IV-2.2.3. Structural and layout requirements

The Alphabeam system facility will have the main following rooms:

- Accelerator room (at level B01);
- ELV room (at level B02);
- High energy beam line (HEBL) room (at level B01);
- Service Control room (at level L00);

- Electrical Cabinet room (at level L01);
- Treatment rooms 1, 2 (at level B01);
- Auxiliary room (at level L00);
- Technical room (at level L00);
- Assembly and Parts Storage room (at level B02);
- Target Storage room (at level L01);
- Treatment Control room 1, 2 (at level B01).

The ion source, the High Voltage (HV) isolation platform, the low energy beamline, the tandem accelerator, the support platform and some of the high energy beamline components will be located in a specialized part of the building called the Accelerator Room (IAT48 in Fig. IV-4).

The floor in the accelerator room will support the HV platform, which contains components collectively called the low energy beam line (LEBL); the tandem accelerator plus ELV vessel (i.e., the power supply attached to the tandem accelerator); and some high energy beam line (HEBL) components. The beamline height in the accelerator room will match the beamline height in the HEBL and treatment rooms, specified as 1.4 meters. The floor has been designed to support all of the dead loads, live loads and any possible out-of-normal conditions. The main equipment that will be supported in the accelerator room is the tandem accelerator with ELV vessel, whose weight will be spread equally over the centre cut-out supports.

There will be a staircase placed in the accelerator room which will allow personnel to traverse from the accelerator room to the ELV room. Electrical cables from the electrical cabinet room (PAT06) on the L01 level will primarily travel to the accelerator room.

The components of the high energy beam line in the HEBL room will be supported directly by the concrete floor for rigidity. Heavier components such as shielding, the Beam Shaping Assembly (BSA) and treatment room components will also be supported directly by the concrete floor for rigidity so that these large loads will not impart deflections or vibrations to the sensitive beam line. Cable and pipe management in the HEBL room will primarily use trenches.

Permanent fiducials will be used with a laser tracker or similar device for alignment of various components. These fiducials will primarily be installed in the accelerator room, HEBL room and the treatment rooms. They are placed on various walls at different elevations and floors.

In order to allow movement and installation of the biggest elements of the Alphabeam system, the freight elevator size and capacity have been properly evaluated and designed; the doorway and door opening in crucial points of the BNCT areas have been duly addressed.

Clinical areas and patients' paths

The BNCT facility layout has been designed in order to separate the paths reserved to patients accessing respectively, the BNCT, the proton therapy and hadron therapy areas; dedicated paths for CNAO personnel have been foreseen, as well.

Patients will enter the Center from the main entrance; they will proceed with the acceptance procedures in the atrium, will move to the first basement floor (B01) using the stairs or the correspondent elevator and will reach the existing waiting room. In case patients on stretchers undergo boron therapy, they will instead use a reserved path starting from outside the building, using a dedicated elevator.

After the treatment, the potentially radioactive patient will stop in the awakening room, if necessary, or in the adjacent warm waiting room before leaving the BNCT areas; exiting the facility, the boron treated patient will follow a different path with respect to the one used for entering, in order to avoid mixing flows of patients.

Since CNAO is also a research centre and the main goal is to also perform research activities in BNCT, dedicated paths for the researchers directed to the experimental treatment room on the first basement floor exist: they will access by means of a dedicated elevator and will reach the room with a dedicated route.

For maintenance operations and purposes, separated routes have been foreseen, as well; actually, technicians will use an elevator, which can also be accessed directly outside of the building at the ground floor, and can also be used for the movement of high-technology spare parts, connecting the different levels, up to the rooftop, where new plants will be located.

IV-2.2.4. Radiation protection

In order to properly address the radiation safety design of the BNCT suite, a wide variety of radiation safety related issues were studied: the prompt radiation production, the shielding effectiveness, the activation of air and the dosimetric effect of the exhaust release. The activation issues for the machine and the building structures (walls, doors, etc..) both short and long term, take into account materials handling, the personnel exposure evaluation and the main decommissioning issues. The radiological effects of emergency scenarios were also considered.

As a general remark, the main challenges to be faced in terms of radiation protection dealt with the fact that CNAO premises are not on a green field: actually, one synchrotron for hadron therapy beams is already installed and operational (seven days a week) and a new additional proton therapy synchrotron will run once the BNCT facility is ready and commissioned.

As a basis for the radiation safety evaluation, the physics undergoing BNCT treatment delivery has been investigated and the corresponding simulations have been performed, according to the CNAO facility layout, used as the basis for the Monte Carlo models used for the study.

First of all, the physics of the radiation production has been studied, starting from neutron generation: neutrons are produced by bombarding natural Li targets with high current proton beams, as described previously in Sections IV-2.1.4–IV-2.1.6. High fidelity, time-dependent, radiative transport models were developed in MCNP6 by TLS' physicists to understand the potential radiation hazards associated with each of these source terms on timescales that spanned a single treatment interval through the end-of-life for the facility. Simultaneously the CNAO Radiation Protection Department ran simulations using the FLUKA Monte Carlo code: the comparison between the two sets of results showed consistency, in order to back the crucial design choices, both in terms of logistic and materials.

After that, the facility rooms' shielding design has been based on the prompt radiation produced by the complete BSA system: neutrons, γ, and X rays, etc. One of the most delicate issues in the design, since a layout with opposing treatment rooms was chosen, is the cross-talk between the two rooms, i.e., one room has to be accessible while the other is in use, or when the machine is on and the beam is stopped on the beam dump or on the diagnostics detector.

Actually, the radiation protection design is based on the evaluation of the effects due to several prompt radiation sources:

- Neutrons or photons produced by the target and backscattered from the Beam Shaping Assembly, when the other treatment room is in use;
- Neutrons and photons produced by the proton beam striking the copper dump, the machine parts, and the photons created by the materials irradiated by the neutron beam.

Moreover, treatment beam neutrons and those scattered out of it impact the facility design, as well as radiation protection planning, due to undesirable activation of the Alphabeam and facility components. Due to the peculiar physical phenomena involved, a careful design and a careful selection of the materials was chosen, in order to match the needs arising from several points of view.

As an example, given the simulation results the finishes of the treatment rooms and high-technology rooms are going to be treated differently from the other areas, for radiation protection needs: the concrete to be employed in these walls and floors will be enriched with a small boron percentage. This choice has a great impact on the shielding thickness, since the boron in the first layer damps the neutron transmission on the concrete and the secondary gamma production, thus reducing the overall transmitted dose rate, and it lowers both the thermal neutron fluence rate in the room and in the walls, reducing both the air activation (mainly due to ^{41}Ar, by the (n,γ) reaction with ^{40}Ar) and the walls' activation.

IV-3. REGULATORY, RESEARCH AND CLINICAL ACTIVITIES

IV-3.1. Regulatory aspects

This section is not legally binding but intends to provide general overview of the regulatory aspects of a BNCT medical device facility as applicable to the Italian situation. The standard to be applied to medical devices is the MDR 2017/745 EU Regulation[1]. The regulation aims to increase product quality and safety standards with a risk–benefit clinical approach to the device. This makes the EU the guarantor of public health and of the smooth functioning of the internal market in the domain of medical devices. Compliance with MDR[1] requirements and any relevant harmonised standards, for this type of medical device, will be certified by a notified body, responsible for the conformity assessment procedures.

The MDR clearly focuses onto the clinical investigation and evaluation, meaning pre-market clinical trials in the conformity assessment, and post-market clinical follow up on CE marked medical devices whose certificate depends on the intended use.

A 'manufacturer' of medical devices is "a natural or legal person natural or legal person who manufactures or fully refurbishes a device or has a device designed, manufactured or fully refurbished, and markets that device under its name or trademark"[1].

As concerns the BNCT device, the manufacturer is TLS, who will issue and keep up to date the technical documentation. According to the MDR[1], as TLS is a US entity, an authorised representative has to be identified in an EU Member State in order to fulfil all manufacturer duties. The technical documentation will bring the device in conformity with the requirements of this Regulation.

---

[1] https://eur-lex.europa.eu/legal-content/EN/TXT/HTML/?uri=CELEX:32017R0745&from=IT

Evidence of compliance with the general safety and performance requirements, laid down in this Regulation, has to be based on clinical data. Compliance with harmonised standards means for manufacturers to demonstrate compliance with the general safety and performance requirements and other legal requirements (e.g., EN 60601-1 Medical Electrical equipment, EN 60601-1-2 Electromagnetic compatibility EMC; EN 62304 Medical device software — Software life cycle processes; IEC radiation protection, EN 10993-1 biocompatibility, etc.). This includes mechanical and/or electrical tests, reliability tests, software verification and validation, performance tests, evaluation of biocompatibility and biological safety, if applicable.

The manufacturer has to demonstrate the conformity according to IEC's standards with preclinical tests. The manufacturer has to contact an accredited laboratory (e.g., in Italy IMQ laboratory[2]), to perform the tests for electrical safety, electromagnetic compatibility, etc. Clinical evaluation has to be based on data from clinical investigations carried out under the responsibility of CNAO Foundation.

Another important step of the authorisation process is the approval of the drug:

- Non-commercial drug, first in humans. In this case, the drug cannot be administered to humans, so that an experimentation of the drug, which, according to Regulation 536/2014, needs to be carried out following several distinct steps. The experimentation process proceeds as follows: laboratory tests, animal tests and, to conclude, human tests;
- Commercial drug. In this case, the drug can be administered to humans since it has already obtained authorization.

In a first phase, it seems that authorising bodies consider the use of the drug and medical device subject to a combined authorisation for a specific pathology. Eventually, the extension of the authorisation could be matter of discussion opening to independent pathways for new drugs and new pathologies applied to the same medical device.

The Ministry of Health evaluates clinical trials with medical devices without the CE marking (Article 62 of Regulation (EU) MDR 2017/745). An application with all the documentation required by the Regulation (Annex XV, chapter II of MDR[1]) has to be submitted to the Ministry of Health (in accordance with Article 70 of MDR[1]), before starting the clinical investigation, to demonstrate conformity.

The clinical trial can be started after a check of the Ministry of Health about the accuracy of the application. Then the Ministry will continue scientific evaluation activities, in order to take possible corrective measures according to Article 76 of MDR[1]. Usually, an application for a clinical investigation is submitted by the Sponsor (CNAO Foundation in this case). The validation of the application can only be carried out with a positive opinion given by an ethics committee nationally recognized (Comitato Etico Pavia of reference for CNAO). Once the clinical results have been collected at the end of the clinical investigation, the manufacturer will be able to complete the technical dossier and to request the CE marking.

Matching the authorization of the boron compound and its use in the BNCT device with the authorization procedure of the medical device itself will be the result of a strong collaboration between manufacturer, clinical trial sponsor and regulatory agencies.

---

[2] https://www.imq.it/en

IV-3.2. Computational dosimetry and treatment planning

Treatment planning for Neutron Capture Therapy (NCT) has a different approach with respect to conventional planning with photons or charged particles. In fact, it exclusively relies on Monte Carlo (MC) simulations for dose calculation, because of the complex nature of neutron interactions with matter.

This topic includes many computational aspects, mainly related to the neutron beam simulation in proximity of the neutron beam aperture. In the particle energy range of interest to NCT, there are many MC transport toolkits having physics models sufficiently accurate. Nonetheless, defining a particle source which can be validated both in space and energy is not trivial. This is due to the particle mixed field characteristics of NCT irradiation. A faithful representation of the initial radiation source is the key for an accurate evaluation of the treatment.

Methods of defining a neutron beam source for NCT treatment planning will be investigated to come out with phase space 'surface source' files that define the characteristics of the radiation source at/near the aperture of the neutron beam to avoid repeating the computationally expensive calculations of neutron transport down the beam line, for future studies. Phase space files will record position, direction, energy, particle type, statistical weight, and history number of simulated particles, which originate from a detailed simulation of the neutron beamline.

For research and intercomparison activities, the radiation transport code MCNP, developed at Los Alamos National Laboratory [IV-12], is planned to be extensively used as a dose calculation engine, since being well validated, relatively easy to use and offering flexibility in source definition and robust variance reduction techniques. Intercomparisons among other radiation transport code (e.g., PHITS [IV-13], Geant4 [IV-14] or FLUKA [IV-15]) are also foreseen to possibly define unique physical settings in phantom geometry first, and patient-like anatomy afterwards. The auxiliary support of Geant4 DNA-package [IV-16] will be also considered to investigate biological damages induced at the cellular and sub-cellular scale, with the aim of developing predictive radiobiological models. The Monte Carlo code to be used in the Alphabeam commercial treatment planning system (TPS) is currently under evaluation.

The purpose of developing this simulation framework is also to define a protocol through which the computational results can be validated with experimental measurements. This can be achieved by comparing between calculation and measurements in two different phantoms, e.g., large rectangular or cylindrical water phantoms and a smaller anthropomorphic phantom, to provide a test bench for the source model and TPS. Rectangular water phantoms allow freedom of position for dosimeters and are frequently equipped with motorized drives for remote positioning of detectors to permit measurement at multiple positions. This ability is especially advantageous for measuring along multiple axes. Anthropomorphic phantoms provide more realistic and clinically relevant conditions to determine the accuracy and calibrate the TPS.

Finally, auxiliary tools might be developed to speed-up monitor unit checks for NCT treatment plans, handling comparisons with in-phantom dose measurements. For example, as demonstrated by JAEA [IV-17], a simple water phantom can be used to conduct an extensive set of measurements to confirm treatment plans; e.g., in-phantom measurements of the neutron flux and dose components along the central axis of the beam, beam profiles as a function of depth, output measurements, and the effect of the sizes of the beam port and phantom on the dose components.

IV-3.3. Experimental and environmental dosimetry

The BNCT irradiation field is complex, since it is constituted by different components, namely thermal and epithermal neutrons, fast neutrons (although this component ought to be minimized, a tail is usually present in the neutron spectrum), prompt γ rays from neutron absorption and inelastic reactions (also from the Li target), and γ rays from the activation of the BSA structural materials.

Both the proton beam current striking the target and the neutron radiation field will be monitored independently and in real time.

The characterization of the irradiation field will be performed with neutron activation techniques (bare and Cd-covered In and Au foils) and a graphite ionization chamber filled with CO_2 for the γ ray component. The fast neutron component will be characterized through neutron spectrometry with superheated emulsions [IV-18], the ACSPECT silicon telescope [IV-19], the DIAMON spectrometer [IV-20] and the NCT-WES spectrometer [IV-21].

In-phantom dosimetry will be performed mainly with thermolumiscent detectors (TLDs) [IV-22] and gel dosimeters [IV-23].

The complex irradiation field will also be characterized through microdosimetric techniques. Microdosimetry measures the stochastic distribution of the energy imparted in a microscopic volume by single events. The measurements of the distributions obtained with and without ^{10}B atoms in the target material allows to characterize the mixed BNCT radiation field, and to determine the fractional contribution to the total absorbed dose due to photon and neutron interactions. Pairwise measurements with and without boron allow to quantify the additional dose fraction due to the boron neutron capture reaction. The relative biological effectiveness (RBE) of the BNCT radiation field can thus be estimated [IV-24–IV-33]. The microdosimetric characterization can be performed at the standard 1 μm site-size, or at smaller diameters down to about 30 nm using advanced tissue equivalent proportional counters [IV-34]. The measurement at site sizes in the nanometre range is interesting because it has been demonstrated that for decreasing target diameters the ratio between the dose mean lineal energy value for the specific radiation quality and that for the reference radiation field, $\bar{y}_D / \bar{y}_{D\text{-}ref}$ becomes closer to the corresponding RBE [IV-35–IV-36]. The microdosimetric characterization of the BNCT radiation field at CNAO will be performed both at the micrometre and at the nanometre level using advanced tissue equivalent proportional counters.

The radiation environmental monitoring for a BNCT is mainly based on the measurement of the photon and neutron components inside and outside the treatment, HEBL, and accelerator rooms. The radiation fields involved inside the rooms are particularly challenging to measure because they are both mixed (photons and neutrons) and very intense (when compared to the usual fields for which standard monitors are conceived). The neutron component is peaked in the epithermal range, for which the typical neutron instruments are usually not optimized. Outside the shielding, the radiation field is completely dominated by photons, with a significant high energy component arising from the nuclear capture reactions. The radiation monitoring systems have to be chosen or designed in order to be able to measure very intense neutron fields, to have a good neutron/photon discrimination capability and an adequate spectral response.

IV-3.4. Boron measurement and clinical dosimetry

Due to its unique cell-level selectivity, BNCT crucially relies on the estimation of ^{10}B-concentration in several compartments (tumour tissue, blood, and other healthy tissues). The ideal method for ^{10}B quantification would be able to perform the measurement at the macroscopic level as well as to be sensitive to the microscopic distribution (cellular and subcellular level); also, it would be non-invasive and capable of producing a result in vivo and in real time [IV-37–IV-38]. Presently, we lack a single technique compliant with all the aforementioned requirements.

Thanks to the long experience of clinical BNCT, the scientific community can count on well benchmarked techniques capable of measuring ^{10}B ppm in different kind of samples and situations [IV-39]. Presently, measurement of ^{10}B-concentrations in patients during a BNCT treatment is performed using atomic spectrometry, choosing among:

- Inductively Coupled Plasma Atomic Emission Spectrometry (ICP-AES) or Inductively Coupled Plasma Optical Emission Spectrometry (ICP-OES);
- Inductively Coupled Plasma Mass Spectrometry (ICP-MS).

The former is by far the most readily available and thus presently the first choice in BNCT [IV-5, IV-40–IV-43].

Atomic spectrometry gives the mean ^{10}B amount in a macroscopic sample (volumes of the order of magnitude of few ml or less) which generally can be assumed as a mixed sample in term of cell subpopulations. ICP-AES/ICP-OES are typically performed on patient's blood samples taken before, during and after neutron irradiation. Thanks to the quick response, the ^{10}B-concentration is used to monitor and correct the beam time prescribed to each patient to better conform the delivered dose to the tumour as estimated by the TPS engine. These evaluations crucially rely also on a pharmacokinetics model of the boronated drug as well as on fixed ratios among the ^{10}B-concentration in blood compared to tumour and healthy tissues.

A very close alternative to the described emission spectrometry is prompt gamma neutron activation analysis, also known as prompt gamma ray spectrometry or prompt gamma ray analysis, based on the detection of the 478 keV γ ray emitted after 94% of ^{10}B neutron capture reactions, due to the initial excited state of the ^{7}Li recoil nucleus [IV-44–IV-45]. The main difference from ICP-AES/OES is the need, in parallel to the treatment neutron beam, of a dedicated thermal neutron beam characterized by a very low background to push downwards the sensitivity and thus to limit as much as possible to measuring time.

The knowledge of ^{10}B-concentrations gained by spectrometry techniques performed during neutron irradiation is typically coupled with an a priori study of boronated drug uptake by each patient performed by PET and a ^{18}F-fluorinate analogue of the boronated drug (generally, BPA) [IV-41, IV-46]. PET imaging is currently implemented in BNCT clinical protocol to evaluate patient eligibility. It is typically performed hours or even days before the actual neutron irradiation and gives important information about the pharmacokinetics of the boronated drug, in particular it is used to quantify the tumour-to-normal tissue and the tumour-to-blood ratios of the ^{10}B-concentrations which will be coupled with the ICP-AES(OES) measurements to adjust the treatment.

For monitoring the bio-distribution of boron carriers in vivo, protocols based on magnetic resonance imaging (MRI) and spectroscopy (MRS) have been proposed, with the aim of exploiting the non-ionizing nature of MR techniques to improve BNCT treatments. In-vivo preclinical studies focused on the detection of ^{19}F-labelled BPA demonstrated that: ^{1}H-MRI ^{19}F-MRI can provide a ^{19}F-FBPA distribution map [IV-47]; the best irradiation time can be identified by the study of the correlation between ^{19}F-MRI and blood ^{19}F-MRS [IV-46]; BNCT efficacy can be improved by L-DOPA preloading [IV-48]. This approach requires the availability of unconventional MR scanner hardware. With the aim of exploiting clinically widely available MR acquisition settings, a technique for BPA quantification based on the detection of the aromatic proton signals of BPA with MRS has been proposed. It was demonstrated to be feasible in a phantom study [IV-49], whereas the implementation in vivo on humans needs to be replicated [IV-50].

At the preclinical level, to support boronated drug development as well as in vitro and in vivo studies, there is a plethora of dedicated techniques capable of pushing our knowledge of ^{10}B-concentration and microscopic spatial distribution towards the single cell uptake. Just to mention a few: neutron autoradiography, charged particle spectroscopy, secondary ion mass spectrometry, electron energy loss spectroscopy, etc. Typically, these techniques are quite time consuming, work off-line, and can eventually require highly sophisticated and expensive instrumentations, thus it is hard to suggest these methods as the first line approach to quantify ^{10}B during patient treatment.

In terms of clinical dosimetry, it has to be stressed that ^{10}B-concentration clearly controls the therapeutic dose distribution (i.e., the boron dose component) but does not act alone. Boron dose comes from the convolution of ^{10}B-concentration at a certain point and the neutron flux, in particular the thermal fraction, at the same position. Presently neutron flux information is mainly determined by activation methods, validated during or immediately after patient irradiation through comparison with the TPS calculation [IV-51]. Dedicated methods are available for the other dose components, such as: the γ dose and the neutron beam component are generally estimated by the twin ionization chamber techniques; the nitrogen dose component estimated by combining activation measurements and TPS calculations. It has to be stressed that all these techniques are applied straightforwardly in phantoms, while they are affected by huge indirect and off-line estimations in case of a patient.

A strategy to mitigate a lot of the limits listed for the described techniques is the development and implementation of a real time BNCT dose monitoring by single photons detection (PG-SPECT) [IV-52]. As already mentioned, a 478 keV γ ray is available to monitor the therapeutic dose due to ^{10}B in tissues, while the 2.2 MeV γ ray produced by the radiative neutron capture reaction on ^{1}H is widely accepted as a signal to map the neutron flux in the irradiated region. In this scenario, an optimized and dedicated single photon detecting system such as a modified SPECT tomograph [IV-53] or a dedicated Compton camera [IV-54] would surround the patient during the irradiation to produce real-time 2D and 3D maps of ^{10}B dose and thermal neutron flux. Although such a system has been suggested for several years, many scientific teams are still working on its development and its practical implementation is still missing in a BNCT clinical facility.

In conclusion, for the practical implementation of BNCT at CNAO, we suggest proceeding by pursuing the three following strategies:

(a) ^{18}F-FBPA PET before BNCT treatment coupled with ICP-AES measurements on the patient's blood samples to know ^{10}B-concentration in tumour and tissues to monitor and optimise the delivered doses;
(b) Further investigating the possible implementation of MRI and MRS techniques for monitoring the bio-distribution of borate compounds in vivo;

(c) Investing in the development of a prototype of dose monitoring system based on PG-SPECT for single photons detection at 478 keV and 2.2 MeV to quantify and image in vivo and in real time the BNCT therapeutic dose as well as the thermal neutron flux.

IV-3.5. Radiobiology

The radiobiological studies foreseen at the CNAO BNCT facility will be focused either on beam/apparatus commissioning to assess the safety and the characteristics of the system, or on preclinical research aiming to enhance the therapeutic gain thanks to the knowledge of BNCT, in particular of AB-BNCT, radiobiology. In fact, in addition to the evident need to deepen the knowledge of the effects of BNCT on tumoral/normal cells/tissues, the relatively short history of AB-BNCT clearly reflects on scarce radiobiological studies, which are fundamental to highlight potential differences and similarities between reactor-based and accelerator-based BNCT effects, as well as to exploit them for a better treatment. For example, the few basic studies performed in vitro so far report RBE values of epithermal neutron beams that are in the range 2.2–2.6 with a Be target [IV-55], and 1.7–1.9 with a Li target [IV-56]. Therefore, it will not be surprising if further differences in cellular and molecular responses will come to light. In the context of comparing different types of facilities, the possibility offered to the Pavia community of being able to easily conduct experiments with beams produced both by a reactor (LENA) and by an accelerator (CNAO) will be unique.

All these experiments will be designed by taking into account both the need of reproducing standard experimental conditions for beam quality assurance, and the future clinical trials that will be conducted at the centre. CNAO's approach for the design of the pre-clinical experimental studies for beam commissioning prior to clinical trials will be, in the first instance, directed towards the assessment of the safety of the CNAO treatment in normal cells/tissues and in animal models through radiobiological standard tests, in order to be able to compare the results with those collected by other BNCT facilities.

Thus, cytotoxicity, clonogenicity and cell viability will be evaluated under different experimental conditions, e.g., with boron carrier only, with neutrons with/without boron carrier, with multiple concentrations of boron carrier, in CHO and V79 hamster cells (as quality assurance references) and in human normal cell lines (e.g., fibroblasts and endothelial cells). In parallel, genotoxicity under the same experimental conditions will be assessed by means of γ-H2AX staining and micronuclei formation and/or SCEs (sister chromatid exchanges). Concerning the in vivo experiments, they will be performed in order to evaluate local and systemic toxicity, in/out-of-field, at different beam depths, after whole body irradiation or local irradiation in the absence and presence of boron compounds. In particular, intestinal crypt regeneration, skin lesions, mucosal lesions and blood toxicities will be reported and compared with those of γ or X ray irradiation, to calculate RBE/CBE values. In this regard, it will be taken into account that the mixed radiation field that characterizes BNCT, where high- and low-LET particles coexist, makes the evaluation of RBE not trivial. Although the conventional approach is based on fixed RBE and CBE values, other possibilities will also be explored, such as the 'photon isoeffective dose approach' [IV-57] and/or the 'Modified Stochastic Microdosimetric Kinetic' model [IV-58].

An important issue that will be taken into consideration is the differences in body size and in body thickness between humans and mice: the first problem will be overcome through the use of a dedicated acryl support that will allow to position the animals in a way to cover the entire irradiation field. The second concern will be circumvented by exploiting the 'mouse three layer model' designed and described by Masutani [IV-59]. The observation periods will be established to investigate both short and long term effects.

The use of small rodents as in vivo animal models will be possible thanks to the availability of the animal facility of the University of Pavia located near to the LENA Neutron Reactor and to the CNAO Center. The facility, other than to conventional animal housing and care, can provide housing of BNCT treated rodents thanks to the availability of a radiobiology laboratory. Only animals included in experimental protocols approved by the animal welfare body of the University of Pavia and authorized by the Ministry of Health will be accepted at the Center.

A second step of the pre-clinical experimental studies will be addressed to the study of the efficacy of the treatment. Therefore, cancer cell killing will be measured by clonogenic cell survival assay in several cancer cell lines selected according to the planned clinical trials, with and without boron carriers. In parallel, the effects on tumour growth will be evaluated by means of tumour growth delay assay in xenograft models of mice with target cancer types.

IV-3.6. Clinical trial procedures

In the era of advanced technology, oncologic treatments are aimed at improving the therapeutic ratio, local tumour control rates and patient survival while reducing treatment related toxicities.

Boron neutron capture therapy may play a major role in this scenario by exploiting the ability to combine drug delivery and different uptake between normal tissue versus cancer cells and irradiation by an epithermal neutron beam, in order to obtain a large dose gradient between organs at risk and tumour [IV-60]. It results in a selective killing of cancer cells and could represent a promising therapy to overcome the eternal dilemma between risks and benefits of a specific treatment [IV-61]. Due to its highly focused radiation, BNCT could be exploited in many contexts, particularly in tumour volumes near critical structures or in case of recurrent disease independently of their resectability and previous radiation.

According to recent literature, BNCT has shown promising results in malignant gliomas, glioblastoma multiforme and recurrent high-grade meningioma [IV-62]. The majority of these trials were with reactor-based epithermal neutron beams, although nowadays especially in Japan, BNCT activities are shifting to accelerators in a hospital setting [IV-60]. Due to the nature of the single accelerator BNCT facility in Italy, CNAO will design non-randomized, non-comparative, prospective, open-label, phase I/II studies to determine safety and activity of BNCT in the treatment of inoperable and/or recurrent anaplastic astrocytomas or glioblastomas following conventional radiation therapy. In this setting, such clinical trials aim to establish BNCT as a conventional clinical treatment both for primary and salvage procedures.

In line with the Japanese experience, BNCT plays a role in treatment of recurrent head and neck cancers, especially in the context of re-irradiation not suitable for carbon ions [IV-61, IV-63–IV-64]. At CNAO, due to the decades of experience in treating recurrent tumours, we aim to perform clinical trials to test the safety and effectiveness of BNCT with regard to its therapeutic ratio as compared with conventional patient selection criteria [IV-65].

We will explore by dose/volume escalation studies safety and efficacy of BNCT in selected oligometastatic cancers [IV-62]. An additional topic of interest will be to investigate a combination of immunotherapy and BNCT [IV-66]. All trials will have to be designed with either ^{10}B-BSH (sodium borocaptate) or ^{10}B-BPA ((2S)-2-amino-3-(4-(^{10}B)dihydroxy(^{10}B)phenyl)propanoic acid) or eventually more selective boronated compounds depending on the request of the relevant national regulatory authority.

At CNAO, trials will be conducted by clinicians with experience in performing phase I/II clinical trials in radiation oncology and with novel drugs, in a close collaboration with industry and regulatory authorities. Multidisciplinary collaboration within a multi-professional staff including radiation oncologists, medical oncologists, radiologists, nuclear medicine physicians, pathologists, surgeons, nurses, biologists and medical physicists will be of paramount importance in designing and carrying on the clinical trials. International collaboration with other BNCT facilities will be encouraged, especially in view of multicentre studies with same protocol and prescription dose.

IV-4. DEVELOPMENT OF NEW BORATE COMPOUNDS

The most important requirements for a BNCT delivery agent are:

- Low intrinsic toxicity;
- High tumour uptake, typically 20–50 μg ^{10}B, with an ideal uptake of 30 μg ^{10}B;
- Low uptake in normal tissue: ideally boron concentrations exceeding 3:1 for tumour-to-normal tissue and tumour-to-blood;
- Relatively quick clearance from normal tissues and blood and persistence in tumour for several hours to allow for neutron irradiation.

In addition to these characteristics, it would also be desirable to add the theranostic delivery agents, i.e., those agents equipped with a specific reporter capable of carrying out the dosimetry of ^{10}B either in situ or in a very short time namely, within the time of the clearance.

Finding the Goldilocks region of perfect delivery agent is a dynamic process and has been refined thanks to the increase in scientific knowledge, progress of molecular biology, nuclear medicine, and organic chemistry. Historically, three different generations of BNCT delivery agents have been identified; the first generation of boron compounds was developed in the early 1950s. Compounds like boric acid, borax, and pentaborates were synthesised, but their concentrations in tumours were observed to be very low compared to that in the brain, and also their accumulation was transient. Ten years later, the second generation of boron compounds appeared. Two efficient and prominent boron compounds, i.e., BPA and BSH had emerged in the 1960s. They had significantly less toxicity and persisted longer in animal tumours than in related molecules. Besides, boron concentration ratios were 4:1 in the tumour/brain as well as tumour/blood. Hence, their satisfactory results in vitro and in vivo allowed authorization for clinical use. ^{10}B-BPA has become by far the most widely used boron delivery agent in clinical and pre-clinical applications of BNCT. Very recently, ^{10}B-BPA was synthetized under GMP protocols and approved in Japan for clinical use under the name of Borofalan (previously known as SPM 011).

The advent of liposome chemistry, nanomedicine, nucleic acid chemistry and the development of monoclonal antibodies, has allowed the development of the third generation of boron compounds. There is a constant need to develop more efficient and selective boron delivery agents. The success of the binary radiotherapeutic modality depends on the selective uptake of a therapeutic dose of the boron-containing agent by cancer cells. Hence, one of the most challenging tasks is the synthesis and development of smart, boron delivery agents for clinical use. Boron clusters that possess high hydrolytic stability are newly emerging tools for BNCT; however, their toxicity and lack of tumour selectivity limit their direct benefits in BNCT. Tumour-targeting agents, such as polyamines, unnatural amino acids, peptides, proteins, nucleosides, sugars, porphyrins, antibodies, liposomes and nanoparticles are being conjugated with either BPA or BSH molecules to develop a better delivery system. Boron compounds have also been encapsulated into the aqueous core of liposomes. Liposomes made of specific lipid components maintain the ability to cross the tumour membrane and reside intracellularly.

BNCT is now at the crossroads: where do we go from here? The logical answer to this question, would be to develop a fourth generation of boron carriers, but that even this falls short. To make BNCT clinically accessible, it is necessary to have ^{10}B-BPA produced according to GMP standards, also with a view to reshoring EU strategy. A fourth generation boron delivery agent could be an oncotropic molecule modified with ^{10}B.

A glimpse on how the fourth generation boron deliveries will look can be traced back to the new synthetic tactics:

- Novel synthetic bioconjugation at cysteine (Cys) residues in peptides and proteins has emerged as a powerful tool in chemistry. Cys-borylation could proceed at room temperature and tolerates a variety of functional groups present in complex polypeptides;
- A simple synthetic approach for various novel boronated protected tryptophans using a regio- and chemoselective – C–H-activation or electrophilic substitution – of 4- and 5-boronated indoles could be in principle employed. Boron-containing heterocycles are important in a variety of applications from drug discovery to materials science. BN/CC isosterism may greatly expand the possibilities. In particular, the application of BN/CC isosterism to arenes provides a vast available chemical space in the aromatic hydrocarbons;
- Click chemistry has applications in nearly all areas of modern chemistry and has had a significant impact on the synthesis and development of radiopharmaceuticals;
- Novel drug delivery systems targeting the comparatively large gaps between capillaries that characterize tumour tissue are also under active development.

The preparation of new borate compounds will be a matter of interdisciplinary collaboration within the CNAO network and a crucial role will be played by TLS that has a large, active group developing next-generation BNCT pharmaceuticals.

IV-5. CONCLUSION

In the bylaws of CNAO Foundation, research is stated as a fundamental pillar, and the decision to create a clinical centre for BNCT is definitely in line with this objective. The construction of the new facility is presently ongoing, and it is expected to be operational in 2024.

The accelerator based BNCT system will be supplied by TLS in the frame of an agreement signed in 2020 between CNAO and TLS. The purpose of the agreement is to install, run, CE certify the TLS technology and launch a clinical trial phase to enrol patients eligible for BNCT applications.

A collaboration between CNAO, University of Pavia, INFN, and Polytechnic of Milan has been set up to pursue the goal of treating patients with this innovative modality. The collaboration is open to international groups to integrate experience and multidisciplinary competence to reach the goal.

Seven areas have been identified and, as described in this article, the work interests: regulatory aspects, radiobiology, computational dosimetry and treatment planning, experimental and environmental dosimetry, boron measurement and clinical dosimetry, clinical trial procedure for bnct, development of new borated compounds.

REFERENCES TO ANNEX IV

[IV-1]   BARTH, R.F., et al., Current status of boron neutron capture therapy of high grade gliomas and recurrent head and neck cancer, Radiat. Oncol. **7** 146 (2012) 1–21.

[IV-2]   KOIVUNORO, H., et al., Boron neutron capture therapy for locally recurrent head and neck squamous cell carcinoma: An analysis of dose response and survival, Radiother. Oncol. **137** (2019) 153–158.

[IV-3]   MIYATAKE, S.I., et al., Boron neutron capture therapy for malignant brain tumours, Neurol. Med. Chir. **56** 7 (2016) 361–371.

[IV-4]   AIHARA, T., et al., BNCT for advanced or recurrent head and neck cancer, Appl. Radiat. Isot. **88** (2014) 12–15.

[IV-5]   WANG, L.W., et al., Clinical trials for treating recurrent head and neck cancer with boron neutron capture therapy using the Tsing–Hua Open Pool Reactor, Cancer Commun. (2018) 37–38.

[IV-6]   ZONTA, A., et al., Clinical lessons from the first applications of BNCT on unresectable liver metastases, J. Phys. Conf. Ser. **41** (2006) 484–495.

[IV-7]   ZONTA, A., et al., Extra corporeal liver BNCT for the treatment of diffuse metastases: what was learned and what is still to be learned, Appl. Radiat. Isot. **67** (2009) 67–70.

[IV-8]   SUZUKI, M., Boron neutron capture therapy (BNCT): a unique role in radiotherapy with a view to entering the accelerator–based BNCT era, Int. J. Clin. Oncol. **25** (2020) 43–50.

[IV-9]   BAYNOV, B., et al., Accelerator based neutron source for the neutron–capture and fast neutron therapy at hospital, Nucl. Instrum. Methods Ser. A **413** (1998) 397–426.

[IV-10]   LEE, C., ZHOU, X–L., Thick target neutron yields for the ^{7}Li(p,n)^{7}Be reaction near threshold, Nucl. Instrum. Methods Ser. B **152** (1999) 1–11.

[IV-11]   INTERNATIONAL ATOMIC ENERGY AGENCY, Current Status of Neutron Capture Therapy, TECDOC 1223, IAEA, Vienna (2001).

[IV-12]   WERNER, C.J., et al., "MCNP6.2 Release Notes", Los Alamos National Laboratory, report LA-UR-18-20808 (2018).

[IV-13]   SATO, T., et al., Features of Particle and Heavy Ion Transport Code System PHITS, Version 3.02, J. Nucl. Sci. Technol. **55** (2018) 684–690.

[IV-14]   AGOSTINELLI, S., et al., GEANT4—a simulation toolkit, Nucl. Instrum. Methods Ser. A **506** 3 (2003) 250–303.

[IV-15]   FERRARI, A., RANFT, J., SALA, P.R., FASSÒ, A., FLUKA: A multi–particle transport code (Program version 2005) (No. CERN–2005–10). CERN (2015).

[IV-16]   INCERTI, S., et al., The GEANT4–DNA project. Int. J. Model. Simul. Sci. Comput. **1** 02 (2010) 157–178.

[IV-17]   JAPAN ATOMIC ENERGY AGENCY, https://www.jaea.go.jp/english/.

[IV-18]   D'ERRICO, F., Radiation dosimetry and spectrometry with superheated emulsions, Nucl. Instrum. Methods Ser. B **184** (2001) 229–254.

[IV-19]   AGOSTEO, S., FAZZI, A., INTROINI, M.V., LORENZOLI, M., POLA, A., A telescope detection system for direct and high resolution spectrometry of intense neutron fields, Radiat. Meas. **85** (2016) 1–17.

[IV-20]   POLA, A., RASTELLI, D., TRECCANI, M., PASQUATO, S., BORTOT, D., DIAMON: a portable, real–time and direction–aware neutron spectrometer for field characterization and dosimetry, Nucl. Instrum. Methods Ser. A **969** (2020) 164078.

[IV-21]   BEDOGNI, R., et al., NCT–WES: A new single moderator directional neutron spectrometer for neutron capture therapy. Experimental validation, EPL **134** (2021) 42001.

[IV-22]   GAMBARINI, G., et al., Determination of gamma dose and thermal neutron fluence in BNCT beams from the TLD–700 glow curve shape, Radiat. Meas. **45** (2010) 640–642.

[IV-23]   GAMBARINI, G., et al., Fricke gel dosimetry in epithermal or thermal neutron beams of a research reactor, Radiat. Phys. Chem. **116** (2015) 21–27.

[IV-24]   ZAMENHOF, R.G., Microdosimetry for boron neutron capture therapy: A review, J. Neuro-Oncol. **33** (1997) 81.

[IV-25]   WUU, C.S., et al., Microdosimetry for boron neutron capture therapy, Radiat. Res. **130** (1992) 355.

[IV-26]   KOTA, C., et al., The use of low pressure tissue equivalent proportional counters for the dosimetry of neutron beams used in BNCT and BNCEFNT, Med. Phys. **27** (2000) 535.

[IV-27]   BURMEISTER, J., et al., Microdosimetric intercomparison of BNCT beams at BNL and MIT, Med. Phys. **30** (2003) 2131.

[IV-28]   HSU, F.Y., et al., Microdosimetry study of THOR BNCT beam using tissue equivalent proportional counter, Appl. Radiat. Isot. **67** (2009) S175.

[IV-29]   MORO, D., et al., BNCT dosimetry performed with a mini twin tissue–equivalent proportional counter (TEPC), Appl. Radiat. Isot. **67** (2009) S171.

[IV-30]   COLAUTTI, P., et al., Microdosimetric measurements in the thermal neutron irradiation facility of LENA reactor, Appl. Radiat. Isot. **88** (2014) 147.

[IV-31]   SATO, T., et al., Microdosimetric modeling of biological effectiveness for boron neutron capture therapy considering intra– and intercellular heterogeneity in ^{10}B-distribution, Sci. Rep. **8** (2018) 988.

[IV-32] HU, N., et al., Evaluation of PHITS for microdosimetry in BNCT to support radiobiological research, Appl. Radiat. Isot. **161** (2020) 109148.

[IV-33] SELVA, A., et al., Microdosimetry of an accelerator based thermal neutron field for boron neutron capture therapy, Appl. Radiat. Isot. **182** (2022) 110144.

[IV-34] BORTOT, D., et al., A novel TEPC for microdosimetry at nanometric level: response against different neutron fields, Radiat. Prot. Dosim. **180** 1–4 (2018) 172–176.

[IV-35] LINDBORG, L., GRINDBORG, J.E., Nanodosimetric results and radiotherapy beams: a clinical application?, Radiat. Prot. Dosim. **70** 1–4 (1997) 541–546.

[IV-36] MAZZUCCONI, D., et al., Nano–microdosimetric investigation at the therapeutic proton irradiation line of CATANA, Radiat. Meas. **123** (2019) 26–33.

[IV-37] WITTING, A., et al., Boron analysis and boron imaging in biological materials for BNCT, Crit. Rev. Oncol. Hematol. **68** (2008) 66–90.

[IV-38] PROBST, T.U., Methods for boron analysis in boron neutron capture therapy (BNCT). A review, Fresenius', J. Anal. Chem. **364** (1999) 391–403.

[IV-39] WITTIG, A., MOSS, R., NAKAGAWA, Y., (Eds), Neutron Capture Therapy – Principles and Applications, Springer (2012).

[IV-40] SHIBATA, Y., et al., Prediction of boron concentration in blood form patients on boron neutron capture therapy, Anticancer Res. **23** (2003) 0250–7005.

[IV-41] SAVOLAINEN, S., et al., Boron neutron capture therapy (BNCT) in Finland: technological and physical prospects after 20 years of experience, Phys. Med. **29** 3 (2013) 233–248.

[IV-42] BUSSE, P.M., et al., "The Harvard–MIT BNCT program", Frontiers in Neutron Capture Therapy, Springer (2001) 37–60.

[IV-43] GONZÁLEZ, S.J., et al., First BNCT treatment of a skin melanoma in Argentina: dosimetric analysis and clinical outcomes, Appl. Radiat. Isot. **61** 5 (2004) 1101–1105.

[IV-44] MATSUMOTO, T., et al., Phantom experiment and calculation for in vivo ^{10}boron analysis by prompt gamma ray spectroscopy, Phys. Med. Biol. **36** 3 (1991) 329–338.

[IV-45] RAAIJMAKERS, C.P.J., et al., Boron neutron capture therapy using prompt gamma–ray analysis, Acta Oncol. **34** (1995) 517–523.

[IV-46] ISHIWATA, K., 4–Borono–2–^{18}F–fluoro–L–phenylalanine PET for boron neutron capture therapy–oriented diagnosis: Overview of a quarter century of research, Ann. Nucl. Med. **33** (2019) 223–236.

[IV-47] PORCARI, P., et al., In vivo ^{19}F MRI and ^{19}F MRS of ^{19}F-labelled boronophenylalanine–fructose complex on a C6 rat glioma model to optimize boron neutron capture therapy (BNCT), Phys. Med. Biol. **53** (2008) 6979–6989.

[IV-48] CAPUANI, S., et al., L–DOPA Preloading increases the uptake of borophenylalanine in C6 glioma rat model: A new strategy to improve BNCT efficacy, Int. J. Radiat. Oncol. Biol. Phys. **72** (2008) 562–567.

[IV-49] TIMONEN, M., et al., Acquisition-weighted MRSI for detection and quantification of BNCT ^{10}B–carrier L-p-boronophenylalanine–fructose complex, a phantom study, J. Radiat. Res. **50** (2009) 435–440.

[IV-50] TIMONEN, M., et al., ^{1}H MRS studies in the Finnish boron neutron capture therapy project: Detection of ^{10}B–carrier, L-p-boronophenylalanine–fructose, Eur. J. Radiol. **56** (2005) 154–159.

[IV-51] JÄRVINEN, H., VOORBRAAK, W. (Eds), Recommendations for the Dosimetry of Boron Neutron Capture Therapy, NRG Report 21425/03.55339/C, Petten, The Netherlands (2003).

[IV-52] KOBAYASHI, T., et al., A noninvasive dose estimation system for clinical BNCT based on PG–SPECT – Conceptual study and fundamental experiments using HPGe and CdTe semiconductor detectors, Med. Phys. **27** 9 (2000) 2124–2132.

[IV-53] MINSKY, D., et al., First tomographic image of neutron capture rate in a BNCT facility, Appl. Radiat. Isot. **69** (2011) 1858–1861.

[IV-54] LEE, T., et al., Monitoring the distribution of prompt gamma rays in boron neutron capture therapy using a multiple–scattering Compton camera: a Monte Carlo simulation study, Nucl. Instrum. Methods Ser. A **798** (2015) 135–139.

[IV-55] ONO, K., Prospects for the new era of boron neutron capture therapy and subjects for the future, Ther. Radiol. Oncol. **2** (2018) 40–46.

[IV-56] IMAMICHI, S.Y., "Evaluation of the BNCT System in National Cancer Center Hospital Using Cells and Mice", 16th International Congress of Radiation Research, 25–29 August 2019, Manchester, UK (2019).

[IV-57] GONZÁLEZ, S., SANTA CRUZ, G., The photon-isoeffective dose in boron neutron capture therapy, Radiat. Res. **178** (2012) 609.

[IV-58] SATO, T., MASUNAGA, S.I., KUMADA, H., HAMADA, N., Microdosimetric modeling of biological effectiveness for boron neutron capture therapy considering intra– and intercellular heterogeneity in ^{10}B-distribution, Sci. Rep. **8** (2018) 988.

[IV-59] MASUTANI, M., "Pre–clinical evaluation of biological effects of AB–BNCT system", IAEA Virtual Technical Meeting on Best Practices in Boron Neutron Capture Therapy (March 2022), IAEA, Vienna (2022).

[IV-60] MATSUMOTO, Y., et al., A critical review of radiation therapy: From particle beam therapy (proton, carbon, and BNCT) to beyond, J. Pers. Med. **11** 8 (2021) 825.

[IV-61] HE, H., et al., The basis and advances in clinical application of boron neutron capture therapy, Radiat. Oncol. **16** 1 (2021) 216.

[IV-62]   MALOUFF, T.D., et al., Boron neutron capture therapy: A review of clinical applications, Front. Oncol. **11** (2021) 601820.

[IV-63]   HIROSE, K., et al., Boron neutron capture therapy using cyclotron–based epithermal neutron source and borofalan (^{10}B) for recurrent or locally advanced head and neck cancer (JHN002): An open–label phase II trial, Radiother. Oncol. **155** (2021) 182–187.

[IV-64]   AIHARA, T., et al., BNCT for advanced or recurrent head and neck cancer, Appl. Radiat. Isot. **88** (2014) 12–15.

[IV-65]   WARD, M.C., et al., Refining patient selection for reirradiation of head and neck squamous carcinoma in the IMRT era: a multi–institution cohort study by the MIRI collaborative, Radiat. Oncol. Bio. Phys. **100** (2018) 586–585.

[IV-66]   TRIVILLIN, V.A., et al., Evaluation of local, regional and abscopal effects of boron neutron capture therapy (BNCT) combined with immunotherapy in an ectopic colon cancer model, Br. J. Radiol. **94** 1128 (2021) 20210593.

THE BNCT PROJECT OF SHONAN KAMAKURA GENERAL HOSPITAL

TOMIO INOUE, SHINICHI GOTOH
Shonan Kamakura General Hospital, 1370-1 Okamoto, Kamakura, Kanagawa 247-8533, Japan

Abstract

The medical corporation, Tokushukai Group, has established the Advanced Medical Center at Shonan Kamakura General Hospital. The purpose is to provide state-of-the-art medical care to the locals. Another goal is to accept patients with advanced or refractory cancer. The facility accepts patients 24 hours a day, 365 days a year. Based on this concept, we introduced two types of particle beam therapy devices: a proton beam therapy device and boron neutron capture therapy (BNCT).

V-1.   INTRODUCTION

Shonan Kamakura General Hospital, located in Kamakura City, Kanagawa Prefecture, provides a 24-hour emergency centre. The Tokushukai Group operates many medical facilities in Japan and overseas. Among them, the core hospital of 71 general hospitals in the group is Shonan Kamakura General Hospital. Currently, our hospital has 658 beds and 2,012 staff. It is also a regional cancer medical cooperation base hospital designated by the Minister of Health, Labor and Welfare. We have more than 2,000 cancer registrations annually. Our hospital is also accredited by JCI, Joint Commission International, which is an international medical quality standard. Furthermore, we have set up the Advanced Medical Center for cancer treatment.

BNCT has been introduced in the Advanced Medical Centers. The Center consists of two TomoTherapy units for XRT, one unit for intracavitary irradiation iridium RALS, one unit for proton therapy. Furthermore, the Department of Nuclear Medicine is also planning internal radionuclide therapy for neuroendocrine tumours with ^{177}Lu. In addition, there is also an oncology centre that provides anti-cancer drug treatment.

V-2.   SYSTEM

The BNCT system is described in Table V-1. The current value of the proton beam is measured in real-time by an independent Faraday cup. We are constantly checking the correlation between this current value and the neutrons generated here. A gamma-ray area monitor is installed behind the ceiling in the treatment room corner as a real-time monitor of fields in the treatment room. This value guides entering and leaving the room immediately after treatment.

TABLE V-1. DESCRIPTION OF THE BNCT SYSTEM

Component	Description
Type of accelerator	Cockcroft-Walton type proton beam accelerator
Number of beamlines	1
Beam energy, current	2.6 MeV, 40 mA (max)
Target materials	Rotating water-cooled Li petal
BSA design	Neutrons can be generated by colliding a proton beam from an accelerator with a rotating Li petal. The neutron deceleration mechanism obtained by this method is mainly deceleration by magnesium fluoride. See Fig. V-1
Collimator diameter	Five: Φ = 8, 11, 14, 17, 20 cm.

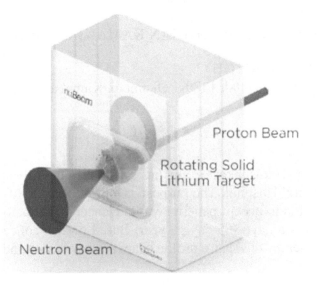

FIG. V-1. Target and neutron beam shaping assembly.

A sliding gantry type CT device and a 6-axis robotic bed are used for patient positioning.

V-3. FACILITY DESIGN AND LAYOUT

The BNCT treatment room is entirely housed on a 14 m × 28 m site. The system consists of three rooms, including an accelerator room, a beam dump room, and a treatment room. The accelerator room has three SF_6 gas tank for accelerator insulation, and the accelerator is a 2.6 MeV Cockcroft-Walton type proton beam accelerator. The beam dump room is an area for removing unnecessary beams. The treatment room is equipped with a solid Li target system and the beam shaping assembly. A sliding gantry type of CT device and a 6-axis robotic couch are used for patient positioning.

The volume of air exchange in the treatment room is 162 m^3, and both the intake and exhaust volumes are set to 6,700 m^3/hour.

The material of choice for construction is boronated polyethylene of 25 mm thickness for the floor, ceiling and walls of the treatment room and the beam dump room. Boronated polyethylene can have two effects: absorption by boron (n, α) reaction and moderation by polyethylene. Therefore, this material is effective in a wide range from medium to high neutron velocities to the thermal range as well. Figure V-2 gives the accelerator, beam dump, and treatment rooms.

V-4. PROJECT

BNCT research is conducted at the Clinical Research Center. With the scientific research grant obtained by Dr. Katsumi Tomiyoshi, we are developing a drug that can replace BPA to maintain a high concentration of boron in the tumour. This is a study to add a compound with a new function to BPA, and is currently conducting synthetic and empirical studies.

Clinically, we plan to start with recurrent head and neck cancer. In the future, we are also considering the treatment of intractable tumours and the treatment of recurrent cases.

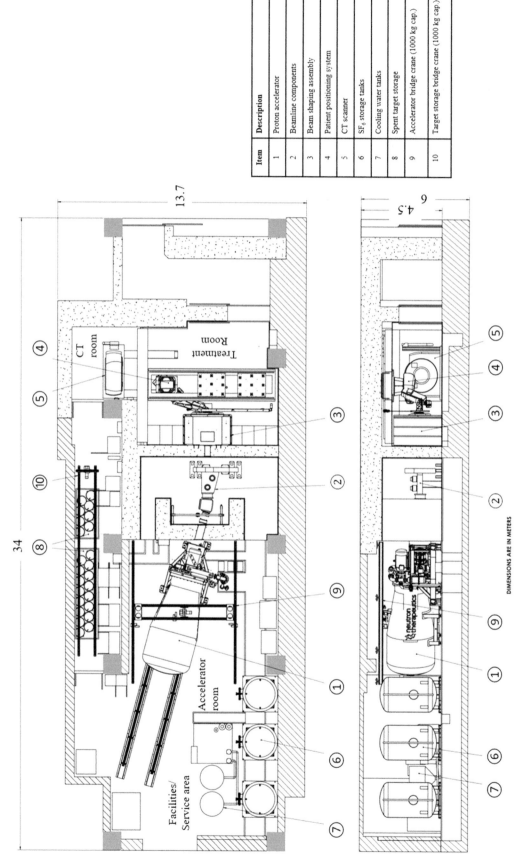

Item	Description
1	Proton accelerator
2	Beamline components
3	Beam shaping assembly
4	Patient positioning system
5	CT scanner
6	SF_6 storage tanks
7	Cooling water tanks
8	Spent target storage
9	Accelerator bridge crane (1000 kg cap.)
10	Target storage bridge crane (1000 kg cap.)

DIMENSIONS ARE IN METERS

FIG. V-2. Layout of the BNCT facility at Shonan Kamakura General Hospital (courtesy of Neutron Therapeutics, Inc.).

V-5. FACILITY OPERATION

Our facility requires a two-hour BPA infusion prior to BNCT irradiation. However, since the dedicated time for the treatment room is the sum of the positioning and irradiation time, it is considered within 60 minutes in one case. Since the actual clinical practice of BNCT has not started yet, the number of treatments per year cannot be predicted, as of March 2022. The staff listing is given in Table V-2.

TABLE V-2. LIST OF STAFF

Position	No.
Medical doctors	3
Medical physicists	3
Radiation therapy technologists	2

Annex VI.

DESCRIPTION OF THE ACCELERATOR-BASED BNCT PROJECT AT THE HELSINKI UNIVERSITY HOSPITAL

HANNA KOIVUNORO[1], LIISA PORRA[2], HEIKKI JOENSUU[2]
[1]Neutron Therapeutics Inc., Danvers, MA, USA
[2]Comprehensive Cancer Center, Helsinki University Hospital and University of Helsinki, Helsinki, Finland

Abstract

Boron neutron capture therapy (BNCT) has a long history in Finland. The BNCT treatments were carried out with a nuclear reactor for over a decade starting in 1999. After the closure of the reactor, Helsinki University Hospital started looking for options to continue BNCT with an accelerator-based device. A collaboration agreement was made with Neutron Therapeutics Inc. to acquire a nuBeam BNCT system. The system is currently under commissioning with the aim to achieve a license to start clinical trials.

VI-1. INTRODUCTION

In the period 1999–2012, 249 patients received over 300 BNCT treatments at the Helsinki University Hospital (HUS) using the FiR 1 research reactor, a 250-kW TRIGA Mark II research reactor at Otaniemi, Espoo, as the neutron source and L-boronophenylalanine–fructose (L-BPA–fr) as the boron carrier [VI-1–VI-4]. The BPA infusion was given intravenously over 2 hours. The BPA dose was escalated from 290 to 450 mg/kg in a clinical trial [VI-3]. The patients were treated either within one of the four phase I/II clinical trials carried out at the centre or as compassionate treatments. The tumour types treated were gliomas (anaplastic astrocytoma or glioblastoma), head and neck cancer (squamous cell carcinomas, adenocystic carcinomas, sarcomas), aggressive meningioma, melanoma, and lymphoma. The FiR 1 reactor was closed for BNCT treatments in 2012. In 2016, the HUS entered into a collaboration with Neutron Therapeutics Inc. (Danvers, MA) to construct an accelerator-based nuBeam BNCT suite in Helsinki. The nuBeam installation started at the HUS cancer centre in 2018 after the reconstruction of the building that houses the accelerator was completed. The neutron therapy system incorporates an image-guided robotic patient positioning system, an in-room CT with a sliding gantry, an on-line neutron beam monitoring system, and a treatment planning system designed for BNCT. The facility is primarily intended for clinical use, but basic and preclinical research is also planned to be performed at the site [VI-5].

VI-2. PROJECT

The first clinical trial at HUS will focus on recurrent head and neck cancer with L-BPA–fr as the boron carrier. In the first study protocol, L-BPA–fr is intended to be administered in an identical manner as was done at the FiR 1 facility, i.e., as a 2-hour intravenous infusion with the neutron irradiation started about 1 hour after completion of the L-BPA–fr infusion when the blood boron concentration had decreased to ~20 mg/g [VI-2]. An interim safety analysis and an early efficacy analysis will be carried out after the first patients have been treated, with one aim to obtain a CE mark for the system. Once the safety of the BPA mediated accelerator based BNCT system has been confirmed, there are plans to expand BNCT to other solid tumours and to combine it with systemic cancer therapy and/or conventional radiotherapy. When needed, ^{18}F-FBPA-PET can be performed at the Cyclotron Unit of the HUS Medical Imaging Center located next to the nuBeam facility on the same campus. The nuBeam system is also planned to be used for testing of novel boron carriers. The nuBeam facility is located ideally for preclinical trials since many medical research laboratories are located on the campus.

VI-2.1. System

VI-2.1.1. Proton accelerator and beam shaping assembly

nuBeam includes a single-ended electrostatic Cockcroft-Walton type proton accelerator. The proton beam energy is 2.6 MeV and the maximum current 40 mA. A beam transport system directs the protons onto a cooled, rotating, solid lithium target. An optimized beam shaping assembly (BSA), containing a MgF_2 moderator, produces a neutron beam designed to be consistent with IAEA TECDOC 1223 [VI-7]. The patient treatment room shielding is composed of concrete, lead, and borated and lithiated plastic. It is also supplemented with a lead beam shutter that slides to cover the nozzle when irradiation is not active. The interchangeable circular beam delimiter sizes range from 8 cm to 20 cm, the nominal size being 14 cm in diameter. The beam delimiters are produced from enriched ^{6}Li-polyethylene. The activated spent lithium targets can be exchanged automatically by the patient positioning robot. The spent targets are stored in shielded enclosures in the radiation storage room at the BNCT facility until the radioactivity has decayed to a level where the targets can be treated as non-radioactive materials.

VI-2.1.2. Real time neutron beam monitoring system

The on-line neutron beam monitoring system consists of ionization chamber type detectors placed inside the BSA, embedded in the beam delimiter material so that they view the epithermal beam near the beam port. At this location the beam monitors are not affected by the presence of the patient. Two redundant neutron monitors are designed to be sensitive primarily to neutron radiation and insensitive to gamma radiation. A third, neutron-insensitive ionization chamber is used to monitor the gamma component of the beam. The neutron beam is automatically terminated based on the neutron fluence measured by the beam monitors. This on-line neutron beam monitoring system provides more reliable measurements of the delivered neutron fluence than relying on calibrated proton current data.

VI-2.1.3. Patient positioning system

The treatment room facilitates a 6-axis robotic patient positioning and image-guiding system, Exacure, manufactured by BEC GmbH (Reutlingen, Germany). In-room imaging is performed with a rail mounted Somatom ConfidenceVR CT scanner (Siemens Healthineers, Erlangen, Germany), located behind neutron shields during the irradiations. Figure VI-1 shows a view of the treatment room. On the treatment day, the patient is positioned onto the robotic couch using an earlier prepared mask and vacuum pillows. The patient's position is automatically adjusted to match the planning images captured just prior to irradiation.

VI-2.1.4. Facility design and layout

The facility layout is illustrated in Fig. VI-2. The bunker walls consist of high-density concrete (3.4 g/cm^3), selected to minimize the wall thickness. The wall thickness is varied according to radiation intensity, with a maximum thickness of 110 cm in the beam-facing wall. Due to substantially lower radiation levels, the walls of the accelerator hall are only 40 cm thick and consist of standard density concrete. The beamline room walls are covered with standard 5% borated plastic for neutron absorption. The treatment room walls are covered with lithiated plastic to minimize scattering and to maximize neutron absorption without creation of γ rays. Neutron absorbing plastic is not needed in the accelerator hall. The treatment room and the accelerator hall are classified as controlled areas with access allowed only for radiation workers. The control room is classified as a supervised area. The classification of the areas is shown in Fig. VI-3.

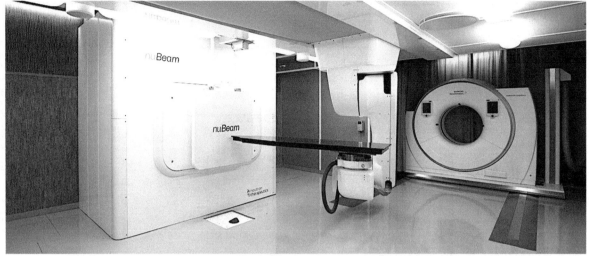

FIG. VI-1. The nuBeam treatment room at the Helsinki University Hospital cancer center illustrating the CT scanner (right), the robotic image-guided patient positioning system (middle), and the beam shaping assembly and the beam shutter (left).

A blood boron measurement laboratory housing an Agilent 5110 instrument (Palo Alto, CA) for the inductively coupled plasma-atomic emission spectrometry (ICP-OES) analysis of blood samples is located on the second floor of the facility, where also the material activation analysis room is located. The activation analysis of the foil and wire dosimeters is performed with a high-purity gamma spectrometer ORTEC HPGe Gamma-Ray Detector (Zoetermeer, The Netherlands). It has an integrated cryo-cooling system, a DSPEC-50 multi-channel analyzer, analysis software (LVis), and an integrated Hidex automatic sample changer.

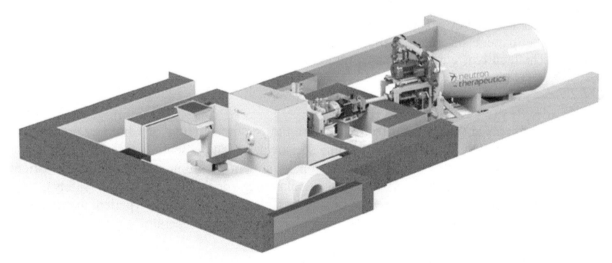

FIG. VI-2. The nuBeam compact accelerator-based neutron suite installed at the Helsinki University Hospital Cancer Center. Bunker size is 20 m × 13 m.

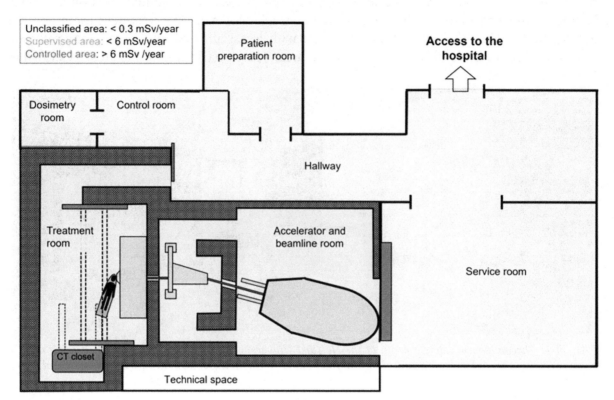

FIG. VI-3. Radiation protection classified areas of the nuBeam facility at the Helsinki University Hospital.

VI-2.2. Costs and management

VI-2.2.1. Financing plan

The Helsinki University Hospital and Neutron Therapeutics Inc. formed a joint project to install a compact accelerator-based neutron source at the Helsinki University Hospital campus area. The Helsinki University hospital has constructed the facility and the supporting laboratories, and Neutron Therapeutics Inc. provides the nuBeam suite including the neutron source, the patient positioning robot, the in-room imaging, and the treatment planning system. The facility is currently in a commissioning phase.

VI-2.2.2. Staff

Patient treatments at the Helsinki facility require the personnel, some of whom do not need to be dedicated for BNCT and may have duties elsewhere. The essential staff for running the facility include an oncologist (patient selection, treatment planning and patient care, patient follow-up), a radiation physicist (treatment planning, dosimetry, radiation safety), two radiographers (patient simulation and positioning, imaging), a nurse (patient care, blood sampling), a chemist (blood boron analysis), and a neutron beam operator.

VI-2.2.3. Treatment capacity

The facility initial license permits 22 hours of neutron irradiation per week, which is sufficient for treating at least 1000 patients annually, each receiving BNCT twice.

VI-2.2.4.   Neutron beam dosimetry

Based on the current recommendations for BNCT dosimetry published by the European project for a Code of Practice for BNCT dosimetry [VI-6] and Ref. [VI-7] and following our previous practice at the FiR 1 reactor facility, the Helsinki nuBeam facility uses the neutron activation method for neutron fluence measurements [VI-8] and the paired-ionization chamber method [VI-9] for neutron and gamma dose measurements. The neutron spectrum is defined in air with a multifoil set as described by Auterinen et al. [VI-10]. The neutron flux is measured in the phantoms with diluted Al-Au and Al-Mn (1 % weight Au/Mn) activation foils and using Al-Mn wires (2.6 % weight Mn) from Goodfellow (Cambridge, UK). The daily quality assurance is performed with an Au-Al and Mn-Al foil pair near the thermal neutron maximum at the depth of 2 cm in a cylindric PMMA phantom, 20 cm in diameter and 24 cm in length (Fig. VI-4). A full characterization of the neutron flux, the photon dose distribution, and the total dose distribution is achieved with the Au-Al and Mn-Al foils (the depth profile), Mn-Al wires (the lateral profiles), and paired ionization chambers (the 3D dose; the photon dose, and the total dose) in a large water phantom (a PTW MP3-PL Hi-Tech 3D Water Phantom) (Fig. VI-4). In addition, end-to-end testing is performed in a small cubic 20 cm wide water phantom to enable image-guided phantom positioning, and 3D measurements of the dose and the neutron flux distributions.

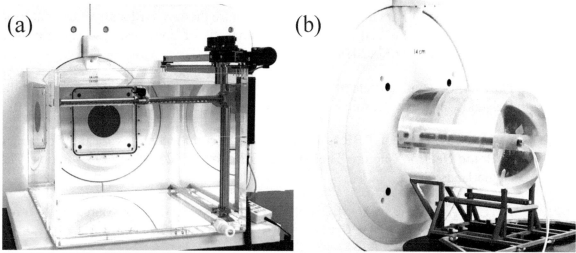

FIG. VI-4. (a) The PTW MP3-PL Hi-Tech 3D Water Phantom applied for full characterization of the dose and the neutron flux distribution. The dimensions of the phantom are 63 cm × 63.2 cm × 52 cm (width × depth × height). (b) the cylindrical (diameter 20 cm, length 24 cm) PMMA phantom used primarily for daily QA and beam model normalization.

VI-2.2.5.   Treatment planning

Treatment planning is based on full Monte Carlo simulation with material assignment deduced from DICOM images (CT, MRI, and PET) of the patient. The dose engine for patient dose calculation was developed by Neutron Therapeutics Inc. on top of the Geant4 MC toolkit [VI-11]. The dose engine is used through the RayStation treatment planning system (Stockholm, Sweden), where the target volumes, organs at risk, and their tissue compositions and boron concentrations, the biological dose calculation model (the RBE factors), and the neutron beam parameters and directions are defined. In addition, RayStation is used for reporting of the doses.

VI-3. SUMMARY

The nuBeam accelerator facility began installation in 2018. The neutron beam commissioning is currently ongoing to obtain a license to carry out clinical trials at the facility. The first clinical trial focuses on recurrent inoperable head and neck cancer. A CE mark is planned to be applied for the nuBeam system after the completion of the first phase of the clinical trial. The nuBeam facility is intended to be used both for patient treatments and preclinical research.

REFERENCES TO ANNEX VI

[VI-1] JOENSUU, H., et al., Boron neutron capture therapy of brain tumors: clinical trials at the Finnish facility using boronophenylalanine, J. Neuro-Oncol. **62** (2003) 123–134.

[VI-2] KANKAANRANTA, L., et al., Boron neutron capture therapy in the treatment of locally recurred head-and-neck cancer: final analysis of a phase I/II trial, Int. J. Radiat. Oncol. Biol. Phys. **99** (2011) 97100.

[VI-3] KANKAANRANTA, L., et al., L-boronophenylalanine-mediated boron neutron capture therapy for malignant glioma progressing after external beam radiation therapy: a phase I study, Int. J. Radiat. Oncol. Biol. Phys. **80** (2011) 369–376.

[VI-4] KOIVUNORO, H., et al., Boron neutron capture therapy for locally recurrent head and neck squamous cell carcinoma: An analysis of dose response and survival, Radiother. Oncol. **137** (2019) 153–158.

[VI-5] PORRA, L., et al., Accelerator-based boron neutron capture therapy facility at the Helsinki University Hospital, Acta Oncol. **61** (2022) 269–273.

[VI-6] JÄRVINEN, H., VOORBRAAK, W. (Eds), Recommendations for the Dosimetry of Boron Neutron Capture Therapy, NRG Report 21425/03.55339/C, Petten, The Netherlands (2003).

[VI-7] INTERNATIONAL ATOMIC ENERGY AGENCY, Current Status of Neutron Capture Therapy, TECDOC 1223, IAEA, Vienna (2001).

[VI-8] AUTERINEN, I., SERÉN, T., UUSI-SIMOLA, J., KOSUNEN, A., SAVOLAINEN, S., A toolkit for epithermal neutron beam characterisation in BNCT, Radiat. Prot. Dosim. **110** (2004) 587–593.

[VI-9] KOSUNEN, A., et al., Twin ionisation chambers for dose determination in phantom in an epithermal neutron beam, Radiat. Prot. Dosim. **81** (1999) 187–194.

[VI-10] AUTERINEN, I., SERÉN, T., ANTTILA, K., KOSUNEN, A., SAVOLAINEN, S., Measurement of free beam neutron spectra at eight BNCT facilities worldwide, Appl. Radiat. Isot. **61** (2004) 1021–1026.

[VI-11] AGOSTINELLI, S., et al., GEANT4 - a simulation toolkit, Nucl. Instrum. Methods Phys. Res. Sect. A **506** 3 (2003) 250–303.

ACCELERATOR-BASED NEUTRON IRRADIATION SYSTEM WITH A LITHIUM TARGET IN THE NATIONAL CANCER CENTER HOSPITAL, JAPAN

H. IGAKI[1], S. NAKAMURA[2], M. NAKAMURA[3] J. ITAMI[4]

[1]Department of Radiation Oncology, National Cancer Center Hospital, Tokyo, Japan.
[2]Division of Radiation Safety, National Cancer Center Hospital, Tokyo, Japan.
[3]Cancer Intelligence Care Systems, Inc. Tokyo, Japan.
[4]Shin-Matsudo Accuracy Radiation Therapy Center, Shin-Matsudo Central General Hospital, Matsudo, Japan

Abstract

The National Cancer Center Hospital is one of the institutes of the National Cancer Center, a Japanese national research and development agency, and a Japanese leading cancer hospital. We have developed an accelerator-based neutron irradiation system, CICS-1, under a contract of joint research between the National Cancer Center and the CICS, Co. Inc. A clinical trial has been ongoing since November 2019 based on the contracted research from CICS, Co. Inc. and StellaPharma J.V. Co., Ltd. The aim of this clinical trial is to get an approval of CICS-1 as a medical device for boron neutron capture therapy (BNCT) from the governmental authority, and to extend the indication of SPM-011 (borofalan ^{10}B) to skin tumours. Our accelerator-based neutron irradiation system clinically implemented a Li target for BNCT for the first time in the world.

VII-1. INTRODUCTION

National Cancer Center Hospital is an institute of the National Cancer Center, a Japanese national research and development agency, the Japanese leading institution for basic cancer research and clinical cancer treatment, and started an accelerator-based BNCT project in 2010, under the research contraction with Cancer Intelligence Care Systems (CICS), Inc. Based on this project, a Li-targeted accelerator-based neutron irradiation system was installed in the Department of Radiation Oncology, National Cancer Center Hospital. After adjustment of the neutron beam, physical and biological evaluation was started in 2016. CICS-1 is the world's first Li-targeted neutron irradiation system for BNCT for clinical use [VII-1]. This system is characterized by a compact beam shaping assembly (BSA) due to the low incident energy of the neutron beam. We have started a clinical trial of BNCT in order to get the governmental approval of CICS-1 and SPM-011 (borofalan ^{10}B) in treating skin tumours (melanoma and angiosarcoma) by BNCT based on the contracted research from CICS and StellaPharma J.V. Co., Ltd. The aim of this clinical trial is to get approval of CICS-1 as a medical device for BNCT and from the governmental authority, and to extend the indication of SPM-011 (borofalan ^{10}B) to skin tumours.

In this article, we introduce our accelerator-based neutron irradiation system of the world's first Li-targeted one, CICS-1.

VII-2. PROJECT

The following sections provide a description of the project.

VII-2.1. System

VII-2.1.1. Type of accelerator

The BNCT system at the National Cancer Center Hospital uses a high current proton RFQ (radio frequency quadrupole) linear accelerator. It consists of an injector system, a low energy beam transport system, an RFQ linac, an RF transmission line, a klystron system including its power supply unit, a high energy beam transport system, quadrupole doublet magnets, a bending magnet, a wobbling magnet, and steering magnets.

VII-2.1.2. Number of beamlines

There is one treatment room, and the beam line is bent 90° downward by the bending magnet, hitting the target on the floor below and producing neutrons. By downsizing the equipment, a vertical beam was realized.

VII-2.1.3. Beam energy, current

The maximum proton beam current is 20 mA and is accelerated up to 2.5 MeV. The average current is more than 10 mA to get enough neutron flux for the treatment.

VII-2.1.4. Target materials

Our equipment uses a solid Li target. Using the vacuum vapour deposition method, we deposited a thin solid Li layer on a conical Cu substrate. The thickness of the Li target along the incident proton angle reaches more than approximately 100 μm. Additionally, a Pd layer is formed between the Li layer and the substrate to prevent blistering.

Neutron production in the $^7Li(p,n)^7Be$ reaction is accompanied by production of the radioactive ^7Be isotope. Beryllium-7, accumulated in the Li layer, has a long half-life of 53.3 days and emits γ rays of 478 keV by radioactive decay. This activation causes unnecessary radiation exposure to maintenance personnel and makes it difficult to replace the target, so a cleaning system was installed. The spent Li layer containing ^7Be is automatically cleaned with water after a specified service period. Next, the aqueous solution containing Li and radioactive ^7Be is discharged through pipes into a decay tank outside the treatment room. These operations are automatically controlled by the Li deposition and cleaning system. After cleaning, the target substrate is removed, and a new target substrate is attached by maintenance personnel.

VII-2.1.5. Beam shaper assembly

The BSA consists of four main parts: a moderator, a neutron reflector, a gamma shield, and a thermal neutron shield. The moderator is sintered MgF_2 that decelerates the generated neutrons to the energy range of epithermal neutrons. The neutron reflector is made of Pb, which efficiently guides neutrons to the irradiation port. The gamma shield is also made of Pb and shields the secondary gamma rays generated during the neutron beam moderation. The thermal neutron shield is a silicone sheet with LiF.

VII-2.1.6. Real time monitor

In our system, the real time monitor used as the monitor unit utilizes the proton beam monitor. The relationships between the neutron and photon yields and proton current have been sufficiently investigated by a non-clinical test over the lifetime of the neutron source [VII-2].

VII-2.1.7. Collimator

The size of the circular beam delimiter is 24 cm in diameter. The collimator for the irradiation port is currently under development.

VII-2.1.8. Patient positioning system

In BNCT, it is necessary to set up the patient's tumour site as close as possible to the irradiation port and precisely at a position where the dose distribution calculated in the treatment plan can be obtained. Since the irradiation port of this device is oriented in the horizontal plane, an irradiation table that can be moved in five axial directions was used under the irradiation port. During treatment, the patient is placed on a removable carbon-fibre tabletop attached to the irradiation table. This tabletop can be attached to the patient couch of the CT scanner. The patient on the tabletop is moved from the CT room to the BNCT treatment room using a stretcher. Sharing the tabletop allows for accurate setup and reduces the burden on the patient.

VII-2.2. Facility design and layout

VII-2.2.1. Facility overview

The accelerator based BNCT system is constructed within an existing hospital (National Cancer Center Hospital, Tokyo, Japan). As a result, medical information from, e.g., PET-CT, PET-MR, etc…, is utilized for BNCT. Therefore, BNCT can be performed in an environment of daily clinical practice.

Our facility for the accelerator based BNCT system can be divided into three sections (an accelerator room, a procedure area before/after BNCT, and a treatment room). Figure VII-1 shows the layout of the accelerator based BNCT system of which the 2.5 MeV RFQ system is installed at the National Cancer Center, Tokyo, Japan. The treatment room for BNCT is located next to the medical linacs used for radiation therapy. The accelerator lies on the floor above the treatment room, delivering the beam vertically downwards. Additionally, the procedure area is nearby the treatment room (CT simulation room, control room, recovery room, etc…).

The internal dimensions of the accelerator room are $20 \times 8 \times 5$ m, and the accelerator is 8 m long. The related equipment, such as power supply, cooling system, maintenance service device, etc…, is also installed in the accelerator room, and no other space is necessary for the accelerator.

In the treatment room, the Li target system, BSA, a treatment couch, and a patient monitoring system are installed. The internal dimensions of the treatment room are $8 \times 7 \times 3.5$ m, and the height of the polyethylene layer is 4 m above the floor. The patient monitoring system has four cameras, a microphone, an emergency button for a patient, and a percutaneous oxygen saturation monitor. Using these devices, the patient's condition is confirmed during BNCT. The irradiation table can be moved on a 5-axis shift, and the off-line image-guided radiotherapy is performed by combining the laser system in the treatment room. The 5-axis shift is sufficient for the system since the neutron beam is isotropically emitted perpendicular to the beam axis.

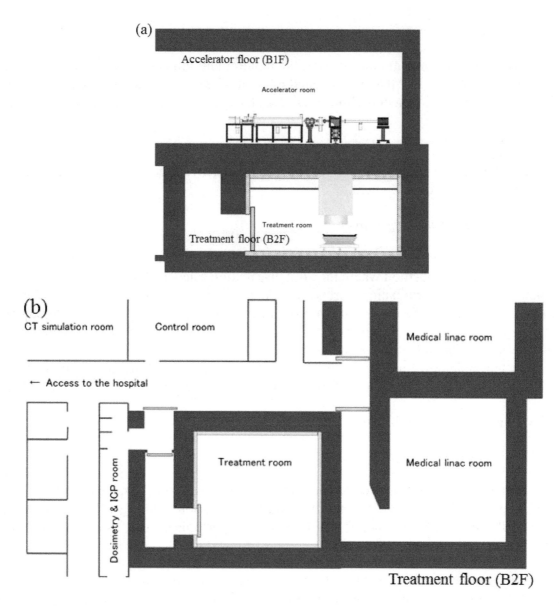

FIG VII-1. Layout of the BNCT system by Cancer Intelligence Care Systems, Inc at the National Cancer Center Hospital [VII-3] (a) Elevation showing the accelerator floor above the treatment floor and (b) Plan view at the level of the treatment floor.

The procedure area has a computed tomography (CT) simulation room, a control room, dosimetry space for quality assurance/control (QA/QC), inductivity coupled plasma optical emission spectrometer for measuring the boron density in the blood of the patient, a treatment planning system, and a waiting room. The CT simulation room is utilized not only for the treatment planning CT, but also the off-line image-guided radiotherapy in BNCT. Hence, the CT simulation room also plays the role of a simulation room. In the control room, the number of required neutrons for BNCT is controlled by the number of protons delivered to the Li target in the system. In the waiting room, drip infusion for the compounds is performed before BNCT and the patient condition is confirmed after BNCT. There is much work in each related occupation (radiation oncologist, medical physicist, radiation technologist, and nurse) before and after BNCT. The time schedule on the treatment day is especially busy. In our facility, it takes approximately four hours for each occupation to perform one treatment.

VII-2.2.2. Radiation zoning

BNCT facilities have to be designed to take into account of the As Low As Reasonably Achievable (ALARA) principle. As a result, a reduction of radiation risk is expected. A BNCT facility needs to consider the dose rate during operation and the shielding requirements for the accelerator based BNCT system.

Our facility is divided into three areas according to the possible risk from radiation and imposing oversight proportional to the risks. The three sections, described above, are divided as follows:

(a) Controlled area: treatment room, accelerator room;
(b) Supervised area: the procedure area before/after BNCT except for the waiting room;
(c) Uncontrolled area: waiting room.

Residual activation of the accelerator based BNCT system, especially in the treatment room, has to be also considered as well as the dose rate during the operation. The treatment room in our facility is mainly shielded by two layers. The two layers consist of a neutron absorbing material and γ ray shielding materials. The material of the first layer is borated polyethylene which absorbs neutrons and suppresses activation, while that of second layer is concrete, which is a shield for gamma-rays. Furthermore, the air in the treatment room may be activated due to neutron exposure. After the neutron exposure, ^{41}Ar production may be expected due to the activation of ^{40}Ar which exists naturally in air. Therefore, the treatment room is under negative pressure with respect to neighbouring spaces, air in the treatment room is exchanged every 348 s, and fresh, temperature-controlled air is supplied to follow each regulation.

VII-2.3. Costs and management

VII-2.3.1. Financial cost

At present, we are conducting a clinical trial aiming the governmental approval as a medical device under the Japanese Pharmaceutical and Medical Device Act. Accordingly, the cost of operation and treatment is difficult to calculate, based on the assumed regular operation of the accelerator.

VII-2.3.2. Staff

The BNCT facility is located within the Department of Radiation Oncology, National Cancer Center Hospital, and adjacent to the conventional medical linear accelerator room for radiotherapy. All staff members of the Department of Radiation Oncology are concurrently working on BNCT with their regular radiotherapy duties. There are nine full-time certified radiation oncologists, one of whom has the certificate of the Japanese Society of Neutron Capture Therapy as of February 2022. Radiation oncologists supervise the overall conditions for BNCT. Dose prescription and treatment decisions are also made by radiation oncologists. Of the four full-time and two part-time medical physicists, two are mainly engaged in BNCT, one of whom has been certified as specialized in radiation oncology. Medical physicists are responsible for treatment planning, dose measurements/verifications, quality assurance/quality control, and implementations for BNCT. They also measure the boron concentration of the patient's blood by ICP. Twenty-two radiotherapy technologists are working in the Department of Radiation Oncology, of whom, six technologists have been certified for radiotherapy technology, and two certified technologists are involved mainly in BNCT. They prepare the patient's fixation and positioning tools, set up in the treatment position in collaboration with medical physicists, and operate a neutron irradiation system during the treatment. One of the five nurses is in charge of caring for the patient during the entire procedure of BNCT including the administration of borofalan (^{10}B) before neutron irradiation. Blood samples are taken from the patient by the attending nurse.

VII-2.3.3. Number of patients per year expected

After obtaining governmental approval of CICS-1, our initial goal is about 100 cases/year and to expand the indications. Because the target cancers are rare, the number of patients will not increase rapidly. Therefore, we also plan to conduct clinical trials to expand the applicable affected areas. Considering patients' quality of life, we believe that the BNCT treatment advocated by CICS can save many patients. If BNCT is recognized as a treatment method for this rare cancer and another in the future, we expect to increase the number of patients further. To this end, we would like to ensure the system's robustness and make it known to patients and physicians.

REFERENCES TO ANNEX VII

[VII-1]   IGAKI, H., et al., Scalp angiosarcoma treated with linear accelerator-based boron neutron capture therapy: A report of two patients, Clin. Transl. Radiat. Oncol. **33** (2022) 128–133.

[VII-2]   NAKAMURA, S., et al., Neutron flux evaluation model provided in the accelerator-based boron neutron capture therapy system employing a solid-state lithium target, Sci. Rep. **11** 1 (2021) 8090.

[VII-3]   CANCER INTELLIGENCE CARE SYSTEMS INC., CICS (2020), https://www.cics.jp/page/english.html.

LINAC-BASED BNCT DEVICE DEVELOPED BY THE iBNCT PROJECT

H. KUMADA[1], K. NAKAI[1], Y. MATSUMOTO[1], S. TANAKA[1], T. SAKAE[1], H. SAKURAI[1], F. NAITO[2], T. KURIHARA[2], T. SUGIMURA[2], M. SATO[2]
[1]Faculty of Medicine, University of Tsukuba, Japan
[2]High Energy Accelerator Research Organization, Japan

Abstract

The iBNCT project, an industry–academia–government collaboration team headed by the University of Tsukuba, was launched in 2011. The project is to develop iBNCT001, a demonstration device of an accelerator-based neutron source applicable to boron neutron capture therapy (BNCT). The accelerator of the iBNCT001 consisting of a radio-frequency quadrupole linac, and a drift tube linac has been adopted. The energy and average current of the protons accelerated by the linac were set to 8 MeV and 5 mA or more, respectively. Beryllium was selected as the neutron target material. iBNCT001 has been completed and has succeeded in generating a neutron beam with sufficient intensity for treatment. Various characteristic measurements were performed to confirm the performance of the neutron beam released from the beam aperture of the device. The measurement results demonstrated that the device generated neutrons with sufficient intensity for BNCT treatment. In 2021 and 2022, a non-clinical study with mice irradiation experiments was conducted to validate the safety and applicability of the treatment. Based on the results of the non-clinical study, we plan to conduct clinical trials in actual patients.

VIII-1. INTRODUCTION

The iBNCT project was launched in 2011 to establish and spread hospital-based BNCT in the near future [VIII-1]. University of Tsukuba, high-energy accelerator research organization (KEK), Japan Atomic Energy Agency (JAEA), local governments, and various manufacturers were involved in the accelerator participated in the project. The project team designed and produced iBNCT001, a demonstration device for an accelerator-based neutron source for BNCT. The project chose Be as the target material of iBNCT001, and a linac was adopted for the accelerator. The linac was designed and produced to accelerate protons of an average current of 5 mA or more at 8 MeV. In 2016, the neutron source succeeded in generating neutrons, although the intensity was low at that time. Subsequently, the device was gradually further improved; in 2020, the average current of the proton beam achieved was approximately 2.1 mA [VIII-2]. The results of various measurements performed under the operating condition of 2.1 mA demonstrated that the device generates neutrons with sufficient intensity for BNCT treatment. The maximum thermal neutron flux in a water phantom was $\sim 1.4 \times 10^9$ cm^{-2}s^{-1}. Based on the physical measurement results, a non-clinical study with mice is being conducted to validate the safety and applicability of the treatment during 2021 and 2022. Based on the results of the non-clinical study, clinical trials for several cancers would be conducted. Details of the device and facility developed by the project and its performance are described below.

VIII-2. PROJECT

The iBNCT project, an industry–academia–government collaboration team headed by University of Tsukuba, was launched in 2011. The University of Tsukuba, KEK, and JAEA have led the project as academic organizations and designed and developed a demonstration device by cooperating with various manufacturers related to accelerator equipment. Ibaraki

Prefecture and Tsukuba City, as local governments, support the activities of the project. The aim of this project is to produce a demonstration device for the accelerator-based BNCT device and to implement clinical trials of BNCT using the demonstration device. Therefore, to achieve this purpose, it is necessary to develop not only an accelerator-based BNCT device but also peripheral devices needed for treatment, such as a treatment planning system (TPS), a patient positioning system, and several radiation monitors. The goals of this project is to obtain regulatory approval for the device developed and to commercialize the device by transferring related technologies to the participating companies.

The concept of the project is 'Realization of BNCT in a stable, safe, and easy manner in a hospital.' In particular, we seek to establish a BNCT that can be completed within a short irradiation time thanks to a high-intensity neutron beam, and to suppress radiation exposure for patients and medical staff by establishing low activation technology for the device.

VIII-2.1. System

VIII-2.1.1. Conceptual design of iBNCT001

In the conceptual design stage of the iBNCT001 device, we first decided to adopt Be for the neutron target. We believe that Be has several advantages as a neutron target for a BNCT neutron source compared to Li, such as a higher melting point and higher heat conductivity. Additionally, 'NeuCure' which is the cyclotron-based BNCT device manufactured by Sumitomo Heavy Industries, Ltd. and registered under the regulatory affairs, is combined with the Be target and has a track record of application in many treatments. Next, we determined the energy of the proton beam irradiating the Be target. When Be is irradiated by protons of energies greater than 13.4 MeV, Be changes into ^{7}Be, which is a radioisotope. Therefore, we determined that the energy of the protons ought to be lower than 13 MeV. In the design stage, we set the proton beam energy to 8 MeV, based on various analyses and evaluations. When Be is irradiated by 8 MeV protons, the neutron energy is 6.1 MeV or less, which is relatively low, and is therefore able to suppress the activation of both Be and several materials in the beam-shaping assembly (BSA). We also discussed the average proton current together with the energy because we had to design a neutron source that could generate sufficient neutron intensity for the treatment. The results of various analyses indicated that a few milliamperes are needed to generate sufficient neutrons for BNCT from the reaction between 8 MeV protons and thin Be. Therefore, the average proton current of the accelerator was set to 5 mA or more.

VIII-2.1.2. Type of accelerator

Regarding the type of the accelerator for iBNCT001 to accelerate protons to 8 MeV and reliably achieve proton currents of 5 mA or more, we have chosen a linac, which consists of a radio-frequency quadrupole linac (RFQ) and a drift tube linac (DTL) [VIII-3]. To ensure large proton currents with a linac, the basic methodologies of the RFQ and DTL adopted in J-PARC [VIII-4], which is a high-performance accelerator-based neutron source installed in Tokai village in Japan, have been applied to iBNCT001. For a klystron, a key component of the linac, the same type in use at J-PARC was adopted. However, one klystron drives both linac tubes of iBNCT001, whereas the J-PARC linac is operated by combining different klystrons for the RFQ and DTL. This contributed to lower cost and space savings for iBNCT001. Regarding peripheral devices for the linac, such as the ion source and cooling system of the linac tubes, dedicated equipment and systems for iBNCT001 have been designed and developed, and finally combined with the linac tubes. Figure VIII-1 shows the actual linac tubes installed in an accelerator room at the iBNCT001 facility. The main specifications of the linac system in 2022 are listed in Table VIII-1. The total length of the linac system including the ion source is approximately 8 m, with a diameter of 1.5 m or less. The linac system was designed and assembled by the KEK group. The values in parentheses in the iBNCT001 column represent the corresponding design target values. The DTL further increased the velocity of the protons from the RFQ, and the energy of the protons finally reached 8 MeV. Accelerated protons are delivered from the linac system to the Be-based target system by passing through a proton beam transport system. As of 2022, the device operates stably, with an average proton current of approximately 2.1 mA.

(a) DTL   (b) RFQ

FIG. VIII-1. DTL (a) and RFQ (b) of the iBNCT001 linac tubes.

TABLE VIII-1. MAIN SPECIFICATIONS OF THE iBNCT001 NEUTRON SOURCE

Items	iBNCT001
Accelerator type	An RFQ and a DTL type linac
Proton energy	8 MeV
Peak proton current	> 28 mA (50 mA)
Proton pulse width	1 ms
RF frequency	324 MHz
Average proton current	2.1 mA (> 5.0 mA)
Target material	Be
Epithermal neutron flux	$> 7.0 \times 10^8$ cm^{-2}·s^{-1} ($> 1.5 \times 10^9$ cm^{-2}·s^{-1})
Beam aperture	Normal: $\Phi = 120, 150$ mm,
	Extended: $\Phi = 100, 120, 150$ mm

Note: The values in parentheses in the iBNCT001 column are the design target values. Φ indicates a diameter.

VIII-2.1.3. Proton beam transport system

The proton beam accelerated to 8 MeV by the linac tube is transported to the Be-based target system by passing through the beam transport system. To decrease the heat load on the Be surface, the 8 MeV protons is expanded to a 13 cm × 13 cm square area just upstream of the Be target using two quadrupoles and two octupole electromagnets. To reduce the heat load on Be, we chose the diffusing method rather than the scanning method for proton irradiation. The extended square proton beam is irradiated onto the surface of Be, and neutrons are emitted.

VIII-2.1.4. Beam energy, current

As described above, the energy of the protons in iBNCT001 was set to 8 MeV, and the linac was designed and produced so that an average proton current of 5 mA or more could be accelerated. As of 2022, the linac can be stably driven with an average proton current of approximately 2.1 mA.

VIII-2.1.5. Target system

As described above, in the iBNCT project, Be was chosen as the target material, and the final proton energy was set at 8 MeV, being lower than that of the leading manufacturer's device. Therefore, the challenges associated with Be targets were increased. Two major challenges have to be addressed during the development. The first issue is to prevent the Be system from melting owing to the large heat load caused by proton beam irradiation. The second issue is developing a target system capable of suppressing blistering in the materials of the target system. To overcome these issues, we devised a Be-based target system consisting of a three-layer structure for the iBNCT001 neutron generation device [VIII-5]. Figure VIII-2 shows an image of the three-layer structure target system. The first layer of the three-layer structure is a Be layer. Its shape is circular with a diameter of 15 cm and a thickness of 0.5 mm. The Bragg peak depth of the proton in Be is approximately 0.55 mm. Thus, to avoid blistering damage caused by accumulation of protons inside the Be, we set the thickness of the Be to 0.5 mm, which is thinner than the penetration range. A Cu heat sink was used for the third layer. To cool the target system from the enormous heat load caused by the high-current proton irradiation, a water channel was placed inside the Cu block. Cooling water in the water channel was circulated at a flow rate of 50 L/min and a high speed of 10 m/s. The second layer, intermediate between the Be layer and the Cu heat sink, is a key layer of the target system. For the material of the second layer, we adopted Pd from candidate materials such as Pd, Nb, Ta, and Ti. This is because the lattice of the material is not easily destroyed even when it contains 10% or more hydrogen. The thickness of the Pd layer was set to 0.5 mm; the same as that of the Be layer. The three materials were firmly bonded via diffusion bonding using the hot isostatic pressing method. The 8 MeV protons do not stop in the Be layer but pass through while releasing neutrons and are finally stopped in the Pd layer. Hence, significant blistering damage can be avoided in the Be while the Pd accumulates the large quantity of protons. The Cu block layer (the third layer) delivers neutrons from upstream layers to a BSA while removing the heat from the proton irradiation.

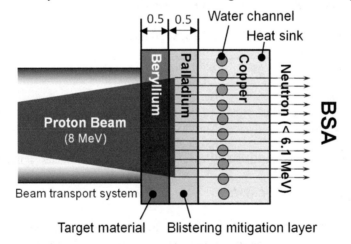

FIG. VIII-2. Schematic of the three-layer structure Be target system at iBNCT.

VIII-2.1.6. Beam shaper assembly

The Be-based target system emits neutrons with a maximum energy of 6.1 MeV following the reaction between the 8 MeV proton beam and Be, then delivers it downstream. The energy of most neutrons passing through the target system is higher than that of the epithermal neutrons, used in the treatment. Thus, the BSA of iBNCT001 plays a role in decreasing the relatively high-energy neutrons to therapeutic epithermal neutrons [VIII-5], as well as emitting them from a beam aperture toward the irradiation field of a patient and minimizing out-of-field leakage beyond the beam aperture.

316

Figure VIII-3 shows a schematic image of the iBNCT001 BSA. First, high-energy neutrons originating from the target system are removed and filtered using an Fe block as a fast-neutron filter. The fast-neutron filter was positioned in close contact to the Be-based target system. Next, the energy of neutrons passed through the fast-neutron filter was further moderated by the moderator block, which was installed just behind the iron block. The moderating material is magnesium fluoride (MgF_2), which is traditionally manufactured as a single crystal; however, it is expensive, and difficult to make into a large block. Thus, we attempted to produce the compound as a sintered body. The success of this development made it possible to manufacture a large block of MgF_2 at low cost, and we could adapt the material to the moderator of our BSA. Two thinner filters are installed behind the MgF_2 moderator block. One is a Cd plate with a thickness of 2 mm that cuts thermal neutrons in the neutrons from the upstream. The other is made of Bi with thickness of 5 mm, which functions as a γ-ray filter. Both types of radiation are unnecessary for treatment. The energy components of the neutrons that pass through the moderator block and several filters are adjusted to be mainly therapeutic epithermal neutrons. A collimator was installed behind the filters. Details of the collimator are explained in the next section.

Meanwhile, regarding shielding, one of the roles of the BSA, radiation leakage (neutrons and γ rays of all energies) outside the beam aperture is shielded as much as possible using several shielding blocks surrounding the moderator and collimator. The shielding materials were mainly polyethylene (PE), LiF + PE, Pb, and concrete. Many blocks of PE and LiF + PE surround the moderator and collimator, and concrete blocks enclose the BSA.

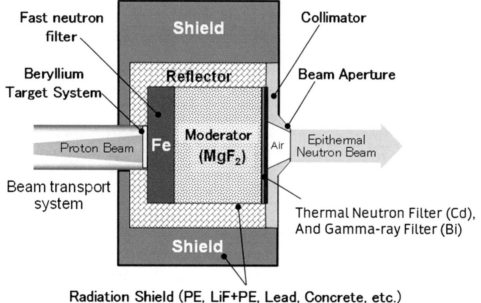

FIG. VIII-3. Schematic of the iBNCT001 BSA with the Be target system.

VIII-2.1.7. Collimator

To provide neutrons to the irradiation field in a patient's body, they are centralized in the beam aperture using the collimator installed behind the γ ray filter. iBNCT001 employs a horizontal irradiation facility. Therefore, patients can be irradiated in both the sitting and lying posture. The collimator consists mainly of Pb and PE blocks, including LiF. Most neutrons from the upstream filters are delivered to the beam aperture by passing through the air beamline. Meanwhile, neutrons that deviate from the beamline and enter the surrounding collimator material (PE and LiF + PE) are scattered within the materials, while some of the neutrons return to the beamline side to increase the neutron intensity at the beam aperture.

317

Several beam apertures made of LiF + PE have been prepared and can be easily exchanged. The optimal size of the beam aperture for each irradiation can be selected and installed by considering the size of the tumour and/or risk organs around the tumour. The appropriate beam aperture used for treatment is determined during treatment planning using the treatment planning system (TPS). In addition, iBNCT001 can be combined with extended beam apertures because of its high neutron intensity and sufficient irradiation time.

In addition, iBNCT001 can be combined with extended beam apertures. As of 2022, three types of extended beam apertures with diameters of 10, 12, and 15 cm were prepared. The distance from the wall to the surface of the beam aperture was extended by 10 cm. In the application of the 10 cm extended beam aperture, the neutron intensity drops to approximately 60%. However, by using an extended beam aperture, interference between the patient and wall can be avoided. This is particularly effective in the irradiation of head and neck cancer and can suppress shoulder interference in patients. Therefore, the patient can receive irradiation in a comfortable posture, even if the irradiation time is long. Consequently, this contributes to the improvement in irradiation accuracy. By applying the extended beam aperture to the treatment, iBNCT001 can generate a neutron beam of sufficient intensity that allows treatment to be completed within an hour. Figure VIII-4 shows a characteristic experiment with a water phantom for the extended beam aperture.

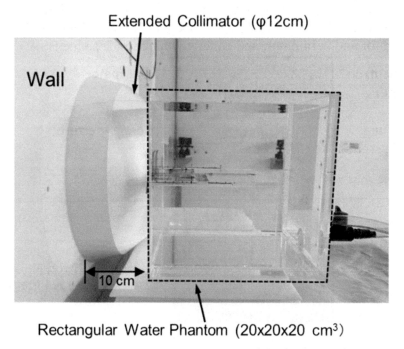

FIG. VIII-4. Experiment for measuring the beam characteristics of an extended beam aperture with a rectangular water phantom.

VIII-2.1.8. Treatment planning system

To perform BNCT, various medical devices related to radiation therapy are required. Therefore, in addition to producing a linac-based neutron source device, the iBNCT project has also developed medical devices such as a TPS, patient positioning system, measurement equipment for boron concentration in blood samples, and radiation monitors. The TPS (the development code name is Tsukuba Plan) has been applied to dose calculations using the Monte Carlo algorithm [VIII-6]. For the Monte Carlo transport code, the particle and heavy ion transport code (PHITS) system, which is being developed and improved by JAEA, has been used [VIII-7]. For the nuclear data library that is needed to perform the Monte Carlo calculation, JENDL-4.0

has been adopted [VIII-8]. The development of the first version of Tsukuba Plan has been completed, and several verifications are being carried out to validate the applicability of the system to treatment planning work in a clinical study with iBNCT001. The verification results demonstrated that Tsukuba Plan has a dosimetry performance that allows dose calculation for treatment planning in BNCT. Tsukuba Plan will be applied in clinical studies using iBNCT001 [VIII-9].

VIII-2.1.9. Patient positioning system

In radiation therapy, including BNCT, positioning of a patient aims to swiftly guide the patient to the irradiation position determined in the treatment plan determined by the TPS and to fix the patient in position accurately. In BNCT, it is possible that the patient's position may change during irradiation because the irradiation time is longer than that of conventional radiation therapy. Therefore, the patient has to be firmly immobilized in the irradiation position to prevent movement. Furthermore, real-time monitoring of the patient's posture and position control during irradiation may be required.

In the iBNCT project, the method of guiding and positioning patients using multi-laser beams was adopted. In this method, before the actual positioning work, many laser beams from the ceiling and both walls of the irradiation room are directed to several points in the irradiation position, such as the nasal apex, eyes, and ears of the patient. During the positioning process, the patient can be guided to the irradiation position by matching each point on the patient to its respective laser beam. The patient is then immobilized in the irradiation position. This method using multiple laser beams is the same as the positioning system applied in the clinical trial performed at the reactor-based BNCT facility at JRR-4, and thus, already has many practical results [VIII-10–VIII-11].

Moreover, the project is also developing another positioning system that can monitor several points on a patient in real time during irradiation. Using motion capture technology with multiple camaras, three-dimensional coordinates of each point on the patient are detected in real time [VIII-12]. When the system is complete, the position of a patient can be monitored in real time during irradiation.

VIII-2.2. Facility design and layout

As mentioned above, the iBNCT001 device is a demonstration device for a linac-based neutron source for BNCT developed by Tsukuba's iBNCT project. The device was installed in a building at the Ibaraki neutron medical research center 'iNMRC' in Tokai village, Japan. Schematic images of the facility layout installed in the iNMRC are shown in Fig. VIII-5. The device needed to be installed in an existing building. Therefore, at the design stage, it was built to fit equipment, beam transport tubes, the BSA, and irradiation rooms that make up the facility according to the layout of each room in the building. Hence, the beam line of the beam transport system for the demonstration device was longer than necessary. The treatment room is located on a different floor from the accelerator room; thus, the layout is slightly complicated. The ion source and both linac tubes (RFQ and DTL) were installed in the accelerator room on the second basement floor. Meanwhile, the BSA with the Be-based target system and the treatment room were installed in a room on the first basement floor. For iBNCT001, the number of treatment rooms with BSA is one, with an area of approximately 20 m^2.

There are other rooms related to the accelerator, such as the cooling device, power generators, and klystron rooms on several floors in the building. Furthermore, other facilities and rooms related to the BNCT research such as the biological laboratory and accelerator control room, are installed on the first floor of the iNMRC building. An animal isolator for housing irradiated mice is also installed in the biological laboratory.

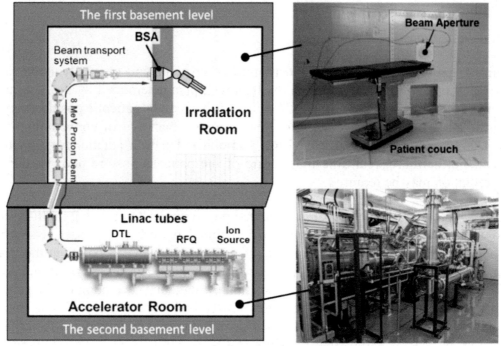

FIG. VIII-5. Layout of the iBNCT001 facility in iNMRC.

VIII-3. COSTS AND MANAGEMENT

VIII-3.1. Financial cost

The iBNCT project was selected as one of the core projects of the 'Tsukuba International Strategic Zone,' which was one of the international strategic zones of Japan in 2011. Strategic zones commit to industrial promotion given an advantage in regulatory standards requirement and financial help from governmental bodies. Hence, iBNCT001 was developed to obtain national competitive funds by utilizing the system of strategic zones. The cost of constructing iBNCT001 was approximately three million US dollars.

Regarding the building where iBNCT001 was installed, Ibaraki Prefecture provided a building for the iBNCT project based on international strategic zones and renovated it so that the demonstration device could be installed. The laboratory, control room, and living rooms have also been improved in the building at the same time.

VIII-3.2. Staff

The University of Tsukuba manages the facility, and KEK, in cooperation with companies that are accustomed to operating radiation facilities, is in charge of operating and maintaining the facility. Two or more staff members are always stationed in the accelerator control room, and the iBNCT001 linac is usually driven by two operators.

ACKNOWLEDGEMENTS

This work was supported by the Tsukuba International Strategic Zone, based on the strategic zones of Japan. The University of Tsukuba Hospital supported the development and operation of the device by collaborating with the project.

REFERENCES TO ANNEX VIII

[VIII-1]   KUMADA, H., et al., Project for the development of the linac based NCT facility in University of Tsukuba, Appl. Radiat. Isot. **88** (2014) 211–215.

[VIII-2]   KUMADA, H., et al., Evaluation of the characteristics of the neutron beam of a linac-based neutron source for boron neutron capture therapy, Appl. Radiat. Isot. **165** (2020) 109246.

[VIII-3]   KUMADA, H., et al., Development of linac-based neutron source for boron neutron capture therapy in University of Tsukuba, Plasma Fusion Res. **13** 2406006 (2018) 1–6.

[VIII-4]   IKEGAMI, M., Beam commissioning and operation of the J-PARC linac, Prog. Theor. Exp. Phys. **2012** (2012) 02B002.

[VIII-5]   KUMDA, H., et al., Development of beryllium-based neutron target system with three-layer structure for accelerator-based neutron source for boron neutron capture therapy, Appl. Radiat. Isot. **160** (2015) 78–83.

[VIII-6]   KUMADA, H., TAKADA, K., SAKURAI, Y., SUZUKI, M., Development of a multimodal Monte Carlo based treatment planning system, Radiat. Protect. Dosim. **180** 1–4 (2018) 286–290.

[VIII-7]   SATO, T., NIITA, K., MATSUDA, N., HASHIMOTO, S., Particle and heavy ion transport code system, PHITS, version 2.52, J. Nucl. Sci. Technol. **50** (2013) 913–923.

[VIII-8]   SHIBATA, K., IWAMOTO, O., NAKAGAWA, T., IWAMOTO, N., JENDL-4.0: a new library for nuclear science and engineering, J. Nucl. Sci. Technol. **48** (2011) 1–30.

[VIII-9]   KUMADA, H., et al., Verification for dose estimation performance of a Monte-Carlo based treatment planning system in University of Tsukuba, Appl. Radiat. Isot. **166** (2020) 109222.

[VIII-10]   KUMADA, H., MATSUMURA, A., NAKAGAWA, Y., Development of the patient setting system for medical irradiation with research reactor, Trans. At. Energy Soc. Jpn. **1** (2002) 59–68.

[VIII-11]   KUMADA, H., et al., Verification of the computational dosimetry system in JAERI (JCDS) for boron neutron capture therapy, Phys. Med. Biol. **49** (2004) 3353–3365.

[VIII-12]   KUMADA, H., et al., Monitoring patient movement with boron neutron capture therapy and motion capture technology, Appl. Radiat. Isot. **163** (2020) 109208.

LINAC-BASED BNCT FACILITY DEVELOPED BY DAWON MEDAX

YOUNG-SOON BAE and the DAWON MEDAX A-BNCT TEAM
BNCT Center, Dawon Medax, 241, Gangnam-daero, Seocho-gu, Seoul, Rep. of Korea

Abstract

Dawon Medax (DM) has developed an epithermal neutron source for boron neutron capture therapy (BNCT) clinics in Korea, 'DM-BNCT'. The proton LINAC accelerator with a Be target, as an epithermal neutron source, is adopted for BNCT development using well-known conventional RF technology that is widely used in modern research accelerators. The proton beam energy and current are 10 MeV and 4 mA (max. average current). The beam shaping assembly (BSA) is designed to provide good neutron beam quality which almost satisfy the IAEA reference values in Ref. [IX-1]. Also, the BSA is designed to have high epithermal neutron energy conversion efficiency, using a figure of merit defined in this paper. The DM-BNCT project is now at the stage of submitting a formal application to the Korean FDA for the license of a human clinical trial for the BNCT facility after the successful completion of animal pre-clinical tests in the fall of 2021.

IX-1. INTRODUCTION

One of the main limitations of BNCT is the lack of appropriate epithermal neutron sources. In the past, the only available neutron beams for BNCT with enough intensity were from research reactors. These are nuclear facilities outside the hospital environment. Such facilities were difficult to adapt to clinical trials. However, in recent decades, high-current, low-energy particle accelerators of various types have been developed. These particle accelerators produce an epithermal neutron beam suitable for BNCT with a specific target system depending on the particle beam energy. At the DM-BNCT facility, a 10 MeV radio-frequency linear accelerator (LINAC) and Be target system has been developed to produce a fast neutron beam via the $^9Be(p,n)^9B$ nuclear reaction [IX-2].

IX-2. PROJECT

Dawonsys Co. and Dawon Medax started the project to develop an accelerator-based boron neutron capture therapy system in 2016 with funding from the Korean government to install an accelerator-based epithermal neutron source at Songdo, Incheon [IX-3]. The neutron source facility was designed and constructed to include the accelerator room, treatment room, control room, the patient preparation and recovery rooms on the underground floor, and the supporting utility and cell laboratory on the ground floor. In this project, an optimized epithermal neutron energy spectrum is obtained satisfying the IAEA reference value for epithermal neutron flux ($> 1 \times 10^9$ $cm^{-2} \cdot s^{-1}$) using a 10-MeV proton radio-frequency linear accelerator technology. The epithermal neutron source system consists of RFQ and DTL accelerators as the main beam accelerators, a duoplasmatron ion source, a beam transport line on which the proton beam profile and current are monitored, and a water-cooled solid Be target for neutron generation from the nuclear reaction $^9Be(p,n)^9B$ inside a beam shaping assembly (BSA). An on-line epithermal neutron beam monitoring system using a $LiCaAlF_6$ scintillator has also been installed. A comprehensive verification and validation testing of the entire system has been conducted including measurement of proton and neutron beam characteristics by licensing authorities for the medical device. Recently, we completed sets of animal pre-clinical test using boron delivery agents of boronophenylalanine (BPA) drug in which the significant therapeutic effect of BNCT has been confirmed and submitted the formal application to the Korean Ministry of Food and Drug Safety to obtain a license for a human clinical-trial of the BNCT facility. A treatment planning system (TPS) has also been developed for BNCT [IX-4] and was used in the pre-clinical test. This annex introduces the DM accelerator-based BNCT facility.

IX-2.1. System

IX-2.1.1. Type of accelerator

The accelerator, which was named 10-MeV Linac-based Pulsed Epithermal Neutron Source (LPENS10), is a linear accelerator using conventional radio-frequency accelerator technology [IX-5–IX-6]. The accelerator consists of an ion source, low energy beam transport (LEBT), 3-m long, 4-vane type Radio Frequency Quadrupole (RFQ), 5-m long Alvarez-type Drift Tube Linac (DTL), and beam transport line (BTL). Figure IX-1 shows the layout of the proton accelerator. The RFQ and DTL are designed to resonate at an RF frequency of 352 MHz. The maximum peak proton beam current of each pulse is 50 mA.

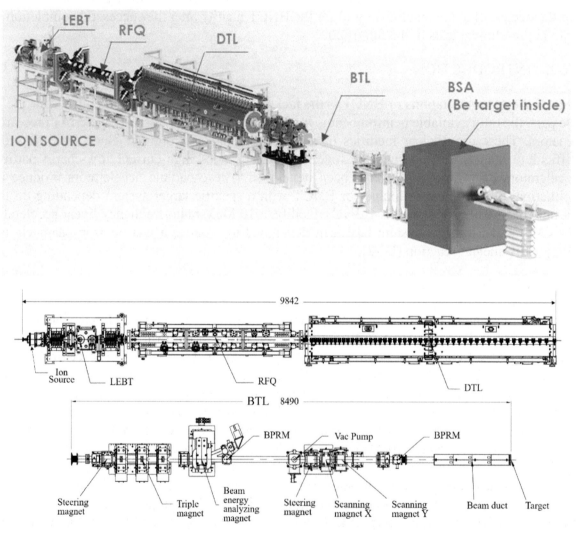

FIG. IX-1. The radio-frequency proton LINAC accelerator and BSA of DM-BNCT.

The proton ion source is a duoplasmatron type. A duoplasmatron ion source is a plasma-based ion source that generates protons from hydrogen gas through arc-discharge. It enables high current proton beam extraction with low emittance (< 0.2 π-mm-mrad in normalized rms unit) which is beneficial for beam matching into the following accelerator. For BNCT treatment of patients, a highly stable ion source is required for constant proton beam current to enable a 1-hour irradiation for the patients at constant neutron flux. The stable proton beam extraction is obtained with the help of a solid-state switching system of plasma arc-discharge and high voltage (50 keV) power supplies. The primary 50 keV proton beam extracted from the ion source is transported through a low energy beam transport equipped with two electromagnetic solenoids to focus the proton beam into the RFQ entrance. The focused and matched proton beam is first accelerated by the RFQ up to 3 MeV, and then it is further accelerated up to 10 MeV by the DTL. The DM-BNCT facility aims for BNCT to become a popular and compact cancer therapy for adoption inside hospitals. Therefore, the final proton beam energy was determined to be 10 MeV in order to ensure sufficient epithermal neutron flux at the exit of the beam port of the BSA. Also, the DTL accelerator is almost directly connected to the RFQ without requiring any additional beam matching part, namely, a Medium Energy Beam Transport (MEBT) between the RFQ and DTL. However, a significant difficulty is to have a duty factor high enough (8–10%) to have a sufficiently high epithermal neutron flux for BNCT treatment. Thus, the linear accelerator system is designed to be operated with a peak beam current of 50 mA and duty factor up to 20%, defined by the maximum pulse duration of 1.3 ms multiplied by the pulse repetition rate of up to 150 Hz. This is indeed a challenging goal in terms of RF breakdown issues inside the RFQ and DTL cavities.

The accelerated beam is transported at 10 MeV to the neutron generating target through the BTL. The main components of the BTL are a triplet quadrupole magnet focusing system and wobbler magnets. The beam is transmitted to the centre of the target 8 m away from the DTL with the use of steering magnets, and the beam size is controlled using the triplet quadrupole magnet system. There are two wobbler magnets (scanning X and Y) in the horizontal and vertical directions of the BTL to rotate the beam pulse onto the target surface in order to reduce the thermal heat flux. The beam alignment is confirmed by measuring the beam profile using the beam profile monitors installed at two different locations: after the triple magnet and before the entrance of the beam duct as shown in Fig. IX-1. The beam loss in the BTL is monitored using a current measurement device, an AC current transformer, installed at the entrance of the beam duct.

IX-2.1.2. Number of beamlines

Currently, only a straight beam line is available. But the BTL line will have a bending magnet that allows directing the ion beam either to treatment rooms at ±45° or straight through.

IX-2.1.3. Beam energy, current

The proton beam energy is 10 MeV with an energy spread of less than 5%. The beam energy is measured using a beam analysing magnet (just a bending magnet) and a single aperture slit. The nominal operation beam current is 2 mA as an average value. It can be varied from 2.3 μA to 2 mA by increasing the pulse width and repetition rate of the ion source's high voltage semiconductor switching system. The plasma discharge of the ion source is controlled to maintain constant peak beam current on the target.

IX-2.1.4. Target materials

The neutron generating target material is Be. The Be target, with rectangular cross section of 10 cm × 10 cm, has two layers: a thin layer of pure Be to generate neutrons via the $^9Be(p,n)^9B$ reaction and a Cu substrate with water cooling micro-channels for efficient heat removal. The Be is brazed onto the Cu substrate. This target is vulnerable to a blistering-limited lifetime, but it will be upgraded to a three-layered target with backing material resistant to hydrogen blistering [IX-7].

IX-2.1.5. Beam shaper assembly

The BSA converts fast neutrons with average energy of 2.8 MeV generated from the Be target to an epithermal neutron beam for clinical applications. It consists of an MgF_2 moderator, a composite reflector of aluminium near the target and lead blocks, and filters for fast neutrons, thermal neutrons, and γ radiation. Figure IX-2 presents the cross-sectional view of the DM-BNCT BSA and collimator. The BSA is movable to investigate radiation from the target and replace the target. The flux of epithermal neutrons is 1.03×10^9 cm^{-2}·s^{-1} at an average current of 2 mA with a 10 MeV proton beam energy, while the contributions of thermal and fast neutrons and γ radiation are acceptable: $\phi_{th}/\phi_{epi} = 1/340$, $D_{fn}/\phi_{epi} = 3.3 \times 10^{-13}$ Gy·cm^2, $D_\gamma/\phi_{epi} = 0.9 \times 10^{-13}$ Gy·cm^2. The epithermal neutron flux per unit proton beam power is 5×10^7 cm^{-2}·s^{-1}·kW^{-1}. The epithermal neutron conversion efficiency of the DM-BNCT BSA is $\phi_{epi}/Y = 2.5 \times 10^{-5}$ cm^{-2}, newly defined as a figure of merit to evaluate the performance of the BSA. Here, Y is the total neutron yield at the Be target integrated over solid angle and energy. Such a neutron beam quality factor gives an approximately 10.0 cm advantage depth (AD) and about 7.7 cm treatable depth (TD) assuming a tumour-to-blood ^{10}B ratio of 3.5 and a ^{10}B-concentration in blood of 18 ppm. The estimation of AD and TD come from the Snyder phantom [IX-8] calculation using the MCNP program [IX-9].

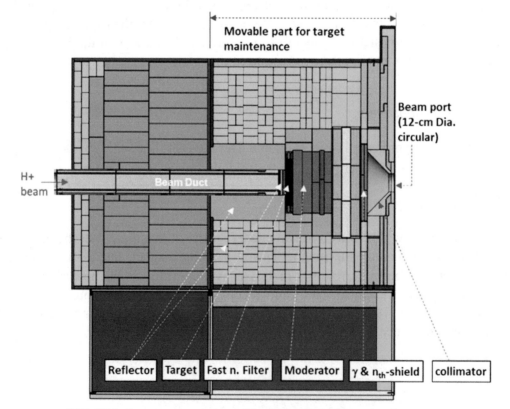

FIG. IX-2. Cross-sectional view of the DM-BNCT BSA and collimator.

IX-2.1.6. Real time monitor

The proton beam current and the epithermal neutron flux at the beam port are monitored in real time. The proton beam current is measured at several different locations in the accelerator using an AC current transformer supplied by the Bergoz company. But the final delivered beam current at the target is measured using a simple resistor circuit and Ohm's law. The detector of the epithermal neutron flux is a $LiCaAlF_6$ crystalline scintillator. It shows good scintillation response under thermal neutrons [IX-10]. The detector is located inside the BSA collimator which is described below. It is calibrated and verified to provide the epithermal neutron flux at the beam aperture port using the gold foil activation method. The neutron and γ radiation levels in the treatment room are also monitored in real time.

IX-2.1.7. Collimator

The DM-BNCT BSA is designed to allow the mainly epithermal energy components of the neutrons pass through the moderator blocks. Epithermal neutrons are centralized in the beam aperture using a collimator installed behind the thermal neutron and γ-ray filters. The collimator mainly consists of lead and boron containing polyethylene blocks. Neutron beams are delivered to the beam aperture at the end of the collimator by passing through a cylindrical cone-shaped air beamline. The cone angle and length are optimized to have the maximum epithermal neutron flux at the beam aperture exit. The beam aperture diameter is 12 cm. Figure IX-3 shows the neutron energy spectrum at the beam aperture exit and the lateral profile of the epithermal neutron flux at certain longitudinal locations. The neutron energy spectrum calculated using MCNP agrees well with multi-foil measurements as shown in Fig. IX-3(a). For the lateral profile of the epithermal neutron flux at the beam port, as shown in Fig. IX-3(b), the flux reduces by about 30% within the beam aperture exit. The peak value of the epithermal neutron flux at the centre of the beam aperture exit decreases by about 10% when the distance from the beam aperture exit is increased by 1 cm.

The installation of an additional beam aperture at the beam port is being planned, to take into account the size and location of tumour and organs at risk around the tumour. The final type of the beam aperture used for the clinical trial will be determined using the treatment planning system.

IX-2.1.8. Treatment planning system

DM-BNCT developed a treatment planning system 'DM-BTPS' for dose calculation. The dose calculation in DM-BTPS is based on the existing seraMC engine code using the Monte Carlo algorithm [IX-11]. But, the original seraMC engine has been modified and upgraded to be adaptive to a modernized parallel computing server and a graphical user interface developed in house for the radiotherapy planning system. The verification of the DM-BTPS during the development are carried out with depth-dose measurement using gold wire in a water phantom and using a paired ion chamber in a tissue-equivalent phantom. The applicability of the system to treatment planning work in a clinical trial is now being validated.

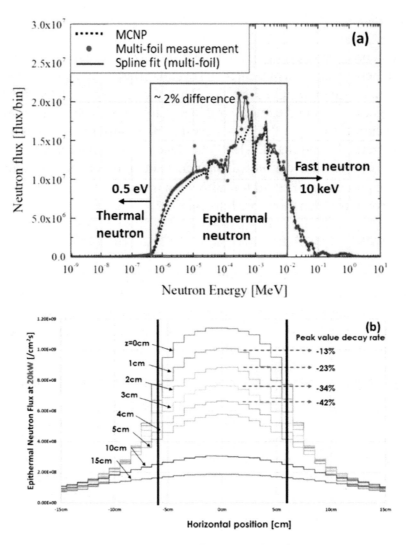

FIG. IX-3. (a) The neutron energy spectrum and (b) the lateral profile of the epithermal neutron flux from the beam exit port out to a distance of 15 cm from the beam port.

IX-2.1.9. Patient positioning system

Since the neutron beam line and the treatment room lie in the horizontal plane, patients can be treated in both sitting and lying postures. Two X ray systems will be used to confirm the position of the patient determined in the treatment plan by the TPS. To fix the patient to the position accurately, a multi-axis rotatable positioning bed and multi-laser beams have been adopted. Once the patient position is fixed, several points on the patient are monitored in real time during treatment using a multiple camera system with motion capture software. It is called the 'motion monitoring system'. Three-dimensional coordinates of each point on the patient can be detected in real time during the treatment so that it can be used to ensure that patient movement does not exceed the acceptable limit determined by the TPS.

IX-2.2. Facility design and layout

Figure IX-4 shows the layout of the DM BNCT facility. The facility consists of an accelerator room, three treatment rooms, and a treatment control room. The neutron source is placed in a radiation-shielded room on the basement floor (Fig. IX-4). The treatment rooms and accelerator control room are on the same floor. The radiation shielding wall is made of concrete. The wall thickness of the accelerator room is 1 m, and the wall between the accelerator room and treatment room is about 2 m. The BSA is installed inside the wall between the accelerator room and treatment room. The width of the facility in the underground floor is 27 m × 34 m and 5 m high. The water pumps and electrical utilities are placed on the first floor. The treatment rooms and accelerator room are maintained with negative air pressure for radiation safety.

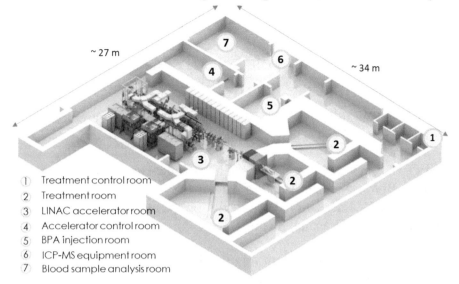

1. Treatment control room
2. Treatment room
3. LINAC accelerator room
4. Accelerator control room
5. BPA injection room
6. ICP-MS equipment room
7. Blood sample analysis room

FIG. IX-4. The layout of the Dawon Medax accelerator based facility.

The facility is equipped with dosimeters, a neutron and γ-ray radiation monitoring system, neutron flux monitors, CCD-cameras in the treatment rooms and accelerator room, high purity germanium γ-ray detector, ICP-MS equipment for boron concentration in blood samples etc.

IX-3. COSTS AND MANAGEMENT

IX-3.1. Financial cost

The main component costs of the facility construction are the LINAC accelerator and BSA. The cost of construction of the building will depend on the number of beamlines and the size of facility determined by the layout of treatment room and the clinical space. The estimated operational and treatment costs are not yet known, but they will not be as high as the cost of the boron delivery agent (BPA).

IX-3.2. Staff

The Dawon Medax BNCT centre has two divisions for the development, operation, and maintenance of the accelerator-based BNCT facility; one is the accelerator physics division and the other is the engineering division. The total number of staff of the two divisions is about two dozen employees including three PhD scientists and many mechanical and electrical engineers. Two or three operators, mainly RF and electrical engineers, are trained for the operation of the facility in a clinical trial. There are also a dozen staff working for the clinical trial and boron delivery drug including two medical doctors.

IX-3.3. Number of patients per year expected

The first phase of the clinical test is planned to start in 2022 with 20–30 patients expected to be treated. But the actual number of patients is under discussion with medical doctors of the hospitals participating in the clinical trial.

ACKNOWLEDGEMENTS

This project was supported by the Ministry of Trade, Industry and Energy.

REFERENCES TO ANNEX IX

[IX-1]   INTERNATIONAL ATOMIC ENERGY AGENCY, Current Status of Neutron Capture Therapy, TECDOC 1223, IAEA, Vienna (2001).

[IX-2]   HAWKESWORTH, M.R., Neutron radiography equipment and methods, At. Energy Rev. **15** (1977) 169–220.

[IX-3]   SEO, H.J., et al., "Current status of non-clinical trials using proton linear accelerator based Boron Neutron Capture Therapy System in Republic of Korea," Plenary lecture, 19th International Congress on Neutron Capture Therapy, September1 October 2021, Granada, Spain.

[IX-4]   YI, J., et al., "Development and validation of a dose calculation code for BNCT TPS based on the seraMC", Technical Meeting on Best Practices in Boron Neutron Capture Therapy, 14–18 March 2022 IAEA, Vienna (2022).

[IX-5]   KIM, D.S., et al., "Development of the accelerator-based Boron Neutron Capture Therapy system for cancer treatment within 1 hour therapeutic time", 17th International Congress on Neutron Capture Therapy, 2–7 October 2016, University of Missouri in Columbia, Missouri, USA (2016).

[IX-6]   LEE, C.H., et al., Status of development and planning activities on CANS in Korea, J. Neutron Res. **23** 2–3 (2021) 127–141.

[IX-7]   YAMAGATA, Y., et al., Development of a neutron generating target for compact neutron sources using low energy proton beams, J. Radioanal. Nucl. Chem. **305** 3 (2015) 787–794.

[IX-8]   TORRES-SÁNCHEZ, P., et al., Optimized beam shaping assembly for a 2.1-MeV proton-accelerator-based neutron source for boron neutron capture therapy, Sci. Rep. **11** (2021) 7576.

[IX-9]   WERNER, C.J., et al. (Eds), MCNP Version 6.2 Release Notes. 2, LA-UR-18-20808 (2018).

[IX-10] YANAGIDA, T., et al., Europium and sodium co-doped $LiCaAlF_6$ scintillator for neutron detection, Appl. Phys. Express **4** (2011) 106401.

[IX-11] NIGG, D., et al., SERA an advanced treatment planning system for neutron therapy and BNCT, Trans. Am. Nucl. Soc. **80** (1999) 223–232.

KANSAI BNCT MEDICAL CENTER OF
OSAKA MEDICAL AND PHARMACEUTICAL UNIVERSITY

KOJI ONO
Kansai BNCT Medical Center, Osaka Medical and Pharmaceutical University,
2-7 Daigaku-machi, Takatsuki-shi, Osaka 569-8686, Japan

Abstract

The Kansai BNCT Medical Center is equipped with the world's first boron neutron capture therapy (BNCT) irradiation system 'NeuCure', which was developed by the BNCT research group of the Institute for Integrated Radiation and Nuclear Science, Kyoto University (KURNS) in collaboration with Sumitomo Heavy Industries and Stellar Pharma, Inc. The unit at our centre is the third since the first unit at Kyoto University. In June 2020, we started BNCT for recurrent or locally advanced inoperable head and neck cancer with insurance coverage. In addition to this clinical activity, we are working on clinical and basic research areas.

X-1. INTRODUCTION

In 2018, the Kansai BNCT Medical Center installed the third unit of the world's first BNCT irradiation device (NeuCure), which was developed by the BNCT research group of KURNS in collaboration with Sumitomo Heavy Industries and Stellar Pharma, and started BNCT for recurrent and locally advanced inoperable head and neck cancer in June 2020. We are carefully exploring the efficacy and adverse events of BNCT with gradual increases in treatment dose. In clinical research, we are conducting a physician-led phase II trial for recurrent high-grade meningioma (WHO II and III). The clinical research on positron emission tomography imaging conducted with 4-borono-2-fluoro-L-phenylalanine (FBPA-PET), which is essential for the advancement of BNCT, is also a very important research task. As for basic research, we are studying the BNCT effects of novel boron compounds in cultured cell systems, the effects on excised tumours after boron drug administration, and micro-distribution of drugs in tissues by α-autoradiography. In medical physics research, where radiation control regulations are less stringent, we are also investigating more efficient irradiation methods and more optimum neutron beam formation. This paper reports on the current status of the centre's facilities, staff, and activities.

X-2. FACILITIES IN OUR CENTRE

X-2.1. System

The BNCT system is described in Table X-1.

TABLE X-1. DESCRIPTION OF THE BNCT SYSTEM

Component	Description
Type of accelerator	Cyclotron
Number of beamlines	1
Beam energy, current	30 MeV, 1 mA
Target materials	Be
BSA design	see Fig. X-1
Collimator	0 cm-long, diameters: 100 mm, 120 mm, 150 mm
	5 cm-long, diameter 120 mm
	10cm-long, diameter 120 mm

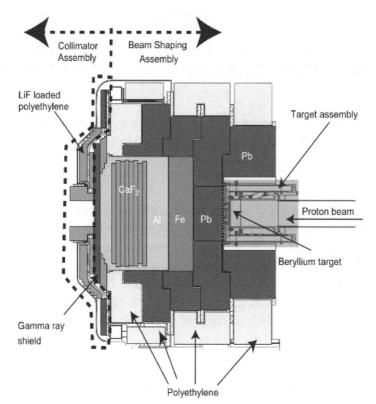

FIG. X-1. The BSA in use at Kansai BNCT Medical Center.

X-2.2. Patient positioning system

The patient positioning system follows the steps below:

(a) Determination of the irradiation target area based on medical images: computed tomography (CT)/magnetic resonance imaging (MRI) and positron emission tomography (PET);

(b) Tentative determination of neutron beam direction and irradiation position by X ray CT simulation;

(c) Creation of an inner mask to fix the affected area (head and neck);

(d) Confirmation of the feasibility of irradiation in the tentatively determined positions in the simulation room by fixing the outer mask, which was further layered on the inner mask, to the collimator;

(e) After the above preliminary preparations, on the day of BNCT, the patient is fixed on a supine bed or a sitting irradiation chair with a collimator in the room next to the neutron irradiation room;

(f) Confirm whether the irradiation field and the irradiation centre are set correctly to the collimator hole by viewing from the upstream side of neutron irradiation;

(g) After making the necessary fine adjustments, the unit is automatically transferred to the irradiation room and fixed to the neutron irradiation port;

(h) Perform the neutron irradiation.

X-2.3. Facility design and layout

The facility is in a 4-storey building of 4017 m² area in total. Regular concrete is used except for the wall between the cyclotron-beam shaper assembly room and the irradiation room (where concrete containing boron is used). Table X-2 gives a description of the layout, and Table X-3 some of the equipment inside.

TABLE X-2. GENERAL DESCRIPTION OF THE FACILITY LAYOUT

Floor	Description of functions
1st basement	Ion source, waste storage etc.
1st floor	Area for performing BNCT as well as a hot laboratory for radioisotope tracer labelling for PET cyclotrons for BNCT and radioisotope production for PET.
	The PET and BNCT areas, including ion source, waste storage etc., are designated as radiation-controlled areas.
	The Hot Laboratory, where tracers for PET are synthesised, is controlled at negative pressure to ensure that no air containing tracers leaks outside the room.
2nd floor	Rooms for BNCT patient examination, boron drug administration, to simulate irradiation position setting and X ray-CT room.
	Rooms for PET-CT examination.
	Air pressure differential and air exchange are maintained in the treatment room: air is drawn in at ambient pressure and then filtered out.
3rd floor	Study spaces for medical doctors, medical physicists, radiological technologists et al.
	Working space for administrative staff of the centre

TABLE X-3. DESCRIPTION OF EQUIPMENT IN THE FACILITY

Equipment	No of units
Spectrometers	
ICP	1
UV spectrometer	1
Dosimeters	
Areas monitors	3
Gamma-ray monitors	3
Neutron monitors	2
Survey meters	4
High purity germanium detector	1
Pocket chambers	Many

X-2.4. Costs and management

The facility is managed by the BNCT Joint Clinical Research Institute, a higher-level organization with jurisdiction over the BNCT centre. A committee known as the Cancer Board, consisting of doctors from related departments at Osaka Medical and Pharmaceutical University, deliberates on the suitability of candidate cases for BNCT. There is a monthly staff meeting attended by all staff. It is expected that up to 200 patients could be treated each year. Table X-4 describes the capital, operational and treatment costs of the centre, Table X-5 gives a list of staff and management positions.

TABLE X-4. COSTS ASSOCIATED WITH THE FACILITY

Cost component	Cost (¥)
Capital	
Facility construction cost (Building involving BNCT and PET facilities)	3.1 billion
Accelerator cost (Cyclotrons for BNCT and RI generation for PET)	3.0 billion
Operational	
Annual operational cost	135 million
Treatment	
Electricity/case	200 thousand
Labour cost for the present all staff/year	130 million
Boron drug (average)/case	1.7 million

TABLE X-5. LIST OF STAFF AND MANAGEMENT

Position	No.
Staff	
Medical doctors	
Radiation Oncologist (radiation oncology specialist, BNCT specialist)	3
Neurosurgeon (neurosurgery specialist, BNCT specialist)	1
Head & Neck Surgeon (otolaryngology specialist, BNCT specialist)	1
Nuclear Medicine Specialist	1
Medical physicists	2
Radiopharmaceutical scientist (pharmacist)	1
Radiological technologists	3
Nurses	3
Administrative staff	6
Management	
Administrative officer of BNCT centre (Director of Joint Clinical Institute)	1
Director of the centre (Chief Professor of Radiation Oncology)	1
Deputy Director (OMPU Hospital Director, Chief Professors of Head and Neck Surgery and Neurosurgery)	3
Assistant Deputy Director	2

Note: The management are part of the BNCT Joint Clinical Research Institute

X-3. PROJECT

The project undertakes both basic and clinical research.

X-3.1. Basic research

X-3.1.1. Development of novel boron compounds

Our centre is not equipped to conduct in vivo studies to examine the effects of BNCT on tumours in animals and the response of normal tissues. However, it is possible to verify the effects in vitro using cultured cells. In addition, it is available to irradiate tumours removed after administration of boron compounds with neutrons and to study their effects by in vitro colony formation assay. Furthermore, we are investigating the microscopic distribution of boron compounds in tumours and normal tissues by autoradiography.

Researchers in the university can use the facilities for these studies. In addition, off-campus researchers can also use the facilities in collaboration with the researchers in the centre.

X-3.1.2. Research in medical physics

Medical physics research that contributes to the improvement of actual treatment techniques is very important. BNCT at the present stage is still in its infancy and many points need to be improved. Since slow neutrons are used in BNCT, it is desirable to place the affected area as close as possible to the collimator. However, in BNCT for head and neck cancer, it is often difficult to bring the collimator and the affected area close to each other due to obstructions such as shoulders. Therefore, we are trying to solve this problem by developing an elongated collimator. Fortunately, when the elongated collimator is attached, the air gap between the beam shaping assembly and the collimator exit becomes larger, and in addition, the leakage of neutrons from the gap between the collimator and the patient's affected area is prevented. As a result, the neutron dose rate at the collimator exit is greatly increased and the irradiation time is significantly reduced.

In BNCT, it is not easy to deliver enough thermal neutrons to the deep region of the body. Of course, if the energy is increased drastically, the flux of neutrons reaching deep in the body will increase, but the neutrons will not be converted into thermal neutrons, and BNCT will not work. Therefore, it is necessary to adjust the neutron energy spectrum delicately. We are developing this method.

X-3.2. Clinical research

In BPA-BNCT, the estimation of tumour boron concentration by ^{18}F-FBPA-PET is important to accurately select the patients who are eligible for the indication. Currently, the Japanese administrative authorities have not yet approved ^{18}F-FBPA-PET as a diagnostic imaging method. Therefore, we are conducting a clinical study of ^{18}F-FBPA-PET in patients with various cancers to search for future indications. Within the framework of this research, we are also performing ^{18}F-FBPA-PET on head and neck cancer patients before BNCT to estimate tumour boron concentration and are using it as basic information for tumour dose calculation. We are planning to analyze the relationship between boron dose and tumour response.

BNCT for recurrent malignant glioma has not yet been approved as a medical treatment by the Japanese government, even though its overall survival is much longer than the experience of clinicians and other reports. One of the reasons for this, officials claim, is that there is no evidence of tumour shrinkage. This is because contrast-enhanced MRI misinterprets pseudo-progression as true progression. As long as the RANO (Response Assessment in Neuro-Oncology) criteria based on contrast-enhanced MRI are used, this problem cannot be resolved. Therefore, we will evaluate the acute response of the tumour by ^{18}F-FBPA-PET and estimate the boron neutron dose for the tumour based on the boron concentration determined by PET, and also clarify the relationship of the response with the dose. We are conducting a clinical study for this purpose.

Malignant meningiomas are the most common radioresistant malignant brain tumours. Furthermore, recurrence is extremely difficult to treat. We are currently conducting a clinical trial to explore the effects of BNCT on such cases. Currently, patient enrolment has been completed and we are observing the progress of the patients after treatment.

REFERENCE TO ANNEX X

[X-1]   TANAKA, H., et al., Experimental verification of beam characteristics for cyclotron-based epithermal neutron source (C-BENS), Appl. Radiat. Isot. **69** 12 (2011) 1642–1645.

Annex XI.

CYCLOTRON-BASED EPITHERMAL NEUTRON SOURCE AT KURNS

H. TANAKA[1], M. SUZUKI[1], T. MITSUMOTO[2], K. ONO[3]
[1]Institute for Integrated Radiation and Nuclear Science, Kyoto University, Japan
[2]Sumitomo Heavy Industries, Ltd., Japan
[3]Kansai BNCT Medical Center, Osaka Medical and Pharmaceutical University, Japan

Abstract

The accelerator neutron source installed at the Institute for Integrated Radiation and Nuclear Science, Kyoto University (KURNS), is composed of a 30-MeV, 1-mA cyclotron, beam transport system, Be target, beam-shaping assembly, and an irradiation system. The device was installed in 2008, and after undergoing physical measurements and non-clinical tests, the world's first clinical test using an accelerator neutron source was conducted in 2012. This paper provides an overview of the cyclotron-based neutron source and facility installed at KURNS.

XI-1. INTRODUCTION

KURNS has conducted clinical research on more than 500 cases of brain cancer, head and neck cancer, malignant pleural mesothelioma, etc. using the Kyoto University Research Reactor [XI-1]. While good clinical results with respect to boron neutron capture therapy (BNCT) can be obtained, it is difficult to install a research reactor in a medical institution; therefore, the development of a neutron source using an accelerator is desired. In the early 2000s, accelerator neutron sources using low-energy protons and Li targets were developed [XI-2]. However, because of the lack of current in the accelerator and the low melting point of the Li target, a neutron source that can be installed in a medical institution to perform BNCT has not yet been realized. KURNS and Sumitomo Heavy Industries, Ltd. (SHI), which possess cyclotron manufacturing technology, have been developing accelerator neutron sources since 2007, focusing on the high generation rate of neutrons generated by the incidence of relatively high-energy protons on Be targets.

XI-1.1. Project

After designing and manufacturing the cyclotron and beam-shaping assembly, a cyclotron-based epithermal neutron source (C-BENS) was installed at KURNS in 2008. After physical measurements using phantoms and non-clinical trials [XI-3–XI-4], the world's first phase I clinical trial for recurrent malignant glioma using an accelerator-based neutron source was conducted in 2012. Before starting phase II clinical trials, a second C-BENS was installed at the Southern Tohoku BNCT Research Center (STBRC) in 2014. Phase II clinical trials were initiated in 2015 [XI-5]. KURNS, Osaka Medical College, and STBRC participated in the clinical trials hosted by SHI and Stella Pharma Corporation, which manufactures ^{10}B-p-borono-phenylalanine (borofalan(^{10}B)). Phase I clinical trials for head and neck cancer also began at KURNS in 2014. Phase II clinical trials were initiated at the STBRC in 2016 [XI-6]. KURNS, Kawasaki Medical College, STBRC, and the National Cancer Hospital Japan participated in the clinical trials led by SHI and Stella Pharma Corporation. After the clinical trials, SHI obtained approval as a medical device for head and neck cancer in March 2020. This system is called 'NeuCure' and is available in Japan.

XI-1.2. System

XI-1.2.1. Type of accelerator

Sumitomo Heavy Industries, Ltd has a history of developing various types of cyclotrons. Based on this experience, SHI has developed a cyclotron called HM-30 that can generate a high-current proton beam. This cyclotron is capable of producing a 30-MeV, 1-mA proton beam. Table XI-1 shows the main characteristics of the HM-30. Figure XI-1 is a photograph of the HM-30. To generate a proton beam of 30 MeV and 1 mA, acceleration of negative hydrogen ions, charge conversion extraction by carbon foil, and vertical injection from an external ion source are adopted. In positive ion acceleration, the beam loss during beam extraction by the deflector becomes large, making it a source of heat and radiation generation. However, it is possible to change the extraction orbit by charge conversion using foil, allowing the extraction of the proton beam with almost 100% efficiency.

TABLE XI-1. THE MAIN CHARACTERISTICS OF THE HM-30 CYCLOTRON

Characteristic	
Accelerated Particle	Negative hydrogen ion
Extraction Energy	30 MeV
Extraction Method	Carbon foil stripping
Nominal Operation Current	1 mA
Magnet Size	3.0 m × 1.6 m × 1.7 m
Weight	60 tons

FIG. XI-1. The HM-30 cyclotron installed at KURNS.

XI-1.2.2. Number of beamlines

The KURNS building was originally planned for use for a purpose other than C-BENS. Only one irradiation room is present. The proton beam extracted from the cyclotron is bent using a bending magnet and guided to the Be target. A magnet is installed in the beamline to form a beam.

XI-1.2.3. Beam energy, current

In HM-30, negative hydrogen ions are accelerated to 30 MeV and can be extracted with a current of 1 mA. To generate a 1-mA proton beam, the ion source can generate negative hydrogen ions with currents of 10 mA or greater.

XI-1.2.4. Target materials

When using protons with energies far exceeding 30 MeV, reactions with heavy metals, such as Ta and W could be used. However, when heavy metals are used, it is expected that it will be difficult to replace the target because the activation of the target is high. Therefore, Be metal was selected as the neutron generation target. It was set to be thinner than 5.8 mm, which is the range of Be for protons with an energy of 30 MeV. Consequently, the protons from the beam do not stay in Be, but rather stop in the cooling water on the back surface, preventing blistering. To reduce the heat load on Be, two magnets are installed on the beamline, and the proton beam can be expanded.

XI-1.2.5. Beam shaper assembly

Figure XI-2 is a schematic layout of the beam-shaping assembly. Irradiation of the Be target with 30-MeV protons produces high-energy neutrons of up to 28 MeV in the forward direction. Lead and Fe are installed behind the target, and high-energy neutrons lose energy via inelastic scattering. Lead has higher threshold energy; therefore, it is installed upstream. Lead is also installed as a reflector material around the moderator. Those neutrons with energy of approximately 1-MeV become epithermal neutrons when passing through Al and CaF_2. Aluminium has valleys in the cross-section near 27 keV and 100 keV; therefore, these energies are transmitted. Since a neutron energy of 100 keV is too high for use in BNCT, such neutrons are attenuated by the resonance cross-section of F. Therefore, the neutron energy spectrum that has passed through this BSA forms a peak near 27 keV, designed to treat deeply sited tumours. Cadmium is installed downstream of Al and CaF_2 to shield thermal neutrons. Lead can be placed behind Cd to shield the γ rays in the treatment beam.

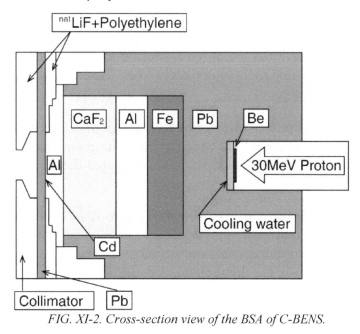

FIG. XI-2. Cross-section view of the BSA of C-BENS.

XI-1.2.6.   Real-time monitor

An ammeter that monitors the proton beam current incident on the target is installed in the beamline upstream of the target. Neutrons are indirectly monitored by determining the relationship between the proton charge and thermal neutron fluence in the water phantom installed after the collimator. Although it is under development, we are proceeding with the testing of a multi-type ionization chamber [XI-7], a real-time neutron monitor that combines a tiny scintillator and an optical quartz fibre [XI-8].

XI-1.2.7.   Collimator

The collimator is made of polyethylene containing LiF and installed behind the Pb gamma shield. Collimators with diameters of 10 cm, 12 cm, and 15 cm are used to reduce the dose outside the irradiation field. The shielding effect was confirmed by conducting an irradiation test using a water phantom to simulate a human body [XI-9]. In the future, a collimator with a maximum diameter of 25 cm will also be available, considering irradiation of the trunk.

XI-1.2.8.   Patient positioning system

A laser pointer is installed on the wall of the irradiation room, and it is possible to correlate it with a marker that matches the position set in the simulation room. Before irradiation, an X ray generator is used for computed radiography imaging to confirm the beam direction with the digitally reconstructed radiographic image created in the treatment plan, and the image can be acquired using an imaging plate.

Since BNCT has to be performed as close as possible to the collimator, an irradiation system that can be set in the sitting position is installed when irradiating the head and neck. In the case of brain tumours, irradiation is often performed in a lying position with an irradiation bed. The patient is fixed using a mask and suction bag to suppress movement during irradiation.

XI-1.3. Facility design and layout

Figure XI-3 shows a cross-sectional view of the first floor in the Innovation Research Laboratory Medical area at KURNS. Since it was originally installed in a space not designed for C-BENS, it is not necessarily an optimized layout. In KURNS, the shielding thickness is sufficient to allow irradiation time for approximately five patients per week. The patient setting is performed in the preparation room.

In addition, the activated air is properly treated by the exhaust duct and exhaust fan. B_4C rubber is installed to reduce the activation of the concrete in the irradiation room. We evaluated the thermal neutron intensity in the irradiation room and reported the activation of the concrete [XI-10].

There is a space for an ICP installation required for clinical BNCT on the third floor, CT for treatment planning on the second floor, and dosimetry devices required for QA. A high-purity germanium detector with background shielding is used to measure the gamma rays emitted by the activation metal foils for QA. The thermoluminescence dosimeter equipment is also located in the preparation room.

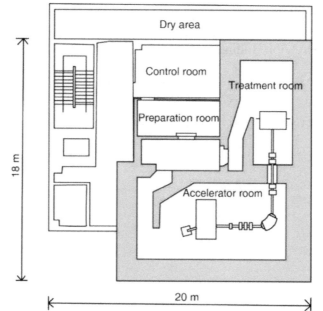

FIG. XI-3. Schematic layout of the facility with C-BENS installed.

REFERENCES TO ANNEX XI

[XI-1]   SAKURAI, Y., KOBAYASHI, T., Characteristics of the KUR Heavy Water Neutron Irradiation Facility as a neutron irradiation field with variable energy spectra, Nucl. Instrum. Methods Phys. Res. Sect. A **453** 3 (2000) 569–596.

[XI-2]   ALLEN, D.A., BEYNON, T.D., GREEN, S., Design for an accelerator-based orthogonal epithermal neutron beam for boron neutron capture therapy, Med. Phys. **26** 1 (1999) 71–76.

[XI-3]   TANAKA, H., et al., Characteristics comparison between a cyclotron-based neutron source and KUR-HWNIF for boron neutron capture therapy, Nucl. Instrum. Methods Phys. Res. Sect. B **267** 11 (2009) 1970–1977.

[XI-4]   TANAKA, H., et al., Experimental verification of beam characteristics for cyclotron-based epithermal neutron source (C-BENS), Appl. Radiat. Isot. **69** 12 (2011) 1642–1645.

[XI-5]   KAWABATA, S., et al., Accelerator-based BNCT for patients with recurrent glioblastoma: a multicenter phase II study, Neuro-Oncol. Adv. **3** 1 (2021) 1–9.

[XI-6]   HIROSE, K., et al., Boron neutron capture therapy using cyclotron-based epithermal neutron source and borofalan (^{10}B) for recurrent or locally advanced head and neck cancer (JHN002): An open-label phase II trial, Radiother. Oncol. **155** (2021) 182–187.

[XI-7]   FUJII, T., et al., Study on optimization of multi-ionization-chamber system for BNCT, Appl. Radiat. Isot. **69** 12 (2011) 1862–1865.

[XI-8]   TANAKA, H., et al., Note: Development of real-time epithermal neutron detector for boron neutron capture therapy, Rev. Sci. Instrum. **88** 5 (2017) 056101.

[XI-9]   TSUKAMOTO, T., et al., A phantom experiment for the evaluation of whole body exposure during BNCT using cyclotron-based epithermal neutron source (C-BENS), Appl. Radiat. Isot. **69** 12 (2011) 1830–1833.

[XI-10]   IMOTO, M., et al., Evaluation for activities of component of Cyclotron-Based Epithermal Neutron Source (C-BENS) and the surface of concrete wall in irradiation room, Appl. Radiat. Isot. **69** 12 (2011) 1646–1648.

BORON COMPOUNDS USED IN THE FIRST CLINICAL TRIALS IN PATIENTS WITH INTRACRANIAL TUMOURS IN THE USA

AGUSTINA MARIANA PORTU[1,2], MARÍA SILVINA OLIVERA[1]

[1]Comisión Nacional de Energía Atómica (CNEA), Av. General Paz 1499, B1650KNA, San Martín, Buenos Aires, Argentina
[2]Consejo Nacional de Investigaciones Científicas y Técnicas (CONICET), Godoy Cruz 2270, Ciudad Autónoma de Buenos Aires, Argentina

Abstract

A summary of the first clinical trials of BNCT for patients with intracranial tumour[1] in the USA from 1951 to 1961 is presented, describing the boron compounds used. Among a considerable amount of boron compounds studied both experimentally and in biodistributions, Sodium tetraborate, sodium pentaborate, *p*-carboxyphenylboronic acid (PCPB) and sodium *closo*-decahydrodecaborate (GB-10) were chosen for clinical trials. The number of patients, the route of administration, the facility and the clinical outcome are reported.

XII-1. INTRODUCTION

The disappointing results of the first trials at Brookhaven National Laboratory (BNL) and Massachusetts Institute of Technology (MIT) were attributed to inadequate penetration of the thermal neutron beams, poor localization of boron in the tumour and/or high boron concentrations in blood, contributing to damage to the vascular endothelium [XII-2]. The results of early trials are summarized in Table XII-1. They involved four compounds:

XII-1.1. Sodium tetraborate ($Na_2B_4O_7 \cdot 10H_2O$)

Borax (sodium tetraborate) was the first compound considered for the treatment of malignant glioma BNCT at BNL. Biodistribution studies ($n = 58$; [XII-3]) and the first series ($n = 10$) of clinical trials were performed between 1951 and 1953 at the Brookhaven Graphite Research Reactor using this compound. For the clinical trials, borax enriched with ^{10}B (96%) in glycerol was administered intravenously (i.v., 19–46 mg·kg^{-1} body weight-bw) 17–40 min prior irradiation to the intact scalp and skull (no craniotomy was performed). No curative effect and painful scalp lesions were observed [XII-4–XII-6].

XII-1.2. Sodium pentaborate ($Na_2B_5O_8 \cdot 4H_2O$)

It was used in the second ($n = 9$) and third ($n = 10$) series of the clinical trials of malignant glioma at the Brookhaven Graphite Research Reactor (approx. 1953–1959) and in the clinical trial performed at the Brookhaven Medical Research Reactor ($n = 17$, 1959–1961). Sodium pentaborate was administered (16–50 mg·kg^{-1} bw) with D-glucose in the molar ratio 2:1. Craniotomy was performed to expose the cortical surface. In some cases, the route of injection was the internal carotid artery, while other patients were infused through a peripheral vein. The

---

[1] Some studies refer to malignant glioma. As the patients were advanced or even terminal, it is believed they could all be considered glioblastoma multiforme. For example, in the third series of the BNL treatments, a patient cerebral sarcoma was treated [XII-1].

irradiation was performed approximately 30 min after the boron compound injection [XII-7]. The outcome was unsuccessful in all of these series [XII-8–XII-10].

XII-1.3. *p*-carboxyphenylboronic acid

At the Massachusetts General Hospital, *p*-carboxyphenylboronic acid (PCPB) was selected among 140 boron compounds developed by Soloway and co-workers after several experimental studies (e.g. Ref. [XII-11]). It was tested in toxicological studies in terminal glioblastoma multiforme (GBM) (or other cerebral diseases) patients [XII-12] and biodistribution studies performed in patients while they were undergoing craniotomy for tumour resection [XII-13]. The compound was synthetized as ^{10}B-enriched PCPB [XII-14] and it was administered (i.v., 15–30 mg·kg^{-1} bw) to 16 patients with advanced glioblastoma multiforme[2] and they were irradiated at open craniotomy, at the MIT nuclear reactor between 1959 and 1961 (first series). No therapeutic effect was observed, and extensive radiation necrosis was observed in some of them. All of the patients were dead within a year [XII-15].

XII-1.4. Sodium *closo*-decahydrodecaborate (GB-10, Na$_2$B$_{10}$H$_{10}$)

GB-10 (CAS 666747-96-2) is a sodium salt of the polyhedral borane anion $(B_{10}H_{10})^{-2}$ [XII-16]. It is a non-toxic, hydrophilic compound that does not cross the blood brain barrier and it is highly stable in aqueous solution [XII-17]. Along with BPA and BSH, it is accepted for human use [XII-18].

Sodium perhydrodecaborate[3] was used in the second series of the Massachusetts General Hospital's BNCT trial. GB-10 proved to be the most soluble, chemically stable and biologically inert of all of the tested compounds [XII-13]. It was used in two patients, who were infused with 30 mg·kg^{-1} bw via i.c. injection. The procedure and outcome were comparable to the first series of Massachusetts General Hospital's patients. Failure of the treatment was not associated with compound toxicity, but with higher levels of boron in vessels than in the tumour [XII-15].

In 2002, Diaz and co-workers performed toxicological studies in 15 patients (GBM or non-small cell lung cancer) using GB-10 (from 5 to 30 mg ^{10}B·kg^{-1}) [XII-19]. These studies were carried out in the frame of boron neutron capture-enhanced fast neutron therapy proposal at the University of Washington.

Note that the bibliography mentions the use of boric acid derivatives in the initial clinical trials (e.g., Refs [XII-14, XII-17, XII-20]), but no specific literature demonstrating this statement could be found. It is not clear if Ref. [XII-12] performed toxicological studies on humans. The first radiobiological studies of BNCT were carried out using boric acid [XII-21]. The pharmacokinetics of boric acid in humans has been explored and reviewed [XII-22].

---

[2] One turned out to be an amelanotic melanoma.
[3] This is the nomenclature used in the bibliography of the 1950s, which refers to the same compound as GB-10 (they have the same IUPAC name). In more recent bibliography, sodium perhydrodecaborate and GB-10 are mentioned indistinctively [XII-14, XII-18].

TABLE XII-1. SUMMARY OF EARLY CLINICAL TRIALS AND OUTCOMES

Period	Reactor	Ser.	n	Pathology	Compound	Solution.	Dose (mg/kg)	Route	Craniotomy	Ref.	Overall effect	MeST (mos.)	Survival range (mos.)
1951–1953		1	10		ST ^{10}B (96%)	glycerine	16–46	i.v.	N	[XII-5]	No effect	97	43–185
?–?	BGRR	2	9	MG[a]	SP	aqueous (D-glucose)	32–50	i.v./i.c.	Y		Scalp lesions due to radiation	147	93–337
?–1959		3	10		SP	aqueous (D-glucose)	26–60	i.c.	Y			96	29–158
1959–1961	BMRR	1	17	MG[b]	SP	aqueous (D-glucose)	?	i.v.	Y			87	
1959–1961	MIT	1	16	GBM[c]	PCPB with ^{10}B		15–30	i.v.	Y	[XII-15]	Extensive radiation necrosis; recurrent tumour; massive intracranial haemorrhage; acute bacterial meningitis	180	<1 year.
1959–1961		2	2	GBM	PCPB		30	i.c	Y				

Note: Ser.: Series ; n : number of participants in trial; Obs.: Observation; MG: malignant glioma; GBM : glioblastoma multiforma; PCPB: sodium perhydrodecaborate (Na$_2$B$_{10}$H$_{10}$); SP: sodium pentaborate; ST: sodium tetraborate; i.v.: intravenous; i.c. intracarotoid; MeST: Median Survival Time; MIH Massachusetts Institute of Technology; BMRR: Brookhaven Medical Research Reactor; BGRR: Brookhaven Graphite Research Reactor.

[a] Ref. [XII-1] reported a case of cerebral sarcoma.
[b] Ref. [XII-6] reported one case of another cerebral tumour.
[c] 1/16 turned out to be amelanotic melanoma.

345

REFERENCES TO ANNEX XII

[XII-1]   FARR, L.E., HAYMAKER, W., CALVO, W., YAMAMOTO, Y.L., LIPPINCOTT, S.W., Neutron-capture therapy in a case of cerebellar sarcoma treated initially with X-radiation – A clinical and histological study, Acta Neuropathol. **1** 1 (1961) 34–55.

[XII-2]   INTERNATIONAL ATOMIC ENERGY AGENCY, Current Status of Neutron Capture Therapy, TECDOC 1223, IAEA, Vienna (2001).

[XII-3]   SWEET, W.H., JAVID, M., The possible use of neutron-capturing isotopes such as boron 10 in the treatment of neoplasms. 1. Intracranial tumors, J. Clin. Invest. **31** 6 (1952) 604–610.

[XII-4]   FARR, L.E., et al., Treatment of glioblastoma multiforme, Am. J. Roentgenol. **71** (1954) 279–291.

[XII-5]   GODWIN, J.T., FARR, L.E., SWEET, W.H., ROBERTSON, J.S., Pathological study of eight patients with glioblastoma multiforme treated by neutroncapture therapy using boron 10, Cancer **8** 3 (1955) 601–615.

[XII-6]   SWEET, W.H., Early history of development of boron neutron capture therapy of tumors, J. Neuro-Oncol. **33** 1–2 (1997) 19–26.

[XII-7]   ARCHAMBEAU, J.O., The effect of increasing exposures of the reaction on the skin of Man, Radiology **94** (1970) 179–187.

[XII-8]   FARR, L.E., "Neutron capture therapy: Years of experimentation – years of reflection", invited lecture presented at the medical department, Brookhaven National Laboratory (1991).

[XII-9]   FARR, L.E., et al., Effect of thermal neutrons on central nervous system. apparent tolerance of central nervous system structures in Man, Arch. Neutrol. **4** (1961) 246–257.

[XII-10]   SLATKIN, D.N., A history of boron neutron capture therapy of brain tumours, Brain **114** 4 (1991) 1609–1629.

[XII-11]   SOLOWAY, A.H., Correlation of drug penetration of brain and chemical structure, Science **128** 3338 (1958) 1572–1574.

[XII-12]   SOLOWAY, A.H., GORDON, D.S., Evaluation of two boron-containing drugs for use in the proposed neutron-capture irradiation of gliomas, J. Neutropathol. **1** (1960) 1–476.

[XII-13]   SWEET, W.H., SOLOWAY, A.H., BROWNELL, G.L., Boron-slow neutron capture therapy of gliomas, Acta Radiol. Ther. Phys. Biol. **1** 2 (1963) 114–121.

[XII-14]   SOLOWAY, A.H., et al., The chemistry of neutron capture therapy, Chem. Rev. **98** 4 (1998) 1515–1562.

[XII-15] ASBURY, A., OJEMAN, R., NIELSEN, S., SWEET, W.H., Neuropathologic study of fourteen cases of malignant brain tumour treated by boron-10 slow neutron capture radiation, Neuropathol. Exp. Neurol. **31** (1972) 278–303.

[XII-16] MUETTERTIES, E.L., BALTHIS, J.H., CHIA, Y.T., KNOTH, W.H., MILLER, H.C., Chemistry of boranes. VIII.l Salts and acids of $B_{10}H_{10}^{-2}$ and $B_{12}H_{12}^{-2}$, Inorg. Chem. **3** 3 (1964) 444–451.

[XII-17] LARAMORE, G.E., SPENCE, A.M., Boron neutron capture therapy (BNCT) for high-grade gliomas of the brain: A cautionary note, Int. J. Radiat. Oncol. Biol. Phys. **36** 1 (1996) 241–246.

[XII-18] HAWTHORNE, M.F., LEE, M.W., A critical assessment of boron target compounds for boron neutron capture therapy, J. Neuro-Oncol. **62** 1 (2003) 33–45.

[XII-19] DIAZ, A., STELZER, K., LARAMORE, G., WIERSEMA, R., "Pharmacology studies of $Na_2{}^{10}B_{10}H_{10}$ (GB-10) in human tumor patients", Research and Development in Neutron Capture Therapy, (Proc. 10th Int. Congr. Neutron Capture Therapy, Essen, Germany, Sept. 8–13, 2002) (2002) 993–999.

[XII-20] LARAMORE, G.E., et al., Boron neutron capture therapy: A mechanism for achieving a concomitant tumor boost in fast neutron radiotherapy, Int. J. Radiat. Oncol. Biol. Phys. **28** 5 (1994) 1135–1142.

[XII-21] KRUGER, P.G., Some biological effects of nuclear disintegration products on neoplastic tissue, Proc. Natl. Acad. Sci. **26** 3 (1940) 181–192.

[XII-22] MURRAY, J.F., A comparative review of the pharmacokinetics of boric acid in rodents and humans, Biol. Trace Elem. Res. **66** (1998) 331–341.

Annex XIII.

TWO-STEP RATE INFUSION OF 4-BORONO-L-PHENYLALANINE (BPA) SOLUTION IN BPA-BNCT (ONO'S METHOD)

KOJI ONO

BNCT Joint Clinical Institute, Osaka Medical and Pharmaceutical University,

2-7 Daigaku-machi, Takatsuki-shi, Osaka 569-8686, Japan

Abstract

In boron neutron capture therapy (BNCT) conducted with 4-borono-L-phenylalanine (BPA) (BPA-BNCT), we examined the infusion rate of BPA, changes in blood ^{10}B-concentration, and the accuracy of estimation of ^{10}B-concentration during neutron irradiation. When 250 or 500 mg/kg body weight BPA was administered at a constant rate in 2 hours, the increment of the blood ^{10}B-concentration slowed down in the middle and became about 50% of the first half in the latter. Neutron irradiation was initiated immediately after the end of the 2-hour administration, and the ^{10}B-concentration at the end of the irradiation had decreased to 62% of that immediately after the end of administration. However, when 500 mg/kg was administered at 200 mg·kg^{-1}·h^{-1} for 2 hours, then slowed down to 100 mg·kg^{-1}·h^{-1}, and neutron irradiation was performed during the slowed-down administration, the ^{10}B-concentration immediately after the end of irradiation could be maintained at 98% of the level immediately after the end of administration. The stable ^{10}B-concentration during irradiation contributes to the reliable application of Compound Biological Effectiveness (CBE) values. Furthermore, radiobiological studies have shown that the biological effects of BNCT are more uniform when irradiated with maintained blood ^{10}B-concentrations.

XIII-1. INTRODUCTION

In BNCT, it was usual to administer ^{10}B-drugs intravenously at a constant rate, followed by neutron irradiation. In this case, the ^{10}B-concentration during irradiation is estimated according to the decay curve accumulated in clinical study. According to this value, the physical dose of the ^{10}B neutron reaction is calculated and the irradiation time is determined. However, a large inter-individual error is inevitable in this estimation. Originally, the CBE values used to multiply the physical dose of the ^{10}B neutron reaction in determining the biological X ray equivalent dose were determined experimentally on the basis of blood ^{10}B-concentrations. In this sense, it is essential to know the exact blood ^{10}B-concentration during neutron irradiation. Furthermore, from another point of view, it is questionable whether the concentration and distribution of ^{10}B-drugs in tumour tissue can be maintained during neutron irradiation after the end of administration. In order to overcome the above problems and questions, I have devised a two-step rate intravenous administration method of BPA, and this paper describes the clinical and experimental data on which it was based.

XIII-2. SEARCH FOR THE BORON CONCENTRATION IN BLOOD IN CLINICAL BPA-BNCT

XIII-2.1. Protocol 1

Between June 2004 and February 2006, two different schedules were employed in the administration of BPA–fructose complex solution in BNCT for recurrent head and neck cancer.

Twelve patients received 250–500 mg/kg of BPA in 1–2 h. Four patients received two doses, so that 16 kinetic data points were accumulated. The blood samples were taken at 5–20 min intervals from the beginning of BPA administration until the end of BNCT, except during neutron irradiation. The ^{10}B-concentration was measured by prompt γ-ray spectrometry. The results are shown in Table XIII-1, which shows a high concentration of ^{10}B in the blood immediately after the administration of BPA and a large decrease at the end of the neutron irradiation in all cases. The ^{10}B-concentration immediately after the end of neutron irradiation was 62% of that immediately before the start of irradiation.

TABLE XIII-1. ^{10}B-CONCENTRATION MEASUREMENTS USING PROMPT GAMMA SPECTROSCOPY BEFORE AND AFTER IRRADIATION USING PROTOCOL 1

Case No.	Disease	Dose of BPA (mg/kg)	^{10}B level before BNCT (ppm)	^{10}B level after BNCT (ppm)	After/Before (%)
1	Parotid Ca.	250	24.5	13.9	56.7
2-1	Maxillary Ca. (Adenoid Cystic Ca.)	250	17.7	12.9	72.9
2-2		250	31.8	11.5	36.2
3	Hypopharyngeal Ca.	250	18.6	11.1	59.7
4	Breast Ca (SCC)	250	22.9	12.5	54.5
5-1	Nasal Ca. (Angiosarcoma)	250	18.7	12.8	68.4
5-2		500	32.9	23.5	71.4
6-1	Maxillary Ca.	500	31.9	21.1	66.1
6-2		500	31.9	17.7	55.5
7	Buccal Ca.	500	26.9	19.8	73.6
8	Tongue Ca.	500	29.1	20.3	69.8
9	Mesophayngeal Ca. (Nodes)	500	28	17	60.7
10	Buccal Ca. (Nodes)	500	34.8	18.6	53.4
11-1	Mesophayngeal Ca.	500	32.1	20	62.3
11-2	Mesophayngeal Ca.	500	26	19.1	73.5
12	Gingeval Ca. (Nodes)	500	30.2	18.2	60.3

Note: Ca. = cancer

Figure XIII-1 shows the changes in the blood ^{10}B-concentration over 2 hours in patients treated with BPA at 500 mg·kg^{-1}·(2h)$^{-1}$. The increase in the blood ^{10}B-concentration is biphasic. The slope of the first half of the curve is approximately twice the slope of the second half.

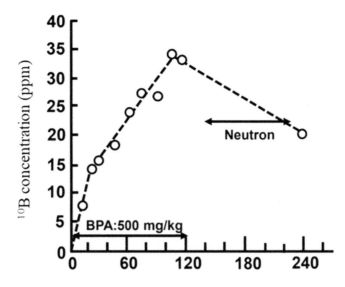

FIG. XIII-1. Concentration of ^{10}B versus time for patients under Protocol 1.

XIII-2.2. Protocol 2

On the basis of the results of Protocol 1, a total of 20 doses of BPA were administered in 15 patients according to the following dosing schedule and the changes in the blood ^{10}B-concentration were studied. The total dose of BPA was fixed at 500 mg/kg, with 200 mg·kg^{-1}·h^{-1} for the first 2 hours and 100 mg·kg^{-1}·h^{-1} for the following 1 hour. ^{10}B-concentrations were determined by prompt γ-ray spectrometry as in Protocol 1. The data for each case are summarized in the Table XIII-2. At the end of the neutron irradiation, blood ^{10}B-concentrations remained on average at 98% of the level immediately before the start of neutron irradiation, when the administration of 400 mg/kg dose was completed. The changes in ^{10}B-concentration during 400 mg/kg dose were similar to those in protocol 1 (Fig. XIII-2). This comparison of two protocols shows that in clinical BPA-BNCT practice with Protocol 2 allows accurate prediction of the ^{10}B neutron reaction dose, as long as the ^{10}B-concentration immediately before irradiation is monitored.

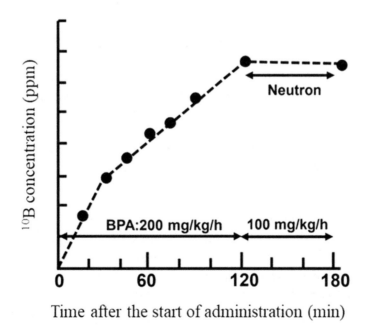

FIG. XIII-2. Concentration of ^{10}B versus time for patients under Protocol 2.

TABLE XIII-2. ^{10}B-CONCENTRATION MEASUREMENTS USING PROMPT GAMMA SPECTROSCOPY BEFORE AND AFTER IRRADIATION USING PROTOCOL 2

Case No.	Disease	Dose of BPA (mg/kg)	^{10}B level before BNCT (ppm)	^{10}B level after BNCT (ppm)	After/Before (%)
1-1	Buccal Ca.	500	32.5	35.8	110
1-2		500	36.1	36.2	100
2	Tongue Ca. (Nodes)	500	20	19.5	97.5
3	Tongue Ca.	500	33.5	31.3	93.4
4	Breast Ca. (Angiosarcoma)	500	24.9	22.7	91.1
5-1	Sebaceous Ca.	500	32.5	31.5	96.9
5-2		500	35	32.5	92.9
6	Breast Ca (SCC)	500	30	30.8	103
7-1	Gingeval Ca.	500	26.3	233	88.5
7-2		500	25.5	25.7	101
8	Hypopharyngeal Ca.	500	22.3	24.4	109
9	Parotid Ca.	500	26.5	23.1	87.2
10	Laryngeal Ca.	500	26	25.8	99.2
11-1	Lip Ca.	500	21.7	26.4	122
11-2		500	23.6	24	102
12	Maxillary Ca.	500	30.7	32.9	107
13	Mesophayngeal Ca. (Nodes)	500	20.2	20.6	102
14-1	Gingeval Ca. (Nodes)	500	19.1	16.7	87.4
14-2		500	19.1	16.7	87.4
15	Malignant Schwanoma	500	29.8	28.9	97

Note: Ca. = cancer

XIII-3. RADIOBIOLOGICAL IMPLICATIONS OF NEUTRON IRRADIATION UNDER CONTINUOUS BPA ADMINISTRATION

XIII-3.1. Consideration from the viewpoint of accurate adaptation of compound biological effectiveness values

Compound biological effectiveness values are experimentally determined values, and they are based on physical doses calculated according to the assumption that ^{10}B is uniformly distributed in the tissues concerned at concentrations equal to those in the blood.

In this case, there is no problem if the ^{10}B-concentrations between the blood and the normal tissue is parallel at an assumed ratio. However, there are some tissues where this is not the case. For example, in the central nervous system, experimental and ^{18}F-FBPA-PET data show that the ^{10}B-concentration of BPA does not parallel the blood ^{10}B-concentration in a certain ratio. Moreover, it has been found that the CBE value calculated on the blood ^{10}B-concentration varies with this ratio [XIII-1]. Therefore, in order to apply CBE values properly, it is desirable to stabilize the blood ^{10}B-concentration at least during neutron irradiation. From this reason, the two-step rate administration method of BPA is desirable.

If the damage to the small vasculatures is the cause of the late adverse events of BNCT, the blood ^{10}B-concentration is the determining factor. For example, necrosis and ulceration, which are late adverse events of the skin, are considered to be examples of this. Even in such cases, stabilization of blood ^{10}B-concentration during neutron irradiation may contribute to accurate estimation of physical dose and biological X ray equivalent dose.

XIII-3.2. The effect of reducing the heterogeneity of ^{10}B-distribution in the tumour

As mentioned in the introduction, BPA is excreted relatively rapidly from the body in the urine. While this may reduce its toxicity, it may also make it difficult to maintain ^{10}B-concentration in the tumour tissue at a level during neutron irradiation and to estimate the ^{10}B neutron reaction dose accurately.

Since BPA can accumulate in tumour cells, it is expected that a certain level of ^{10}B-concentration will be maintained, but as the ^{10}B-concentration in the blood decreases, the concentration in the tumour stroma will also decrease, and a decline of the cell-killing effect will finally follow. The following experiments were carried out to elucidate this point. Bilateral nephrectomy was performed prior to BPA injection in order to inhibit urinary excretion of BPA and to keep the level of ^{10}B-concentration in the tumour, including the concentration in the stroma. The tumour used in the experiment was an SCCVII tumour. Tumour bearing mice without nephrectomy were prepared as controls. After administration of BPA, the tumours were irradiated with neutrons, removed, and single cell suspensions were prepared by mechanical and enzymatic treatment. Tumour cells were cultured in the medium containing cytochalasin B for 48 hours. Then, cells in culture dishes were fixed with Carnoy's solution, stained by PI, and observed under a fluorescence microscope. The binucleated cells, indicating that they had passed the mitotic phase, were classified into two fractions according to the presence or absence of micronuclei in the cytoplasm.

In a previous study by the author, it was found that the proportion of cells without micronuclei corresponded to the surviving cell fraction assayed by the colony formation [XIII-2–XIII-3]. The relationship between the proportion of cells without micronuclei and the neutron dose would be a biphasic curve if tumour cells that had not taken up BPA were included.

The results of the experiment are shown in Fig. XIII-3. Maintaining ^{10}B-concentration in the blood to be constant by nephrectomy greatly reduced the biphasic nature of the effect and made it more uniform.

Although maintaining a constant interstitial ^{10}B-concentration does not completely eliminate the difference in the BNCT effect between on the cells that take up BPA and those that do not. The significant decrease in heterogeneity suggests that a similar effect can be expected in clinical BPA-BNCT.

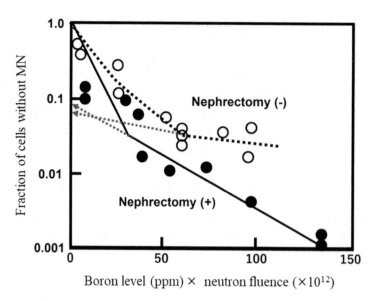

FIG. XIII-3. Fractions of cells without micronuclei (MN) for experiments with (+) and without (-) nephrectomy.

XIII-4. CONCLUSION.

By modifying the BPA infusion schedule, a high and stable blood ^{10}B-concentration during neutron irradiation was achieved. This facilitated treatment planning before BNCT, and the treatment time required to deliver the desired dose to the tumour could be accurately and quickly determined during neutron irradiation.

In addition, maintaining a stable blood ^{10}B-concentration is expected to improve the intratumoural microdistribution of BPA, leading to the enhanced BNCT effects on tumours.

Note: This method was first presented at ICNCT-2006 at Takamatsu, and since then it is usually employed in clinical study and practice in Japan.

REFERENCES TO ANNEX XIII

[XIII-1]   ONO, K., An analysis of the structure of the compound biological effectiveness factor, J. Radiat. Res. **57** Suppl. 1 (2016) i83–i89.

[XIII-2]   ONO, K., MASUNAGA, S., AKABOSHI, M., AKUTA, K., Estimation of initial slope of cell survival curve following irradiation from micronucleus frequency in cytokinesis blocked cells, Radiat. Res. **138** (1994) s101–s104.

[XIII-3]   ONO, K., et al., The combined effect of boronophenylalanine and borocaptate in boron neutron capture therapy for SCCVII tumors in mice, Int. J. Radiat. Oncol. Biol. Phys. **43** 2 (1999) 431–436.

DEVELOPMENT OF TUMOUR-SELECTIVE BORON AGENTS FOR PRECISION MEDICINE BASED ON BORON NEUTRON CAPTURE THERAPY

ATSUSHI FUJIMURA[1,2], KAZUYO IGAWA[1], HIROYUKI MICHIUE[1]

[1]Neutron Therapy Research Center, Okayama University, Kita-ku, Okayama, 700-8558, Japan
[2]Department of Cellular Physiology, Okayama University Graduate School of Medicine, Dentistry, and Pharmaceutical Sciences, Kita-ku, Okayama, 700-8558, Japan.

Abstract

In March 2020, medical devices and drugs related to boronophenylalanine (BPA)-boron neutron capture therapy (BNCT), a combination of chemotherapy and radiotherapy, were approved for the treatment of unresectable locally advanced or locally recurrent head and neck cancer in Japan. BPA-BNCT is a therapy that is expected to play a role in the future of precision medicine, which is the concept of optimizing therapeutic efficacy by analyzing the genetics, lifestyle, and/or environment of individual patients. The use of genetic traits to guide selection of appropriate treatments is widely accepted in cancer therapy, and it is the reason for the success of many molecular-targeted drugs. BPA is a boron agent that is taken up by cells via amino acid transporters, many of which are highly expressed by cancer cells and contribute to the cell specificity of BPA-BNCT. However, boron delivery by BPA is not suitable for all cancers, and new boron formulations that are taken up through different mechanisms will be necessary. In this review, we discuss the points to consider when developing boron agents suitable for BNCT, using the sodium borocaptate formulation being developed at Okayama University as an example.

XIV-1.1. INTRODUCTION

Boron neutron capture therapy (BNCT) is a cancer treatment that kills cells using high linear energy transfer particles (^{7}Li and ^{4}He nuclei) produced by the nuclear reaction $^{10}B + n \rightarrow {}^7Li + {}^4He$ when ^{10}B is irradiated with neutrons [XIV-1]. Based on this principle, two points are important for BNCT to become standardized as a cancer treatment method: securing a source of neutron beams and developing cancer cell-selective boron delivery. In March 2020, an accelerator-based BNCT treatment system using boronophenylalanine (BPA) (Sumitomo Heavy Industries, Tokyo, Japan; Stellapharma, Osaka, Japan) was listed on the Japanese National Health Insurance drug price list, thus making hospital-based BNCT treatment a practical option. The system is currently approved for the treatment of unresectable locally advanced or locally recurrent head and neck cancer, but the indications are expected to be expanded to include refractory cancers, especially malignant brain tumours, and other cancer types that occur in areas where functional preservation is required.

With the advent of hospital-based treatment systems, BNCT is becoming a standard treatment option for cancer. The main challenges for its further expansion and refinement are (1) how to determine the efficacy of BNCT, and (2) how to adapt the approach for cancers and/or patients whose pathological profiles are unsuitable for BPA-mediated BNCT. In this review, we will discuss some ongoing research in these areas.

XIV-2. PATIENT STRATIFICATION FOR BPA-BNCT

When possible, careful patient stratification plays a large role in ensuring the success of cancer treatments including with BNCT. The boron preparation BPA used in BNCT system is a boron-adduct analogue of tyrosine/phenylalanine and it is thought to be taken up and accumulate within cancer cells via amino acid transporters such as the L-type amino acid transporters LAT-1 and LAT-2 and the $ATB^{0,+}$ transporter. These transporters are aberrantly upregulated in many cancers [XIV-2]; indeed, expression of LAT-1 (encoded by the solute carrier family 7 member 5 gene, SLC7A5), is highly upregulated in head and neck cancers, thus providing a potential mechanism for the efficacy of BPA-BNCT in its originally approved indication [XIV-3–XIV-4]. Similarly, melanoma cells express high levels of LAT-1 and are considered likely to take up BPA, although whether or not BNCT is an appropriate local treatment method remains unclear. In malignant brain tumours such as glioblastoma (GBM), the expression level of LAT-1 depends on the disease subtype [XIV-5]. Because GBM is essentially confined to the intracranial region, it is expected that BNCT would be of significant benefit if GBM patients can be appropriately selected [XIV-6–XIV-7].

One potential method for patient stratification for BPA-BNCT is positron emission tomography (PET) with the diagnostic candidate compound ^{18}F-fluoroboronophenylalanine (^{18}F-FBPA). This imaging modality can provide a semi-quantitative estimate of FBPA distribution throughout the body, thereby identifying patients whose tumours are mostly likely to take up BPA. Importantly, FBPA has been reported to behave similarly to BPA in vivo and to accumulate in many tumour types [XIV-8–XIV-9]. In addition, depending on the cancer type and site, the timing between BPA administration and subsequent irradiation may be a factor in determining the optimal treatment strategy.

XIV-3. DEVELOPMENT OF PRODUCTS BEYOND BORONOPHENYLALANINE

The use of BPA will certainly accelerate implementation of BNCT, but other options have to be provided for those patients who cannot tolerate treatment with BPA. Cross-sectional analysis of data obtained to date shows that, although LAT-1 expression is high in some cancers such as head and neck, other cancers may not upregulate the transporter, or may even express it at lower levels than normal tissues. Thus, there is an urgent need to develop boron preparations that are preferentially taken up by cancer cells through a mechanism distinct from that of BPA. Sodium borocaptate (BSH) is a promising and well-researched boron preparation for BNCT. Kawabata et al. reported that addition of a combined BPA/BSH-BNCT protocol to conventional radiotherapy resulted in favourable therapeutic effects in glioblastoma [XIV-10]. In vitro cellular experiments found that BSH alone is rarely taken up by tumour cells, whereas in vivo animal experiments suggested that it is taken up by glial fibrillary acidic protein-positive cells [XIV-11]. Although the mechanism of BSH uptake into these cells is unclear, this finding supports the demonstration by Kawabata et al. that BSH may enhance the efficacy of BPA-BNCT.

Against this backdrop of potential drug development, various attempts have been made to promote the cell membrane permeability and subcellular localization of BSH. Several delivery methods, such as chemical bonding to polyarginine, have shown promising experimental results [XIV-12], but many issues, including the safety and stability of BSH, will need to be resolved before it can be translated to clinical applications. To solve these problems, we recently developed a delivery system in which self-assembling peptide A6K (3-D Matrix, Tokyo, Japan) and BSH are simply mixed together, leading to formation of a spherical complex about 100 nm in diameter that behaves like a peptosome. We found that the complexes are efficiently taken up by glioblastoma cells both in vitro and in vivo [XIV-13]. We identified a certain cell membrane surface protein that serves as an anchor for cellular uptake of the A6K/BSH complex, and we have also developed an immunodetection method to determine the suitability of boron delivery by A6K/BSH by analyzing histopathological specimens. Interestingly, our analysis suggests that BSH accumulates in nucleoli with RNA-related proteins [XIV-14]. By measuring the expression levels of these proteins, which mediate A6K/BSH uptake and subcellular localization, across cancer types, we have demonstrated that A6K/BSH may be a promising agent for patients who would not be eligible for BPA-BNCT; for example, because their tumours express low levels of LAT-1. Nevertheless, the physical properties of BSH have yet to be elucidated in detail, and this information will be crucial to the development of future BSH formulations for BNCT.

XIV-4. FUTURE PERSPECTIVES FOR BNCT-BASED PRECISION MEDICINE

The development of powerful new genomics approaches has ushered in a new era in precision medicine for cancer. Classification of cancers based on gene expression profiles will continue to allow subdivision into multiple groups with different genetic characteristics, even if the primary lesions are the same. In addition, the improvement in cost, time, and accuracy of next-generation sequencers has opened up a range of possible diagnostic and therapeutic approaches for individual patients, thus broadening the application of precision medicine. The gold standard for establishing the efficacy and safety of new treatments is through Phase I, II, and III clinical trials, but this incremental approach is often difficult in the field of cancer, where treatment modalities may change rapidly. Therefore, in recent years, Basket trials have emerged as a useful method to evaluate a single targeted therapy against multiple diseases with a common biomarker [XIV-15]. To date, Basket trials have evaluated the therapeutic effect of molecular-targeted drugs on multiple cancers that share the same gene mutations and amplifications. These trials are extremely useful because they can test the safety and efficacy for multiple cancers in a short time, and are particularly important for rare cancers, which may be difficult to examine in classical trials due to low patient numbers [XIV-16].

In the case of BNCT, the therapeutic dose is directly related to the magnitude of boron uptake and its selectivity for cancer versus normal tissue. Thus, we believe that a BNCT clinical study that specifically assesses boron drug uptake is most appropriately undertaken as a Basket trial. For example, the uptake of BPA using ^{18}F-FBPA-PET will facilitate recruitment of patients who will obtain the most benefit from BPA-BNCT. In this example, patient inclusion criteria would include high uptake of boron drug in the tumour, low uptake in normal tissue within the irradiated area, and no deleterious effects on important normal tissues/organs in the irradiated area. In some cases, it may be necessary to consider side effects and reduce the standard neutron dose to mitigate the severity of such reactions. We also believe that safe and effective new boron drugs can be developed to support BNCT-mediated precision medicine by employing drugs that have established favourable clinical safety profiles. Since BNCT involves both chemotherapy and radiation therapy, it represents a new model for clinical trial design that provides maximum effect in a short period of time while minimizing side effects.

XIV-5. CONCLUSION

Patient stratification for BNCT with BPA or other boron agents will benefit from imaging modalities such as PET with ^{18}F-FBPA, which can determine the cancer's propensity for boron uptake. For BNCT to continue to offer hope to cancer patients, it will be important to develop new boron agents that are taken up through mechanisms distinct from that utilized by BPA. In addition, we have to deepen our understanding of the molecular characteristics that affect the pharmacokinetics and pharmacodynamics of such boron drugs.

ACKNOWLEDGMENTS

We thank Anne M. O'Rourke, PhD, from Edanz Group (https://jp.edanz.com/ac) for editing a draft of this manuscript.

REFERENCES TO ANNEX XIV

[XIV-1]   SAUERWEIN, W., Principles and history of neutron capture therapy, Strahlenther. Onkol. **169** (2013) 1–6.

[XIV-2]   WONGTHAI, P., et al., Boronophenylalanine, a boron delivery agent for boron neutron capture therapy, is transported by $ATB^{0,+}$, LAT1 and LAT2, Cancer Sci. **106** (2015) 279–286.

[XIV-3]   LIM, D., QUAH, D.S., LEECH, M., MARIGNOL, L., Clinical potential of boron neutron capture therapy for locally recurrent inoperable previously irradiated head and neck cancer, Appl. Radiat. Isot. **106** (2015) 237–241.

[XIV-4]   The Human Protein Atlas, https://www.proteinatlas.org/ENSG00000103257-SLC7A5/pathology.

[XIV-5]   GlioVis, Data Visualization Tools for Brain Tumor Datasets, http://gliovis.bioinfo.cnio.es.

[XIV-6]   MIYATAKE, S., et al., Survival benefit of boron neutron capture therapy for recurrent malignant gliomas, J. Neuro-Oncol. **91** (2009) 199–206.

[XIV-7]   KAWABATA, S., et al., Survival benefit from boron neutron capture therapy for newly diagnosed glioblastoma patients, Radiat. Isot. **67** (2009) S15–S18.

[XIV-8]   MENICHETTI, L., et al., Positron emission tomography and [^{18}F]BPA: a perspective application to assess tumour extraction of boron in BNCT, Appl. Radiat. Isot. **67** (2009) S351–S354.

[XIV-9]   WATANABE, T., et al., Comparison of the pharmacokinetics between L-BPA and L-FBPA using the same administration dose and protocol: a validation study for theranostic approach using [^{18}F]-L-FBPA positron emission tomography in boron neutron capture therapy, BMC Cancer **16** (2016) 859.

[XIV-10]   KAWABATA, S., et al., Survival benefit from boron neutron capture therapy for newly diagnosed glioblastoma patients, Radiat. Isot. **67** (2009) S15–S18.

[XIV-11]   NEUMANN, M., BERGMANN, M., GABEL, D., Cell type selective accumulation of mercaptoundecahydro-*closo*-dodecaborate (BSH) in glioblastoma multiforme, Acta Neurochir. (Wien) **145** (2003) 971975.

[XIV-12]   IGUCHI, Y., et al., Tumor-specific delivery of BSH-3R for boron neutron capture therapy and positron emission tomography imaging in a mouse brain tumor model, Biomaterials **56** (2015) 1017.

[XIV-13]   MICHIUE, H., et al., Self-assembling A6K peptide nanotubes as a mercaptoundecahydrododecaborate (BSH) delivery system for boron neutron capture therapy (BNCT), J. Control. Release **330** (2021) 788–796.

[XIV-14] FUJIMURA, A., et al., In vitro studies to define the cell-surface and intracellular targets of polyarginine-conjugated sodium borocaptate as a potential delivery agent for boron neutron capture therapy, Cells **9** (2020) 2149.

[XIV-15] CUNANAN, K.M., et al., Basket trials in oncology: a trade-off between complexity and efficiency, J. Clin. Oncol. **35** (2017) 271–273.

[XIV-16] HYMAN, M.D., et al., HER kinase inhibition in patients with HER2- and HER3-mutant cancers, Nature **554** (2018) 189–194.

REGULATORY STEPS FOR CLINICAL INVESTIGATION WITH AN ACCELERATOR IN EUROPE

SANDRO ROSSI, MARIO CIOCCA, ANGELICA FACOETTI, ARIANNA SERRA
CNAO Foundation
Strada Campeggi, 53
27100, Pavia, Italy

Abstract

This article intends to describe the regulatory steps for clinical investigation according to Regulation (EU) MDR 2017/745, which "aims to ensure the smooth functioning of the internal market as regards medical devices, taking as a base a high level of protection of health for patients and users, and taking into account the enterprises that are active in this sector. At the same time, this Regulation sets high standards of quality and safety for medical devices in order to meet common safety concerns as regards such products. Both objectives are being pursued simultaneously and are inseparably linked whilst one not being secondary to the other" [XV-1].

XV-1. INTRODUCTION

XV-1.1. EU Medical Device Regulation (MDR) 2017/745 - Medical Devices discipline

CE marked Medical Devices have to follow the General Safety and Performance Requirements, in accordance with Annex I of the Regulation. Another important aspect of the new MDR is the Clinical Evaluation that the Manufacturer has to carry out in order to demonstrate the clinical benefit and safety of the Medical Device for the patient and user.

In addition, the clinical evaluation has to include:

- A summary of the risk and benefit analysis, including information regarding known or foreseeable risks and any undesirable effects;
- An indication of the residual risks to be taken into account and further investigated in the clinical evaluation;
- A demonstration that all known and predictable risks and any undesirable side-effects are minimized and are acceptable, compared to the assessed benefits for the patient and/or user.

Final compliance checks with the General Safety and Performance Requirements, according to its intended use, has to be performed following clinical data that provide clinical evidence.

The MDR distinguishes between 'pre-market clinical trials' in the context of clinical evaluation for assessment of conformity and 'post-market clinical trials' on CE marked medical devices, according to the intended use or in the context of the clinical evaluation with different purposes from the conformity assessment.

A device can only be placed on the market or put into service if it complies with the MDR. The manufacturer could use harmonized standards to demonstrate compliance with the requirements of the MDR. The responsible manufacturer defines the intended use and risk classes of the medical device, according to Annex VIII of MDR 2017/745 [XV-1]. The manufacturer has to select the competent Notified Body on similar products, accredited according to Regulation (EU) 2017/745 [XV-1]. The Notified Body can issue a CE Certificate, based on the evaluation of the technical documentation provided by the Manufacturer and all the designs and the construction activities of the medical device implemented by the Manufacturer and managed through its own Company Quality Management System. The Manufacturer issues the EC declaration of conformity, following the issue of the CE Certificate.

XV-2. SAFETY REQUIREMENTS

According to Annex I of Regulation (EU) 2017/745 [XV-1], the device under investigation has to be compliant with the applicable general safety and performance requirements and, with regard to those aspects, every precaution has to be taken to protect the health and safety of the subjects.

Compliance with harmonised standards is a means for manufacturers to demonstrate compliance with the general safety and performance requirements and other legal requirements (e.g. EN 60601-1 Medical Electrical equipment, EN 60601-1-2 Electromagnetic compatibility EMC; EN 62304 Medical device software – Software life cycle processes; IEC radiation protection, EN 10993-1 biocompatibility, etc...). This includes mechanical and/or electrical tests, reliability tests, software verification and validation, performance tests, evaluation of biocompatibility and biological safety, if applicable.

If the manufacturer is not able to demonstrate conformity according to IEC's standards, the manufacturer has to contact an accredited laboratory (e.g., in Italy, IMQ laboratory[1]), to perform the tests for electrical safety, electromagnetic compatibility, etc....

XV-3. DOSIMETRY SAFETY

XV-3.1. Dosimetric commissioning of the system

Within the approval process of the boron neutron capture therapy (BNCT) facility as a medical device by the national regulatory authorities, the dosimetric commissioning of the system represents a crucial aspect. In this perspective, two main tasks can be identified: the experimental verification of the irradiation field and microdosimetry of the mixed radiation fields for BNCT Quality Assurance purposes.

XV-3.2. Experimental verification of the neutron field

Experiments are strongly suggested to verify both the relative spatial distribution and the absolute value of the neutron flux against Monte Carlo simulations. Suggested experimental procedure includes the use of a cubic water phantom (acrylic case with a typical external size around 200 mm side and 3-mm wall thickness) placed in front of the beam shaping assembly, as well as measurements of the neutron flux on the beam central axis at different depths along beam direction. The utilization of a small scintillator neutron detector coupled with an optical fibre, accurately moved inside the water phantom, is strongly suggested for relative measurements [XV-2]. To extract the pure thermal neutron component, thus discriminating it from the epithermal one, the cadmium subtraction method can be adopted. For absolute flux determination, the gold foil activation method is preferred, as reported in detail in Ref. [XV-2].

XV-3.3. Microdosimetry for quality assurance

Since BNCT relies upon the production of high-LET fragments (He and Li nuclei) which deposit their energy locally, desirably within tumour cells, microdosimetry can represent a valuable tool to investigate experimentally the effects of mixed radiation fields at the cellular level. The capability and reliability of different miniature detectors to measure the energy

---

[1] https://www.imq.it/en

deposited in micron-sized sensitive volumes with similar dimensions as biological cells have already been shown in a complex scenario involving mixed radiation fields: preliminary experiences with silicon-based detectors, synthetic microdiamonds and mini-tissue-equivalent proportional counters, reported for clinical carbon ion beams [XV-3–XV-4], appear promising also in the context of BNCT. In particular, as recently shown by Ref. [XV-5], microdosimetry using silicon detectors in a water phantom, with ^{10}B uniformly distributed in water at a concentration around 25 parts-per-million (typical value in the tumour for BNCT), can be successfully applied to BNCT, once properly optimized.

XV-4. RADIOBIOLOGY SAFETY

Radiobiological commissioning of particle beams with or without ^{10}B is a mandatory step before the clinical use of the machine, to ensure safe and accurate treatments. A biological system already well standardised for RBE (relative biological effectiveness) determination, which has been shown to be sensitive and reproducible, has to be used to determine RBE of neutron beam (hydrogen dose), which is specific for each accelerator based BNCT (AB-BNCT) system. For this reason, in vitro clonogenic survival assays with cell lines already well characterized from a radiobiological point of view (e.g., V79, CHO cells) and cells of specific cancer types for planned clinical trials need to be performed. In particular, studies measuring the biological effects in an entire irradiation field to test the homogeneity of the field are fundamental for a proper beam characterization, in terms of quality assurance and efficacy. Cytotoxicity can be measured with cancer and normal cells in the absence and presence of boron compounds at multiple concentrations.

In addition, the safety assessment ought to be carried out by evaluating in vivo effects such as survival, skin lesions, mucosal lesions, blood toxicity and intestinal crypt regeneration in mice, after whole body irradiation or local irradiation in the absence and presence of boron compounds. In particular, intestinal crypt survival is still considered to be the best biological system for in vivo quality assurance in non-conventional radiation therapy of new clinical beams [XV-6]. These studies need to be planned and the experimental set-up designed, taking also into consideration the different body size and thickness of mice compared to humans. The observation periods of these effect can be set to investigate both short and long term effects. In addition, the experimental determination of RBEs, both in vitro and in vivo, will assist accurate comparisons of the neutron beams at the different BNCT facilities used in clinical trials. Finally, for efficacy evaluation studies, the effects on tumour growth delay can be evaluated using mice xenograft models with clinical target cancer types and compared with local X ray irradiation.

XV-5. PRE-MARKET CLINICAL TRIALS

The main actors of a clinical investigation are [XV-1]:

- 'Sponsor' "means any individual, company, institution or organisation which takes responsibility for the initiation for the management and setting up of the financing of the clinical investigation." "Where the sponsor of a clinical investigation is not established in the Union, that sponsor has to ensure that a natural or legal person is established in the Union as its legal representative";
- 'Investigator' "means an individual responsible for the conduct of a clinical investigation at a clinical investigation site". "The investigator shall be a person exercising a profession which is recognised in the Member State concerned as qualifying for the role of investigator on account of having the necessary scientific knowledge and experience in patient care. Other personnel involved in conducting a clinical investigation shall be

suitably qualified, by education, training or experience in the relevant medical field and in clinical research methodology, to perform their tasks.";

- 'Subject' "means an individual who participates in a clinical investigation." In many cases, 'subject' is a synonym of 'patient';
- 'Competent authority';
- 'Ethics committee'.

The following terms are used as defined in the Medical Device Regulation (EU) 2017/745 (MDR) Article 2 [XV-1]:

- 'Clinical investigation' "means any systematic investigation involving one or more human subjects, undertaken to assess the safety or performance of a device";
- 'Clinical investigation plan' "means a document that describes the rationale, objectives, design, methodology, monitoring, statistical considerations, organisation and conduct of a clinical investigation";
- 'Clinical data' "means information concerning safety or performance that is generated from the use of a device and is sourced from the following:
  - clinical investigation(s) of the device concerned,
  - clinical investigation(s) or other studies reported in scientific literature, of a device for which equivalence to the device in question can be demonstrated,
  - reports published in peer reviewed scientific literature on other clinical experience of either the device in question or a device for which equivalence to the device in question can be demonstrated,
  - clinically relevant information coming from post-market surveillance, in particular the post-market clinical follow-up;"
- 'Clinical evidence' "means clinical data and clinical evaluation results pertaining to a device of a sufficient amount and quality to allow a qualified assessment of whether the device is safe and achieves the intended clinical benefit(s), when used as intended by the manufacturer".

XV-5.1. General requirements regarding clinical investigations conducted to demonstrate conformity of devices

"The rules on clinical investigations ought to be in line with well-established international guidance in this field, such as the international standard ISO 14155:2020 [XV-7] on good clinical practice for clinical investigations of medical devices for human subjects, so as to make it easier for the results of clinical investigations conducted in the Union to be accepted as documentation outside the Union, and conversely for the results of clinical investigations conducted outside the Union in accordance with international guidelines to be accepted within the Union." [XV-1].

"Clinical investigations shall be designed, authorised, conducted, recorded and reported in accordance with the provisions of Article and of Articles 63 to 80 of MDR 2017/745, the acts adopted pursuant to Article 81, and Annex XV of MDR 2017/745, where carried out as part of the clinical evaluation for conformity assessment purposes, for one or more of the following purposes:

(a) to establish and verify that, under normal conditions of use, a device is designed, manufactured and packaged in such a way that it is suitable for one or more of the specific purposes listed in point (1) of Article 2, and achieves the performance intended as specified by its manufacturer;

(b)   to establish and verify the clinical benefits of a device as specified by its manufacturer;

(c)   to establish and verify the clinical safety of the device and to determine any undesirable side-effects, under normal conditions of use of the device, and assess whether they constitute acceptable risks when weighed against the benefits to be achieved by the device." [XV-1]

A clinical investigation can only be carried out if all the following conditions are met [XV-1]:

(a)   "...authorisation issued by the Member State(s) in which the clinical investigation is to be conducted...";

(b)   "an ethics committee, set up in accordance with national law, has not issued a negative opinion in relation to the clinical investigation, which is valid for that entire Member State under its national law";

(c)   "the sponsor, or its legal representative or contact person, is established in the Union";

(d)   "vulnerable populations and individuals are adequately protected...";

(e)   "the anticipated benefits to the subjects or to public health justify the foreseeable risks and inconveniences and compliance with this condition is constantly monitored";

(f)   "the subject or, if the subject is not able to give his or her informed consent, his or her designated legal representative, has given informed consent...";

(g)   "the subject or, where the subject is not able to give informed consent, his or her legally designated representative, has been provided with the contact details of an entity where further information can be received in case of need";

(h)   "the rights of the subject to physical and mental integrity, to privacy and to the protection of the data concerning him or her in accordance with Directive 95/46/EC [XV-8] are safeguarded";

(i)   "the clinical investigation has been designed to involve as little pain, discomfort, fear and any other foreseeable risk as possible for the subjects, and both the risk threshold and the degree of distress are specifically defined in the clinical investigation plan and constantly monitored";

(j)   "the medical care provided to the subjects is the responsibility of an appropriately qualified medical doctor...";

(k)   "no undue influence, including that of a financial nature, is exerted on the subject, or, where applicable, on his or her legally designated representatives, to participate in the clinical investigation";

(l)   "the investigational device(s) in question conform(s) to the applicable general safety and performance requirements set out in Annex I [XV-6] apart from the aspects covered by the clinical investigation and that, with regard to those aspects, every precaution has been taken to protect the health and safety of the subjects";

(m) the device under investigation complies with the dosimetry characterization and radiobiological safety.

"The sponsor of a clinical investigation shall submit an application to the Member State(s) in which the clinical investigation is to be conducted ... accompanied by the documentation referred to in Chapter II of Annex XV" [XV-1].

At the end of the years devoted to experimentation and once the clinical results have been collected, the manufacturer will be able to complete the Technical Dossier and the CE marking could be requested.

XV-6. DRUG AUTHORIZATION

Another important step of the authorisation process for treatment with BNCT is the approval of the drug. The boron drug has to be authorized by European Medicines Agency or by the national agency, for example, in Italy, Agenzia Italiana del Farmaco.

For information on the boron drug authorization process two scenarios are possible, depending on whether the drug in question is or is not a commercial drug:

- 'Non-commercial drug'. In this case, the drug cannot be administered to humans, so that an experimentation of the drug, which, according to Regulation 536/2014, need to be carried out following several distinct steps. The experimentation process proceeds as follows: laboratory tests, animal tests and, to conclude, human tests;
- 'Commercial drug'. In this case, the drug can be administered to humans, since it has already obtained authorization.

In a first phase, it seems that authorising bodies consider the use of drug and medical device subject to a combined authorisation for a specific pathology. Eventually, the extension of the authorisation could be matter of discussion opening to independent pathways for new drugs and new pathologies applied to the same medical device.

REFERENCES TO ANNEX XV

[XV-1]   EUROPEAN UNION REGULATION (EU) 2017/745 OF THE EUROPEAN PARLIAMENT AND OF THE COUNCIL of 5 April 2017 on medical devices, amending Directive 2001/83/EC, Regulation (EC) No 178/2002 and Regulation (EC) No 1223/2009 and repealing Council Directives 90/385/EEC and 93/42/EEC (2017).

[XV-2]   WATANABE, K., et al., First experimental verification of the neutron field of Nagoya University Accelerator-driven neutron source for boron neutron capture therapy, Appl. Radiat. Isot. **168** (2020) 109553.

[XV-3]   COLAUTTI, P., et al., Microdosimetric study at the CNAO active-scanning carbon-ion beam, Radiat. Prot. Dosim. **180** 1–4 (2018) 157–161.

[XV-4]   MAGRIN, G., et al., Microdosimetric characterization of clinical carbon-ion beams using synthetic diamond detectors and spectral conversion methods, Med. Phys. **47** 2 (2020) 713–721.

[XV-5]   VOHRADSKY, J., GUATELLI, S., DAVIS, J.A., TRAN, L.T., ROSENFELD, A.B., Evaluation of silicon based microdosimetry for boron neutron capture therapy quality assurance, Phys. Med. **66** (2019) 8–14.

[XV-6]   FACOETTI, A., et al., In vivo radiobiological assessment of the new clinical carbon ion beams at CNAO, Radiat. Prot. Dosim. **166** 1–4 (2015) 379–382. GUEULETTE, J., et al., Intestinal crypt regeneration in mice: a biological system for quality assurance in non-conventional radiation therapy, Radiother. Oncol. **73** Suppl. 2 (2004) S148–S154.

[XV-7]   INTERNATIONAL ORGANIZATION FOR STANDARDIZATION, Clinical investigation of medical devices for human subjects — Good clinical practice. ISO 14155:2020 (2020).

[XV-8]   EUROPEAN COMMUNITY, Directive 95/46/EC of the European Parliament and of the Council of 24 October 1995 on the protection of individuals with regard to the processing of personal data and on the free movement of such data (1995).

Annex XVI.

THE APPROVAL PATH FOR CLINICAL BNCT IN JAPAN

S. KATAOKA, Y. KIKUCHI, T. UMEHARA, S. MASUI
Medical and Advanced Equipment Unit, Industrial Equipment Division, Sumitomo Heavy Industries, Ltd., ThinkPark Tower, 2-1-1, Osaki, Shinagawa-ku, Tokyo 141-6025, Japan

Abstract

This article describes the regulatory environment in Japan and how 'BNCT' (the physical equipment, the software and the pharmaceutical combined) became approved for clinical use for unresectable, locally advanced or locally recurrent head and neck cancer.

XVI-1. INTRODUCTION

In 2020, an accelerator based BNCT (AB-BNCT) system and a dose calculation program for BNCT were approved as medical devices in Japan together with the approval of a boron drug, boronophenylalanine (BPA). The intended use is to treat unresectable, locally advanced or locally recurrent head and neck cancer. The devices and the drug were the first commercialized BNCT system in the world. In general, a new medical device and a new drug carrying major risks are required to perform a clinical trial for commercialization. This is a requirement in common with many countries where high risk medical products are sought to market. However, one country's regulatory processes may not exactly match those in other countries.

BNCT has some latitude in the neutron beam characteristics, which depend on the system design, to achieve an acceptable clinical result. In particular, the neutron energy spectrum varies a little from system to system. It is possible that a clinical outcome in BNCT has an inextricable link to a specific device and a specific drug: it may differ to some degree if a different combination were selected. Therefore, BNCT products look like combination products, because the device and the drug will be used together in the same clinical trial and the efficacy of the treatment is interdependent. The U.S. Food & Drug Administration defines combination products as therapeutic and diagnostic products that combine drugs, devices, and/or biologic products. From a regulatory standpoint, it does not seem self-evident at present whether the BNCT system is a combination product. Devices and drugs have different pathways to approval and are regulated under different types of regulatory authorities or different departments in the relevant agency. An applicant intending to market their BNCT system for the first time is strongly suggested to consult with their regulatory authority for guidance and advice as early as the non-clinical study phase. Accelerator-based systems are on the way to propel commercialization of BNCT, but the review process has yet to be fully established because of the lack of precedents of complete systems and the few installations in the world.

The other important component is software, especially the treatment planning system (TPS). According to the definition of Software as a Medical Device (SaMD) provided by the International Medical Device Regulators Forum[1], SaMD is "software intended to be used for one or more medical purposes that perform these purposes without being part of a hardware medical device". During the clinical trial performed in Japan, the Japanese regulation was revised and replaced by the Pharmaceutical and Medical Device Agency (PMDA[2]), and SaMD was legislated in 2014. Consequently, all TPSs for BNCT became classified as SaMD. In the past, the TPS 'Simulation Environment for Radiotherapy Applications' (SERA) was widely

---

[1] https://www.imdrf.org/
[2] https://www.pmda.go.jp/english/index.html

used for treatment planning for reactor based BNCT. SERA is a suite of software modules including geometric modelling from medical imaging such as CT and/or MRI, three-dimensional Monte-Carlo radiation transport, dose computation, and an integrated graphical user interface. SERA was used in the clinical trial for the first medical device registration in Japan; however, development of SERA has not been continued and the software was not being maintained by any commercial supplier. Therefore, it was inappropriate for SERA to become classified as SaMD in the current regulatory framework, and, for commercialization of BNCT, a new TPS needed to be developed. The new TPS was reviewed simultaneously with the equipment and the drug for its equivalency in comparison with SERA.

The Japanese regulatory authority, the PMDA, has four measures to provide faster access to safer and more effective medical devices and drugs on the market. Those regulatory mechanisms are 'Priority Review', 'Rapid Review', 'Breakthrough Therapy', and 'SAKIGAKE'. The SAKIGAKE[3,4] designation system was created in 2013 under the Japanese government's policy for pioneering the practical application of innovative medical products [XVI-1]. Similar regulatory pathways are, for example, 'Breakthrough Therapy' in the United States and 'Priority Medicines' in the European Union. Per the applicants' requests, the Japanese government designated the BNCT equipment, the dose calculation program, and the drug for BNCT as 'SAKIGAKE' in 2017, and the review processes were expedited.

PMDA's decision was that the system under review was not a combination product, but the use of the designated drug was mandatory. The approved indication was restricted only to the cancer that was treated in the clinical trial. In the review process, PMDA vigilantly gave attention to the validity of the criteria in safety, and the standards and references from the following organizations were referred to:

- International Electrotechnical Commission (IEC);
- International Organization for Standardization (ISO);
- Japanese Industrial Standards (JIS);
- International Atomic Energy Agency (TECDOC-1223 [XVI-2]);
- The applicant's internal standards for particle therapy equipment and/or their modification optimized for BNCT.

Basically, the framework of the review was not dissimilar to the normal kind of review, including, but not limited to, safety, efficacy, and quality. However, BNCT differs greatly from conventional radiotherapy and particle therapy in some respects, such as neutron dose measurement and the unique dose calculation caused by the mechanism of BNCT. Taking these specifics into consideration, below were the items heavily weighted in the review:

(a) Performance and appropriateness of the specification limits:
   (i) Repeatability and stability of calibration of dose monitoring system;
   (ii) Linearity of dose monitoring system;
   (iii) Depth–dose curve;
   (iv) Peak dose;
   (v) Positioning reproducibility of the treatment bed;
   (vi) Measurement precision of the charge monitors of charged particle beam;
   (vii) Irradiation field size;
   (viii) Time of continuous proton beam irradiation;
   (ix) Dose distribution calculation functions;

---

[3] https://www.mhlw.go.jp/english/policy/health-medical/pharmaceuticals/dl/140729-01-01.pdf
[4] https://www.mhlw.go.jp/english/policy/health-medical/pharmaceuticals/dl/140729-01-02.pdf

(x)   Dose calculation algorithm.
(b) Device safety:
  (i) Physicochemical properties;
  (ii)   Electrical safety and electromagnetic compatibility;
  (iii)   Biological safety;
  (iv)   Radiation safety;
  (v)   Mechanical safety;
  (vi)   Stability and durability;
  (vii)   Studies to support the safety of dose engine.
(c) Other design verification studies:
  (i)   Other performance studies;
  (ii)   Other safety studies;
  (iii)   Studies to support the efficacy of the device;
  (iv)   Studies to support usage method of the system.
(d) Conformity to the requirements specified in the relevant act;
(e) Risk management;
(f) Manufacturing process;
(g) Clinical Data (Phase I and Phase II) or Alternative Data Accepted by PMDA;
(h) Plan for Post-marketing Surveillance etc.

Based on PMDA's review and suggestion, the Minister of Health, Labour and Welfare made the final decision on the approvals. During the first experience of the regulatory process, four salient points emerged from the Japanese case:

(a) The equipment and the drug needed to be approved simultaneously in pairs, using data taken from the same clinical trial;
(b) The new TPS was reviewed and approved by comparing with the existing TPS, and also in pairs with the equipment and the drug;
(c) The medical device application is reviewed indication by indication, and a clinical trial is necessary for each indication;
(d) Post-marketing surveillance is mandatory.

In Japan, many of the devices and the drugs that receive approvals will be also covered shortly thereafter by national health insurance, which may differentiate the Japanese review process from that of other countries.

REFERENCES TO ANNEX XVI

[XVI-1]   MINISTRY OF HEALTH, LABOUR AND WELFARE (Japan) Strategy of SAKIGAKE, SAKIGAKE PT, 17 June 2014 (2014).

[XVI-2]   INTERNATIONAL ATOMIC ENERGY AGENCY, Current Status of Neutron Capture Therapy, TECDOC 1223, IAEA, Vienna (2001).

Annex XVII.

APPLICATION FOR MARKETING APPROVAL OF
STEBORONINE 9000 mg/300 ml FOR INFUSION

YOSHIYA IGUCHI, HIDEKI NAKASHIMA, YOSHIMITSU KATAKUSE, TOMOYUKI ASANO
Sakai R&D Center, Stella Pharma Corporation, Japan

Abstract

Steboronine is a drug containing borofalan, a phenylalanine derivative (BPA)[1], isotopically enriched in ^{10}B. Steboronine (active substance borofalan (^{10}B)) made by Stella Pharma Corporation was approved as a new drug in Japan by the Pharmaceuticals and Medical Devices Agency (PMDA), the regulatory authority, in March 2020 [XVII-1][2]. This article gives an outline of the data requirements for a new drug submission for BNCT within Japan.

XVII-1. INTRODUCTION

The basic data concerning Steboronine are summarized in Table XVII-1.

TABLE XVII-1. BASIC DATA DESCRIBING STEBORONINE [XVII-1][2]

Brand Name	Steboronine 9000 mg/300 mL for infusion
INN	Borofalan (^{10}B)
Applicant	Stella Pharma Corporation
Dosage Form/Strength	An injectable solution containing 9000 mg of borofalan (^{10}B) per bag (300 mL)
Chemical Structure	
Molecular formula	$C_9H_{12}{}^{10}BNO_4$
Molecular weight	208.21
Chemical name	(S)-2-Amino-3-[4-(^{10}B)dihydroxyboranylphenyl]propanoic acid
Dosage and Administration	Borofalan (^{10}B) is administered as an intravenous infusion 2 hours at the rate of 200 mg·kg^{-1}·h^{-1}, followed by irradiation of neutron beams to the cancer area for adults. During the irradiation, borofalan (^{10}B) is intravenously infused at 100 mg·kg^{-1}·h^{-1}.

XVII-2. DRUG SUBSTANCE: BOROFALAN (^{10}B)

XVII-2.1. Characterization

"The drug substance consists of white crystals or a crystalline powder. The general properties of the drug substance, including description, solubility, hygroscopicity, melting point, thermal analysis, dissociation constant, and partition coefficient have been determined.

The chemical structure of the drug substance has been elucidated by elemental analysis, mass spectrometry, ultraviolet/visible spectrophotometry (UV/VIS), infrared spectrophotometry (IR), nuclear magnetic resonance spectroscopy (NMR: ^{1}H-NMR and ^{13}C-NMR), and single crystal X ray diffractometry" [XVII-1].

---

[1] BPA was registered with the International Nonproprietary Name (INN) as borofalan (^{10}B).
See https://www.who.int/teams/health-product-and-policy-standards/inn
[2] PMDA deliberation: https://www.pmda.go.jp/files/000237990.pdf

XVII-2.2. Control of drug substance

Each manufacturing process of the drug substance was established as a Good Manufacturing Practice (GMP) process and validated. Impurities which occurred in the manufacturing process and quality assurance period were identified following ICH[3] guideline Q3A[4] for controlling impurities in active substances [XVII-2].

The approved "specifications for the drug substance include content, description, identification (UV/VIS and IR), purity (clarity and color of solution, chloride, heavy metals, arsenic, related substances (by liquid chromatography – LC), optical isomers (LC), residual solvents (gas chromatography – GC), ^{10}B abundance ratio (inductively coupled plasma mass spectrometry), loss on drying, and assay (LC). Borofalan(^{10}B) which is more than 99% ^{10}B-enriched has been used for development study and as a commercial product in Japan" [XVII-1].

XVII-2.3. Stability of drug substance

The stability test for borofalan(^{10}B) is conducted following the long-term stability requirement in ICH guideline Q1A[5] which is stability testing of new drug substances [XVII-3]. Table XVII-2 shows the storage conditions and terms for stability tests for the drug substance. A photostability testing also conducted following Q1B[6] and results showed the drug substance was photostable [XVII-4].

TABLE XVII-2. STABILITY TESTS OF THE DRUG SUBSTANCE

Test	Temperature	Humidity	Storage period
Long term testing	25 °C	60% RH	60 months
Accelerated testing	40 °C	75% RH	6 months

XVII-3.   DRUG PRODUCT: STEBORONINE 9000 mg/300 ml FOR INFUSION

XVII-3.1. Description and composition of drug product and formulation development

"The drug product is a liquid for injection containing 9000 mg of the drug substance packed in a soft bag (300 mL). The drug product contains excipients D-sorbitol as a solubilizer for borofalan(^{10}B), stabilizing agent and pH adjuster, and water for injection" [XVII-1].

XVII-3.2. Control of drug product

Each manufacturing process for the drug product was established as a GMP process and validated. Impurities which were contained in the active substance and occurred in the manufacturing process and quality assurance period were identified following the ICH guideline Q3B[7] for controlling impurities in drug product [XVII-5]. In addition, as Steboronine is an infusion drug product, it is necessary to establish sterilization assurance.

The approved "specifications for the drug product consist of content, description, identification (ninhydrin reaction, borate, and UV/VIS), osmotic pressure ratio, pH, purity (related substances [LC] and optical isomers [LC]), bacterial endotoxin, extractable volume, foreign insoluble matter, insoluble particulate matter, sterility, and assay (LC)" [XVII-1].

---

[3] ICH: The International Council for Harmonisation of Technical Requirements for Pharmaceuticals for Human Use
[4] https://database.ich.org/sites/default/files/Q3A%28R2%29%20Guideline.pdf
[5] https://database.ich.org/sites/default/files/Q1A%28R2%29%20Guideline.pdf
[6] https://database.ich.org/sites/default/files/Q1B%20Guideline.pdf
[7] https://database.ich.org/sites/default/files/Q3B%28R2%29%20Guideline.pdf

XVII-3.3. Stability of drug product

The stability test for Steboronine was conducted following the long-term stability requirement in ICH guideline Q1A [XVII-3] for new drug products. Table XVII-3 shows the storage conditions and terms for stability tests for the drug product. A photostability testing also conducted following Q1B [XVII-4] and results showed the drug substance was photostable.

TABLE XVII-3. STABILITY TESTS OF DRUG PRODUCT

Test	Temperature	Humidity	Storage period
Long term testing	5 °C	–	36 months
Accelerated testing	25 °C	60% RH	6 months

XVII-4. CTD-BASED NEW DRUG APPLICATION

To apply for approval of boron drugs as medicines, it is necessary to configure the Common Technical Document (CTD)[8] and submit it to the regulatory authorities. The composition of the CTD is harmonized by ICH and shown in the M4[9,10] guideline [XVII-6–XVII-7]. It addresses the organization of the information to be presented in registration applications for new pharmaceuticals. All of the quality data is the results by using Steboronine which is formulated more than 99% ^{10}B enriched Borofalan(^{10}B) as drug substance showed Tables XVII-4–XVII-6.

TABLE XVII-4. CTD COMPOSITION OF DRUG SUBSTANCE

Section	Title
3.2.S.1	General information
3.2.S.1.1	Nomenclature
3.2.S.1.2	Structure
3.2.S.1.3	General properties
3.2.S.2	Manufacture
3.2.S.2.1	Manufacturer(s)
3.2.S.2.2	Description of manufacturing process and process controls
3.2.S.2.3	Control of materials
3.2.S.2.4	Controls of critical steps and intermediates
3.2.S.2.5	Process validation and/or evaluation
3.2.S.2.6	Manufacturing process development
3.2.S.3	Characterisation
3.2.S.3.1	Elucidation of structure and other characteristics
3.2.S.3.2	Impurities
3.2.S.4	Control of the drug substance
3.2.S.4.1	Specification(s)
3.2.S.4.2	Analytical procedures
3.2.S.4.3	Validation of analytical procedures
3.2.S.4.4	Batch analyses
3.2.S.4.5	Justification of specification(s)
3.2.S.5	Reference standards or materials
3.2.S.6	Container closure system
3.2.S.7	Stability
3.2.S.7.1	Stability summary and conclusions
3.2.S.7.2	Post-approval stability protocol and stability commitment
3.2.S.7.3	Stability data

---

[8] https://www.ich.org/page/ctd
[9] https://database.ich.org/sites/default/files/M4S_R2_Guideline.pdf
[10] https://database.ich.org/sites/default/files/M4S_Q%26As_R2_Q%26As.pdf

TABLE XVII-5. CTD COMPOSITION OF DRUG PRODUCT

Section	Title
3.2.P.1	Description and composition of the drug product
3.2.P.2	Pharmaceutical development
3.2.P.2.1	Components of the drug substance
3.2.P.2.1.1	Drug substance
3.2.P.2.1.2	Excipients
3.2.P.2.2	Drug product
3.2.P.2.2.1	Formulation development
3.2.P.2.2.2	Overages
3.2.P.2.2.3	Physicochemical and biological products
3.2.P.2.3	Manufacturing process development
3.2.P.2.4	Container closure system
3.2.P.2.5	Microbiological attributes
3.2.P.2.6	Compatibility
3.2.P.3	Manufacture
3.2.P.3.1	Manufacturer(s)
3.2.P.3.2	Batch formula
3.2.P.3.3	Description of manufacturing process and process controls
3.2.P.3.4	Controls of critical steps and intermediates
3.2.P.3.5	Process validation and/or evaluation
3.2.P.4	Control of excipients
3.2.P.4.1	Specifications
3.2.P.4.2	Analytical procedures
3.2.P.4.3	Validation of analytical procedures
3.2.P.4.4	Justification of specifications
3.2.P.4.5	Excipients of animal or human origin
3.2.P.4.6	Novel excipients
3.2.P.5	Control of the drug product
3.2.P.5.1	Specifications
3.2.P.5.2	Analytical procedures
3.2.P.5.3	Validation of analytical procedures
3.2.P.5.4	Batch analyses
3.2.P.5.5	Characterisation of impurities
3.2.P.5.6	Justification of specification(s)
3.2.P.6	Reference standards or materials
3.2.P.7	Container closure system
3.2.P.8	Stability
3.2.P.8.1	Stability summary and conclusion
3.2.P.8.2	Post-approval stability protocol and stability commitment
3.2.P.8.3	Stability data

TABLE XVII-6. CTD COMPOSITION OF QUALITY PART APPENDICES

Section	Title
3.2.A.1	Facilities and equipment
3.2.A.2	Adventitious agents safety evaluation
3.2.A.3	Excipients
3.2.R	Regional information

Safety and efficacy data also ought to be prepared according to guideline of the CTD. All of the non-clinical study reports indicated by Table XVII-7 need to be shown in the CTD. In Japan, single dose toxicity data were necessary both for the boron drug and for BNCT irradiation.

TABLE XVII-7. CTD COMPOSITION OF SAFETY DATA

Section	Title
4.2.1	Pharmacology
4.2.1.1	Primary pharmacodynamics
4.2.1.2	Secondary pharmacodynamics
4.2.1.3	Safety pharmacology
4.2.1.4	Pharmacodynamic drug interactions
4.2.2	Pharmacokinetics
4.2.2.1	Analytical methods and validation reports (if separate reports are available)
4.2.2.2	Absorption
4.2.2.3	Distribution
4.2.2.4	Metabolism
4.2.2.5	Excretion
4.2.2.6	Pharmacokinetic drug interactions (nonclinical)
4.2.2.7	Other pharmacokinetic studies
4.2.3	Toxicology (Drug and BNCT)
4.2.3.1	Single-dose toxicity (in order by species, by route)
4.2.3.2	Repeat-dose toxicity (in order by species, by route, by duration; including supportive toxicokinetics evaluations)
4.2.3.3	Genotoxicity
4.2.3.3.1	In vitro
4.2.3.3.2	In vivo (including supportive toxicokinetics evaluations)
4.2.3.4	Carcinogenicity (including supportive toxicokinetics evaluations)
4.2.3.4.1	Long-term studies (in order by species; including range-finding studies that cannot appropriately be included under repeat-dose toxicity or pharmacokinetics)
4.2.3.4.2	Short- or medium-term studies (including range-finding studies that cannot appropriately be included under repeat-dose toxicity or pharmacokinetics)
4.2.3.4.3	Other studies
4.2.3.5	Reproductive and developmental toxicity (including range-finding studies and supportive toxicokinetics evaluations) (If modified study designs are used, the following sub-headings need to be modified accordingly.)
4.2.3.5.1	Fertility and early embryonic development
4.2.3.5.2	Embryo-fetal development
4.2.3.5.3	Prenatal and postnatal development, including maternal function
4.2.3.5.4	Studies in which the offspring (juvenile animals) are dosed and/or further evaluated.
4.2.3.6	Local tolerance
4.2.3.7	Other toxicity studies (if available)
4.2.3.7.1	Antigenicity
4.2.3.7.2	Immunotoxicity
4.2.3.7.3	Mechanistic studies (if not included elsewhere)
4.2.3.7.4	Dependence
4.2.3.7.5	Metabolites
4.2.3.7.6	Impurities
4.2.3.7.7	Other

REFERENCES TO ANNEX XVII

[XVII-1]   PHARMACEUTICALS AND MEDICAL DEVICES AGENCY, Report on the Deliberation Results, Steboronine 9000 mg/300 mL for Infusion (English translation), February 6, 2020, Tokyo, Japan (2020).

[XVII-2]   INTERNATIONAL CONFERENCE ON HARMONISATION OF TECHNICAL REQUIREMENTS FOR REGISTRATION OF PHARMACEUTICALS FOR HUMAN USE, ICH Harmonised Tripartite Guideline: Impurities in New Drug Substances Q3A(R2), Current Step 4 version, dated 25 October 2006, ICH, Geneva (2006).

[XVII-3]   INTERNATIONAL CONFERENCE ON HARMONISATION OF TECHNICAL REQUIREMENTS FOR REGISTRATION OF PHARMACEUTICALS FOR HUMAN USE, ICH Harmonised Tripartite Guideline: Stability Testing of New Drug Substances and Products Q1A(R2), Current Step 4 version dated 6 February 2003, ICH, Geneva (2003).

[XVII-4]   INTERNATIONAL CONFERENCE ON HARMONISATION OF TECHNICAL REQUIREMENTS FOR REGISTRATION OF PHARMACEUTICALS FOR HUMAN USE, ICH Harmonised Tripartite Guideline: Stability Testing: Photostability Testing of New Drug Substances and Products Q1B, Current Step 4 version dated 6 November 1996, ICH, Geneva (1996).

[XVII-5]   INTERNATIONAL CONFERENCE ON HARMONISATION OF TECHNICAL REQUIREMENTS FOR REGISTRATION OF PHARMACEUTICALS FOR HUMAN USE, ICH Harmonised Tripartite Guideline: Impurities in New Drug Products Q3B(R2), Current Step 4 version dated 2 June 2006, ICH, Geneva (2006).

[XVII-6]   INTERNATIONAL CONFERENCE ON HARMONISATION OF TECHNICAL REQUIREMENTS FOR REGISTRATION OF PHARMACEUTICALS FOR HUMAN USE, ICH Harmonised Tripartite Guideline: Stability Testing: The Common Technical Document for the Registration of Pharmaceuticals for Human Use: Safety – M4S(R2) Nonclinical Overview and Nonclinical Summaries of Module 2 Organisation of Module 4, Current Step 4 version dated 20 December 2002, ICH, Geneva (2002).

[XVII-7]   INTERNATIONAL CONFERENCE ON HARMONISATION OF TECHNICAL REQUIREMENTS FOR REGISTRATION OF PHARMACEUTICALS FOR HUMAN USE, M4S Implementation Working Group Questions & Answers (R4) Current version dated November 11, 2003, ICH, Geneva (2003).

Annex XVIII.

PRACTICAL GUIDELINES FOR PRECLINICAL EXPERIMENTS FOR ACCELERATOR-BASED BNCT SYSTEMS

SHOJI IMAMICHI[1, 2, 3], YOSHITAKA MATSUMOTO[4, 5], SHIN-ICHIRO MASUNAGA[6, 7], MITSUKO MASUTANI[1,2]

[1]Division of BNCT, EPOC, National Cancer Center, Tokyo, Japan
[2]Department of Molecular and Genomic Biomedicine, CBMM, Nagasaki University Graduate School of Biomedical Sciences, Japan
[3]Radiation Oncology Research Laboratory Center, Institute for Integrated Radiation and Nuclear Science, Kyoto University, Japan
[4]Department of Radiation Oncology, Clinical Medicine, Faculty of Medicine, University of Tsukuba, Japan
[5]Proton Medical Research Center, University of Tsukuba Hospital, Japan
[6]Research Center of Boron Neutron Capture Therapy, Osaka Prefecture University
[7]Particle Radiation Biology, Division of Radiation Life Science, Institute for Integrated Radiation and Nuclear Science, Kyoto University

Abstract

This article is intended to guide practical planning of preclinical tests before conducting clinical trials for accelerator-based boron neutron capture therapy (AB-BNCT). Pre-planning of clinical trials for treatment of specific types of cancers with AB-BNCT is necessary to make preclinical evaluation plans for discussion with the regulatory authorities. For different types of AB-BNCT systems, the assessment of safety and effectiveness and evaluations may need to be different. In this article, the current guidelines of set up conditions for AB-BNCT systems and necessary considerations are discussed.

XVIII-1.   INTRODUCTION

In Japan, an AB-BNCT system based around a cyclotron with a Be target was approved for medical treatment in 2020 [XVIII-1], and clinical trials for another AB-BNCT system consisting of a linac with a solid Li target have been started [XVIII-2]. An AB-BNCT system consisting of a linac with a Be target has also been developed and preclinical tests are ongoing [XVIII-3–XVIII-5]. Preclinical tests have to be scheduled before clinical trials of AB-BNCT systems; i.e., safety and efficacy of AB-BNCT systems need to be proven using cell based and animal experiments. Although the regulatory levels for AB-BNCT systems differ among countries, based on experience and information, some practical guidelines reflecting current knowledge and experiments for preclinical assessment are summarized in this article.

For cell based and xenograft experiments, it is preferable to use cells of specific cancer types for the planned clinical trials. Determination of the values for relative biological effectiveness (RBE) of the neutron beam (hydrogen dose), which is specific for each AB-BNCT system, needs to be carried out if a new AB-BNCT system is being developed [XVIII-6]. A safety evaluation study needs to include evaluation of cytotoxicity and genotoxicity, and assessment of in vivo effects using animal models, such as mice. The biological effects need to be compared with those of γ or X ray irradiation to calculate RBE values [XVIII-7]. The observation periods are suggested to be defined to investigate both short and long term effects. Efficacy evaluation studies ought to include clonogenic cell survival assays of several types of cancer cells. Effects on tumour growth can be evaluated using xenograft models of mice with target cancer types.

The final evaluation with preclinical tests ought to be carried out using drugs produced under GMP (good manufacturing practice) and close-to-GLP (good laboratory practice) level facilities and conditions. After irradiation with the AB-BNCT system, the long-term observation of mice at irradiation facilities would often be difficult because of the regulations concerning radioactive materials and other laws. Therefore, it may be necessary to confirm that the activation levels are low and to transfer the irradiated animals to outside facilities. Consultation with the relevant regulatory authorities for full planning is unavoidable.

XVIII-2. SETUP CONDITIONS FOR BIOLOGICAL EVALUATIONS

XVIII-2.1. Preparation for preclinical assessments for accelerator systems

AB-BNCT systems have already been installed in medical facilities, and it is important for preclinical assessments for AB-BNCT systems that the facilities required for cell culture, radioisotope production/handling, and animal facilities, as well as a γ or X ray irradiation system for animals are all available. Collaboration with the physical assessment team and radiation oncologists will be also essential. These points need to be confirmed before preparing for preclinical assessments.

XVIII-2.2. Dose calculation for biology experiments with accelerator systems

In order to characterize the beam of an AB-BNCT system it is advisable to conduct biological dosimetry for validation of the physical dosimetry evaluations.

The traditional way to calculate the bioequivalent dose (hereinafter referred to as the equivalent dose) is by multiplying each calculated absorbed dose component by the corresponding RBE values. The equivalent dose of the boron component in a certain tissue is calculated by multiplying the absorbed dose per ppm boron by the tissue boron concentration and by the CBE (compound biological effectiveness) value, which is the RBE factor for boron dose for a certain boron drug specified for each tissue. The boron concentration in tissues ought to be calculated from the blood concentration and tissue/blood concentration ratio determined prior to the experiments (see also Ref. [XVIII-8]).

$$D \text{ (Gy-Eq)} = D_B \times \text{RBE}_B + D_N \times \text{RBE}_N + D_H \times \text{RBE}_H + D_\gamma \times \text{RBE}_\gamma \quad \text{(XVIII-1)}$$
$$D_B \text{ (Gy)} = (7.43 \times B(\text{ppm})) \times f \times 10^{-14} \quad \text{(XVIII-2)}$$
$$D_N \text{ (Gy)} = (6.78 \times N) \times f \times 10^{-14} \quad \text{(XVIII-3)}$$

where $f = \int \phi(t) \cdot dt$: fluence of thermal neutrons (cm^{-2})

The CBE values have been determined for individual boron compounds, boronophenylalanine (BPA) and sodium borocaptate (BSH), and for each tissue, mostly from in vivo rodent experiments (Table XVIII-1). The CBE values are required to convert the boron dose, which accounts for most of the dose, to the photon equivalent dose. The boron dose is derived from the reaction between thermal neutrons and boron nuclei. Therefore, extrapolation of the values obtained from reactor based BNCT can be used. However, the concept of CBE was generated with a given geometry and boron dose in order to minimize the occurrence of adverse events on normal tissues, especially for late adverse events, or to estimate the effect on tumour tissues for each endpoint. Thus, CBE factors have been measured not from in vitro cultured cell experiments but using certain BNCT systems and in vivo tests in tissues. Therefore, the CBE factors will not be exact for different AB-BNCT systems and can be applicable only within limitations.

In practice, the same CBE value can be used for different AB-BNCT systems as for reactor based BNCT if the boron drug (e.g., BPA) is the same. However, if the endpoints and AB-BNCT irradiation set up conditions are different, some levels of uncertainty will inevitably exist.

For γ rays, because there are few differences among AB-BNCT systems, almost the same RBE values can be utilized. In contrast, since the spectral weightings of fast, epithermal, and thermal neutrons are different among the different AB-BNCT systems, the RBE value for neutron (hydrogen) dose needs to be measured in individual systems prior to the preclinical tests. The RBE values used for reactor based BNCT cannot be used as they are for AB-BNCT. Combined biological effects of primary or secondary γ rays and neutrons need also to be considered individually in each AB-BNCT system. These values obtained by physical measurements ought to be validated in the biological evaluation.

Since the neutron characteristics and γ ray contamination rates are different between reactor based and the accelerator based BNCT systems, direct extrapolation of biological effects will not be possible. Limited extrapolation of biological effects may be possible among similar AB-BNCT systems. However, the precise extrapolation of biological effects will not be possible among different types of AB-BNCT systems. To date, the biological effects of each dose radiation component have been evaluated individually in some reports [XVIII-9–XVIII-11]. By extending this approach, limited extrapolation may become possible by evaluating the contribution of each dose component.

TABLE XVIII-1. RBE FOR NEUTRON DOSE AND CBE FACTORS FOR BPA AND BSH

Radiation	Tumour	Brain	Skin	Mucosa [XVIII-12, XVIII-13]	Lung [XVIII-14]	Liver (hepatocytes) [XVIII-15]
Neutrons						
Thermal neutron	3.0	3.0	3.0	3.0	3.0	3.0
Epithermal neutron	3.0	3.0	3.0	3.0	3.0	3.0
$^{10}B(n, \alpha)^7Li$						
BPA	3.8	0.32 + N/B × 1.65	2.5	4.9	0.32 + N/B × 1.80	4.3
BSH	2.5	0.4	0.8	0.3		0.9
γ ray	1.0	1.0	1.0	1.0	1.0	1.0

Note: Data are taken from Ref. [XVIII-6]. N/B is the ratio of boron concentration in normal skin/blood.

XVIII-2.3. For typical system geometries of accelerator systems

Current experience implies that the minimum desirable epithermal neutron flux would be 10^9 cm^{-2} s^{-1}. Beams of 5×10^8 cm^{-2} s^{-1} have been used but may result in rather long irradiation times. The AB-BNCT systems currently being developed for clinical applications utilize two types of targets, Be and Li, which can produce epithermal neutron fluxes of suitable strength. The biological effects of each type may reflect the individual dose components and differences in dose rates. The irradiation geometry for biological testing, such as the use of a bolus of ~2 cm thickness to increase the thermal neutron fluence at the irradiation site, at the front and back of the cell samples or animals, will also affect the biological response and need to be considered sufficiently beforehand at the physical evaluation stage.

XVIII-2.3.1. Setting up for different beam directions of accelerator systems

Depending on the position of the beam port, direction of the beam, and irradiation room, set up conditions for biological testing will be different and need to be prepared accordingly.

XVIII-2.3.2. For a horizontal beam

The horizontal beam geometry has been tested for various types of reactor based BNCT systems and these set up conditions can be applied in biology experiments. Acrylic plates or water equivalent materials can be used, if necessary, to maximize neutron flux. The dose on the top, lateral and bottom sides of the beam port at the irradiation position for a given irradiation field will be affected by the set-up materials and equipment (Fig. XVIII-1).

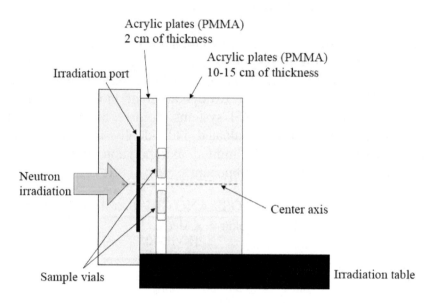

FIG. XVIII-1. The set up for horizontal neutron beam irradiation.

XVIII-2.3.3. For a vertical beam

Some of the considerations for a vertical irradiation system include:

- The biological effects ought to be evaluated at the centre, inside, and outside of the irradiation field in a symmetrical manner for the surrounding set up geometry, matching physical evaluations (Fig. XVIII-2);
- To maximize the thermal neutron flux and to stabilize the irradiation dose compositions, either acrylic plates or water equivalent materials can be placed under the samples to be irradiated. The acrylic plates need to be 10–15 cm in thickness (with more than 10 cm thickness, the reflected thermal neutron flux is maximized). The acrylic plate size has to be larger than the irradiation field (port) diameter. To prevent leakage of the neutron beam, the bolus and samples, cells and animals need to be tightly spaced. Spaces between the samples and equipment can be filled with soft water-equivalent bolus materials, available from vendors, to obtain a reproducible irradiation geometry;
- Installation of acrylic plates behind the samples will also be useful for stabilization of beam compositions. Neutron reflections from the floor need to also be controlled;
- Mice and rats can be positioned in non-constrained natural body positions.

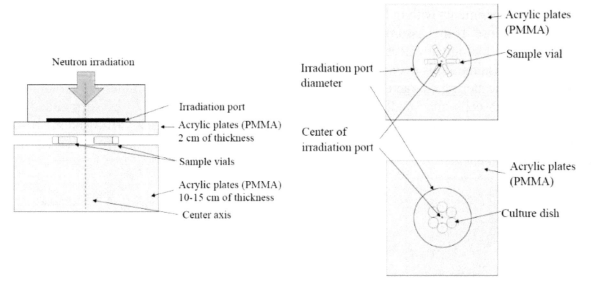

FIG. XVIII-2. The set up for vertical neutron beam irradiation.

XVIII-2.4. Considerations for set-ups for different accelerator systems

XVIII-2.4.1.   Cyclotron with a beryllium target and horizontal beam

The proton energy and current are important factors for neutron beam generation from a Be target, and one can classify such systems as a 'high-energy type' with a proton energy of 30 MeV and a 'medium-energy type' with a proton energy of 8 MeV and a beam current of 1 mA. Since the primary neutron energy is high, it is necessary to slow the neutrons down to create an epithermal neutron beam ($0.6 \, \text{eV} < E_\text{n} \leq 10 \, \text{keV}$) suitable for BNCT, and it is essential to evaluate the γ ray dose at a certain depth generated at the accumulated time (depth dose evaluation), because the γ ray dose rate contribution increases. With a suitable beam shaper assembly, it is possible to form a neutron beam of good directionality with a peak in the epithermal energy region, although it is accompanied by γ ray contamination. The currently reported irradiation field of the NeuCure AB-BNCT system with a Be target on a cyclotron is 10–15 cm in diameter.

XVIII-2.4.2.   The geometry for cells and animal experiments

In order to minimize the non-uniformity of the dose distribution in irradiated samples, the sample arrangement needs to be as uniform as possible and point-symmetrical with respect to the beam centre as shown in Fig. XVIII-1. In addition, regarding the posture of the experimental animal to be irradiated, consideration needs to be given to avoid an unnatural posture such as an inverted position and to arrange the animal in as natural a posture as much as possible.

XVIII-2.4.3.   Linac with a beryllium target and horizontal beam

For a medium-energy type linear accelerator, with a proton energy of 8 MeV and a beam current of 10 mA, it is also possible to form a fairly directional beam with a peak in the epithermal energy region, although the beam also contains γ ray contamination.

XVIII-2.4.4.   The geometry for cells and animal experiments

In biological experiments with radiation, it is extremely important to ensure the uniformity and stability of the dose. If this assumption cannot be maintained, the horizontal axis of the graph representing the biological effect will be uncertain. In biological experiments with neutron radiation, this assumption is even more important than for other types of radiation, because of the diffuse and scattering characteristics of neutron radiation. Therefore, in neutron irradiation experiments with cells and small animals, the irradiated sample ought to be surrounded by water-equivalent materials (polyethylene, acrylic, etc.), as shown in Figs XVIII-3 and XVIII-4, in order to collect the scattered components as much as possible and to minimize the variability between experiments.

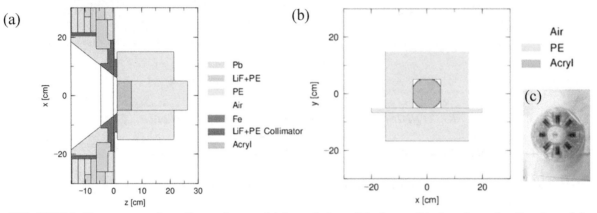

FIG. XVIII-3. The geometry for cell experiments. (a) lateral view of the beam, (b) view down the direction of the beam, (c) photo of cell irradiation tube holder.

The handling of irradiated samples varies from country to country and from organization to organization, and it is important to adapt the system to the situation. For example, in the mouse irradiation container shown in Fig. XVIII-4b, the container is hermetically sealed to prevent leakage and dispersion of the radioactive materials to the outside. The air and vaporized anaesthesia that flow through the tube enable stable irradiation and guarantee that the experimental animals will not awaken during irradiation. Please refer to this as an example for such experiments.

XVIII-2.4.5.   Linac with a solid lithium target and vertical beam

A solid Li target permits use of a low energy proton beam that can yield a primary neutron energy of 1 MeV or less. This means that the fast neutron component of the beam is small. An example of an AB-BNCT system that uses a Li target is the CICS-1 linac at the National Cancer Center hospital, Japan, which is built in a vertical beam geometry (Fig. XVIII-2). Features of this system include a low fast neutron component, low degree of activation, but poor neutron beam directionality. The activation by thermal neutron rays also needs to be considered. A gradual decrease in neutron beam output that is dependent on irradiation time occurs for this AB-BNCT system type with solid Li target and linac [XVIII-2]. Therefore, the irradiation time will be calculated for each experiment based on the proton current and a slight dose rate change needs to be considered in the evaluation of biological effects. This system uses a wobbling beam, and the biological effects need to be evaluated at the centre, inside and outside of the irradiation field in a symmetrical manner as shown in Fig. XVIII-2.

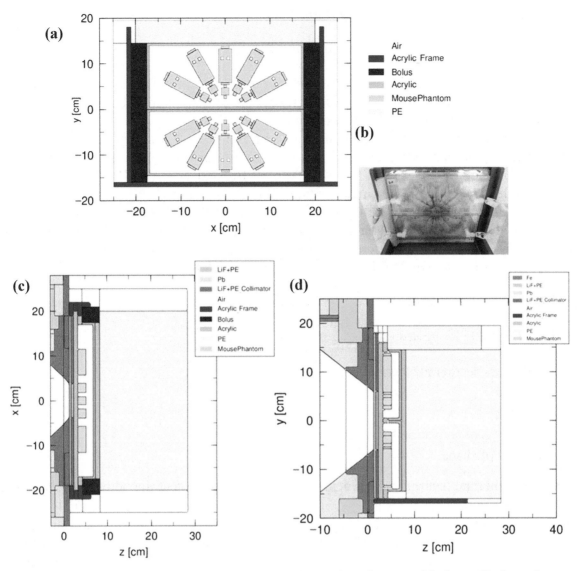

FIG. XVIII-4. The geometry for small animal experiments. (a) view from direction of the beam (b) photo of mouse irradiation holder, (c) view from the top of the beam, (d) lateral view of the beam.

XVIII-2.5. Effect of neutron beam properties on biological effects

The following contributions need to be considered:

- Neutron spectrum: Since the RBE values of thermal, epithermal, and fast neutrons and γ rays are different, the cell-killing effect depends on the mixing ratio of each radiation constituent;

- γ ray contamination rate: When the ratios of thermal, epithermal, and fast neutrons change, the γ ray contamination rate changes. Therefore, for the same reason as above, the cell-killing effect and the cell survival rate change. Other biological effects such as the anti-tumour effect will be similarly affected. Therefore, the evaluation needs to be performed at several beam-depth points using water-equivalent bolus materials;

- Dose rate: Gamma rays, which are low LET radiation, show a lower cell-killing effect as the dose rate decreases. In contrast, the dose rate dependence of the cell-killing effect of thermal, epithermal, and fast neutrons (high LET radiation) is reported to be small but may begin to contribute when a longer irradiation is carried out [XVIII-16].

XVIII-2.6. Requirements for set up conditions

XVIII-2.6.1. Activation of cells and animals

The activation levels of cells and animals needs to be tested to confirm that they are below the limits required by Ref. [XVIII-17] or other regulation levels. Freshly dead animals or tissues can be used for the evaluation of activation levels. Only after this confirmation, can the irradiated cells and animals be safely transferred to outside facilities for further analysis and observations. The induced radionuclides in mice after irradiation with an AB-BNCT system with a Li target were ^{24}Na, ^{38}Cl , ^{42}K, ^{56}Mn, ^{80m}Br, and ^{82}Br [XVIII-18–XVIII-19]. The activation levels of mice for the major radionuclides, ^{24}Na and ^{38}Cl, have to be lower than the regulation limit after neutron beam irradiation by any AB-BNCT system [XVIII-18–XVIII-19]. The induced activity and the saturated activity, expressed using $Bq \cdot g^{-1} \cdot mC^{-1}$ and $Bq \cdot g^{-1} \cdot A^{-1}$, respectively, can be estimated considering the mass of the samples, and the proton current of the accelerator in any AB-BNCT system [XVIII-18–XVIII-19].

XVIII-2.6.2. Shielding materials for neutron beams

The requirements for neutron shielding materials include:

- High cross section for thermal neutrons;
- Thinness;
- Solidity;
- Low levels of activation from neutron irradiation;
- Stability of shape.

The use of thermal neutron beam shielding is necessary to protect the body of the mouse and other animal models. A thermal neutron reacts with ^{10}B to produce an α particle and a ^{7}Li nucleus, both short range particles. Therefore, the maximization of the thermal neutron flux as well as the shielding have to be fully considered in experiments. Some substances with small atomic number (e.g., ^{1}H, ^{12}C) may be used as moderators. For example, acrylic plates, a paraffin block, and a water equivalent bolus can be used. For maximization of the thermal neutron component, it is necessary to set up the irradiation using materials appropriate to the particular beam characteristics of the AB-BNCT system.

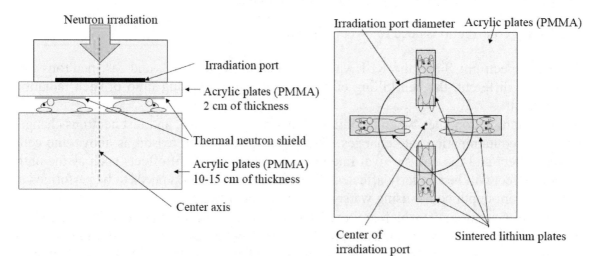

FIG. XVIII-5. Use of shielding materials for neutron beam irradiation of mice.

The thermal neutron shielding plates need to be constructed of materials with high cross-sections. Plates made of ^{6}Li-enriched resins have been used in reactor facilities [XVIII-20–XVIII-22], and sintered Li plate (Nikkei Sangyo Co., Ltd., Nippon Light Metal Group) were used at the National Cancer Center Hospital in Japan [XVIII-23]. These plates can be approximately 5 mm thick. If the posterior end of the hind legs are to be locally irradiated, the head and body may be covered with a plate (Fig. XVIII-5). A box type thermal neutron shield using sintered Li plates has also been designed for local irradiation of mice. However, with this design, extensive attenuation of the thermal neutron flux could occur and the estimated irradiation time to achieve typical planned doses could become unsuitably long.

XVIII-2.6.3. *Requirement and consideration for evaluation of boron drugs*

In order to characterize the beam of AB-BNCT systems, it is advisable to conduct biological dosimetry after sufficient physical dosimetry evaluations have been performed. The initial investigation of the biological effects of known boron drugs, such as BPA and BSH, in cell-based assays will be useful before testing newly developed boron drugs of interest.

The evaluation of biological effects will be first carried out with cell based tests by comparison with the known boron drugs. Before in vivo animal studies start, the absorption, distribution, metabolism, excretion, and toxicity test data of boron drugs need to be available. In vivo animal model studies need to verify the differences between clinically relevant blood boron levels and boron levels in tumours/normal tissues following administration of the boron drugs. For the estimation of boron concentration in animals during treatment with certain boron drugs, it is necessary to pre-determine the boron concentration changes in blood and target tissues after administration at several doses. The boron concentration will be measured by ICP-AES, ICP-MS, or prompt γ ray analysis. Then these values can be used to estimate boron doses in the target tissues in that animal model when irradiation experiments with the BNCT system are performed.

XVIII-2.6.4. *Endpoint evaluation in comparison with γ or X ray irradiation*

Since BNCT is also one of the modalities of radiotherapy, there is no endpoint that is specific to BNCT, and the same endpoint as for general radiotherapy can be set. It is also possible to compare biological effects with other treatments by using the same endpoint.

As an example, in vitro studies employing BNCT may include the following.

- Intracellular boron accumulation test;
- Evaluation of intracellular boron drug localization;
- Cell susceptibility to the boron drug;
- Cytotoxicity test of the boron drug, such as micronucleus assay (one of the genetic toxicity tests) and other DNA damage markers after neutron beam irradiation;
- Irradiation field dose and leakage dose evaluation test using cells and animals;
- Effects of neutron irradiation on mouse survival and haematological tests for short term and late effects;
- Depth–dose evaluation tests using cells and animals after neutron beam irradiation;
- Effects of neutron irradiation on risk tissues for short term and late effects;
- Verification test of boron drug administration routes in animal models, such as mice;
- Dose-dependent BNCT antitumour effect verification test.

Both cell based and animal experiment facilities will be installed in a controlled area of the facility, and experiments will be conducted to handle radioactive substances including living organisms and measuring instruments according to the regulations in force. In most cases, it is necessary to have prior discussions with the regulatory authorities.

XVIII-3.   EXPERIMENTAL PLANNING

XVIII-3.1. Safety evaluation study

Safety evaluation may be carried out with both cell based and animal based studies. Cell based studies can include cell survival assays and genotoxicity tests such as micronucleus formation assay with and without boron drug treatments and neutron beam irradiation. Early endpoints for animal experiments could be haematological testing, effects on mucosa, skin reactions, damage in intestinal crypts. Late endpoints could include cataract, fibrosis, fertility, and cancer development.

XVIII-3.1.1.   RBE evaluation of the neutron beams of AB-BNCT systems

The determination of RBE of the neutron beam for the individual AB-BNCT systems is necessary for the assessment of biological effects. According to the equation:

$$D \text{ (Gy-Eq)} = D_B \times RBE_B + D_N \times RBE_N + D_H \times RBE_H + D_\gamma \times RBE_\gamma \tag{XVIII-4}$$

When the thermal component in the epithermal neutron beam of the AB-BNCT system is very small, this can be approximated to:

$$D \text{ (Gy-Eq)} = D_H \times RBE_H + D_\gamma \times RBE_\gamma \tag{XVIII-5}$$

Then, by measurement of D_H and D_γ, and by setting $RBE_\gamma = 1.0$, the value of RBE_H can be calculated. The dose rate effect of the epithermal neutron beam from an AB-BNCT system has not yet been sufficiently characterized and this may affect the value of RBE_H.

Usually, the colony formation assay (clonogenic survival assay) is conducted using several kinds of cancer cell lines. For Kyoto University Reactor, the value of RBE_H is reported to be approximately 3.0.

XVIII-3.1.2.   Cytotoxicity evaluation

Cell survival rate will be measured by the colony formation method (comparison with dose that reduces cell survival rate to 10%, D_{10}, and γ ray irradiation, D_0, value) as previously conducted for BNCT systems on reactors. Other cell survival estimation methods have been recently examined and shown to be limitedly applicable for AB-BNCT systems.

XVIII-3.1.3.   Genotoxicity evaluation

The frequency of occurrence of chromosomal abnormalities, such as the frequency of micronuclei formation and frequencies of sister chromatid exchanges can be tested as an initial response.

XVIII-3.1.4.   *Assessment of in vivo effects*

For example, using mice, survival, weight changes, skin lesions, mucosal lesions, haematological status, and intestinal crypt regeneration can be evaluated after whole body or local neutron beam irradiation (with or without administration of boron drugs) in comparison with γ or X ray irradiation. Oral mucosa lesions may be evaluated similarly to the survival of mice after neutron beam irradiation after treatment with boron drugs and mock-treatment [XVIII-12, XVIII-24]. Severe oral mucosa lesions cause feeding difficulties. Haematological status may be monitored by white blood cell numbers, platelet numbers, and red blood cell numbers. If a transient decrease of these numbers is expected, it may be advisable to extend the observation period until complete recovery. Skin lesions may be monitored in parallel by skin reaction scores over time [XVIII-25]. Intestinal crypt regeneration can be used to evaluate an early biological effect at day 3.5–4.0. For intestinal crypt regeneration assessment, the effect of dose ranges of approximately 10–20 Gy-eq can be evaluated, but methods for the measurement of lower ranges of doses has not yet been developed. Administration by bolus injection or infusion of boron drugs can be carried out using intravenous, intraperitoneal, or subcutaneous routes, depending on the drugs [XVIII-25–XVIII-26].

XVIII-3.1.5.   *Assessment of in- and out-of-field and beam depth direction biological effects*

Cell-based evaluation, and subsequent animal model testing may be conducted for the assessment of in- and out-of-field as well as beam depth direction biological effects. When in vivo effects are analyzed, it is advisable to monitor the skin dose as well for each assessment because the skin dose may become the limiting dose if thermal neutrons are a major component of the neutron beam.

XVIII-3.1.6.   *For evaluation of in- and out-of-field effects*

As mentioned above, the assessment can be carried out using cells and animal models, such as mouse, based on physical dose evaluation results. To overcome the difference in the body size between mice and humans and measure biological effects in the total irradiation field, an example of set up models for a vertical beam port is shown in Fig. XVIII-6. For a horizontal beam, this method may be also applicable for biological evaluation of in- and out-of-field positions.

XVIII-3.1.7.   *For evaluation of beam-depth direction effect*

The set-up to overcome the differences in the body thickness of mice and humans and measure biological effects of beam-depth direction with a vertical beam port is shown in Fig. XVIII-7. The number of mice in an array may be increased or decreased depending on the beam conditions. For a horizontal beam, this method may be also applicable for biological evaluation of beam-depth direction.

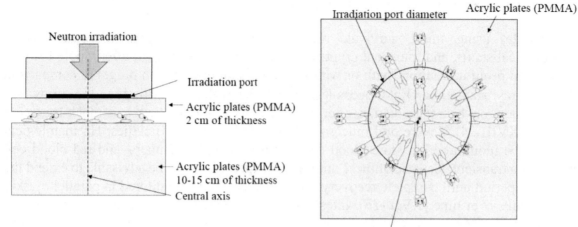

FIG. XVIII-6. The set up for evaluation of in- and out-of-field irradiation.

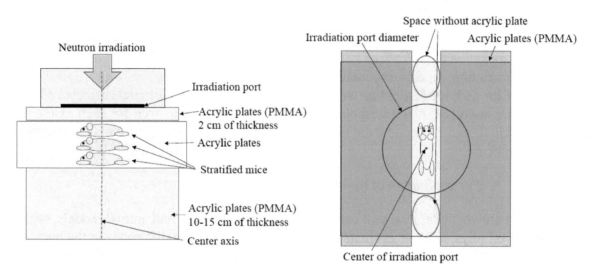

FIG. XVIII-7. The set up for evaluation of beam-depth direction.

XVIII-3.2. Efficacy evaluation study

XVIII-3.2.1. Cancer cell killing in the absence and presence of boron drugs

If planned preclinical testing is intended to support performing clinical trials, cells of the type of the target cancer and normal cells of the corresponding risk organs need to be included in the assessment.

XVIII-3.2.2. Effects on tumour growth

For the assessment of the anti-tumour effect, the tumour size is measured over time in a subcutaneous tumour graft model. If the tumour diameter can be significantly suppressed by observation at 2–4 weeks after irradiation, for example, it is judged that there is an antitumour effect. In both in vitro and in vivo studies, the response in the acute phase is examined, and the final tumour control effect may not be confirmed. The TCD_{50} (tumour control dose 50) assay may be also used to evaluate the tumour control effect.

XVIII-4.   OTHER CONSIDERATIONS

It is desirable to conduct tests to assess normal tissue responses in medium-sized animals whose metabolism, anatomy, and pharmacokinetics are closer to humans. However, issues of clinical facility constraints, animal rearing within controlled areas, and difficulties in maintaining positions during treatment, etc. are usually present. Due to these problems, careful preparation is required for implementation of medium-sized animal models. Instead, the efficacy and safety evaluation for humans have been estimated from the test results of rodents, such as mice, as far as possible, taking into consideration the clinical results of reactor based BNCT. Interpretation of antitumour effects also requires analysis that takes into account the heterogeneity of boron drug distribution within the tumours, their metabolism in the bodies and the characteristics of tumours and normal tissues.

Due to the limited number of AB-BNCT facilities that can be used for basic investigations, few academic papers have been published. On the other hand, many papers on animal experiments using reactor based BNCT systems have been reported, which can be referred to when the difference in beam characteristics is examined.

To minimize animal experiments and optimize evaluation conditions, three-dimensional culture systems, ex vivo assessment strategies, or human iPS cell derived assessment systems are being developed and expected to become useful in the near future.

The progress of development of AB-BNCT systems is rapid. Therefore, the practical guidelines for preclinical assessment are expected to be improved accordingly.

ACKNOWLEDGEMENTS

We deeply thank J. Itami, H. Igaki, S. Nakamura, Y. Imahori, M. Nakamura, M. Suzuki, A. Matsumura, A. Schwint and members of the JSNCT and the ISNCT for kind advice for summarizing this article.

REFERENCES TO ANNEX XVIII

[XVIII-1]   HIROSE, K., et al., Boron neutron capture therapy using cyclotron-based epithermal neutron source and borofalan (^{10}B) for recurrent or locally advanced head and neck cancer (JHN002): An open-label phase II trial, Radiother. Oncol. **155** (2021) 182–187.

[XVIII-2]   NAKAMURA, S., et al., Neutron flux evaluation model provided in the accelerator-based boron neutron capture therapy system employing a solid-state lithium target, Sci. Rep. **11** 1 (2021) 8090.

[XVIII-3]   KUMADA, H., et al., Project for the development of the linac based NCT facility in University of Tsukuba, Appl. Radiat. Isot. **88** (2014) 211–215.

[XVIII-4]   KUMADA, H., et al., Evaluation of the characteristics of the neutron beam of a linac-based neutron source for boron neutron capture therapy, Appl. Radiat. Isot. **165** (2020) 109246.

[XVIII-5]   SUZUKI, M., Boron neutron capture therapy (BNCT): a unique role in radiotherapy with a view to entering the accelerator-based BNCT era, Int. J. Clin. Oncol. **25** 1 (2020) 43–50.

[XVIII-6] ONO, K., Prospects for the new era of boron neutron capture therapy and subjects for the future, Ther. Radiol. Oncol. **2** (2018) 40–46.

[XVIII-7] PEDROSA-RIVERA, M., et al., Thermal neutron relative biological effectiveness factors for boron neutron capture therapy from in vitro irradiations. Cells **9** 10 (2020) 2144.

[XVIII-8] INTERNATIONAL ATOMIC ENERGY AGENCY, "Radiobiology: dose components in BNCT", Current Status of Neutron Capture Therapy, TECDOC 1223, IAEA, Vienna (2001) Ch 6.2.

[XVIII-9] BORTOLUSSI, S., et al., Understanding the potentiality of accelerator based-boron neutron capture therapy for osteosarcoma: dosimetry assessment based on the reported clinical experience, Radiat. Oncol. **12** 1 (2017) 130.

[XVIII-10] SATO, E., et al., Radiobiological response of U251MG, CHO-K1 and V79 cell lines to accelerator-based boron neutron capture therapy, J. Radiat. Res. **59** 2 (2018) 101–107.

[XVIII-11] SATO, T., et al., Microdosimetric modeling of biological effectiveness for boron neutron capture therapy considering intra- and intercellular heterogeneity in ^{10}B-distribution, Sci. Rep. **8** 1 (2018) 988.

[XVIII-12] MORRIS, G.M., et al., Boron microlocalization in oral mucosal tissue: implications for boron neutron capture therapy, Br. J. Cancer **82** 11 (2000) 1764–1771.

[XVIII-13] CODERRE, J.A., et al., The effects of boron neutron capture irradiation on oral mucosa: evaluation using a rat tongue model, Radiat. Res. **152** 2 (1999) 113–118.

[XVIII-14] KIGER, J.L., et al., Functional and histological changes in rat lung after boron neutron capture therapy, Radiat. Res. **170** 1 (2008) 60–69.

[XVIII-15] SUZUKI, M., et al., The effects of boron neutron capture therapy on liver tumours and normal hepatocytes in mice, Jpn. J. Cancer Res. **91** 10 (2000) 1058–1064.

[XVIII-16] MATSUYA, Y., et al., A model for estimating dose-rate effects on cell-killing of human melanoma after boron neutron capture therapy, Cells **9** 5 (2020) 1117.

[XVIII-17] INTERNATIONAL ATOMIC ENERGY AGENCY, Radiation Protection and Safety of Radiation Sources: International Basic Safety Standards, General Safety Requirements Part 3, No. GSR Part 3, IAEA, Vienna (2014).

[XVIII-18] PROTTI, N., et al., Gamma residual radioactivity measurements on rats and mice irradiated in the thermal column of a TRIGA Mark II reactor for BNCT, Health Phys. **107** 6 (2014) 534–541.

[XVIII-19] NAKAMURA, S., et al., Evaluation of radioactivity in the bodies of mice induced by neutron exposure from an epi-thermal neutron source of an accelerator-based boron neutron capture therapy system, Proc. Jpn. Acad. Ser. B Phys. Biol. Sci. **93** 10 (2017) 821–831.

[XVIII-20] TORU, K.K., Development of LiF tile neutron shield and measurement of tritium release from it, KURRI Tech. Rep. KURRI TR-198, Kyoto (1980).

[XVIII-21] SUZUKI, M., et al., A preliminary experimental study of boron neutron capture therapy for malignant tumours spreading in thoracic cavity, Jpn. J. Clin. Oncol. **3** 4 (2007) 245–249.

[XVIII-22] KOGA, J., et al., Measurement of γ rays from ^{6}LiF tile as an inner wall of a neutron-decay detector, J. Instrum. **16** 2 (2021) P02001.

[XVIII-23] IMAMICHI, S., et al., Evaluation of the biological effectiveness in cells and mice for BNCT system in National Cancer Center Hospital, Proc. Jpn. Acad. Ser. B **93** 10 821–831.

[XVIII-24] KREIMANN, E.L., et al., The hamster cheek pouch as a model of oral cancer for boron neutron capture therapy studies: selective delivery of boron by boronophenylalanine, Cancer Res. **61** 24 (2001) 8775–8781.

[XVIII-25] FUKUDA, H., et al., Boron neutron capture therapy of malignant melanoma using ^{10}B-paraboronophenylalanine with special reference to evaluation of radiation dose and damage to the normal skin, Radiat. Res. **138** 3 (1994) 435–442.

[XVIII-26] MASUNAGA, S., et al., The dependency of compound biological effectiveness factors on the type and the concentration of administered neutron capture agents in boron neutron capture therapy, Springerplus **3** (2014) 128.

THE USE OF TARGET AND TREATMENT VOLUMES IN BNCT

YUAN-HAO LIU[1,2,3,4], YI-CHIAO TENG[1,5], YI-WEI CHEN[6], LING-WEI WANG[6], YAN-HUNG WU[6]

[1]Neuboron Therapy System Ltd., Xiamen; [2]Xiamen Humanity Hospital, Xiamen; [3]Neuboron Medtech Ltd., Nanjing; [4]Nanjing University of Aeronautics and Astronautics, Nanjing; [5]National Tsing Hua University, Hsinchu; [6]Taipei Veterans General Hospital, Taipei

Abstract

This describes the use of gross tumour volume (GTV), clinical target volume (CTV) planning target volume (PTV), and Planning organ at Risk Volume (PRV) in boron neutron capture therapy (BNCT), and how to apply them appropriately according to the fundamentals of BNCT. Rather than using the ordinary GTV, we propose the use of GTV_{B10} for a better dose estimation and insight into the clinical outcome and its prediction. For BNCT, CTV is not mandatory since its margin does not contain a significant number of ^{10}B atoms; if this term is not used appropriately, the dose calculation may be incorrect due to the neutron suppression. A PTV is applied to the collimator opening setting to ensure that GTV_{B10} is in the irradiation field, also to make a compensation to the instrumental and positioning uncertainty. PRV provides a conservative dose evaluation for Organ at Risk (OAR); if the PRV has a higher boron uptake, the neutron suppression has to be carefully evaluated and handled. In this annex, we also present a dose calculation frame for CTV, PTV, and PRV, if they are used in the BNCT treatment plan.

XIX-1. INTRODUCTION

From our experiences in clinical practice, the conventional definition of gross tumour volume (GTV), clinical target volume (CTV) and planning target volume (PTV) in conventional radiotherapy as indicated in ICRU-50 [XIX-1] is not completely complied with in BNCT: some modification might be needed to follow the spirit of these terms. In addition, the use of planning organ at risk volume (PRV) [XIX-2] needs additional care. These result from the biological targeting characteristic of BNCT attributed to the use of targeting boron carriers (drugs). The presence of ^{10}B atoms within the target region is essential for the creation of therapeutic boron dose. Therefore, such a property has to be carefully taken into account when defining and delineating the GTV, CTV, PTV, and PRV in BNCT. The following introduces the concept of boron dose, and then discuss and suggest a way of using GTV, CTV, PTV, and PRV in BNCT.

XIX-2. THE THERAPEUTIC DOSE – BORON DOSE

The therapeutic dose of BNCT is the boron dose, D_{boron}[1]. It is delivered to the 'target region' as an integration result of thermal neutrons and ^{10}B atoms distributed in the target volume, V_{ROI}, over an irradiation time, t, which can be defined as follows:

$$D_{boron,physical}(Gy) = \frac{Q \int_{V_{ROI}} \int_t \int_E N_{B10}(x,y,z,t)\phi_n(x,y,z,E,t)\sigma_{B10}(E)dEdtdV}{V_{ROI} \times \rho_{ROI}} \quad (XIX-1)$$

---

[1] The tumour dose or the normal tissue dose is actually the sum of different dose components, including boron dose, neutron dose, nitrogen dose, hydrogen dose, photon dose, etc. However, the main therapeutic dose is the boron dose.

where x, y, and z stand for the position in the space; E is the neutron energy (MeV), and t is the irradiation time (s); N_{B10} is the number of ^{10}B atoms; ϕ_n stands for neutron flux (cm$^{-2}\cdot$s^{-1}), and σ_{B10} is the ^{10}B capture cross section (barn); Q is the reaction energy release per boron neutron capture reaction; ρ_{ROI} is the physical density of the ROI (region of interest) material.

Note that we use absorbed dose instead of the biological weighted dose for convenience. From Eq. (XIX-1), a volume without ^{10}B atoms will not receive the therapeutic boron dose. From this background knowledge, we can discuss the use of GTV, CTV and PTV in BNCT accordingly.

XIX-3. GROSS TUMOUR VOLUME

Generally, the patient geometry and anatomy information are acquired using CT or MRI scans, which do not contain any information regarding ^{10}B distribution. Hence, the GTV defined in conventional radiotherapy using anatomical images cannot correctly reflect the boron dose information. Using multimodal images in this situation is strongly suggested. The definition of GTV is hereby suggested as follows for BNCT:

GTV$_{anatomy}$ (or GTV$_{CT,MRI}$)	Gross tumour volume defined according to anatomy medical images (sometimes with the aid of functional images such as PET or SPECT);
GTV$_{B10}$ (or GTV$_{PET}$)	Gross tumour volume defined according to functional images that reflects the ^{10}B uptake within the tumour (e.g., ^{18}F-FBPA [XIX-3] PET scan). This is similar to the concept of biological target volume (BTV) [XIX-4].

In BNCT practice [XIX-5–XIX-9] in Taiwan, GTV is the only applied term in BNCT clinical treatment; most of the time, GTV$_{MRI}$ was used, and in some cases GTV$_{CT}$ was used. The term of GTV$_{B10}$ is a more advanced term considering the volume which really receives the boron dose. Theoretically, GTV$_{anatomy}$ is larger than or equal to GTV$_{B10}$, because some parts of the tumour may not absorb boron carriers, which explains recurrence in some BNCT treated tumours. In the following, we present three real cases using ^{18}F-FBPA PET scans.

In therapeutic evaluation, the volumetric dose information, which strongly relates to the therapeutic outcome, is GTV$_{B10}$. Figure XIX-1 shows an ideal uptake of boron carriers within the tumour, therefore the GTV$_{B10}$ is equal to the GTV$_{anatomy}$. However, in Figures XIX-2 and XIX-3, the GTV$_{B10}$ is clearly smaller than the GTV$_{anatomy}$; consequently, additional attention is required while making the treatment plan. Readers may naturally raise a question when using GTV$_{B10}$ – how to deal with the rest of the tumour volume where there is no or less ^{10}B uptake? The solution for these cases may be a combination of conventional radiotherapy or charged particle beam to perform dose painting, i.e., to give a boost shot to the cold spots/volume[2] [XIX-10]. Thus, the use of GTV$_{anatomy}$ and GTV$_{B10}$ are both needed, and the volumetric dose information in both defined GTVs has to be calculated and reported separately (i.e., doses of GTV$_{B10}$, GTV$_{anatomy}$, and GTV$_{anatomy} \not\subseteq$ GTV$_{B10}$ have to be reported).

---

[2] The cold spots of GTV$_{anatomy}$ have to be further confirmed by biopsy or other medical examination modalities, for the sake of understanding whether there are tumour cells or not. If there are, or are suspected to be, tumour cells, additional radiation dose (or CXT) can be applied to cover the cold spots. On the other contrary, if there are no observable tumour cells, additional dose boost may not be applied.

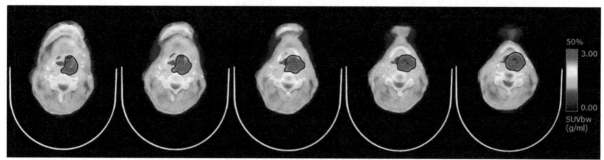

FIG. XIX-1. A uniform uptake of boron carriers in a squamous cell carcinoma, which was scanned using ^{18}F-FBPA PET. The $GTV_{anatomy}$ is equal to the GTV_{B10}.

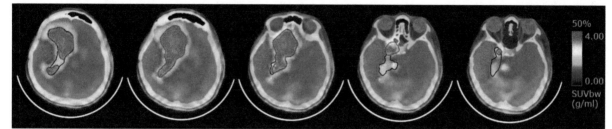

FIG. XIX-2. ^{18}F-FBPA PET scan of a large GBM shows a non-uniform uptake in the tumour lesion. The region delineated by the red line is GTV_{B10} and the region delineated using the dark blue line is $GTV_{anatomy}$ defined by MRI.

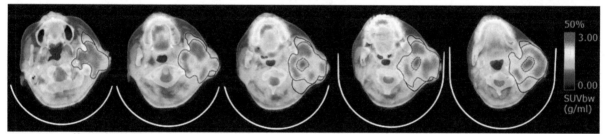

FIG. XIX-3. ^{18}F-FBPA PET scan of a large head and neck tumour where the boron uptake uniformity is poor. The region delineated using the dashed red line is a necrosis volume or a hypoxia region, where there is no boron uptake. The region delineated using the dark blue line is the typical $GTV_{anatomy}$ defined using MRI. Apparently, the $GTV_{anatomy}$ is much bigger than the GTV_{B10}.

Nonetheless, some facilities may not have access to ^{18}F-FBPA PET scanning and cannot obtain the ^{10}B distribution information. In that case, a fixed tumour-to-normal-tissue ratio (TNR) is used, and only the $GTV_{anatomy}$ is defined. For the sake of good tumour control, the strategy of maximizing tumour dose has to be conducted; that is to say, using the normal tissue tolerance dose as the dose constraint rather than tumour prescribed dose. In addition, when a second irradiation is foreseen, the TNR obtained in the patient is normally lower, and the dose performance of the second irradiation may be poorer than the first irradiation. Thus, maximizing the tumour dose under the tolerance limit of normal tissue is worth a consideration.

XIX-4. CLINICAL TARGET VOLUME

One uses CTV to deal with the potential infiltration of tumour cells at the boundary which cannot be clearly identified by medical images. In conventional radiotherapy, one can deliver additional dose from the radiation beam to cover the margin, i.e., CTV; however, in BNCT, this is a more complicated issue.

Two fundamental questions of using CTV in BNCT arise as follows:
(a) Is CTV mandatory in BNCT?
(b) If it is used, how does one calculate its boron dose?

For the first question, our answer is no; it is not mandatory nor needed for BNCT if you have sufficient information regarding the in vivo boron spatial distribution. Conventionally, the CTV will be prescribed a certain amount of dose in order to eliminate the infiltrating tumour cells and obtain better tumour control. However, this is not straightforward in BNCT. The boron uptake within the margin of the CTV is very poor or none, and therefore the boron dose rate is very low; accordingly, Eq. (XIX-1) shows that when there are no ^{10}B atoms, there is no therapeutic dose. Therefore, the use of CTV in BNCT may not benefit patient, but cause confusion in dose calculation, dose contouring, and dose delivery. Next, we will discuss the two questions thoroughly.

A proposed method to deal with the lack of ^{10}B atoms in the CTV is artificially adding ^{10}B atoms to the margin during neutron transport calculation, which is certainly not correct. Furthermore, such a fake uptake will lead to a disturbance (i.e., neutron flux suppression, or the so-called self-shielding/self-absorption effect) of the neutron flux within the GTV$_{B10}$, and the calculated dose in the tumour could be wrong. The correct way of calculating the dose in the margin is using the exact boron uptake in that place where the boron dose rate will be low. The whole situation will lead to another important issue – dose contouring and dose delivery.

One has to pay attention to the 'dose contouring and delivery' of the margin of the CTV. Unlike the conventional way, it may be dangerous to intentionally achieve a certain dose prescription to the CTV; the additional dose to the target region could only be achieved by the means of beam dose (i.e., doses of neutron and γ rays), which is harmful to normal tissues. Delivering additional dose to the margin using the neutron beam will increase undesired dose to normal tissues because the beam does not differentiate normal tissues and tumour tissues; the differentiation is done by the targeting drug, i.e., ^{10}B-carrier. For example, Figure XIX-4 shows the CTV (orange line) defined by using MRI images, and the GTV$_{B10}$ (red line) defined by an ^{18}F-FBPA PET scan, of a brain stem tumour patient. If one forces the delivery of additional dose to the CTV around the GTV$_{B10}$, the normal brain may be damaged.

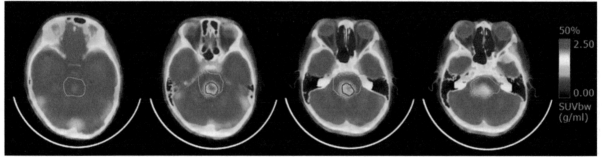

FIG. XIX-4. The CTV defined by MRI images (orange line), and GTV$_{B10}$ (red line) of a brain stem tumour.

It is important to stress that if active tumour cells exist in the boundary region, the boron carrier will do the job of targeting and these cells will be comprised in the GTV$_{B10}$; if there is no ^{10}B, there is no possibility of delivering a selective dose by BNCT. The use of CTV in BNCT is therefore not necessary, but BNCT may still use CTV for the sake of dose summation in combining with other radiation therapy. Keep in mind, if CTV is used in BNCT, it is not advised to use its dose as a dose constraint. The difference in CTV usage between conventional radiotherapy and BNCT needs to be clearly addressed.

Nevertheless, if a functional image is not available (e.g., ^{18}F-FBPA PET scan) and GTV$_{B10}$ is not applicable, one has to follow the conventional way to define CTV and calculate the dose accordingly using a pseudo ^{10}B concentration[3, 4] (do not add additional ^{10}B atoms into CTV margin during the neutron transport calculation but do the math afterward). Figure XIX-5 gives an illustration of this idea.

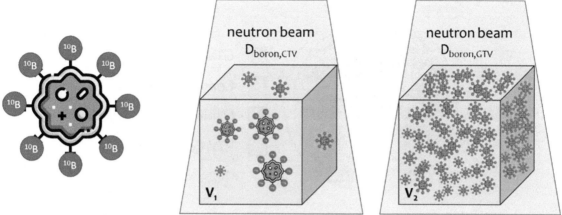

(a) tumor cell targeted by ^{10}B carriers   (b) infiltrated region with few tumor cells   (c) Region with abundant tumor cells

FIG. XIX-5. Illustration and explanation of the concept of using a pseudo boron concentration in estimating the CTV boron dose, assuming a targeting boron carrier (drug), which has a uniformly targeting performance to every tumour cell in the lesion as well as in the infiltrated region, is used on a patient. The tumour lesion is irradiated by a uniform neutron beam, and V_1 has the same volume and mass as V_2. V_1 is located at the infiltrated area between the GTV and normal tissue, i.e., CTV margin; V_2 is located at the boundary between the GTV and the CTV, but still within the GTV volume. It is known that the absorbed dose of the CTV, $D_{boron,CTV}$ is much smaller than that of GTV, $D_{boron,GTV}$. However, the probability of destroying the tumour cells is similar, or equal, in V_1 and V_2. The illustration presents the inapplicability of the macroscopic dose concept in the CTV margin of BNCT. However, the dose response obtained from the nearby GTV could be a good reference, if the targeting/binding rates of the boron carriers to the tumour cells are the same in V_1 and V_2. The TCP (the tumour control probability, which is proportional to the probability of destroying tumour cells) is generally established on the macroscopic dose, therefore one could use a pseudo boron concentration (same as V_2) to mimic the dose response in the CTV (i.e., V_1).

XIX-5. PLANNING TARGET VOLUME

PTV is applied to compensate the uncertainties arising from patient positioning, beam delivery setup, organ movement, and so on. However, as explained in previous sections, there are no or fewer ^{10}B atoms in the margin added outside the GTV$_{B10}$. Therefore, delivering additional dose to an additional margin of CTV will not contribute to the therapeutic benefits, but instead it would increase the risk to OARs. Thus, if one intends to use PTV in BNCT, it ought to be defined according to GTV$_{B10}$ or GTV$_{anatomy}$ rather than CTV. However, as stated in Section XIX-4, the dose of PTV is not suitable for use as a dose constraint either. Thus, we propose to apply the concept of PTV to the collimator opening size; that is to say, add an additional margin to the collimator opening in order to compensate the abovementioned uncertainties. Figure XIX-6 provides a simple illustration. It has to be borne in mind that the

---

[3] The use of CBE for GTV$_{anatomy \notin B10}$ and the margin of CTV has to be further discussed and studied. For the cases in which GTV$_{B10}$ is available, we suggest use of normal tissue CBE for GTV$_{anatomy \notin B10}$ and the margin of CTV for a conservative purpose. For the cases without GTV$_{B10}$, we suggest use of the same CBE value as the GTV for CTV, as well as PTV (will be discussed in the subsequent section).

[4] Due to the cellular targeting property of boron carriers, the concept of 'volumetric dose' in CTV margin may have to be modified; in other words, the TCP (tumour control probability) of CTV margin may be quite different from the GTV because the dose to the infiltrated tumour cells may have different radiobiological response. The dose needed to kill a tumour cell is small, but the key point is whether or not there is a sufficient number of ^{10}B atoms inside the tumour cell. Therefore, use of a pseudo boron concentration in the boron dose calculation for the margin of CTV could be useful in assessing the TCP of the target region. What happens there is a microscale event, but one may use the macroscale dose obtained from neighbourhood tissue (i.e., GTV) to mimic the response.

tumour is targeted by boron carriers (drugs), and the selectivity of dose delivery to tumour is done by the boron carriers. If the tumour movement is within a uniform neutron field, the dose to tumour will not change during motion, i.e., the integration of Eq. (XIX-1) is the same everywhere within the neutron field. Hence, it is important to keep the tumour motion within the beam field; that is to say, the beam field needs to be slightly larger than the tumour.

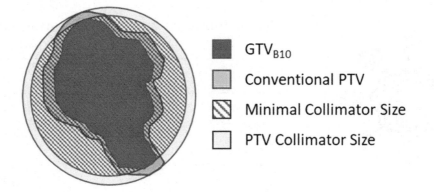

FIG. XIX-6. Illustration of the delineation of PTV and the corresponding collimator size.

Therefore, we do not suggest the use of PTV in BNCT as in other radiotherapies; instead, we suggest adding an additional margin to the collimator size to ensure the coverage of GTV_{B10}. In fact, the movement (≤ 10 mm) of GTV_{B10} in the perpendicular plane of the incident neutron beam direction will not result in significant dose difference (~1%) [XIX-11] when we use a beam opening size equal to the largest diameter of GTV_{B10}. This is because of neutron scattering in the body's tissues. As a result, a slightly larger margin (5–10 mm, depending on the uniformity of the beam) for the collimator size is enough to compensate the uncertainties arising from the definition of PTV.

Nonetheless, if a functional image is not available (e.g., a PET scan) and GTV_{B10} is not applicable, one has to follow the conventional way to define a PTV and calculate the dose in the same way as described in Section XIX-4.

XIX-6. PLANNING ORGAN AT RISK VOLUME

A PRV is proposed to reduce the interobserver errors for the contouring and provides a conservative dose evaluation for OAR; in addition, a larger volume will improve the statistics of the dose simulated by neutron transport due to more events recorded in the ROI. However, one point has to be brought to the reader's attention: if the PRV with a higher boron uptake (e.g., mucosa) is very close to the tumour, or is located on the path between the neutron beam and tumour, the neutron suppression due to the additional ^{10}B atoms has to be carefully evaluated. One may use two-step calculations to avoid the disturbance of introducing PRV into the calculation; the first-step calculation uses PRV to obtain the OAR dose, and tumour dose is obtained from the second-step calculation which does not use PRV. Another method to correct such an inference is using an additional margin to the designated OAR and assign a virtual boron concentration to its boron dose tally (as described in the last paragraph of Section XIX-4).

XIX-7. SUMMARY

The use of GTV, CTV, PTV, and PRV in BNCT needs to be carefully considered. Two important concepts need to be kept in mind:

(a) Where there is no boron atom, there is no therapeutic boron dose; thus, we do not suggest prescribing the dose to CTV and PTV, but only to calculate their doses for the consideration of a combined therapy with another radiation therapy modality. The dose coverage of tumour parts that do not absorb boron can be obtained by dose painting by a successive boost of conventional radiotherapy or particle therapy;

(b) If a dose prescription is required for CTV or PTV, the dose can be calculated by using the method suggested in this article; one ought never to artificially change the boron concentration in the CTV and PTV margins, which will lead to a distortion of dose calculation during neutron transport; the boron dose could be calculated by using a pseudo boron concentration later.

We strongly suggest using functional imaging to determine GTV_{B10}. The synthesis of ^{18}F-FBPA by nucleophilic reaction [XIX-12] has been a recent breakthrough and will be available for other facilities in a few years. Irradiating the patient without the spatial distribution information of ^{10}B atoms could be far less effective, and the clinical outcome is not predictable.

Table XIX-1 shows the use of GTV, CTV, and PTV in two different scenarios; in the first one, the functional image is available, i.e., GTV_{B10} is defined, and in the second a fixed TNR for the tumour volume is used. It remains a debate among practitioners about the necessity of CTV and PTV, as well as how to delineate them. Nonetheless, Table XIX-1 provides the basic concept of how it can be processed.

ACKNOWLEDGEMENT

We would like to thank Dr. Silva Bortolussi of University of Pavia, for her valuable comments and kind discussion in preparing this article. This work was supported by the Science and Technology Major Project of Xiamen Municipal Bureau of Science and Technology, Grant No. 3502720201031.

TABLE XIX-1. THE DELINEATION OF GTV, CTV, PTV, AND THEIR DOSE CALCULATIONS

Term		Scenarios	
		Boron-10 spatial information is available	Boron-10 spatial information is not available
TNR		Use the functional image to obtain the ^{10}B spatial distribution.	Use a fixed empirical TNR.
GTV	Delineation	$GTV_{anatomy}$ and GTV_{B10} has to be defined.	Define $GTV_{anatomy}$
	Dose Calculation	Use TNR(V) to calculate the dose, and use the dose of GTV_{B10} as prescribed dose.	Use the fixed TNR to calculate the dose. It is suggested to use the OAR dose as dose constraints, rather than the GTV prescribed dose.
CTV	Delineation	Define CTV according to the $GTV_{anatomy}$[a]	Define CTV according to the $GTV_{anatomy}$
	Dose Calculation	Use the normal tissue CBE for the margin and use the normal tissue boron concentration. DO NOT use the same dose description condition of GTV_{B10} for CTV.	Use the same CBE as defined in GTV but use the normal tissue boron concentration; designate a pseudo boron concentration to the tally in the neutron transport calculation, and obtain the boron dose for the margin of CTV.
PTV	Delineation	PTV is defined only for the determination of collimator size. It is suggested to define PTV according to the $GTV_{anatomy}$[b], because the neutron scattering will compensate the movement in the perpendicular plane of the neutron beam incident direction.	Define PTV according to the CTV
	Dose Calculation	Not calculated.	Same as CTV

Note:

[a] There is not yet a consensus on the baseline for delineating of CTV. Some suggest the use of $GTV_{anatomy}$, and others suggest the use of GTV_{B10}.

[b] There is not yet a consensus on the baseline for delineating PTV either. Some prefer to use GTV_{B10}, and others the use of $GTV_{anatomy}$, while there are some debates about using CTV. The debate is about the necessity of exposing more normal tissue to the neutron beam (i.e., protecting normal tissue) versus better tumour control.

REFERENCES TO ANNEX XIX

[XIX-1] INTERNATIONAL COMMISSION ON RADIATION UNITS AND MEASUREMENTS, Prescribing, Recording and Reporting Photon Beam Therapy, International Commission on Radiation Units and Measurements, ICRU Report 50, Washington, D.C., USA (1993).

[XIX-2] INTERNATIONAL COMMISSION ON RADIATION UNITS AND MEASUREMENTS, Prescribing, Recording and Reporting Photon Beam Therapy (Supplement to ICRU Report 50), ICRU Report 62, Washington, D.C., USA (1999).

[XIX-3] CHEN, Y.-W., et al., Using precise boron neutron capture therapy as a salvage treatment for pediatric patients with recurrent brain tumors, Ther. Radiol. Oncol. **4** (2020) 109105.

[XIX-4] CHEN, Y.-W., et al., Compassionate treatment of brainstem tumors with boron neutron capture therapy: A case series, Life **12** 4 (2022) 566.

[XIX-5] CHEN, Y.-W., et al., Salvage boron neutron capture therapy for malignant brain tumor patients in compliance with emergency and compassionate use: Evaluation of 34 cases in Taiwan, Biology (Basel) **10** 4 (2021) 3334.

[XIX-6] HE, J., et al., Nucleophilic radiosynthesis of boron neutron capture therapy-oriented PET probe [^{18}F]FBPA using aryldiboron precursors, Chem. Commun. **57** 71 (2021) 8953–8956.

[XIX-7] LEE, J.-C., et al., Preliminary dosimetric study on feasibility of multi-beam boron neutron capture therapy in patients with diffuse intrinsic pontine glioma without craniotomy, PLOS ONE **12** 6 (2017) e0180461.

[XIX-8] LEE, J.C., et al., The dosimetric impact of shifts in patient positioning during boron neutron capture therapy for brain tumors, Biomed. Res. Int. **2018** (2018) 5826174.

[XIX-9] LING, C.C., et al., Towards multidimensional radiotherapy (MD-CRT): biological imaging and biological conformality, Int. J. Radiat. Oncol. Biol. Phys. **47** 3 (2000) 551–560.

[XIX-10] WANG, L.W., et al., Fractionated BNCT for locally recurrent head and neck cancer: experience from a phase I/II clinical trial at Tsing Hua Open-Pool Reactor, Appl. Radiat. Isot. **88** (2014) 23–27.

[XIX-11] WANG, L.W., et al., Clinical trials for treating recurrent head and neck cancer with boron neutron capture therapy using the Tsing-Hua Open Pool Reactor, Cancer Commun. (Lond.) **38** 1 (2018) 37.

[XIX-12] WATANABE, T., et al., Comparison of the pharmacokinetics between L-BPA and L-FBPA using the same administration dose and protocol: a validation study for the theranostic approach using [^{18}F]-L-FBPA positron emission tomography in boron neutron capture therapy, BMC Cancer **16** 1 (2016) 859.

ACCELERATOR BNCT FOR PATIENTS WITH RECURRENT HEAD AND NECK CANCER: CASE REPORT

TERUHITO AIHARA[1,2], MASAAKI HIGASHINO[2], NAONORI HU[1], RYO KAWATA[2], KEIJI NIHEI[1], KOJI ONO[1]

[1]Kansai BNCT Medical Center
[2]Department of Otolaryngology Head and Neck Surgery, Osaka Medical College, Takatsuki, Japan

Abstract

This article describes the first case of recurrent mesopharyngeal carcinoma treated with accelerator based boron neutron capture therapy at our institution.

XX-1. INTRODUCTION

Boron neutron capture therapy (BNCT) is a form of particle beam radiation therapy, which functions by using an α particle and a ^{7}Li nucleus that are generated by a neutron capture reaction on a ^{10}B nucleus present in the boron carrier compound that has been selectively taken up into the tumour tissue. This treatment is relatively safe because it selectively destroys boron-incorporated tumour cells through the high linear energy transfer particles. Advanced/recurrent head and neck cancers are frequently radio-/chemo-resistant, grow extensively, and require a wide resection including surrounding normal tissues. To avoid severe impairment of head and neck structures, it is necessary to explore a new treatment modality. Mishima first proposed employing BNCT for malignant melanomas utilizing the specific melanin synthesis activity of melanoma cells [XX-1]. Kato et al. [XX-2] began the first BNCT therapy for recurrent parotid gland carcinoma and reported excellent preliminary results. From these encouraging results, many years' experience of BNCT for head and neck cancer, and the trend toward emphasizing the quality of life after treatment, we also began treating patients with BNCT [XX-3–XX-5]. BNCT for head and neck cancer was officially approved by the Japanese government in March 2020 and covered by national health insurance in June 2020. In this article, we present a case study of our first patient with recurrent head and neck cancer who was treated by accelerator BNCT.

XX-2. PATIENT AND METHODS

The patient was a 73-year-old man diagnosed with recurrent oropharyngeal cancer (rT2N0M0, squamous cell carcinoma, P16 negative). Two years earlier, this patient underwent radiation therapy alone (70 Gy / 35 Fr) for oropharyngeal cancer (cT4N1M0).

He underwent a CT scan for treatment planning that was performed with the 'NeuCure dose engine' software program. Intravenous administration of Borofalan (^{10}B) (500 mg/kg bw) was performed for 3 hours using the 2–2–1 method [XX-6] and the irradiation time was determined from the ^{10}B-concentration measured from the blood sample taken just before irradiation. The patient was irradiated with epithermal neutrons produce by the BNCT treatment system NeuCure at a beam current of 1 mA for 41 minutes. The calculated mucosa tissue dose and tumour dose ranged from 2.2 to 11.3 Gy-eq and 24.2 to 40.3 (mean dose of 31.3) Gy-eq, respectively (Fig. XX-1).

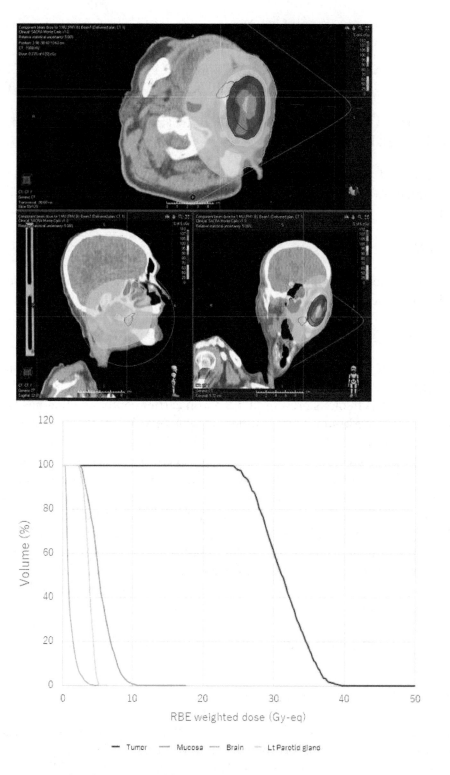

(a)

(b)

FIG. XX-1. Neutron distribution (a) and dose volume histogram (b) for this patient.

XX-3. RESULTS

Almost complete regression was obtained approximately two months after BNCT treatment, without further treatment. The complete regression in the tumour has continued for 22 months up to the time of writing without chronic complications in the form of skin ulcer or xerostomia (Fig. XX-2). The stress from the treatment did not cause any aggravation to the patient's dementia.

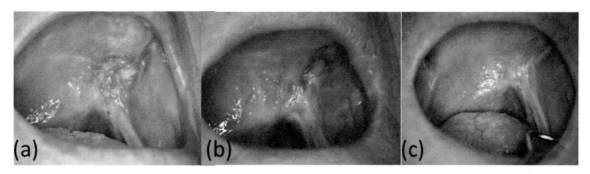

FIG. XX-2. Local findings of this patient before (a), 5 months after (b), and 12 months after (c) BNCT. The tumour disappeared 22 months after BNCT.

XX-4. DISCUSSION

The patient was offered rescue surgery by his doctor, but his family refused due to his dementia. The BNCT treatment was quick and no further aggravation to his illness was observed.

In this way, BNCT can be safe and expected to have a strong therapeutic effect. However, it is not effective for all head and neck cancer cases. No therapeutic effect can be expected in cases where boron compounds do not accumulate in the tumour or in cases where the tumour is deep seated and sufficient neutron flux cannot be delivered. It is necessary for a BNCT specialist to clarify what kind of cases are suitable for BNCT and establishing the efficacy and safety of BNCT for head and neck cancer is an important mission for us.

XX-5. CONCLUSION

We introduced the first case treated with the accelerator BNCT treatment system NeuCure. The therapeutic effect of this case is good, indicating that BNCT is a potential treatment method for patients with head and neck cancer. In the future, in addition to the clinical experience, we believe that a detailed investigation of radiobiology and physics is necessary to further improve the clinical outcome.

REFERENCES TO ANNEX XX

[XX-1]   MISHIMA, Y., Neutron capture treatment of malignant melanoma using ^{10}B-chlorpromazine, Pigment Cell Res. **1** (1973) 215–221.

[XX-2]   KATO, I., et al., Effectiveness of BNCT for recurrent head and neck malignancies, Appl. Radiat. Isot. **61** (2004) 1069–1073.

[XX-3]   AIHARA, T., et al., First clinical case of boron neutron capture therapy for head and neck malignancies utilizing ^{18}F-BPA PET, Head Neck **28** (2006) 850–855.

[XX-4]   AIHARA, T., et al., BNCT for advanced or recurrent head and neck cancer, Appl. Radiat. Isot. **88** (2014) 12–15.

[XX-5]   AIHARA, T., et al., Boron neutron capture therapy for advanced salivary gland carcinoma in head and neck, Int. J. Clin. Oncol. **19** (2014) 37–44.

[XX-6]   ONO, K., et al., "Measurement of BPA in the blood by fluorometry", (Proc. 13th International Congr. on Neutron Capture Therapy, 2–7 November 2008, Florence, Italy, ZONTA A., ALTIERI, S., ROVEDA, L., BARTH, R., Eds) ENEA, Italy (2008) 174.

LIST OF ABBREVIATIONS

AAPM	American Association of Physicists in Medicine
AB-BNCT	Accelerator based BNCT
AD	Advantage Depth
ALARA	As Low As Reasonably Achievable
BGO	Bismuth Germanate
BGRR	Brookhaven Graphite Research Reactor
BMRR	Brookhaven Medical Research Reactor
BNCT	Boron Neutron Capture Therapy
BNL	Brookhaven National Laboratory
BPA	4-borono-L-phenylalanine (this specific isomer, unless specified otherwise)
BPA-BNCT	BNCT conducted using BPA
BPA–fr	BPA complexed with D-fructose
BRN	Brain Radiation Necrosis
BSA	Beam Shaper Assembly
BSH	Mercapto-undecahydro-*closo*-dodecaborate
BSS	Bonner Sphere Spectrometer
BTL	Beam Transport Line
bw	Body Weight
CBE	Compound Biological Effectiveness
C-BENS	Cyclotron-Based Epithermal Neutron Source
CE	Conformité Européene (marking)
CICS	Cancer Intelligence Care Systems
CLD	Chord Length Distribution
CMm	Cutaneous Malignant melanoma
CNAO	Center for Oncological Hadrontherapy
CPE	Charged Particle Equilibrium
CR	Complete Response
CT	Computed Tomography
CTD	Common Technical Document
CTV	Clinical Target Volume
CZT	Cadmium Zinc Telluride
DM	Dawon Medax
DNA	Deoxyribonucleic Acid
DSB	Double Strand Break
DTL	Drift Tube Linac
EC	European Commission
EELS	Electron Energy Loss Spectroscopy
EGFR	Epidermal Growth Factor Receptor
ELV	Electrostatic Vessel
ENDF	Evaluated Nuclear Data File
ERP	Equipment Reference Point
ESQ	Electrostatic Quadrupole
EXFOR	Experimental Nuclear Reaction Data
FBPA	4-borono-2-fluoro-L-phenylalanine

FBPA–fr	FBPA complexed with D-fructose
FBPA-PET	PET conducted with FBPA
FDG	2-deoxy-2-fluoro-D-glucose
F-DOPA	6-fluorol-L-dopa
FFPE	Formalin-Fixed Paraffin-Embedded
FLT	*O*-(2-fluoroethyl)-L-tyrosine
FNT	Fast Neutron Therapy
GAGG	Gadolinium Aluminium Gallium Garnet
GB-10	Sodium decahydrodecaborate
GBM	Glioblastoma Multiforma
GLP	Good Laboratory Practice
GMP	Good Manufacturing Practice
GTV	Gross Tumour Volume
HEBL	High Energy Beam Line
HGM	High Grade Meningioma
HUS	Helsinki University Hospital
H&N	Head and Neck cancer
i.c.	IntraCarotoid injection
ICH	International Council for Harmonisation of Technical Requirements for Pharmaceuticals for Human Use
ICP	Inductively Coupled Plasma
ICP-AES	ICP-Atomic Emission Spectroscopy (synonym of ICP-OES)
ICP-MS	ICP-Mass Spectroscopy
ICP-OES	ICP-Optical Emission Spectroscopy (synonym of ICP-AES)
ICRP	International Commission on Radiological Protection
ICRU	International Commission on Radiation Units and Measurements
IEC	International Electrotechnical Commission
IGH	ImmunoGlobulin Heavy chain
INFN	Istituto Nazionale di Fisica Nucleare
IsoE	IsoEffective dose
ITV	Internal Target Volume
IUPAC	International Union of Pure and Applied Chemistry
i.v.	IntraVenous injection
JANIS	JAva-based Nuclear Information Software
JCDS	Japan Atomic Energy Agency Computational Dosimetry System
JIS	Japanese Industrial Standards
JENDL	Japanese Evaluated Nuclear Data Library
KERMA	Kinetic Energy Released per unit MAss
KUR	Kyoto University research Reactor
KURNS	Institute for Integrated Radiation and Nuclear Science, Kyoto University
L/N	Lesion/Normal ratio
LAT-1 (or 2)	Large neutral Amino acid Transporter 1 (or 2)
L-DOPA	L-3,4-dihydroxyphenylalanine
LEBL	Low Energy Beam Line
LEBT	Low Energy Beam Transport

LET	Linear Energy Transfer
LIBS	Laser-Induced Breakdown Spectroscopy
LRPFS	LocoRegional Progression-Free Survival
mAb	Monoclonal Antibody
MCNP	Monte Carlo N-Particle
MDR	Medical Device Regulation
MEBT	Medium Energy Beam Transport
MeST	Median Survival Time
MG	Malignant Glioma
MGMT	*O*6-methylguanine-DNA-methyltransferase
MIT	Massachusetts Institute of Technology
mOS	median Overall Survival
MRI	Magnetic Resonance Imaging
MRS	Magnetic Resonance Spectroscopy
MTV	Metabolic Tumour Volume
MU	Monitor Unit
N/B	Normal tissue to Blood concentration ratio
NCR	Neutron Capture Radiography
NCT	Neutron Capture Therapy
nSCC	non-Squamous Cell Carcinoma
NTCP	Normal Tissue Complication Probability
NTD	Neutron Track Detector
OAR	Organs at Risk
OS	Overall Survival
P-cell	Proliferative cell
PANS	Prompt Alpha Neutron Spectroscopy
PCPB	*p*-carboxyphenylboronic acid
PE	Poly(Ethylene)
PEEM	PhotoEmission Electron Microscopy
PEPT (1 or 2)	PEPtide Transporter (1 or 2)
PET	Positron Emission Tomography
PGNA	Prompt Gamma Neutron Activation analysis
PG-SPECT	Prompt Gamma-SPECT
PHITS	Particle and Heavy Ion Transport System
PMDA	Pharmaceutical and Medical Devices Agency
PR	Partial Response
PRV	Planning organ at Risk Volume
PTFE	Poly(TetraFluoroEthylene)
PTV	Planning Target Volume
PVE	Partial Volume Effect
QA	Quality Assurance
QC	Quality Control
Q-cell	Quiescent cell
RANO	Response Assessment in Neuro-Oncology
RBE	Relative Biological Effectiveness

RF	RadioFrequency
RFQ	RadioFrequency Quadrupole
RNA	RiboNucleic Acid
ROI	Region Of Interest
SaMD	Software as a Medical Device
SCC	Squamous Cell Carcinoma
SERA	Simulation Environment for Radiotherapy Applications
SHI	Sumitomo Heavy Industries
SIMS	Secondary Ion Mass Spectroscopy
SLD	Sub-Lethal Damage
SNMS	Secondary Neutral Mass Spectroscopy
SPECT	Single Photon Emission Computed Tomography
SSB	Single Strand Break
STBRC	Southern Tohoku BNCT Research Center
SUV	Standard Uptake Value
T/B	Tumour/Blood concentration ratio
TCD_{50}	Tumour Control Dose 50
TCP	Tumour Control Probability
TD	Treatable Depth
TL	Temporal Lobe
TLD	ThermoLuminiscent Detector
TLS	TAE Life Sciences
TME	Tumour MicroEnvironment
T/N	Tumour/Normal tissue concentration ratio
TPS	Treatment Planning System
VITA	Vacuum Insulated Tandem Accelerator
WHO	World Health Organization
XRT	X Ray Therapy

CONTRIBUTORS TO DRAFTING AND REVIEW

Ahmed, M.	National Institutes of Health, USA
Alberti, D.	University of Torino, Italy
Altieri, S.	University of Pavia, Italy
Asano, T.	Stella Pharma Corporation, Japan
Auterinen, I.	VTT, Finland
Bedogni, R.	INFN, Italy
Belyakov, O.	International Atomic Energy Agency
Besnard-Vauterin, C.	International Atomic Energy Agency
Bortolussi, S.	University of Pavia, Italy INFN, Italy
Busser, B.	Grenoble-Alpes University, France Institut universitaire de France, France
Capala, J.	National Institutes of Health, USA
Chen, Y.-W.	Taipei Veterans General Hospital, Taiwan, China
Chou, F.-I.	National Tsing Hua University, Taiwan, China
Ciraj Bjelac, O.	International Atomic Energy Agency
Cruikshanck, G.	University of Birmingham, UK
Dagrosa, A.	CNEA, Argentina
Deagostino, A.	University of Torino, Italy
Geninatti Crich, S.	University of Torino, Italy
Giammarile, F.	International Atomic Energy Agency
González, S.	CNEA, Argentina
Green, S.	University Hospitals Birmingham, UK
Gryziński, M.	National Centre for Nuclear Research, Poland
Hatazawa, J.	Osaka University, Japan
Hiratsuka, J.	Kawasaki University of Medical Welfare, Japan
Hirose, K.	Southern Tohoku BNCT Research Center, Japan

Holmberg, O.	International Atomic Energy Agency
Hu, N.	Kansai BNCT Medical Center, Japan
Igaki, H.	National Cancer Center Hospital, Japan
Igawa, K.	University of Okayama, Japan
Imamichi, S.	National Cancer Center, Japan
Itami, J.	Shin-Matsudo Central General Hospital, Japan
Jalilian, A.	International Atomic Energy Agency
Kankaanranta, L.	Helsinki University Hospital, Finland
Katsumi, H.	Southern Tohoku BNCT Research Center, Japan
Kim, W.	Dawon Medax Corporation, Republic of Korea
Kiyanagi, Y.	Nagoya University, Japan
Koivunoro, H.	Neutron Therapeutics, Finland
Kondo, N.	Kyoto University, Japan
Kreiner, A.	CNEA, Argentina
Kumada, H.	University of Tsukuba, Japan
Liu, Y.-H.	Neuboron Medtech Ltd, China
Lockyer, N.P.	University of Manchester, UK
Masui, S.	Sumitomo Heavy Industries, Japan
Masunaga, S.	Osaka Prefecture University, Japan
Masutani, M.	Nagasaki University, Japan
Matsumura, A.	Ibaraki Prefectural University of Health Sciences, Japan
Mauri, P.L.	CNR-ITB, Italy
Mavric, H.	International Atomic Energy Agency
Mitsumoto, T.	Sumitomo Heavy Industries Ltd., Japan
Miyatake, S.-I.	Osaka Medical and Pharmaceutical University, Japan
Monti, V.	INFN, Italy
Monti Hughes, A.	CNEA, Argentina

Motto-Ros, V.	ILM, France
Msimang, Z.	International Atomic Energy Agency
Murata, I.	Osaka University, Japan
Nakamura, H.	Tokyo Institute of Technology, Japan
Nakamura, M.	Cancer Intelligence Care Systems, Inc., Japan
Nakamura, S.	National Cancer Center Hospital, Japan
Nievas, S.	CNEA, Argentina
Nigg, D.	Idaho National Laboratory, USA
Olivera, M.	CNEA, Argentina
Ono, K.	Osaka Medical and Pharmaceutical University, Japan
Pan, M.	Western University, Canada
Porra, L.	Helsinki University Hospital, Finland
Porras, J.I.	University of Granada, Spain
Portu, A.M.	CNEA, Argentina
Postuma, I.	INFN, Italy
Protti, N.	University of Pavia, Italy
Quah, D.	National Cancer Center, Singapore
Quintana, J.	CNEA, Argentina
Ridikas, D.	International Atomic Energy Agency
Rossi, S.	CNAO, Italy
Saint Martin, G.	CNEA, Argentina
Sakurai, Y.	Kyoto University, Japan
Sancey Galliot, L.	Grenoble-Alpes University, France
Santa Cruz, G.	CNEA, Argentina
Santos, E.	IN2P3, France
Schwint, A.	CNEA, Argentina
Seo, H.-J.	Dawon Medax Corporation, Republic of Korea